Modern Methods in Scientific Computing and Applications

NATO Science Series

A Series presenting the results of scientific meetings supported under the NATO Science Programme.

The Series is published by IOS Press, Amsterdam, and Kluwer Academic Publishers in conjunction with the NATO Scientific Affairs Division

Sub-Series

I. Life and Behavioural Sciences	IOS Press
II. Mathematics, Physics and Chemistry	Kluwer Academic Publishers
III. Computer and Systems Science	IOS Press
IV. Earth and Environmental Sciences	Kluwer Academic Publishers
V. Science and Technology Policy	IOS Press

The NATO Science Series continues the series of books published formerly as the NATO ASI Series.

The NATO Science Programme offers support for collaboration in civil science between scientists of countries of the Euro-Atlantic Partnership Council. The types of scientific meeting generally supported are "Advanced Study Institutes" and "Advanced Research Workshops", although other types of meeting are supported from time to time. The NATO Science Series collects together the results of these meetings. The meetings are co-organized bij scientists from NATO countries and scientists from NATO's Partner countries – countries of the CIS and Central and Eastern Europe.

Advanced Study Institutes are high-level tutorial courses offering in-depth study of latest advances in a field.
Advanced Research Workshops are expert meetings aimed at critical assessment of a field, and identification of directions for future action.

As a consequence of the restructuring of the NATO Science Programme in 1999, the NATO Science Series has been re-organised and there are currently Five Sub-series as noted above. Please consult the following web sites for information on previous volumes published in the Series, as well as details of earlier Sub-series.

http://www.nato.int/science
http://www.wkap.nl
http://www.iospress.nl
http://www.wtv-books.de/nato-pco.htm

Series II: Mathematics, Physics and Chemistry – Vol. 75

Modern Methods in Scientific Computing and Applications

edited by

Anne Bourlioux

Département de Mathématiques et de Statistique,
Université de Montréal,
Montréal, Québec, Canada

Martin J. Gander

Department of Mathematics and Statistics,
McGill University,
Montréal, Québec, Canada

and

Technical Editor

Gert Sabidussi

Département de Mathématiques et de Statistique,
Université de Montréal,
Montréal, Québec, Canada

Kluwer Academic Publishers

Dordrecht / Boston / London

Published in cooperation with NATO Scientific Affairs Division

Proceedings of the NATO Advanced Study Institute and Séminaire de Mathématiques Supérieures on Modern Methods in Scientific Computing and Applications
Montréal, Québec, Canada
9–20 July 2001

A C.I.P. Catalogue record for this book is available from the Library of Congress.

ISBN 1-4020-0781-7

Published by Kluwer Academic Publishers,
P.O. Box 17, 3300 AA Dordrecht, The Netherlands.

Sold and distributed in North, Central and South America
by Kluwer Academic Publishers,
101 Philip Drive, Norwell, MA 02061, U.S.A.

In all other countries, sold and distributed
by Kluwer Academic Publishers,
P.O. Box 322, 3300 AH Dordrecht, The Netherlands.

Printed on acid-free paper

Printed in the Netherlands.

Table of Contents

Preface

When we first heard in the spring of 2000 that the Séminaire de mathématiques supérieures (SMS) was interested in devoting its session of the summer of 2001—its 40th—to scientific computing the idea of taking on the organizational work seemed to us somewhat remote. More immediate things were on our minds: one of us was about to go on leave to the Courant Institute, the other preparing for a research summer in Paris. But the more we learned about the possibilities of such a seminar, the support for the organization and also the great history of the SMS, the more we grew attached to the project. The topics we planned to cover were intended to span a wide range of theoretical and practical tools for solving problems in image processing, thin films, mathematical finance, electrical engineering, moving interfaces, and combustion. These applications alone show how wide the influence of scientific computing has become over the last two decades: almost any area of science and engineering is greatly influenced by simulations, and the SMS workshop in this field came very timely. We decided to organize the workshop in pairs of speakers for each of the eight topics we had chosen, and we invited the leading experts worldwide in these fields. We were very fortunate that every speaker we invited accepted to come, so the program could be realized as planned.

Four of the eight sessions focused on the techniques of scientific computing: the domain decomposition session with lectures by Laurence Halpern and Frédéric Nataf from France introduced the audience to absorbing boundary conditions and one-way operators, which are key ingredients for fast domain decomposition methods. The new methods proposed were optimized Schwarz methods and a new preconditioner for non-symmetric Schur methods. The multigrid session, given by Gundolf Haase and Ulrich Langer from Austria, provided an introduction to the convergence analysis of multi-grid methods and a new multi-grid technique for the three-dimensional Maxwell equations with impressive performance. Algebraic multi-grid techniques were also presented. The lectures in the session on dynamical systems, given by Tony Humphries and Andrew Stuart from England, started with continuous and discrete dynamical systems, and then went on to stochastic dynamical systems with an analysis of numerical methods for the stochastic case. Finally, the matrix analysis session, led by Zhaojun Bai and Gene Golub from the US, focused on reduced order model techniques for extremely large systems of ordinary differential equations, using a very elegant link to Krylov methods, estimates for entries of functions of matrices through quadrature, and relating orthogonal polynomials to Lanczos, closing the circle.

The other four sessions combined techniques with concrete applications. The session on stochastic differential equations given by George Papanicolaou and Ronnie Sircar from the US started with an overview of limit theorems and the law of large numbers with the key techniques of asymptotic analysis for applications in mathematical finance. In particular, a new model was presented which can reproduce the skewness of the curve of implied volatilities formerly known as the smile curve. The session on image processing dealt with two major techniques: Karol Mikula from Slovakia showed how nonlinear diffusion equations based on the Perona-Malik equation can be used in image processing. The original instability in this approach was corrected with a convolution, and impressive image sequences of a beating heart illustrated the effectiveness of this approach. A full convergence proof for the nonlinear discretization was also shown. The second technique, presented by Jamie Sethian from the US, was level set methods with an almost unreal way to reconstruct missing image parts that are completed by the human brain when we look at an image. The case of the "missing triangle" is particularly spectacular. The third session, on thin films, was presented by Andrea Bertozzi and Thomas Witelski from the US. They showed how asymptotic analysis is essential in capturing phenomena of rupturing and fingering, and how even one-dimensional flows can give insight into the behavior of thin films. The forth session, given by Anne Bourlioux from Canada and Panagiotis Souganidis from the US, introduced asymptotic analysis for combustion problems, and concrete computations which illustrate how high resolution results can be obtained by using the asymptotic results in regions close to flame fronts, where otherwise the required high resolution would be computationally infeasible.

While the sixteen speakers came from six different countries, the total of 75 participants of the ASI come from twenty different countries. In absolute term most came from the US (18), but in relative terms (with respect to the population of the country) most came from Slovakia (7). We are very thankful to Gert Sabidussi and Ghislaine David for the flawless organization of this international event; nobody could have done a better job! We are also grateful for NATO's generous support through its Advanced Study Institutes Program, and to the Centre de Recherches Mathématiques (CRM) and the Université de Montréal for the local support.

The 40th session of the SMS/NATO ASI on Scientific Computing and Applications showed impressively how active the field of scientific computing is today and what tremendous impact it has in science and engineering. Scientific computing has become the third discipline together with mathematical modeling and physical experiments to explore our world and advance science.

The editors

1
2
3
4
5
6
7
8
9
10
11
12
13
14
15
16
17
18
19
20
21
22
23
24
25
26
27
28
29
30
31
32
33
34
35
36
37
38
39
40
41
42
43
44
45
46
47
48
49
50
51
52
53
54
55
56
57
58
59

Key to group picture

1 Kriva
2 Makovicka
3 Sevcovic
4 Meskauskas
5 Kulikov
6 Barbosa
7 Carlini
8 Abrukov
9 Sircar
10 Petrovskaya
11 Ju
12 Matveevna
13 Rezulski
14 Padhi
15 Čunderlik
16 Chuang
17 Kim
18 Bowen
19 Youbissi
20 Ahn
21 Boulet
22 Saas
23 Sabidussi
24 Desautels
25 Lee
26 David J.-L.
27 Jang
28 Horak
29 Kozubek
30 Slavova
31 Ungvarsky
32 Draganescu
33 Eyriboyun
34 Beuchler
35 Of
36 Izsak
37 Mul
38 Handlovicova
39 Bystricky
40 Dienesberger
41 Wabro
42 Reitzinger
43 Vignesvaran
44 Rojas
45 Chalupecky
46 Praks
47 David G.
48 Humphries
49 Halpern
50 Papanicolaou
51 Mikula
52 Souganidis
53 Bourlioux
54 Bertozzi
55 Witelski
56 Nataf
57 Langer
58 Haase
59 Gander

Participants

Victor ABRUKOV
Physical-Technical Faculty
Chuvash State University
428 015 Cheboksary
Russia
victor@chuvsu.ru

André BARBOSA
c/o Helder Coelho
Departamento de Informática
Universidade de Lisboa
Bloco C5 — Piso 1 — Campo Grande
1700 Lisboa
Portugal
andre.dias.barbosa@clix.pt

Sven BEUCHLER
Faculty of Mathematics
TU-Chemnitz
Reichenhainer Str. 39/41
09107 Chemnitz
Germany
sven.beuchler
@mathematik.tu.chemntiz.de

Jean-Philippe BOULET
Dép. de mathématiques et de statistique
Université Laval
Québec, QC, G1K 7P4
Canada
jpboulet@mat.ulaval.ca

Yves BOURGAULT
Dép. de mathématiques et de statistique
Université d'Ottawa
585 King Edward
P.O. Box 450, Station A
Ottawa, ON, K1N 6N5
Canada
ybourg@mathstat.uottawa.ca

Robert BYSTRICKY
Dept. Mathematics — Fac. Civil Engineering
Slovak University of Technology
Radlinskeho 11
813 68 Bratislava
Slovakia
rb@svf.stuba.sk

Elisabetta CARLINI
Dipartimento di Matematica
Università "La Sapienza" di Roma 1
P.le Aldo Moro, 2
00185 Roma
Italy
carlini@giove.mat.uniroma1.it

Vladimir CHALUPECKY
Department of Mathematics
Fac. Nuclear Sci. & Phys. Engineering
Czech Technical University
Trojanova 13
120 00 Praha 2
Czech Republic
chalupecky@km1.fjfi.cvut.cz

Yao-li CHUANG
Department of Physics
Duke University
Durham, NC 27708
USA
chuang@phy.duke.edu

Robert CUNDERLIK
Department of Theoretical Geodesy
Slovak University of Technology
Radlinskeho 11
813 68 Bratislava
Slovakia
cunderli@svf.stuba.sk

Marc-André DESAUTELS
Dép. de mathématiques et de statistique
Université de Montréal
CP 6128, succ. Centre-ville
Montréal, QC, H3C 3J7
Canada
desautel@dms.umontreal.ca

Victorita DOLEAN
CMAP École Polytechnique
91128 Palaiseau Cédex
France
dolean@cmapx.polytechnique.fr

Andrei DRAGANESCU
Department of Mathematics
University of Chicago
5734 S. University Ave.
Chicago, IL 60637
USA
draga@math.uchicago.edu

Mustafa EYRIBOYUN
Mechanical Engineering Department
Zonguldak Karaelmas University
67100 Zonguldak
Turkey
eboyun@karaelmas.edu.tr

Martin J. GANDER
Department of Mathematics & Statistics
McGill University
805 Sherbrooke ouest
Montréal, QC, H3A 2K6
Canada
mgander@math.mcgill.ca

John GREER
Department of Mathematics
Duke University
Box 90320
Durham, NC 27708–0320
USA
jbg33@math.duke.edu

Robert GUÉNETTE
Dép. de mathématiques et de statistique
Université Laval
Québec, QC, G1K 7P4
Canada
robert.guenette@mat.ulaval.ca

Aissa GUESMIA
IRMA, UFR Mathématique & Informatique
Université Louis Pasteur (Strasbourg I)
7, rue René Descartes
67084 Strasbourg Cédex
France
guesmia@math.u-strasbg.fr

Angela HANDLOVICOVA
Dept. Mathematics — Fac. Civil Engineering
Slovak University of Technology
Radlinskeho 11
813 68 Bratislava
Slovakia
angela@svf.stuba.sk

David HORAK
Department of Applied Mathematics
VSB-Technical University of Ostrava
Tr. 17. Listopadu 15
708 33 Ostrava-Poruba
Czech Republic
david.horak@vsb.cz

Ferenc IZSÁK
Department of Applied Analysis
Eötvös Loránd University
Kecskeméti u. 10–12
H–1053 Budapest
Hungary
bizsu@cs.elte.hu

Ning JU
Department of Mathematics
Duke University
Box 90320
Durham, NC 27708–0320
USA
ju@math.duke.edu

Boualem KHOUIDER
Courant Institute of Mathematical Sciences
New York University
251 Mercer St.
New York, NY 10012–1185
USA
khouider@cims.nyu.edu

Johanna KIENESBERGER
Inst. f. Anal. & Comput. Mathematics
Johannes Kepler University of Linz
Altenberger Str. 69
4040 Linz
Austria
kienes@numa.uni-linz.ac.at

Tomás KOZUBEK
Department of Applied Mathematics
VSB-Technical University of Ostrava
Tr. 17. Listopadu 15
708 33 Ostrava-Poruba
Czech Republic
tomas@kozubek@vsb.cz

Zuzana KRIVA
Dept. Mathematics — Fac. Civil Engineering
Slovak University of Technology
Radlinskeho 11
813 68 Bratislava
kriva@svf.stuba.sk

Gennady KULIKOV
Faculty of Mechanics and Mathematics
Ulyanovsk State University
L. Tolstoy Str. 42
432 700 Ulyanovsk
Russia
kulgyu@mci.ulsu.ru

Jindrich MAKOVICKA
Department of Mathematics
Fac. Nuclear Sci. & Physical Engineering
Czech Technical University
Trojanova 13
120 00 Prague 2
Czech Republic
makovick@kmlinux.fjfi.cvut.cz

Inessa MATVEEVA
Sobolev Institute of Mathematics
pr. Koptyuga 4
630090 Novosibirsk
Russia
matveeva@math.nsc.ru

Tadas MESKAUSKAS
CERFACS
42, ave. Gaspard Coriolis
31057 Toulouse Cédex 1
France
meska@cerfacs.fr

Elena MUL
Inst. Computer Information & Technologies
Ternopil Acad. of National Economy
3 Peremoga Square
46004 Ternopil
Ukraine
elena_mul@lycos.com

Günther OF
Mathematisches Institut A
Universität Stuttgart
Paffenwaldring 57
70569 Stuttgart
Germany
ofgr@cip.mathematik.uni-stuttgart.de

Gouri S. PADHI
Dept. Mechanical & Aeronautical Engineer.
University of Limerick
Limerick
Republic of Ireland
gouri@padhi@ul.ie

George PAPANICOLAOU
Department of Mathematics
Stanford University
Stanford, CA 94305–2125
USA
papanico@math.stanford.edu

Natalia PETROVSKAYA
Keldysh Inst. Applied Mathematics, RAS
Boeing Operations International, Inc.
Gazetny per 17/9
103009 Moscow
Russia
natalia.petrovskaya@boeing.com

Roger PIERRE
Dép. de mathématiques et de statistique
Université Laval
Québec, QC, G1K 7P4
Canada
rpierre@mat.ulaval.ca

Pavel PRAKS
Department of Applied Mathematics
VSB-Technical University of Ostrava
Tr. 17. Listopadu 15
708 33 Ostrava-Poruba
Czech Republic
pavel.praks@vsb.cz

Tanvir RAHMAN
Department of Physics
McGill University
3600 University
Montréal, QC, H3A 2K6
Canada
tanvir@physics.mcgill.ca

Stefan REITZINGER
Special Research Program SFB F013
University of Linz
Freistaedterstr. 313
4040 Linz
Austria
reitz@sfb013.uni-linz.ac.at

Michal REZULSKI
Institute of Telecommunications
Warsaw Technical University
15/19 Nowowiejska St.
00665 Warszawa
Poland
rezulski@tele.pw.edu.pl

Marielba ROJAS
CERFACS
42, ave. Gaspard Coriolis
31057 Toulouse Cédex 1
France
mrojas@cerfacs.fr

Laurent SAAS
Institut Français du Pétrole
4, ave. Bois Préau
92852 Rueil Malmaison Cédex
France
laurent.saas@ifp.fr

Daniel SEVCOVIC
Institute of Applied Mathematics
Faculty of Mathematics & Physics
Comenius University
Mlynska dolina
842 48 Bratislava
Slovakia
sevcovic@fmph.uniba.sk

Angela SLAVOVA
Institute of Mathematics & Informatics
Bulgarian Academy of Sciences
Acad. G. Bonchev Str., bl. 8
1113 Sofia
Bulgaria
slavova@math.bas.bg

Amik ST-CYR
Dép. de mathématiques et de statistique
Université de Montréal
CP 6128, succ. Centre-ville
Montréal, QC, H3C 3J7
Canada
stcyr@dms.umontreal.ca

Gábor STIKKEL
Department of Numerical Analysis
Eötvös Loránd University
Pázmány Péter Sétány 1/D
H–1117 Budapest
Hungary
stiko@cs.elte.hu

Romy TOUMA
Dép. de mathématiques et de statistique
Université de Montréal
CP 6128, succ. Centre-ville
Montréal, QC, H3C 3J7
Canada
touma@dms.umontreal.ca

Cyril UNGVARSKY
Inst. Applied Mathematics — F.M.P.I.
Comenius University
842 15 Bratislava
Slovakia
cyril.ungvarsky@fmph.uniba.sk

Ratnam VIGNESVARAN
School of Mathematical Sciences
University of Sussex
Falmer-Brighton BN1 9QH
UK
r.vignesvaran@sussex.ac.uk

Oleg VOLKOV
Dép. de mathématiques et de statistique
Université de Montréal
CP 6128, succ. Centre-ville
Montréal, QC, H3C 3J7
Canada
volkov@dms.umontreal.ca

Markus WABRO
Inst. f. Anal. & Comput. Mathematics
Johannes Kepler University of Linz
Altenberger Str. 69
4040 Linz
Austria
markus.wabro@numa.uni-linz.ac.at

Fabien M. YOUBISSI
Dép. de mathématiques et de statistique
Université Laval
Québec, QC, G1K 7P4
Canada
fyoubi@mat.ulaval.ca

Bence ZAUPPER
Institute of Mathematics
Eötvös Loránd University
Múzeum krt 6–8
H–1088 Budapest
Hungary
benco@freemail.hu

Contributors

Zhaojung BAI
Department of Computer Science
University of California
One Shields Avenue
Davis, CA 95616
USA
bai@cs.ucdavis.edu

Andrea BERTOZZI
Department of Mathematics
Duke University
Box 90320
Durham, NC 27708–0320
USA
bertozzi@math.duke.edu

Anne BOURLIOUX
Dép. de mathématiques et de statistique
Université de Montréal
CP 6128 — succ. Centre-ville
Montréal, QC, H3C 3J7
Canada
bourliou@dms.umontreal.ca

Mark BOWEN
Department of Mathematics
Duke University
Box 90320
Durham, NC 27708–0320
USA
bowen@math.duke.edu

Gene H. GOLUB
Computer Science Department
Stanford University
Stanford, CA 94305–9025
USA
golub@sccm.stanford.edu

Gundolf HAASE
Inst. Analysis & Computational Math.
Johannes Kepler University
Altenberger Str. 69
4040 Linz
Austria
ghaase@numa.uni-linz.ac.at

Laurence HALPERN
Mathématiques — Institut Galilée
Université Paris 13
99, avenue Jean-Baptiste Clément
93430 Villetaneuse
France
halpern@math.univ-paris13.fr

Anthony HUMPHRIES
School of Mathematical Sciences
University of Sussex
Falmer-Brighton BN1 9QH
UK
a.r.humphries@sussex.ac.uk

Mattias JONSSON
Department of Mathematics
University of Michigan
Ann Arbor, MI 48109–1109
USA
mattiasj@umich.edu

Ulrich LANGER
Inst. Analysis & Computational Math.
Johannes Kepler University
Altenberger Str. 69
4040 Linz
Austria
ulanger@numa.uni-linz.ac.at

Karol MIKULA
Department of Mathematics
Fac. of Civil Engineering
Slovak University of Technology
Randlinskeho 11
813 68 Bratislava
Slovakia
mikula@vox.svf.stuba.sk

Frédéric NATAF
Centre de Mathématiques Appliquées
École Polytechnique
91128 Palaiseau Cédex
France
nataf@cmapx.polytechnique.fr

Adib RAHMOUNI
Mathématiques — Institut Galilée
Université Paris 13
99, avenue Jean-Baptiste Clément
93430 Villetaneuse
France
rahmouni@math.univ-paris13.fr

James A. SETHIAN
Department of Mathematics
University of California
Berkeley, CA 94720–3840
USA
sethian@math.berkeley.edu

Ronnie SIRCAR
Dept. of Operations Research
and Financial Engineering
Princeton University
Princeton, NJ 08544
USA
sircar@princeton.edu

Panagiotis SOUGANIDIS
Department of Mathematics
University of Texas at Austin
Austin, TX 78712–1082
USA
souganid@ma.utexas.edu

Andrew STUART
Mathematical Institute
Warwick University
Coventry CV4 7AL
UK
stuart@maths.warwick.ac.uk

Thomas WITELSKI
Department of Mathematics
Duke University
Box 90320
Durham, NC 27708–0320
USA
witelski@math.duke.edu

Computation of large-scale quadratic forms and transfer functions using the theory of moments, quadrature and Padé approximation

Zhaojun BAI

Department of Computer Science
University of California
Davis, CA 95616
USA

Gene GOLUB

Department of Computer Science
Stanford University
Stanford, CA 94305
USA

Abstract

Large-scale problems in scientific and engineering computing often require solutions involving large-scale matrices. In this paper, we survey numerical techniques for solving a variety of large-scale matrix computation problems, such as computing the entries and trace of the inverse of a matrix, computing the determinant of a matrix, and computing the transfer function of a linear dynamical system.

Most of these matrix computation problems can be cast as problems of computing quadratic forms $u^T f(A)u$ involving a matrix functional $f(A)$. It can then be transformed into a Riemann-Stieltjes integral, which brings the theory of moments, orthogonal polynomials and the Lanczos process into the picture. For computing the transfer function, we focus on numerical techniques based on Padé approximation via the Lanczos process and the moment-matching property. We will also discuss issues related to the development of efficient numerical algorithms, including using Monte Carlo simulation.

1 Introduction

In the 1973 paper entitled "*Some Modified Eigenvalue Problems*" by Golub [34], numerical calculation of several matrix computation problems are considered. These include finding constrained eigenvalue problems, determining the eigenvalues of a matrix which is modified by a matrix of rank one, solving constrained and total least squares problems. All these problems require some manipulation before the standard algorithms may be used. Many of these problems have been further studied and widely applied, such as the total least squares problems [59].

In this paper, we survey a collection of "new" modified matrix computation problems. These include computing the entries of the inverse of a matrix, $(A^{-1})_{ij}$, the determinant,

A. Bourlioux and M.J. Gander (eds.), Modern Methods in Scientific Computing and Applications, 1–30.

$\det(A)$, the trace of the inverse of a matrix, $\mathrm{tr}(A^{-1})$, and the transfer function, $h(s) = l^T(I-sA)^{-1}r$. More generally, we are concerned with the problem of computing the quadratic form $u^T f(A)u$ or bilinear form $u^T f(A)v$ involving a matrix functional $f(A)$. The matrix A in question is typically large and sparse. We will discuss those iterative methods in which the matrix A in question is only referenced via matrix-vector products.

The necessity of solving these problems appears in many applications. For example, let $\hat{x}$ be an approximation of the exact solution x of a linear system of equations $Ax = b$, where we assume that A is symmetric positive definite. In order to obtain error bounds of the approximation, we consider the A-norm of the forward error vector $e = x - \hat{x}$:

$$\|e\|_A^2 = e^T Ae = r^T A^{-1}AA^{-1}r = r^T A^{-1}r,$$

where r is the residual vector $r = b - A\hat{x}$. Therefore, the problem becomes to obtain computable bounds for the quadratic form $r^T A^{-1}r$. It is sometimes also of interest to compute the l_2-norm of the error e, $\|e\|_2$. Then it is easy to see that one needs to compute the quadratic form $\|e\|_2^2 = r^T A^{-2}r$. Bounds for the error of linear systems of equations have been studied extensively in [21, 35, 36, 17].

There are a number of applications where it is required to compute the trace of the inverse, $\mathrm{tr}(A^{-1})$, and the determinant, $\det(A)$, of a very large and sparse matrix A, such as in the theory of fractals [56, 63] and lattice Quantum Chromodynamics (QCD) [57, 24, 31].

The solution of a linear least squares problem by the generalized cross-validation technique involves the computation of the trace of the matrix $I - K(K^TK + m\lambda I)^{-1}K$ for estimating a regularization parameter λ, where K is an $m \times n$ matrix, $m \geq n$ [38]. The quadratic forms $u^T \exp(\sqrt{-1}At)u$, $u^T \exp(-\beta A)u$ and $u^T(\omega I - A)^{-1}u$ appear in the computation of magnetic resonance spectra, quantum statistical mechanics and a variety of other problems in chemical physics [50]. In molecular dynamics, the total energy of an electronic structure requires the calculation of a partial eigenvalue sum of a generalized symmetric definite eigenvalue problem. In [6], it is shown that this sum can be obtained through the computation of the trace of the matrix $A(I + \exp((A - \tau)/\kappa)^{-1}$, where τ and κ are parameters. Many other origins and applications of the computation of the quadratic form are listed in [8, 25].

The problem of transfer function computing arises from the steady-state analysis of a linear dynamical system. A successful approximation of the transfer function also leads to a very desirable and potentially powerful reduced-order model of the original full-order system. We will discuss motivations and applications of transfer function computing in Section 4.

2 Quadratic forms

Given an $N \times N$ real symmetric matrix A and a smooth function f such that the matrix function $f(A)$ is defined, the problem of quadratic form computing is to evaluate the quadratic form

$$u^T f(A)u, \tag{2.1}$$

or computing tight lower and upper bounds γ_ℓ and γ_u of the quadratic form $u^T f(A)u$:

$$\gamma_\ell \leq u^T f(A)u \leq \gamma_u. \tag{2.2}$$

Here u is a given real column vector of length N. Without loss of generality, one may assume $u^T u = 1$.

The quadratic form computing problem was first proposed in [21] for bounding the error of linear systems of equations. It has been further studied in [35, 37, 5] and extended to other applications. In this section, we will first discuss the main idea of the approach, and show that the problem of quadratic form computing can be transformed into a Riemann-Stieltjes integral, and then use Gauss-type quadrature rules to approximate the integral, which brings the theory of moments and orthogonal polynomials, and the underlying Lanczos process into the picture.

We note that the quadratic form (2.1) can be generalized to the block case, namely, to evaluate the matrix quadratic form

$$U^T f(A)U,$$

where U is a block of N-vectors. The problem of computing a bilinear form $u^T f(A)v$, where $u \neq v$, can be reduced to the problem of quadratic form computing by the polarization expression:

$$u^T f(A)v = \frac{1}{4}(y^T f(A)y - z^T f(A)z),$$

where $y = u + v$ and $z = u - v$.

2.1 Main idea

Let us now go through the main idea. Since A is symmetric, the eigen-decomposition of A is given by $A = Q^T \Lambda Q$, where Q is an orthogonal eigenvector matrix and Λ is a diagonal matrix with increasingly ordered diagonal elements λ_i, the eigenvalues. Then the quadratic form $u^T f(A)u$ can be written as

$$u^T f(A)u = u^T Q^T f(\Lambda) Q u = \tilde{u}^T f(\Lambda)\tilde{u} = \sum_{i=1}^{N} f(\lambda_i)\tilde{u}_i^2,$$

where $\tilde{u} = (\tilde{u}_i) \equiv Qu$. The last sum can be interpreted as a Riemann-Stieltjes integral

$$u^T f(A)u = \int_a^b f(\lambda)\, d\mu(\lambda), \tag{2.3}$$

where the measure $\mu(\lambda)$ is a piecewise constant function and defined by

$$\mu(\lambda) = \begin{cases} 0, & \text{if } \lambda < a \leq \lambda_1 \\ \sum_{j=1}^{i} \tilde{u}_j^2, & \text{if } \lambda_i \leq \lambda < \lambda_{i+1} \\ \sum_{j=1}^{N} \tilde{u}_j^2, & \text{if } b \leq \lambda_N \leq \lambda \end{cases},$$

and a and b are the lower and upper bounds of the eigenvalues λ_i.

To obtain an estimate for the Riemann-Stieltjes integral (2.3), one can use Gauss-type quadrature rules [33, 22]:

$$I_n[f] = \sum_{j=1}^{n} \omega_j f(\theta_j) + \sum_{k=1}^{m} \rho_k f(\tau_k), \tag{2.4}$$

where the weights $\{\omega_j\}$ and $\{\rho_k\}$ and the nodes $\{\theta_j\}$ are unknown and to be determined. The nodes $\{\tau_k\}$ are prescribed. If $m = 0$, then it is the well-known Gauss rule. If $m = 1$ and $\tau_1 = a$ or $\tau_1 = b$, it is the Gauss-Radau rule. The Gauss-Lobatto rule is obtained for $m = 2$ and $\tau_1 = a$ and $\tau_2 = b$.

The accuracy of the Gauss-type quadrature rules may be obtained by an estimation of the remainder $R_n[f]$:

$$R_n[f] = \int_a^b f(\lambda)\, d\mu(\lambda) - I_n[f].$$

For the Gauss quadrature rule, the remainder $R_n[f]$ can be written as

$$R_n[f] = \frac{f^{(2n)}(\eta)}{(2n)!} \int_a^b \left[\prod_{i=1}^n (\lambda - \theta_i)\right]^2 d\mu(\lambda), \tag{2.5}$$

for some $\eta \in (a, b)$. If the sign of $R_n[f]$ can be determined, then the quadrature $I_n[f]$ is a lower bound (if $R_n[f] > 0$) or an upper lower bound (if $R_n[f] < 0$) of the quadratic form $u^T f(A)u$. Similarly, for the Gauss-Radau rule, the remainder $R_n[f]$ is given by

$$R_n[f] = \frac{f^{(2n+1)}(\eta)}{(2n+1)!} \int_a^b (\lambda - \tau_1) \left[\prod_{i=1}^n (\lambda - \theta_i)\right]^2 d\mu(\lambda), \tag{2.6}$$

for some $\eta \in (a, b)$, and for the Gauss-Lobatto rule it is

$$R_n[f] = \frac{f^{(2n+2)}(\eta)}{(2n+2)!} \int_a^b (\lambda - \tau_1)(\lambda - \tau_2) \left[\prod_{i=1}^n (\lambda - \theta_i)\right]^2 d\mu(\lambda). \tag{2.7}$$

for some $\eta \in (a, b)$.

2.2 Moments

In defining Gaussian quadratures, we choose the weights ω_j and nodes θ_j so that $2n$ moments are calculated exactly, i.e.,

$$\mu_r = \int_a^b \lambda^r\, d\mu(\lambda) = \sum_{j=1}^n \omega_j \theta_j^r$$

for $r = 0, 1, 2, \ldots, 2n - 1$. A different perspective on this statement is that the summation on the right is of the form of a solution to a difference equation. In particular, let $n = 2$; the moments μ_r satisfy a second-order difference equation of the form

$$\alpha\mu_r + \beta\mu_{r-1} + \gamma\mu_{r-2} = 0$$

for certain coefficients α, β and γ. The nodes θ_j are the roots of the characteristic polynomial

$$\alpha\theta^2 + \beta\theta + \gamma = 0.$$

We can use this fact to bound the quadratic form. In particular, note that the first three moments $\mu_0 = \mathrm{tr}(A^0) = N$, $\mu_1 = \mathrm{tr}(A)$ and $\mu_2 = \mathrm{tr}(A^2) = \|A\|_F^2$ can be easily calculated. By using these three moments, it is shown in [7] that for a symmetric positive definite matrix A, we have the following computable bounds for the trace of the inverse $\mathrm{tr}(A^{-1})$:

$$\begin{bmatrix}\mu_1 & \mu_0\end{bmatrix}\begin{bmatrix}\mu_2 & \mu_1\\ b^2 & b\end{bmatrix}^{-1}\begin{bmatrix}\mu_0\\ 1\end{bmatrix} \leq \mathrm{tr}(A^{-1}) \leq \begin{bmatrix}\mu_1 & \mu_0\end{bmatrix}\begin{bmatrix}\mu_2 & \mu_1\\ a^2 & a\end{bmatrix}^{-1}\begin{bmatrix}\mu_0\\ 1\end{bmatrix},$$

where a and b are the lower and upper bounds of the eigenvalues of A, $a > 0$. Furthermore, by the identity

$$\ln(\det(A)) = \mathrm{tr}(\ln A), \tag{2.8}$$

we can also use the first three moments of A to obtain lower and upper bounds of $\ln(\det(A))$:

$$\begin{bmatrix}\ln a & \ln \underline{t}\end{bmatrix}\begin{bmatrix}a & \underline{t}\\ a^2 & \underline{t}^2\end{bmatrix}^{-1}\begin{bmatrix}\mu_1\\ \mu_2\end{bmatrix} \leq \ln(\det(A)) \leq \begin{bmatrix}\ln b & \ln \bar{t}\end{bmatrix}\begin{bmatrix}b & \bar{t}\\ b^2 & \bar{t}^2\end{bmatrix}^{-1}\begin{bmatrix}\mu_1\\ \mu_2\end{bmatrix}$$

where $\underline{t} = (a\mu_1 - \mu_2)/(a\mu_0 - \mu_1)$ and $\bar{t} = (b\mu_1 - \mu_2)/(b\mu_0 - \mu_1)$.

2.3 Orthogonal polynomials and symmetric Lanczos process

Before we discuss how the weights and the nodes are calculated in the Gauss-type quadratures, let us briefly review the theory of orthogonal polynomials and symmetric Lanczos process. For a given nondecreasing measure function $\mu(\lambda)$, it is well-known [58] that a sequence of polynomials $\{p_j(\lambda)\}$ can be constructed via a three-term recurrence

$$\beta_j p_j(\lambda) = (\lambda - \alpha_j)p_{j-1}(\lambda) - \beta_{j-1}p_{j-2}(\lambda),$$

for $j = 1, 2, \ldots, n$ with $p_{-1}(\lambda) \equiv 0$ and $p_0(\lambda) \equiv 1$. Furthermore, they are orthonormal with respect to the measure $\mu(\lambda)$:

$$(p_i, p_j) = \int_a^b p_i(\lambda)p_j(\lambda)\, d\mu(\lambda) = \begin{cases}1, & \text{if } i = j,\\ 0, & \text{if } i \neq j,\end{cases}$$

where one assumes the normalization condition $\int d\mu = 1$ (i.e., $u^T u = 1$). Writing the three-term recurrence in matrix form, we have

$$\lambda p(\lambda) = T_n p(\lambda) + \beta_n p_n(\lambda) e_n,$$

where

$$p(\lambda)^T = [p_0(\lambda), p_1(\lambda), \ldots, p_{n-1}(\lambda)],$$

and

$$T_n = \begin{bmatrix}\alpha_1 & \beta_1 & & \\ \beta_1 & \alpha_2 & \ddots & \\ & \ddots & \ddots & \beta_{n-1}\\ & & \beta_{n-1} & \alpha_n\end{bmatrix}.$$

The following classical symmetric Lanczos process [46] is an elegant way to compute the recurrence coefficients α_j and β_j.

Symmetric Lanczos process Let A be a real symmetric matrix, u a real vector with $u^T u = 1$. Then the following procedure computes the tridiagonal matrix T_n and the orthonormal Lanczos vectors q_j.

```
Let q_0 = 0, β_0 = 0, and q_1 = u
α_1 = q_1^T A q_1
For j = 1, 2, ..., n,
    r_j = A q_j − α_j q_j − β_{j−1} q_j
    β_j = ||r_j||_2
    q_{j+1} = r_j / β_j
    α_{j+1} = q_{j+1}^T A q_{j+1}
```

Let $Q_n = [q_1, q_2, \ldots, q_n]$. Then the symmetric Lanczos process can be written in compact matrix form

$$AQ_n = Q_n T_n + \beta_n q_{n+1} e_n^T,$$

and $Q_n^T Q_n = I$ and $Q_n^T q_{n+1} = 0$. The Lanczos vectors q_j produced by the symmetric Lanczos process and the orthonormal polynomials $\{p_j(\lambda)\}$ are connected by the following:

$$q_j = p_{j-1}(A)u$$

for $j = 1, \ldots, n$. Note that the eigenvalues of T_n are the roots of the polynomial $p_n(\lambda)$.

2.4 Basic algorithms

Gauss-Lanczos (GL) algorithm Using the eigen-decomposition of T_n:

$$T_n = S_n D_n S_n^T,$$

where $D_n = \mathrm{diag}(\theta_1, \theta_2, \ldots, \theta_n)$ and $S_n^T S_n = I_n$. Then for the Gauss quadrature rule, the eigenvalues θ_i of T_n (which are the zeros of the polynomial $p_n(\lambda)$) are the desired nodes. The weights ω_j are the squares of the first elements of the normalized eigenvectors of T_n, i.e., $\omega_j = (e_1^T S_n e_j)^2$.

By combining the Gauss quadrature and the symmetric Lanczos process, we have an algorithm for computing an estimate $I_n[f]$ of the quadrature form $u^T f(A)u$. We refer to it as the Gauss-Lanczos (GL) algorithm.

From the expression (2.5) of the remainder $R_n[f]$, we can determine whether the approximation $I_n[f]$ is a lower or upper bound of the quadratic form $u^T f(A)u$. For example, if $f^{(2n)}(\eta) > 0$, then $I_n[f]$ is a lower bound ℓ_l.

We note that it is not always necessary to explicitly compute the eigenvalues and the first components of eigenvectors of the tridiagonal matrix T_n for obtaining the estimation $I_n[f]$. Indeed, by the fact that $\omega_j = (e_1^T S_n e_j)^2$, the Gauss rule can be written in the form

$$I_n[f] = \sum_{k=1}^{n} \omega_k f(\theta_k) = e_1^T S_n f(D_n) S_n^T e_1 = e_1^T f(T_n) e_1. \tag{2.9}$$

Therefore, if the $(1,1)$ entry of $f(T_n)$ can be easily computed, for example, $f(\lambda) = 1/\lambda$, then the computation of the eigenvalues and eigenvectors of T_n can be avoided. This is the case when applying this technique to error estimation of an iterative method for solving linear system of equations [36, 17].

By exploiting the relation between the Lanczos vectors q_j and orthogonal polynomials p_j, it can be shown that the remainder $R_n[f]$ can be written as

$$R_n[f] = \frac{f^{(2n)}(\eta)}{(2n)!}\beta_1^2\beta_2^2\cdots\beta_n^2$$

for some $\eta \in (a,b)$. Therefore, if $f^{(2n)}(\eta)$ can be estimated or bounded, the error can be easily obtained with little additional cost [16, 25].

Gauss-Radau-Lanczos (GRL) algorithm For the Gauss-Radau and Gauss-Lobatto rules, the nodes $\{\theta_j\}$, $\{\tau_k\}$ and weights $\{\omega_j\}, \{\rho_j\}$ come from eigenvalues and the squares of the first elements of the normalized eigenvectors of an adjusted tridiagonal matrix of T_n, which has the prescribed eigenvalues a and/or b.

To implement the Gauss-Radau rule with the prescribed node $\tau_1 = a$ or $\tau_1 = b$, the above GL algorithm needs to be slightly modified. For example, with $\tau_1 = a$, we need to extend the matrix T_n to

$$\widetilde{T}_{n+1} = \begin{bmatrix} T_n & \beta_n e_n \\ \beta_n e_n^T & \phi \end{bmatrix}. \tag{2.10}$$

Here the parameter ϕ is chosen such that $\tau_1 = a$ is an eigenvalue of $\widetilde{T}_{n+1}$. From [34], it is known that

$$\phi = a + \delta_n,$$

where δ_n is the last component of the solution δ of the tridiagonal system

$$(T_n - aI)\delta = \beta_n^2 e_n.$$

The eigenvalues and the squares of the first components of orthonormal eigenvectors of $\widetilde{T}_{n+1}$ are the nodes and weights of the Gauss-Radau rule.

By combining the Gauss-Radau quadrature and the symmetric Lanczos process, we have an algorithm for computing an estimate $I_n[f]$ of the quadrature form $u^T f(A)u$. We refer to it as the Gauss-Radau-Lanczos (GRL) algorithm.

If $f^{(2n+1)}(\eta) < 0$, then $\widetilde{I}_n[f]$ (with b as a prescribed eigenvalue of $\widetilde{T}_{n+1}$) is a lower bound ℓ_l of the quantity $u^T f(A)u$. $\widetilde{I}_n[f]$ (with a as a prescribed eigenvalue of $\widetilde{T}_{n+1}$) is an upper bound ℓ_u.

Similar to the GL algorithm, it is not always necessary to compute the eigenvalues and the first components of eigenvectors of the tridiagonal matrix $\widetilde{T}_{n+1}$ in order to obtain $\widetilde{I}_n[f]$. In fact, one can show that

$$\widetilde{I}_n[f] = \sum_{k=1}^{n} \omega_k f(\theta_k) + \rho_1 f(\tau_1) = e_1^T f(\widetilde{T}_{n+1})e_1. \tag{2.11}$$

Therefore, if the $(1,1)$ entry of $f(\widetilde{T}_{n+1})$ can be easily computed, for example, $f(\lambda) = 1/\lambda$, we can directly compute $e_1^T f(\widetilde{T}_{n+1})e_1$.

Gauss-Lobatto-Lanczos (GLL) algorithm To implement the Gauss-Lobatto rule, T_n computed in the GL algorithm is updated to

$$\widehat{T}_{n+2} = \begin{bmatrix} T_{n+1} & \psi e_{n+1} \\ \psi e_{n+1}^T & \phi \end{bmatrix}. \tag{2.12}$$

Here the parameters ϕ and ψ are chosen so that a and b are eigenvalues of $\widehat{T}_{n+2}$. Again, from [34], it is known that

$$\phi = \frac{b\delta_{n+1} - a\mu_{n+1}}{\delta_{n+1} - \mu_{n+1}} \quad \text{and} \quad \psi^2 = \frac{b-a}{\delta_{n+1} - \mu_{n+1}},$$

where δ_n and μ_n are the last components of the solutions δ and μ of the tridiagonal systems

$$(T_{n+1} - aI)\delta = e_{n+1} \quad \text{and} \quad (T_{n+1} - bI)\mu = e_{n+1}.$$

The eigenvalues and the squares of the first components of eigenvectors of $\widehat{T}_{n+2}$ are the nodes and weights of the Gauss-Lobatto rule. Moreover, if $f^{(2n+2)}(\eta) > 0$, then $\widehat{I}_n[f]$ is an upper bound ℓ_u of $u^T f(A)u$.

By combining the Gauss-Lobatto quadrature and the symmetric Lanczos process, we have an algorithm for computing an estimation $\widehat{I}_n[f]$ of the quadrature form $u^T f(A)u$. We refer to it as the Gauss-Lobatto-Lanczos (GLL) algorithm.

Similarly to (2.11), we have

$$\widehat{I}_n[f] = \sum_{k=1}^{n} \omega_k f(\theta_k) + \rho_1 f(\tau_1) + \rho_2 f(\tau_2) = e_1^T f(\widehat{T}_{n+2}) e_1. \tag{2.13}$$

2.5 Pseudo-code, software and numerical examples

We have now established all basic algorithms we need to compute the quadratic form $u^T f(A)u$ by applying Gauss, Gauss-Radau and Gauss-Lobatto rules. These algorithms are summarized in the following pseudo-code.

GL, GRL, GLL algorithms Let A be a real symmetric matrix, u a real vector with $u^T u = 1$. f is a given smooth function. Then the following algorithms compute the estimates $I_n[f]$, $\widetilde{I}_n[f]$ and $\widehat{I}_n[f]$ of the quadratic form $u^T f(A)u$ by using the Gauss, Gauss-Radau, and Gauss-Lobatto rules via Lanczos process.

Let $q_0 = 0$, $\beta_0 = 0$, and $q_1 = u$
$\alpha_1 = q_1^T A q_1$
For $j = 1, 2, \ldots, n$,
  $r_j = Aq_j - \alpha_j q_j - \beta_{j-1} q_j$
  $\beta_j = \|r_j\|_2$
  $q_{j+1} = r_j / \beta_j$
  $\alpha_{j+1} = q_{j+1}^T A q_{j+1}$
For GRL algorithm, update T_n according to (2.10)
For GLL algorithm, update T_n according to (2.12)
Compute $I_n[f]$, $\widetilde{I}_n[f]$ and $\widehat{I}_n[f]$ according to (2.9), (2.11) and (2.13)

We note that the "For" loop in the above algorithms is the standard symmetric Lanczos process as described above. The matrix A in question is only referenced in the form of the matrix-vector product. The symmetric Lanczos process can be implemented with only three N-vectors in the fast memory. This is the major storage requirement for the algorithm. These are attractive features for large-scale problems.

For the Gauss-Radau and Gauss-Lobatto rules, we need to have the estimates of a and b as the extreme eigenvalues λ_1 and λ_N of A. Numerical experiments show that more steps of the Lanczos process may be required with poor estimates of a and b. One needs to weigh the cost of using a sophisticated method to obtain good estimates of the extreme eigenvalues against the cost of additional Lanczos iterations. Gershgorin circles can be used to estimate a and b. This is usually sufficient for use in the Gauss-Radau and Gauss-Lobatto rules.

An *ad hoc* choice for determining the number of Lanczos iterations n is to use

$$|I_n[f] - I_{n-1}[f]| \leq \epsilon\, |I_n[f]|,$$

where ϵ is a prescribed tolerance value. This criterion removes the restriction to supply the number of iterations *a priori*. It implies that

$$|I[f] - I_n[f]| \leq |I[f] - I_{n-1}[f]| + \epsilon\, |I_n[f]|.$$

Therefore, the iteration stops if the error is no longer decreasing or is decreasing too slowly.

A Matlab toolset, called QUADFORM, is developed in [25] to implement all GL, GRL and GLL algorithms.

Example 1 This is a symmetric positive definite matrix from the transverse vibration of a vicsek fractal Hamiltonian (VFH). The fractal is recursively constructed. The details are described in [5]. Table 1 shows the numerical results for estimating a few diagonal elements of the inverse of the matrix of dimension $N = 3125$. The parameters a and b are computed by Gershgorin circles. We used $\epsilon = 10^{-4}$ for determining the number of Lanczos iterations in the GRL algorithm. The last two columns are the number of iterations and the approximate values of the conjugate gradient (CG) method.

Table 1: VFH matrix, $e_i^T A^{-1} e_i$ computed by GRL and CG methods.

$(A^{-1})_{ii}$	GRL					CG	
i	iter1	lower bound γ_l	iter2	upper bound γ_u	$\gamma_u - \gamma_l$	iter	$(A^{-1})_{ii}$
1	12	$9.4801416e-01$	19	$9.4801776e-01$	$3.60e-06$	24	$9.4801e-1$
100	11	$1.1005253e+00$	20	$1.1005302e+00$	$4.90e-06$	24	$1.1005e+0$
2000	11	$1.1003685e+00$	19	$1.1003786e+00$	$1.01e-05$	23	$1.1004e+0$
3125	10	$6.4400234e-01$	16	$6.4401100e-01$	$8.66e-06$	21	$6.4400e-1$

Example 2 This is a 1711×1711 matrix from geophysical logs of oil wells data analysis [48]. It is a symmetric positive definite matrix of condition number on the order of 10^7. In this example, we first apply equilibration to improve the condition number. Specifically, the GRL algorithm is applied to the matrix DAD, where $D = \mathrm{diag}(a_{ii}^{-1/2})$. Note that $e_i^T A^{-1} e_i = (De_i)^T (DAD)^{-1} (De_i)$. Table 2 shows the numerical results of GRL and CG methods for estimating few diagonal elements of the inverse of A.

Table 2: Oil Wells matrix, $e_i^T A^{-1} e_i$ computed by GRL and CG methods.

$(A^{-1})_{ii}$	GRL					CG	
i	iter1	lower bound γ_l	iter2	upper bound γ_l	$\gamma_u - \gamma_l$	iter	$(A^{-1})_{ii}$
1	62	$5.2589E-1$	115	$5.4968E-1$	$2.38E-2$	iter	$(A^{-1})_{ii}$
400	244	$7.3332E-2$	344	$7.3605E-2$	$2.73E-4$	335	$5.3138E-1$
900	89	$8.5561E+0$	218	$8.6671E+0$	$1.11E-1$	452	$7.3549E-2$
1400	190	$1.0137E+0$	343	$1.0236E+0$	$9.90E-3$	498	$1.0215E+0$
1711	36	$3.6961E-2$	68	$3.7213E-2$	$2.52E-4$	237	$3.7015E-2$

3 Monte Carlo simulation

In this section, we discuss a Monte Carlo approach for estimating the trace of a matrix function, $\mathrm{tr}(f(A))$. For the task of computing the trace of the inverse of A, we simply take the function $f(\lambda) = 1/\lambda$. For computing the determinant, $\det(A)$, of a symmetric positive definite matrix A, by the identity (2.8), we take $f(\lambda) = \ln(\lambda)$. Instead of applying the GL, GRL or GLL algorithm n times for each diagonal element $e_i^T f(A) e_i$ of $f(A)$, a Monte Carlo approach only applies it m times to obtain an unbiased estimator of the trace of $f(A)$, where in general $m \ll N$. The saving in computational costs could be very significant. Probabilistic confidence bounds for the unbiased estimator can also be obtained. An alternative Monte Carlo approach for computing the trace is presented [12].

Our Monte Carlo approach is based on the following basic property due to [42, 24].

3.1 Proposition *Let H be an $n \times n$ symmetric matrix with $\mathrm{tr}(H) \neq 0$. Let V be the discrete random variable which takes the values 1 and -1 each with probability 0.5 and let z be a vector of n independent samples from V. Then $z^T H z$ is an unbiased estimator of $\mathrm{tr}(H)$, i.e.,*

$$\mathrm{E}(z^T H z) = \mathrm{tr}(H),$$

and

$$\mathrm{Var}(z^T H z) = 2 \sum_{i \neq j} h_{ij}^2.$$

In practice, we take m random sample vectors z_i as described in Proposition 3.1, and then use the GL algorithm to obtain an unbiased estimator of $\mathrm{tr}(f(A))$,

$$\mathrm{E}(z_i^T f(A) z_i) = \mathrm{tr}(f(A)),$$

for $i = 1, 2, \ldots, m$. By using the GRL algorithm, we can obtain a lower bound L_i and an upper bound U_i of the quantity $z_i^T f(A) z_i$:

$$L_i \leq z_i^T f(A) z_i \leq U_i, \tag{3.1}$$

By taking the mean of the m computed lower and upper bounds L_i and U_i, we have

$$\frac{1}{m} \sum_{i=1}^{m} L_i \leq \frac{1}{m} \sum_{i=1}^{m} z_i^T f(A) z_i \leq \frac{1}{m} \sum_{i=1}^{m} U_i. \tag{3.2}$$

It is expected that with a suitable sample size m, Monte Carlo yields good bounds for the quantity $\mathrm{tr}(f(A))$.

To quantitatively assess the quality of such estimation, we can turn to confidence bounds of the estimation. In other words, we can find an interval so that the exact value of $\mathrm{tr}(f(A))$ is in the interval with probability p, where $0 < p < 1$. Hoeffding's exponential inequality in probability theory can be immediately used to derive such confidence bounds [54]. Specifically, let $w_i = z_i^T f(A) z_i - \mathrm{tr}(f(A))$. Since z_i are taken as independent random vectors, w_i are independent random variables. From Proposition 3.1, w_i has zero means (i.e. $E(w_i) = 0$). Furthermore, from (3.1), we also know that w_i has bounded ranges

$$L_{\min} - \mathrm{tr}(f(A)) \leq w_i \leq U_{\max} - \mathrm{tr}(f(A))$$

for all i, $1 \leq i \leq m$, where $U_{\max} = \max\{U_i\}$ and $L_{\min} = \min\{L_i\}$. By Hoeffding's inequality, we have the following probabilistic bounds for the mean of m samples $z^T i f(A) z_i$,

$$P\left(\left|\frac{1}{m}\sum_{i=1}^{m} z_i^T f(A) z_i - \mathrm{tr}(f(A))\right| \geq \frac{\eta}{m}\right) \leq 2\exp\left(\frac{-2\eta^2}{d}\right), \tag{3.3}$$

where $d = m(U_{\max} - L_{\min})^2$ and $\eta > 0$ is a tolerance value, which is related to the probability in the right-hand side of the inequality. In other words, inequality (3.3) tells us that

$$P\left(\frac{1}{m}\sum_{i=1}^{m} z_i^T f(A) z_i - \frac{\eta}{m} < \mathrm{tr}(f(A)) < \frac{1}{m}\sum_{i=1}^{m} z_i^T f(A) z_i + \frac{\eta}{m}\right) > 1 - 2\exp\left(\frac{-2\eta^2}{d}\right).$$

Then from the bounds (3.2), we have

$$P\left(\frac{1}{m}\sum_{i=1}^{m} L_i - \frac{\eta}{m} < \mathrm{tr}(f(A)) < \frac{1}{m}\sum_{i=1}^{m} U_i + \frac{\eta}{m}\right) > 1 - 2\exp\left(\frac{-2\eta^2}{d}\right). \tag{3.4}$$

Therefore, we conclude that the trace of $f(A)$ is in the interval

$$\left(\frac{1}{m}\sum_{i=1}^{m} L_i - \frac{\eta}{m}, \frac{1}{m}\sum_{i=1}^{m} U_i + \frac{\eta}{m}\right)$$

with probability $1 - 2\exp(-2\eta^2/d)$.

If we specify the probability p in (3.4), i.e. $p = 1 - 2\exp(-2\eta^2/d)$, then solving this equality for η/m, yields

$$\frac{\eta}{m} = \sqrt{-\frac{1}{2m}(U_{\max} - L_{\min})^2 \ln\left(\frac{1-p}{2}\right)}. \tag{3.5}$$

Since $(U_{\max} - L_{\min})^2$ is bounded by $2N^2\|f(A)\|_2^2$, we see that with a fixed value of p, $\eta/m \to 0$ as $m \to \infty$, i.e., the confidence interval is essentially given by means of the computed bounds.

We now have a Monte Carlo algorithm which computes an unbiased estimator of $\mathrm{tr}(f(A))$. The algorithm also returns a confidence interval with user specified probability. We note that L_i and U_i are generally very sharp bounds of $z_i^T f(A) z_i$. It would be ideal if we could have a sharp confidence interval, i.e., η/m is small. However, from equation (3.5), we may have to choose a quite large number of samples m. It would be too expensive. Instead, we generally choose a fixed number of samples m and the probability p to compute the corresponding confidence interval. Here is the algorithm based on the Monte Carlo approach.

Monte Carlo algorithm Suppose A is symmetric positive definite. Let m be a chosen number of samples. Then the following algorithm computes (a) an unbiased estimator I_p of the quantity $\mathrm{tr}(f(A))$, and (b) a confidence interval (L_p, U_p) such that $\mathrm{tr}(f(A)) \in (L_p, U_p)$ with a user-specified probability p, where $0 < p < 1$.

For $j = 1, 2, \cdots, m$
  Generate n-vector z_j with uniformly distributed elements in $(0, 1)$.
  For $i = 1 : n$, if $z_j(i) < 0.5$, then $z_j(i) = -1$, otherwise, $z_j(i) = 1$.
  Apply GL algorithm to obtain an estimator I_j of $z_j^T f(A) z_j$
  Apply GRL algorithm to obtain the bounds (L_j, U_j) of $z_j^T f(A) z_j$
  $L_{\min} = \min\{L_{\min}, L_j\}$
  $U_{\max} = \max\{U_{\max}, U_j\}$
  $\eta^2 = -0.5j(U_{\max} - L_{\min})^2 \ln(\frac{1-p}{2})$
  $L_p(j) = \frac{1}{j}\sum_{i=1}^{j} L_i - \frac{\eta}{j}$
  $U_p(j) = \frac{1}{j}\sum_{i=1}^{j} U_i + \frac{\eta}{j}$
End
$I_p = \frac{1}{m}\sum_{j=1}^{m} I_j$
$L_p = L_p(m)$
$U_p = L_p(m)$

A couple of improvements of Monte Carlo simulation for computing the trace of a matrix function have been proposed recently. In [62], it is proposed to use a low discrepancy sampling method for a better choice of sample vectors z_j to improve the convergence rate. One can also develop a variance reduction technique via control regression. The essential idea is to apply the first few easy-to-compute moments of the matrices A and T_n to minimize the variance of the estimates I_j via regression. The following example includes a preliminary result of variance reduction.

Example 3 This is a consistent mass matrix from a regular $n_x \times n_y$ grid of 8-node (serendipity) elements. It is from Higham's test matrix collection available in Matlab's `gallery`. The order of the matrix is $N = 3n_x n_y + 2n_x + 2n_y + 1$. We choose $n_x = n_y = 12$ and then $N = 481$. Numerical results of a Monte Carlo simulation of $\mathrm{tr}(A^{-1})$ and variance reduction are plotted in Fig. 1. Solid horizontal lines in the first top two plots are the exact value $\mathrm{tr}(A^{-1})$. In the top plot, the solid plus and dash circle lines are the estimates by GL algorithm and improved ones with variance reduction for 30 different random samples z_j, respectively. In the middle plot, the solid plus and dash circle lines are the means of the GL estimates and improved ones. The bottom plot is the variances before (solid plus) and after (dash circle) applying the variance reduction technique via control regression.

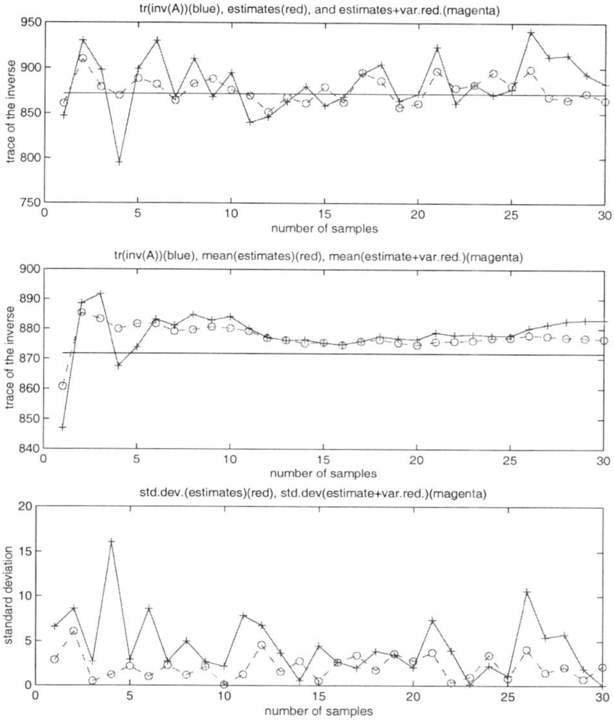

Figure 1: Monte Carlo simulation of $\mathrm{tr}(A^{-1})$.

4 Transfer functions

4.1 Linear dynamical systems and transfer functions

A continuous time-invariant lumped multi-input multi-output linear dynamical system is of the form

$$\left.\begin{aligned} C\dot{x}(t) + Gx(t) &= B\,u(t) \\ y(t) &= L^T x(t) \end{aligned}\right\} \tag{4.1}$$

with initial condition $x(0) = x_0$. Here t is the time variable, $x(t) \in \mathcal{R}^N$ is a state vector, $u(t) \in \mathcal{R}^m$ the input excitation vector, and $y(t) \in \mathcal{R}^p$ the output measurement vector. $C, G \in \mathcal{R}^{N\times N}$ are system matrices, $B \in \mathcal{R}^{N\times m}$ and $L \in \mathcal{R}^{N\times p}$ are input and output distribution arrays, respectively. N is the state space dimension and m and p are the number of inputs and outputs, respectively. In most practical cases, we can assume that m and p are much smaller than N and $m \geq p$.

Linear systems arise in many applications, such as network circuits with linear elements [60], structural dynamics analysis with only lumped mass and stiffness elements [19, 20], linearization of a nonlinear system around an equilibrium point [27], and semi-discretization with respect to spatial variables of a time-dependent differential-integral equations [55, 61].

The matrices C and G in (4.1) are allowed to be singular, and we only assume that the pencil $G + sC$ is *regular*, i.e., the matrix $G + sC$ is singular only for a finite number of values $s \in \mathcal{C}$. The assumption that $G + sC$ is regular is satisfied for all applications we are concerned with that lead to systems of the form (4.1). In addition, C and G in (4.1) are general nonsymmetric matrices. However, in some important applications, C and G are symmetric, and possibly positive definite or positive semidefinite. Note that when C is singular, the first equation in (4.1) is a first-order system of linear differential-algebraic equations. The corresponding linear system is called a descriptor system or a singular system.

The linear system of the form (4.1) is often referred to as the representation of the system in the time domain or in the state space. Equivalently, one can also represent the system in the frequency domain via a Laplace transform. Recall that for a vector-valued function $f(t)$, the Laplace transform of $f(t)$ is defined by

$$F(s) := \mathcal{L}\{f(t)\} = \int_0^\infty f(t)e^{-st}\,dt, \quad s \in \mathcal{C}. \tag{4.2}$$

The physically meaningful values of the complex variable s are $s = \mathrm{i}\omega$, where $\omega \geq 0$ is referred to as the frequency. Taking the Laplace transform of the system (4.1), we obtain the following frequency domain formulation of the system:

$$\left.\begin{aligned} sCX(s) + GX(s) &= B\,U(s) \\ Y(s) &= L^T X(s) \end{aligned}\right\} \tag{4.3}$$

where $X(s)$, $Y(s)$ and $U(s)$ represent the Laplace transform of $x(t)$, $y(t)$ and $u(t)$, respectively. For simplicity, we assume that we have zero initial conditions $x(0) = x_0 = 0$ and $u(0) = 0$.

Eliminating the variable $X(s)$ in (4.3), we see that the input $U(s)$ and the output $Y(s)$ in the frequency domain are related by the following $p \times m$ matrix-valued rational function

$$H(s) = L^T(G + sC)^{-1}B. \tag{4.4}$$

$H(s)$ is known as the *transfer function* or *Laplace-domain impulse response* of the linear system (4.1).

Steady-state analysis, also called frequency response analysis, is to determine the frequency responses $H(\mathrm{i}\omega)$ of the system to external steady-state oscillatory (i.e., sinusoidal) excitation.

Linear dynamical systems have been studied extensively, especially for the case $C = I$, for example, see [44]. Numerous techniques have been developed for performing various analyses of the system. One of the primary computational challenges we are confronted with today is the large state dimension N of the system (4.1). For example, in circuit simulation and structural dynamics applications, N could be as large as 10^6. In addition, the differential equations in the system (4.1) are often stiff from multi-energy and multi-scaling simulation. The system may be required to be analyzed repeatedly for different excitation inputs $u(t)$.

For the sake of simplicity, in the rest of this section we mostly confine our discussion to single-input single-output systems, i.e., $p = m = 1$. We will use lower case letters b and l to denote the input and output distribution vectors, instead of the capital letters B and L.

4.2 Eigensystem methods

Let us first review eigensystem methods as an introduction to compute the transfer function. To compute $H(s)$ about a selected expansion point s_0, let us set

$$A = -(G + s_0C)^{-1}C \quad \text{and} \quad r = (G + s_0C)^{-1}b,$$

where we assume that $G + s_0C$ is nonsingular. Then $H(s)$ can be cast as

$$H(s) = l^T((G + s_0C) + (s - s_0)C)^{-1}b = l^T(I - (s - s_0)A)^{-1}r. \tag{4.5}$$

In other words, we reduce the representation of the transfer function $H(s)$ using only one matrix A. Assume that the matrix A is diagonalizable,

$$A = S\Lambda S^{-1} = S \cdot \mathrm{diag}(\lambda_1, \lambda_2, \ldots, \lambda_N) \cdot S^{-1}.$$

Let $f = S^T l = (f_j)$ and $g = S^{-1}r = (g_j)$, then the transfer function $H(s)$ can be expressed as a partial-fraction expansion,

$$H(s) = f^T(I - (s - s_0)\Lambda)^{-1}g = \sum_{j=1}^{N} \frac{f_j g_j}{1 - (s - s_0)\lambda_j} = \rho_\infty + \sum_{\lambda_j \neq 0} \frac{\kappa_j}{s - p_j}. \tag{4.6}$$

This is known as the *pole-residue representation*, $p_j = s_0 + 1/\lambda_j$ are the *poles* of the system[1], $\kappa_j = -f_j g_j/\lambda_j$ are the *residues*, and $\rho_\infty = \sum_{\lambda_j=0} f_j g_j$ is a constant, which corresponds to the poles at infinity (or zero eigenvalues). Note that it costs $O(N^3)$ operations to diagonalize A, and only $O(N)$ operations to evaluate the transfer function $H(s)$ for each given point s.

Unfortunately, in practice, diagonalization of A is prohibitive when it is ill-conditioned or too large. As a remedy for the possible ill-conditioning of diagonalization, we may use the numerically stable Schur decomposition. Let $A = QTQ^T$ be the Schur decomposition of A. Then

$$H(s) = l^T(I - (s - s_0)A)^{-1}r = (Q^T l)^T(I - (s - s_0)T)^{-1}(Q^T r).$$

Now, it costs $O(N^2)$ to evaluate the transfer function $H(s)$ at each given point s. Alternatively, we can also use the Hessenberg decomposition as suggested in [47].

To reduce the cost of diagonalizing A or computing its decomposition in Schur or Hessenberg form for large N, we may use partial eigen-decomposition. This is also referred to as the *modal superposition method*, for example, see [19]. By examining the pole-residue representation (4.6), it is easy to see that the motivation of this approach comes from the fact that only a few poles (and associated eigenvalues) around the region of frequencies of interest are necessary for the approximation of $H(s)$. Those poles are called the dominant poles. Therefore, to study the steady-state response to an input of the form $u(t) = \tilde{u}e^{i\omega t}$, where $\tilde{u}$ is a constant vector, we express the solution as $x(t) = S_k v(\omega)e^{i\omega t}$, where S_k contains k selected modal shapes (eigenvectors) of the matrix pair $\{C, G\}$ needed to retain all the

[1] By a simple exercise, it can be shown that the definition of poles and residues of the system is independent of the choice of the expansion point s_0.

modes whose resonant frequencies lie within the range of input excitation frequencies. Then one may solve the system

$$(\mathrm{i}\omega\, S_k^T C S_k + S_k^T G S_k) v(\omega) = S_k^T B \tilde{u} \tag{4.7}$$

for $v(\omega)$. Once the selected dominant poles and their corresponding modal shapes S_k are computed, the problem of computing the steady-state response is reduced to solving the $k \times k$ system (4.7). In practice, it is typical that only a relatively small number of the modal shapes is necessary, i.e., $k \ll N$. The problem of finding a few modal shapes S_k within a certain frequency range is one of the well-known algebraic eigenvalue problems in numerical linear algebra [3].

4.3 Padé approximation and moment-matching

Note that the transfer function $H(s)$ of (4.4) is a rational function. More precisely, $H(s) \in \mathcal{R}_{N-1,N}$, where N is the state-space dimension of (4.1).[2] The Taylor series expansion of $H(s)$ of (4.5) about s_0 is given by

$$\begin{aligned} H(s) &= l^T (I - (s - s_0)A)^{-1} r \\ &= l^T r + (l^T A r)(s - s_0) + (l^T A^2 r)(s - s_0)^2 + \cdots \\ &= m_0 + m_1 (s - s_0) + m_2 (s - s_0)^2 + \cdots , \end{aligned} \tag{4.8}$$

where $m_j = l^T A^j r$ for $j = 0, 1, 2, \ldots$, are called *moments* about s_0. Since our primary concern is large state-space dimension N, we seek to approximate $H(s)$ by a rational function $H_n(s) \in \mathcal{R}_{n-1,n}$ over the range of frequencies of interest, where $n \leq N$. A natural choice of such a rational function is a Padé approximation. A function $H_n(s) \in \mathcal{R}_{n-1,n}$ is said to be an nth *Padé approximant* of $H(s)$ about the expansion point s_0 if it matches with the moments of $H(s)$ as far as possible. Precisely, it is required that

$$H(s) = H_n(s) + O((s - s_0)^{2n}). \tag{4.9}$$

For a thorough treatment of Padé approximants, we refer the reader to [11]. Note that equation (4.9) presents $2n$ conditions on the $2n$ degrees of freedom that describe any function $H_n(s) \in \mathcal{R}_{n-1,n}$. Specifically, let

$$H_n(s) = \frac{P_{n-1}(s)}{Q_n(s)} = \frac{a_{n-1}s^{n-1} + \cdots + a_1 s + a_0}{b_n s^n + b_{n-1}s^{n-1} + \cdots + b_1 s + 1}, \tag{4.10}$$

where b_0 is chosen to be equal to 1, which eliminates an arbitrary multiplicative factor in the definition of $H_n(s)$. Then the coefficients $\{a_j\}$ and $\{b_j\}$ of the polynomials $P_{n-1}(s)$ and $Q_n(s)$ can be computed as follows. Multiplying by $Q_n(s)$ on both sides of (4.9) yields

$$H(s) Q_n(s) = P_{n-1}(s) + O((s - s_0)^{2n}). \tag{4.11}$$

[2] $\mathcal{R}_{m,n}$ denotes the set of rational functions with real numerator polynomial of degree at most m and real denominator polynomial of degree at most n.

Comparing the first n $(s-s_0)^k$-terms of (4.11) for $k=0,1,\ldots,n-1$ shows that the coefficients $\{b_j\}$ of the denominator polynomial $Q_n(s)$ satisfy the following system of simultaneous equations:

$$\begin{bmatrix} m_0 & m_1 & \cdots & m_{n-1} \\ m_1 & m_2 & \cdots & m_n \\ \vdots & \vdots & \cdots & \vdots \\ m_{n-1} & m_n & \cdots & m_{2n-2} \end{bmatrix} \begin{bmatrix} b_n \\ b_{n-1} \\ \vdots \\ b_1 \end{bmatrix} = - \begin{bmatrix} m_n \\ m_{n+1} \\ \vdots \\ m_{2n-1} \end{bmatrix}. \qquad (4.12)$$

The coefficient matrix of (4.12) is called the *Hankel matrix*, denoted as M_n. Once the coefficients $\{b_j\}$ are computed, then by comparing the second n $(s-s_0)^k$-terms of (4.11) for $k=n,n+1,\ldots,2n-1$, we see that the coefficients $\{a_j\}$ of the numerator polynomial $P_{n-1}(s)$ can be computed according to

$$\begin{aligned} a_0 &= m_0 \\ a_1 &= m_0 b_1 + m_1 \\ &\vdots \\ a_{n-1} &= m_0 b_{n-1} + m_1 b_{n-2} + \cdots + m_{n-2} b_1 + m_{n-1}. \end{aligned}$$

It is clear that $H_n(s)$ defines a unique nth Padé approximant of $H(s)$ if, and only if, the Hankel matrix M_n is nonsingular. We will assume that this is the case for all n.

This formulates the framework of the asymptotic waveform evaluation (AWE) techniques as they are known in circuit simulation, first presented in [53] around 1990. The manuscript [18] has a complete treatment of the AWE technique and its variants. A survey of the Padé techniques for model reduction of linear systems is also presented in the earlier work [15]. It is well-known that in practice the Hankel matrix M_n is generally extremely ill-conditioned. Therefore, the computation of Padé approximants using explicit moments is inherently numerically unstable. Indeed, this approach can be used only for very small values of n, such as $n \leq 20$, even with some sophisticated schemes to improve the conditioning of the underlying Hankel matrix M_n. As a result, the approximation range of a computed Padé approximant is limited to only a narrow frequency range around the selected expansion point s_0. A large number of expansion points is generally required for the approximation of the transfer function $H(s)$ over a broad frequency range of interest. Since for each expansion point s_0, one has to be concerned with the cost of applying the matrix $A = -(G+s_0C)^{-1}C$, which is generally the most expensive part of the overall computational cost, one would like to use as few expansion points as possible by increasing the order n of Padé approximants with a selected expansion point s_0. Fortunately, numerical difficulties associated with explicit moments can be remedied by exploiting the well-known connection between the Padé approximants and the Lanczos process. We will discuss this connection in the next section.

4.4 Krylov subspaces and the Lanczos process

A Krylov subspace is a subspace spanned by a sequence of vectors generated by a given matrix and a vector as follows. Given a matrix A and a starting vector r, the nth Krylov subspace $\mathcal{K}_n(A,r)$ is spanned by a sequence of n column vectors:

$$\mathcal{K}_n(A,r) = \operatorname{span}\{r, Ar, A^2r, \ldots, A^{n-1}r\}.$$

This is sometimes called the right Krylov subspace. When the matrix A is nonsymmetric, there is a left Krylov subspace generated by A^T and a starting vector l defined by

$$\mathcal{K}_n(A^T, l) = \text{span}\{l, A^T l, (A^T)^2 l, \dots, (A^T)^{n-1} l\}.$$

Note that the first $2n$ moments $\{m_j\}$ of $H(s)$ in (4.8), which define the Hankel matrix M_n in the Padé approximant (4.12), are connected with Krylov subspaces through computing the inner products between the left and right Krylov sequences:

$$m_{2j} = ((A^T)^j l)^T \cdot (A^j b)^T, \qquad m_{2j+1} = ((A^T)^j l)^T \cdot (A^{j+1} b)^T,$$

for $j = 1, 2, \dots, n-1$. Therefore, loosely speaking, the left and right Krylov subspaces contain the desired information of moments, but the vectors $\{A^j r\}$ and $\{(A^T)^j l\}$ are unsuitable as basis vectors. The remedy is to construct more suitable basis vectors:

$$\{v_1, v_2, \dots, v_n\} \quad \text{and} \quad \{w_1, w_2, \dots, w_n\},$$

such that they span the same desired Krylov subspaces, specifically,

$$\mathcal{K}_n(A, r) = \text{span}\{v_1, v_2, \dots, v_n\} \quad \text{and} \quad \mathcal{K}_n(A^T, l) = \text{span}\{w_1, w_2, \dots, w_n\}.$$

It is well known that the nonsymmetric Lanczos process is an elegant way to generate the desired basis vectors of two Krylov subspaces [46]. Given a matrix A, a right starting vector r and a left starting vector l, the nonsymmetric Lanczos process generates the desired basis vectors $\{v_i\}$ and $\{w_i\}$, known as the *Lanczos vectors.* Moreover, these Lanczos vectors are constructed to be biorthogonal

$$w_j^T v_k = 0, \quad \text{for all} \quad j \neq k. \tag{4.13}$$

The Lanczos vectors can be generated by two three-term recurrences. These recurrences can be stated compactly in matrix form as follows:

$$\begin{aligned} AV_n &= V_n T_n + \rho_{n+1} v_{n+1} e_n^T, \\ A^T W_n &= W_n \widetilde{T}_n + \eta_{n+1} w_{n+1} e_n^T, \end{aligned}$$

where T_n and $\widetilde{T}_n$ are the tridiagonal matrices

$$T_n = \begin{bmatrix} \alpha_1 & \beta_2 & & \\ \rho_2 & \alpha_2 & \ddots & \\ & \ddots & \ddots & \beta_n \\ & & \rho_n & \alpha_n \end{bmatrix}, \qquad \widetilde{T}_n^T = \begin{bmatrix} \alpha_1 & \gamma_2 & & \\ \eta_2 & \alpha_2 & \ddots & \\ & \ddots & \ddots & \gamma_n \\ & & \eta_n & \alpha_n \end{bmatrix}$$

and they are related by a diagonal similarity transformation $\widetilde{T}_n^T = D_n T_n D_n^{-1}$, where $D_n = W_n^T V_n = \text{diag}(\delta_1, \delta_2, \dots, \delta_k)$. The projection of the matrix A onto $\mathcal{K}_n(A, r)$ and orthogonally to $\mathcal{K}_n(A^T, l)$ is represented by

$$W_n^T A V_n = D_n T_n.$$

If the nonsymmetric Lanczos process is carried to the end with N being the last step, then it can be viewed as a means of tridiagonalizing A by a similarity transformation:

$$V_N^{-1} A V_N = T_N, \tag{4.14}$$

where T_N is a tridiagonal matrix, with T_n as its $n \times 2n$ leading principal submatrix, $n \leq N$. The Lanczos vectors are determined up to a scaling. We use the scaling $\|v_j\|_2 = \|w_j\|_2 = 1$ for all j.

An algorithm template for the basic nonsymmetric Lanczos process is presented as the following:

> **Nonsymmetric Lanczos process** Let A be a real nonsymmetric matrix, and let r and l be real vectors. Then the following procedure computes the tridiagonal matrices T_n and $\widetilde{T}_n$, and the biorthogonal Lanczos vectors v_k and w_k.

$\rho_1 = \|r\|_2$
$\eta_1 = \|l\|_2$
$v_1 = r/\rho_1$
$w_1 = l/\eta_1$
For $k = 1, 2, \ldots, n$
  $\delta_k = w_k^T v_k$
  $\alpha_k = w_k^T A v_k / \delta_k$
  $\beta_k = (\delta_k/\delta_{k-1})\eta_k$
  $\gamma_k = (\delta_k/\delta_{k-1})\rho_k$
  $v = Av_k - v_k\alpha_k - v_{k-1}\beta_k$
  $w = A^T w_k - w_k\alpha_k - w_{k-1}\gamma_k$
  $\rho_{k+1} = \|v\|_2$
  $\eta_{k+1} = \|w\|_2$
  $v_{k+1} = v/\rho_{k+1}$
  $w_{k+1} = w/\eta_{k+1}$

We note that the nonsymmetric Lanczos process could stop prematurely due to $\delta_k = 0$ (or $\delta_k \approx 0$ considering the finite precision arithmetic). This is called *breakdown*. Our assumption of the nonsingularity of the Hankel matrix M_n guarantees that no breakdown occurs, see [52]. In practice, the problem is curable by a variant of the nonsymmetric Lanczos process, for example, a look-ahead scheme is proposed in [29]. An implementation of the nonsymmetric Lanczos process with a look-ahead scheme to overcome the breakdown can be found in QMRPACK [30].

4.5 Padé approximation using the Lanczos process

Let us first consider the nonsymmetric Lanczos process as a process for tridiagonalizing the matrix A. Then by (4.14), the transfer function $H(s)$ of the original system (4.1) can be rewritten as

$$H(s) = (l^T r)\, e_1^T (I - (s - s_0)T_N)^{-1} e_1 = (l^T r)\, \frac{\det(I - (s - s_0)T_N')}{\det(I - (s - s_0)T_N)}, \tag{4.15}$$

where T'_N is an $(N-1)\times(N-1)$ matrix obtained by deleting the first row and column of T_N. Note that for the second equality, we have used the following Cauchy-Binet theorem to the matrix $I-(s-s_0)T_N$:

$$(I-(s-s_0)T_N)\cdot \operatorname{adj}(I-(s-s_0)T_N) = \det(I-(s-s_0)T_N)\cdot I,$$

where adj(X) stands for the classical adjoint matrix made up of the $(N-1)\times(N-1)$ cofactors of X. Expression (4.15) is called the *zero-pole representation*. It is clear that the poles of $H(s)$ can be computed from the eigenvalues of the $N\times N$ tridiagonal matrix T_N and the zeros of $H(s)$ from the eigenvalues of the $(N-1)\times(N-1)$ tridiagonal matrix T'_N. More precisely, the poles are given by $p_j = s_0 + 1/\lambda_j$, $\lambda_j\in\lambda(T_N)$, and the zeros by $z_j = s_0+1/\lambda'_j$, $\lambda'_j\in\lambda(T'_N)$.

Now, let us turn to large-scale linear systems where the order N of the matrix A is too large to fully tridiagonalize, and where the Lanczos process terminates at n $(\le N)$. Then it is natural to define an n-th reduced-order approximation of the transfer function $H(s)$ as

$$H_n(s) = (l^Tr)\, e_1^T(I-(s-s_0)T_n)^{-1}e_1, \tag{4.16}$$

where T_n is the $n\times n$ leading principal submatrix of T_N, as generated by the first n steps of the nonsymmetric Lanczos process. In analogy to (4.15), we have the zero-pole representation of $H_n(s)$:

$$H_n(s) = (l^Tr)\,\frac{\det(I-(s-s_0)T'_n)}{\det(I-(s-s_0)T_n)}, \tag{4.17}$$

where T'_n is an $(n-1)\times(n-1)$ matrix obtained by deleting the first row and column of T_n.

Now, the question is: what is $H_n(s)$? The answer, which seems surprising to many first-time readers, is that $H_n(s)$ is the Padé approximation of $H(s)$ as computed by using explicit moments in Section 4.3. To show this, let us first recall the following proposition, which was originally developed in [64] for a convergence analysis of the Lanczos algorithm for eigenvalue problems.

4.1 Proposition *If T_n is the $n\times n$ leading principal submatrix of T_N, where $n\le N$, then for any $0\le j\le 2n-1$,*

$$e_1^T T_N^j e_1 = e_1^T T_n^j e_1$$

and for $j=2n$,

$$e_1^T T_N^{2n} e_1 = e_1^T T_n^{2n} e_1 + \beta_2\beta_3\cdots\beta_n\beta_{n+1}\cdot\rho_2\rho_3\cdots\rho_n\rho_{n+1}.$$

A verification of this proposition can be easily carried out by induction. By Proposition 4.1, we immediately see that the first $2n$ moments of $H(s)$ and $H_n(s)$ are matched:

$$m_j = l^TA^jr = (l^Tr)e_1^TT_N^je_1 = (l^Tr)e_1^TT_n^je_1 = \widehat{m}_j \tag{4.18}$$

for $j=0,1,\ldots,2n-1$. Furthermore, by Taylor expansions of $H(s)$ and $H_n(s)$ about s_0 and (4.18), we have

$$H(s) = H_n(s) + (l^Tr)\left(\prod_{j=2}^{n+1}\beta_j\prod_{j=2}^{n+1}\rho_j\right)(s-s_0)^{2n} + O\left((s-s_0)^{2n+1}\right).$$

Therefore, we conclude that $H_n(s)$ is a Padé approximant of $H(s)$.

This Lanczos-Padé connection at least goes back to [39] and [40]. The work of [26, 32] advocates the use of the Lanczos-Padé connection instead of the mathematically equivalent, but numerically unstable AWE method [53] in the circuit simulation community. The Lanczos-based Padé approximation method has become known as the PVL (Padé Via Lanczos) method, as coined in [26]. An overview of various Krylov methods and their applications in model reduction for state-space control models in control system theory is presented in [13]. The presentation style here partially follows the work of [10]. In the following, we present two examples, one from circuit simulation and one from structural dynamics, as empirical validation of the efficiency of the PVL method. We note that in both cases, we only use one expansion point s_0 over the entire range of frequencies of interest. However, the degree of the underlying Padé approximants constructed via the Lanczos process is as high as 60, which seems to be an impossible mission by using explicit moment-matching as discussed in Section 4.3.

Example 4 This example demonstrates the efficiency of the PVL method for a popular circuit problem, which simulates a lumped element network generated by a 3-D electromagnetic problem modeled via the partial element equivalent circuit (PEEC) model [18, 26]. The PEEC model is obtained by appropriate discretizations of the boundary integral formulation of Maxwell's equations for the electric and magnetic fields at any point in a conductor [55]. The order of the system matrices C and G is 306. To capture the dynamic behavior of the transfer function $H(\mathrm{i}\omega)$ over the broad frequency range $[\omega_{\min}, \omega_{\max}] = [1, 5 \times 10^9]$, it is necessary to evaluate $H(s)$ at a large number of frequency points. We used a total of 1001 frequency points. On the left of Fig. 2, we plot the absolute values of $H(\mathrm{i}\omega)$ and the Padé approximant $H_{60}(\mathrm{i}\omega)$ of order 60 generated by the PVL method with only a single expansion $s_0 = 2\pi \times 10^9$. Note that it is nearly indistinguishable from the curve of $|H(\mathrm{i}\omega)|$. The right plot of Fig. 2 is the relative error between $H(\mathrm{i}\omega)$ and $H_{60}(\mathrm{i}\omega)$.

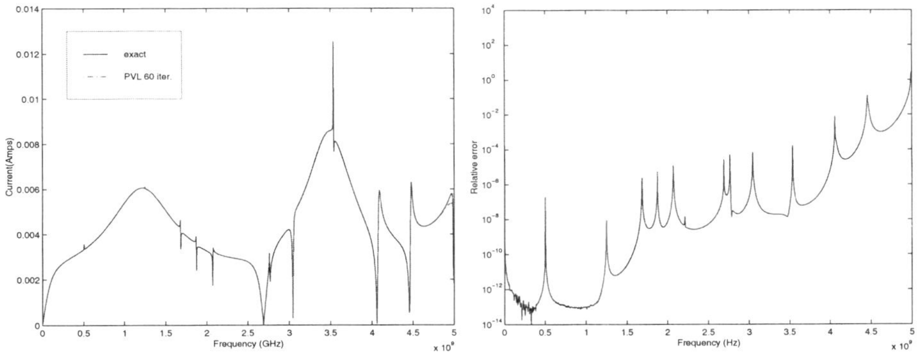

Figure 2: PEEC example, $|H(\mathrm{i}\omega)|$ and PVL $|H_{60}(\mathrm{i}\omega)|$ (left) and relative error $|H(\mathrm{i}\omega) - H_{60}(\mathrm{i}\omega)|/|H(\mathrm{i}\omega)|$ (right).

Example 5 This example is from dynamics analysis of automobile brakes, extracted from

MSC/NASTRAN, a finite element analysis software for structural dynamics [45]. The order of the mass matrix M and stiffness matrix K is 834. The transfer function is of the form $H(\mathrm{i}\omega) = l^T(K-\omega^2 M)^{-1}b$. The expansion point is chosen as $s_0 = 0$. A total of 501 frequency points is evaluated between 0 and 10000 Hz. The left plot of Fig. 3 shows the magnitudes of the original transfer function $H(\mathrm{i}\omega)$ and the reduced-order transfer function $H_{45}(\mathrm{i}\omega)$ after 45 PVL iterations. The right plot of Fig. 3 shows the relative error between $H(\mathrm{i}\omega)$ and $H_{45}(\mathrm{i}\omega)$.

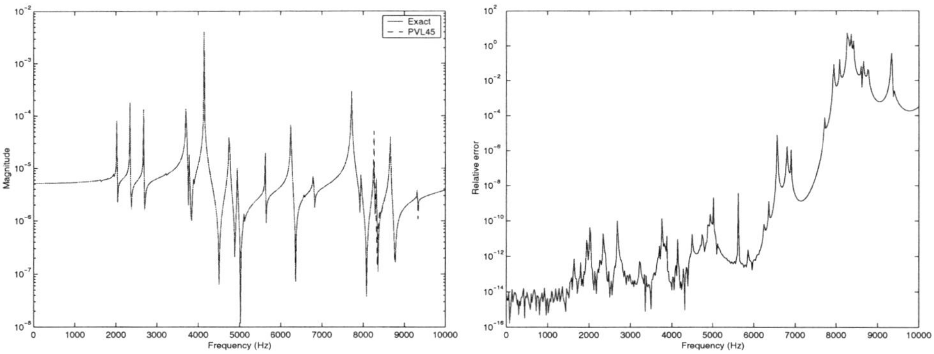

Figure 3: Automobile brake example, $|H(\mathrm{i}\omega)|$ and PVL $|H_{45}(\mathrm{i}\omega)|$ (left) and relative error $|H(\mathrm{i}\omega) - H_{45}(\mathrm{i}\omega)|/|H(\mathrm{i}\omega)|$ (right).

4.6 Error estimation

An important question associated with the PVL method is how to determine the order n of a Padé approximant $H_n(s)$, or equivalently, the number of steps of the Lanczos process in order to achieve a desired accuracy of the approximation. In [9], through an algebraic derivation, it is shown that the forward error between the full-order transfer function $H(s)$ and the reduced-order transfer function $H_n(s)$ is given by

$$H(s) - H_n(s) = (l^T r)\left(\frac{\rho_{n+1}\eta_{n+1}}{\delta_n}\right)[\sigma^2 \tau_{n1}(\sigma)\tau_{1n}(\sigma)]\gamma_{n+1}(\sigma), \tag{4.19}$$

where $\sigma = s - s_0$, $\tau_{1n}(\sigma) = e_1^T(I-\sigma T_n)^{-1}e_n$, $\tau_{n1}(\sigma) = e_n^T(I-\sigma T_n)^{-1}e_1$, and $\gamma_{n+1}(\sigma) = w_{n+1}^T(I-\sigma A)^{-1}v_{n+1}$. From (4.19), we see that there are essentially two factors to determine the forward error of the PVL method, namely $\sigma^2\tau_{n1}(\sigma)\tau_{1n}(\sigma)$ and $\gamma_{n+1}(\sigma)$. Numerous numerical experiments indicate that the first factor, which can be easily computed during the PVL approximation, is the primary contributor to the convergence of the PVL approximation, while the second factor tends to be steady when n increases. Note that $\tau_{n1}(\sigma)$ and $\tau_{1n}(\sigma)$ are the $(1,n)$ and $(n,1)$ elements of the inverse of the tridiagonal matrix $I-\sigma T_n$. This is in agreement with the rapid decay phenomenon observed in the inverse of a band matrix [49]. Fig. 4 shows typical convergence behavior of the factor $|\sigma^2\tau_{n1}(\sigma)\tau_{1n}(\sigma)|$ for a fixed σ. The direct computation of the second factor $\gamma_{n+1}(\sigma)$ would cost just as much as computing the original transfer function. It is advocated that $w_{n+1}^T A v_{n+1}$ be used as an estimation of the factor $\gamma_{n+1}(\sigma)$ near convergence. With this observation, it is possible to implement the PVL method with an adaptive stopping criteria to determine the required number of Lanczos iterations, see [9]. Related work for error estimation can be found in [43, 41] and recently in [51].

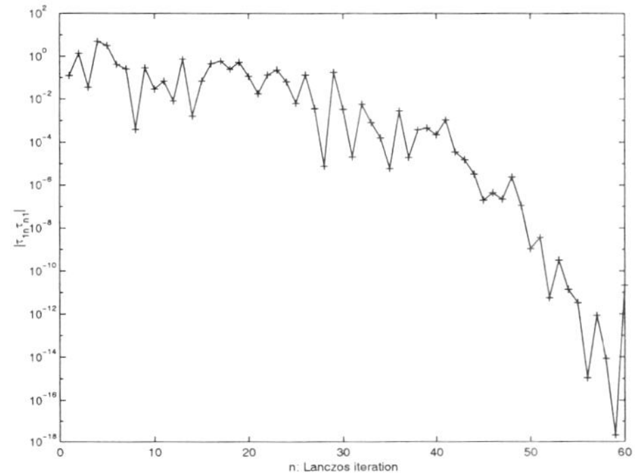

Figure 4: Convergence of $|\tau_{n1}(\sigma)\tau_{1n}(\sigma)|$ for a fixed $\sigma = s - s_0$.

More efficient and accurate error estimations of the PVL approximation and its extension to the other moment-matching based Krylov techniques warrant further study. One alternative approach is to use the technique of backward error analysis. By some algebraic derivation, it can be shown that the reduced-order transfer function $H_n(s)$ of (4.16) can be interpreted as the exact transfer function of a perturbed full-order system. Specifically,

$$H_n(s) = (l^T r) \cdot e_1^T (I - (s - s_0)T_n)^{-1} e_1 = l^T [I - (s - s_0)(A + F_n)]^{-1} r,$$

where

$$F_n = -\frac{1}{\delta_n} \begin{bmatrix} v_n & v_{n+1} \end{bmatrix} \begin{bmatrix} 0 & \eta_{n+1} \\ \rho_{n+1} & 0 \end{bmatrix} \begin{bmatrix} w_n^T \\ w_{n+1}^T \end{bmatrix}.$$

Therefore, one may use $\|F_n\|$ for monitoring convergence. However, it is observed that this is generally a conservative monitor and often does not indicate practical convergence. An open problem is to find an optimal normwise relative backward error

$$\eta(\epsilon) = \min\{\epsilon : l^T \left[I - (s - s_0)(A + F_n)\right]^{-1} r = H_n(s), \ \|F_n\| \leq \epsilon \|A\|\}.$$

With this optimal backward error and perturbation analysis of the transfer function $H(s)$, one may be able to derive a more efficient error estimation scheme.

5 Reduced-order modeling

The Padé approximation of the transfer function $H(s)$ using the Lanczos process naturally leads to an efficient method for reduced-order modeling of large-scale linear dynamical systems. The desired attributes of a reduced-order model include replacing the full-order system by a system of the same type but with a much smaller state-space dimension such that it has an admissible error between the full-order and reduced-order models. Furthermore, the reduced-order model should also preserve essential properties of the full-order system. Such a reduced-order model would let designers efficiently analyze and synthesize the dynamical behavior of the original system within a tight design cycle. Specifically, given the linear dynamical system (4.1), we want to find a reduced-order linear system of the same form

$$\left.\begin{aligned} C_n\dot{z}(t) + G_n z(t) &= B_n u(t) \\ \tilde{y}(t) &= L_n^T z(t) \end{aligned}\right\} \tag{5.1}$$

where $z(t) \in \mathcal{R}^n$, $C_n, G_n \in \mathcal{R}^{n\times n}$, $B_n \in \mathcal{R}^{n\times m}$, $L_n \in \mathcal{R}^{n\times p}$, and $\tilde{y}(t) \in \mathcal{R}^p$. The state-space dimension n of (5.1) should generally be much smaller than the state-space dimension N of (4.1), i.e., $n \ll N$. Meanwhile, the output $\tilde{y}(t)$ of (5.1) approximates the output $y(t)$ of (4.1) in accordance with some criteria for all u in the class of admissible input functions. Furthermore, the reduced-order system (5.1) should preserve essential properties of the full-order system (4.1), such as stability and passivity. We refer to [1, 14] for the definitions of these properties.

Note that the $p \times m$ matrix-valued transfer function of the reduced-order model (4.1) is given by

$$H_n(s) = L_n^T(G_n + sC_n)^{-1}B_n.$$

Hence, for the steady-state analysis in the frequency domain, the objectives of constructing a reduced-order model (5.1) include that the reduced-order transfer function $H_n(s)$ should be an approximation of the transfer function $H(s)$ of the full-order model over the frequency range of interest with an admissible error, and that $H_n(s)$ preserves essential properties of $H(s)$.

We now show how to construct a reduced-order model of the linear system (4.1) in the time domain for transient analysis. With a selected expansion point s_0 as for the steady-state analysis, the linear system (4.1) under the so-called "shift-and-invert" transformation becomes

$$\left.\begin{aligned} -A\dot{x}(t) + (I + s_0A)x(t) &= ru(t) \\ y(t) &= l^T x(t) \end{aligned}\right\}$$

where $A = -(G + s_0C)^{-1}C$ and $r = (G + s_0C)^{-1}b$. Let V_n be the Lanczos vectors generated by the nonsymmetric Lanczos process with matrix A and starting vectors r and l as discussed in Section 4.4. Then considering the approximation of the state vector $x(t)$ by another state vector, constrained to stay in the subspace spanned by the columns of V_n, namely,

$$x(t) \approx V_n z(t) \quad \text{for some} \quad z(t) \in \mathcal{R}^N,$$

yields an over-determined linear system with respect to the state variable $z(t)$:

$$\left.\begin{aligned} -AV_n\dot{z}(t) + (I + s_0A)V_n z(t) &= ru(t) \\ \tilde{y}(t) &= l^T V_n z(t). \end{aligned}\right\}$$

After left-multiplying the first equation by W_n^T, we have

$$\left.\begin{aligned} -W_n^T AV_n\dot{z}(t) + W_n^T(I + s_0A)V_n z(t) &= W_n^T ru(t) \\ \tilde{y}(t) &= l^T V_n z(t). \end{aligned}\right\}$$

Then an n-th reduced-order model of the linear system (4.1) in the time domain is naturally defined as

$$\left.\begin{aligned} C_n\dot{z}(t) + G_n z(t) &= r_n u(t) \\ \tilde{y}(t) &= l_n^T z(t) \end{aligned}\right\} \tag{5.2}$$

where $C_n = -W_n^T A V_n$, $G_n = W_n^T(I + s_0 A)V_n$, $r_n = W_n^T r$ and $l_n = V_n^T l$. By using the governing equations of the nonsymmetric Lanczos process presented in Section 4.4, the quadruples (C_n, G_n, r_n, l_n) can be simply expressed as $C_n = -T_n$, $G_n = (I_n - s_0 T_n)$, $r_n = \rho_1 e_1$, and $l_n = \eta_1 \delta_1 e_1$.

Example 6 Fig. 5 shows transient analysis of a small RLC network presented in [18, p.29]. The system matrices C and G have order 11. An input excitation $u(t)$ of 0.1 ns rise/fall and 0.3 ns duration was simulated. The convergence for orders 2 and 4 of the reduced-order models in the time domain is shown in Fig. 5. The expansion point is chosen to be $s_0 = \pi \times 10^9$.

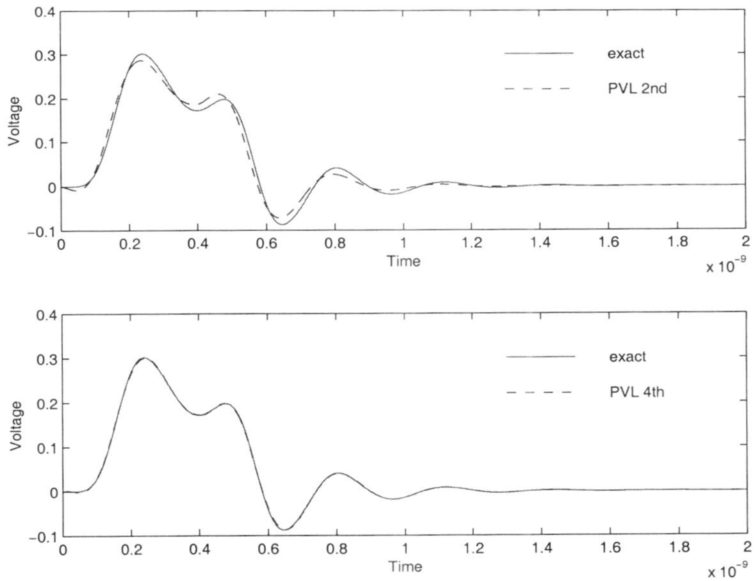

Figure 5: RLC network transient responses: 2nd and 4th order PVL approximation.

The continual and pressing need for accurately and efficiently simulating dynamical behavior of complex physical systems arising from computational science and engineering applications has led to increasingly large and complex models. Reduced-order modeling techniques play an indispensable role in providing an efficient computational prototyping tool to replace such a large-scale model by an approximate smaller model, which is capable of capturing dynamical behavior and preserving essential properties of the larger one.

An accurate and effective reduced-order model can be applied for steady state analysis, transient analysis, or sensitivity analysis of such a system. As a result, it can significantly reduce design time and allow for aggressive design strategies. Such a computational prototyping tool would let designers try "what-if" experiments in hours instead of days.

A myriad of reduced-order modeling methods has been presented in various fields. We can categorize most of these methods into two classes. The first comprises techniques based on the optimization of a reduced-order model according to a suitably chosen criterion. The second class consists of methods that preserve exactly a limited number of parameters of the original model. The work of De Villemagne and Skelton [23] in 1987 provides a survey of early work on these methods. Over the past several years, Krylov-subspace-based techniques, such as the one presented in Section 4, have emerged as one of the most powerful tools for reduced-order modeling of large-scale systems. We refer the reader to recent surveys [27, 28, 2, 4] for further study on the topic.

Acknowledgments We are grateful to the organizers A. Bourlioux, M. Gander, K. Mikula and G. Sabidussi for their invitation and hospitality. Z. Bai is supported in part by NSF grant ACI-9813362 and DOE grant DE-FG03-94ER25219, and in part by a MICRO project (#00–005) from the University of California and MSC.software Corporation. G. Golub is supported in part by NSF grant CCR9971010.

References

[1] B. D. O. Anderson, A system theory criterion for positive real matrices, *SIAM J. Control.* **5** (1967), 171–182.

[2] A. C. Antoulas and D. C. Sorensen, Approximation of large-scale dynamical systems: An overview, Technical report, Electrical and Computer Engineering, Rice University, Houston, TX, Feb. 2001.

[3] Z. Bai, J. Demmel, J. Dongarra, A. Ruhe, and H. van der Vorst, eds., *Templates for the Solution of Algebraic Eigenvalue Problems: A Practical Guide*, SIAM, Philadelphia, 2000.

[4] Z. Bai, Krylov subspace techniques for reduced-order modeling of large-scale dynamical systems, to appear in *Appl. Numer. Math.* (2002).

[5] Z. Bai, M. Fahey, and G. Golub, Some large scale matrix computation problems, *J. Comput. Appl. Math.* **74** (1996), 71–89.

[6] Z. Bai, M. Fahey, G. Golub, E. Menon, and M. Richter, Computing partial eigenvalue sum in electronic structure calculations, Scientific Computing and Computational Mathematics Program SCCM–98–03, Stanford University, 1998.

[7] Z. Bai and G. Golub, Bounds for the trace of the inverse and the determinant of symmetric positive definite matrices, *Ann. Numer. Math.* **4** (1997), 29–38.

[8] Z. Bai and G. Golub, Some unusual matrix eigenvalue problems, in: *Proceedings VECPAR'98—Third Internat. Conf. for Vector and Parallel Processing* (J. Palma,

J. Dongarra, and V. Hernandez, eds.), Lecture Notes in Comput. Sci. **1573**, Springer, 1999, 4–19.

[9] Z. Bai, R. D. Slone, W. T. Smith, and Q. Ye, Error bound for reduced system model by Padé approximation via the Lanczos process, *IEEE Trans. Computer-Aided Design* **18** (1999), 133–141.

[10] Z. Bai and Q. Ye, Error estimation of the Padé approximation of transfer functions via the Lanczos process, *Electron. Trans. Numer. Anal.* **7** (1998), 1–17.

[11] G. A. Baker, Jr. and P. Graves-Morris, *Padé Approximants*, Cambridge University Press, 1996.

[12] R. P. Barry and R. K. Pace, Monte Carlo estimates of the log determinant of large sparse matrices, *Linear Algebra Appl.* **289** (1999), 41–54.

[13] D. L. Boley, Krylov subspace methods on state-space control models, *Circuits Systems Signal Process.* **13** (1994), 733–758.

[14] S. Boyd, L. El Ghaoui, E. Feron, and V. Balakrishnan, *Linear Matrix Inequalities in System and Control Theory*, Classics Appl. Math. **15**, SIAM, Philadelphia, 1994.

[15] A. Bultheel and M. van Barvel, Padé techniques for model reduction in linear system theory: a survey, *J. Comput. Appl. Math.* **14** (1986), 401–438.

[16] D. Calvetti, G. H. Golub, and L. Reichel, A computable error bound for matrix functionals, *J. Comput. Appl. Math.* **103** (1999), 301–306.

[17] D. Calvetti, S. Morigi, L. Reichel, and F. Sgallari, Computable error bounds and estimates for the conjugate gradient method, *Numer. Algorithms* **25** (2000), 75–88.

[18] E. Chiprout and M. S. Nakhla, *Asymptotic Waveform Evaluation*, Kluwer, Dordrecht, 1994.

[19] R. W. Clough and J. Penzien, *Dynamics of Structures*, McGraw-Hill, New York, 1975.

[20] R. R. Craig, Jr., *Structural Dynamics: An Introduction to Computer Methods*, Wiley, New York, 1981.

[21] G. Dahlquist, S. C. Eisenstat, and G. H. Golub, Bounds for the error of linear systems of equations using the theory of moments, *J. Math. Anal. Appl.* **37** (1972), 151–166.

[22] P. Davis and P. Rabinowitz, *Methods of Numerical Integration*, Academic Press, New York, 1984.

[23] C. De Villemagne and R. E. Skelton, Model reductions using a projection formulation, *Internat. J. Control* **46** (1987), 2141–2169.

[24] S. Dong and K. Liu, Stochastic estimation with z_2 noise, *Phys. Lett. B* **328** (1994), 130–136.

[25] M. Fahey, Numerical computation of quadratic forms involving large scale matrix functions, Ph.D. thesis, University of Kentucky, 1998.

[26] P. Feldman and R. W. Freund, Efficient linear circuit analysis by Padé approximation via the Lanczos process, *IEEE Trans. Computer-Aided Design* **14** (1995), 639–649.

[27] R. W. Freund, Reduced-order modeling techniques based on Krylov subspaces and their use in circuit simulation, in: *Applied and Computational Control, Signals, and Circuits*, Vol. 1 (B. N. Datta, ed.), Birkhäuser, Boston, 1999, 435–498.

[28] R. W. Freund, Krylov-subspace methods for reduced-order modeling in circuit simulation, *J. Comput. Appl. Math.* **123** (2000), 395–421.

[29] R. W. Freund, M. H. Gutknecht, and N. M. Nachtigal, An implementation of the look-ahead Lanczos algorithm for non-Hermitian matrices, *SIAM J. Sci. Comput.* **14** (1993), 137–158.

[30] R. W. Freund and N. M. Nachtigal, QMRPACK: a package of QMR algorithms, *ACM Trans. Math. Software* **22** (1996), 46–77.

[31] A. Frommer, T. Lippert, B. Medeke, and K. Schilling, eds., *Numerical Challenges in Lattice Quantum Chromodynamics*, Lecture Notes in Comput. Sci. and Engrg. **15**, Springer, Berlin, 2000.

[32] K. Gallivan, E. Grimme, and P. Van Dooren, Asymptotic waveform evaluation via a Lanczos method, *Appl. Math. Lett.* **7** (1994), 75–80.

[33] W. Gautschi, A survey of Gauss-Christoffel quadrature formulae, in: *E. B. Christoffel—the Influence of His Work on Mathematics and the Physical Sciences* (P. L. Bultzer and F. Feher, eds.), Birkhäuser, Boston, 1981, 73–157.

[34] G. Golub, Some modified matrix eigenvalue problems, *SIAM Rev.* **15** (1973), 318–334.

[35] G. Golub and G. Meurant, Matrics, moments and quadrature, in: *Proc. 15th Dundee Conference, June 1993* (D. F. Griffiths and G. A. Watson, eds.), Longman Sci. Tech., Harlow, 1994.

[36] G. Golub and G. Meurant, Matrices, moments and quadrature II: How to compute the norm of the error iterative methods, *BIT* **37** (1997), 687–705.

[37] G. Golub and Z. Strakoš, Estimates in quadratic formulas, *Numer. Algorithms* **8** (1994), 241–268.

[38] G. Golub and U. Von Matt, Generalized cross-validation for large scale problems, *J. Comput. Graph. Statist.* **6** (1997), 1–34.

[39] W. B. Gragg, Matrix interpretations and applications of the continued fraction algorithm, *Rocky Mountain J. Math.* **5** (1974), 213–225.

[40] W. B. Gragg and A. Lindquist, On the partial realization problem, *Linear Algebra Appl.* **50** (1983), 227–319.

[41] E. Grimme, Krylov projection methods for model reduction, Ph.D. thesis, University of Illinois at Urbana-Champaign, 1997.

[42] M. Hutchinson, A stochastic estimator of the trace of the influence matrix for laplacian smoothing splines, *Comm. Statist. Simulation Comput.* **18** (3) (1989), 1059–1076.

[43] I. M. Jaimoukha and E. M. Kasenally, Oblique projection methods for large scale model reduction, *SIAM J. Matrix Anal. Appl.* **16** (1997), 602–627.

[44] T. Kailath, *Linear Systems*, Prentice-Hall, New York, 1980.

[45] L. Komzsik, *MSC/NASTRAN, Numerical Methods User's Guide, Version 70.5*, MacNeal-Schwendler Corp., Los Angeles, 1998.

[46] C. Lanczos, An iteration method for the solution of the eigenvalue problem of linear differential and integral operators, *J. Res. Nat. Bur. Standards* **45** (1950), 225–280.

[47] A. J. Laub, Efficient calculation of frequency response matrices from state space models, *ACM Trans. Math. Software* **12** (1986), 26–33.

[48] H. Madrid, private communication, 1995.

[49] G. Meurant, A review of the inverse of symmetric tridiagonal and block tridiagonal matrices, *SIAM J. Math. Anal. Appl.* **13** (1992), 707–728.

[50] H.-D. Meyer and S. Pal, A band-Lanczos method for computing matrix elements of a resolvent, *J. Chem. Phys.* **91** (1989), 6195–6204.

[51] A. Odabasioglu, M. Celik, and L. T. Pileggi, Practical considerations for passive reduction of RLC circuits, in: *Proc. Internat. Conf. on Computer-Aided Design*, 1999, 214–219.

[52] B. Parlett, Reduction to tridiagonal form and minimal realizations, *SIAM J. Math. Anal. Appl.* **13** (2) (1992), 567–593.

[53] L. T. Pillage and R. A. Rohrer, Asymptotic waveform evaluation for timing analysis, *IEEE Trans. Computer-Aided Design* **9** (1990), 353–366.

[54] D. Pollard, *Convergence of Stochastic Processes*, Springer Ser. Statist. **15**, Springer, Berlin, 1984.

[55] A. E. Ruehli, Equivalent circuit models for three-dimensional multiconductor systems, *IEEE Trans. Microwave Theory and Tech.* **22** (1974), 216–221.

[56] B. Sapoval, Th. Gobron, and A. Margolina, Vibrations of fractal drums, *Phys. Rev. Lett.* **67** (21) (1991), 2974–2977.

[57] J. C. Sexton and D. H. Weingarten, The numerical estimation of the error induced by the valence approximation, *Nuclear Phys. B Proc. Suppl.* (1994), xx.

[58] G. Szegö, *Orthogonal Polynomials*, 4th ed., Amer. Math. Soc. Colloq. Publ. **23**, Amer. Math. Soc., Providence, RI, 1975.

[59] S. Van Huffel and J. Vandewalle, *The Total Least Squares Problem: Computational Aspects and Analysis*, SIAM, Philadelphia, PA, 1991.

[60] J. Vlach and K. Singhal, *Computer Methods for Circuit Analysis and Design*, Van Nostrand, New York, 1994.

[61] K. Willcox, J. Peraire, and J. White, An Arnoldi approach for generalization of reduced-order models for turbomachinery, FDRL TR–99–1, Fluid Dynamics Research Lab., Massachusetts Institute of Technology, 1999.

[62] M. N. Wong, Finding $\mathrm{tr}(A^{-1})$ for large A by low discrepancy sampling, Presentation at the Fourth International Conference on Monte Carlo and Quasi-Monte Carlo Methods in Scientific Computing, Hong Kong, China, 2000.

[63] S. Y. Wu, J. A. Cocks, and C. S. Jayanthi, An accelerated inversion algorithm using the resolvent matrix method, *Comput. Phys. Comm.* **71** (1992), 125–133.

[64] Q. Ye, A convergence analysis for nonsymmetric Lanczos algorithms, *Math. Comp.* **56** (1991), 677–691.

Thin film dynamics: theory and applications

Andrea L. BERTOZZI

and

Mark BOWEN

Department of Mathematics
Duke University
Durham, NC 27708
USA

Abstract

This paper is based on a series of four lectures, by the first author, on thin films and moving contact lines. Section 1 presents an overview of the moving contact line problem and introduces the lubrication approximation. Section 2 summarizes results for positivity preserving schemes. Section 3 discusses the problem of films driven by thermal gradients with an opposing gravitational force. Such systems yield complex dynamics featuring undercompressive shocks. We conclude in Section 4 with a discussion of dewetting films.

1 Thin films and moving contact lines

The motion of a liquid under the influence of surface tension is a phenomenon we experience every day when we take a shower, drink a cup of coffee, or turn on the windshield wipers. All of these real world situations involve not only the motion of the liquid and the surrounding air, but also their interaction with adjacent hard surfaces. As we know from waxing our cars and cooking with non-stick cookware, the dynamics of a fluid coating a solid surface depend heavily on the *surface chemistry*. Many industrial processes, ranging from spin coating of microchips to de-icing of airplane wings, rely on the ability to control these interactions.

An ongoing challenge is to explain the underlying physics of the motion of a *contact line* (see Fig. 1), a triple juncture of the air/liquid, liquid/solid, and air/solid interfaces. When the system is at rest, the three interfacial energies, determined by the energy per unit area γ_S ($S = al$, ls, or as), are in balance and an equilibrium contact angle θ satisfies Young's law [37]

$$\gamma_{as} - \gamma_{ls} - \gamma_{al}\cos\theta = 0. \tag{1.1}$$

However, the dynamically evolving contact line requires much more subtle modeling involving the interaction between multiple length scales due to the underlying fluid dynamics.

The classical theory of fluids [2] tells us that when a viscous fluid meets a solid boundary, the correct model for the boundary is a "no-slip" condition on the liquid/solid interface[1].

[1]See [50] pp. 676–680 for a comprehensive discussion of the history of the "no-slip" boundary condition.

A. Bourlioux and M.J. Gander (eds.), Modern Methods in Scientific Computing and Applications, 31–79.

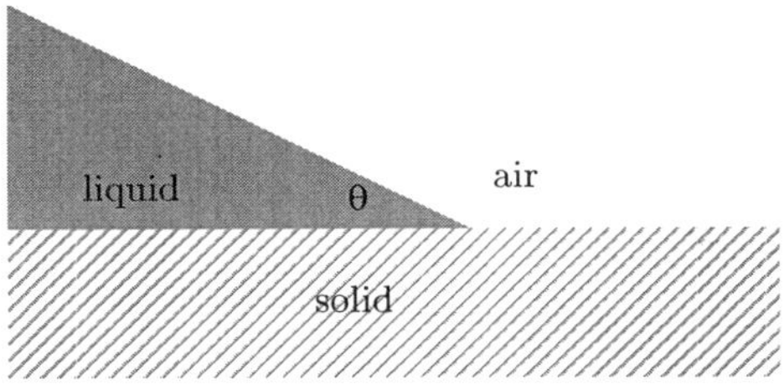

Figure 1: Simple trigonometry is all that is needed to derive Young's law (1.1)—the equation satisfied by the equilibrium contact angle. However, in cases like complete wetting or driven contact lines, a dynamic model is required. Physicists have been fascinated by this problem for decades.

Mathematically, this requires setting the fluid velocity to be zero on the solid boundary. While this makes sense for the *Navier-Stokes equations* (describing the motion of an incompressible fluid) in bounded domains, its relevance to the moving contact line problem is dubious. The difficulty is that a moving contact line coupled with a no-slip condition on the liquid solid interface results in a multi-valued velocity field at the contact line. The velocity field is not well defined because the limit along the solid surface is zero, while the limit along the air/liquid interface (which is moving) is non-zero. Regardless of the liquid/air interface model, the tangential component of force exerted by the fluid on the solid diverges whenever the velocity is multi-valued at the contact line [59, 39]. The fact that such a paradox exists is hardly surprising. We should not expect to find a self-consistent *universal* hydrodynamic model that does not incorporate the surface chemistry.

Several models have been proposed to study the motion of moving contact lines. All of them involve adding an additional effect on a microscopic length-scale. The two that we discuss in the next paragraphs are (1) weakening the no-slip boundary condition via a slip condition effective at small scales, and (2) incorporating the effect of long range *van der Waals* forces (described below) between the liquid and solid. Perhaps the simplest context in which to test these theories is that of thin viscous coating flows. For such problems, a lubrication approximation simplifies bulk flow fluid dynamics, as given by the Navier-Stokes equations, to a single equation, relating the depth averaged horizontal fluid velocity to the shape of the air/liquid interface.

The *Navier slip* condition[2] near the liquid solid interface demands that the velocity at the interface be proportional to its normal derivative:

$$v = \frac{\partial v}{\partial z} k(h), \qquad z = 0. \tag{1.2}$$

Here v is the horizontal fluid velocity and z is the variable in the direction normal to the solid surface. The parameter $k(h)$ is a slip parameter and can depend on the thickness h of the film. We can think of condition (1.2) as a generalization of the standard "no-slip" condition which corresponds to $k \equiv 0$ in (1.2).

[2] A generalization of the "no-slip" condition in which the fluid is allowed to slip tangentially along the solid boundary.

Another choice for removing the singularity due to no-slip is to include microscopic scale forces, in the form of long range *van der Waals* (VW) interactions between the liquid and solid surfaces, near the contact line. The most systematic way to do this is via an additional body force in the fluid of the form (see the review articles [37] sec. II. D and [78] sec. II. E)

$$\Pi(h) = A_D h^{D-5}, \tag{1.3}$$

where D is the "dimension" of the substrate[3]. Here A_D is the Hamaker constant, which depends on the dimension and also on the strength of the interaction between the liquid and solid particles. The sign of A_D depends on whether the VW forces are attractive or repulsive. The particular power in (1.3) comes from assuming that the particles in the fluid interact with particles in the solid substrate via a "non-retarded" potential. See the references listed above and the references therein for more discussion of this and other potentials.

1.1 The basic equation

In its simplest form, the lubrication approximation can be derived from a systematic rescaling and asymptotic expansion of the Navier-Stokes equations in the limit of vanishing *capillary number* C_a and *Reynolds number* R_e, two dimensionless parameters given by

$$C_a = \frac{3\nu V}{\gamma}, \qquad R_e = \frac{\rho V h_0}{\nu}, \tag{1.4}$$

where ν is viscosity, γ is surface tension, ρ is the density of fluid, h_0 a characteristic height of the film, and V is a characteristic velocity of the film. In this case, when the slip model (1.2) is used, the nondimensional lubrication equation

$$\begin{aligned} h_t + \nabla \cdot (f(h)\nabla\Delta h) = 0, \\ f(h) = h^3 + b^{3-p}h^p \end{aligned} \tag{1.5}$$

results. Here b represents a dimensionless slip length. Greenspan [52] derives (1.5) with $b > 0$, "singular slip" $p = 1 > 0$ and $\Pi(h) = 0$. Haley and Miksis [57] consider this same model with different integer values of $p = 0, 1, 2$. The VW model (1.3) results in a similar equation with a second order term that includes the effect of $\Pi(h)$.

The remainder of this section discusses basic mathematical properties of equations of the form (1.5). Section 2 discusses numerical schemes for this problem. Section 3 considers the problem when convective terms are included. Section 4 addresses the dewetting problem and pattern formation, in which effects like (1.3) are included.

The engineering and applied mathematics literature has had much discussion of the most appropriate boundary conditions for an equation of the form (1.5) near the vicinity of a contact line, especially in the context of a slip model (see [78], sections II. B and V and references therein. See also Hocking *J. Fluid Mech.* **239**, 1992). During the past eight years, mathematicians working in the field of nonlinear PDEs have also become interested in this problem from a more analytical point of view.

Equation (1.5) is a fourth order degenerate diffusion equation. Diffusion equations arise in models of physical problems in which some quantity spreads or smears out. The classical

[3] I.e. $D = 2$ for a film on a two-dimensional surface and $D = 1$ for a flow in the plane, bounded by a line.

example is the heat equation $u_t = \nabla \cdot k \nabla u$, where the diffusion coefficient k controls the rate at which heat "diffuses" through the medium. In order to understand the contact line models proposed for equation (1.5) we need to understand something about *degenerate diffusion equations*, where k depends on the solution h, so that $h \to 0$ as $k \to 0$. A well-known family of second order degenerate diffusion equations are the *porous media equations* [79]:

$$h_t - \Delta(\Phi(h)) = 0, \tag{1.6}$$

where $\Phi'(h) > 0$ for $h > 0$, and $\Phi(h) \sim h^m$ as $h \to 0$. Here, $m > 1$ makes the equation degenerate.

The Laplacian operator Δ endows the evolution equation (1.6) with some important properties:

(1) *Instantaneous smoothing* of the solution in regions of positive h, since the equation is uniformly parabolic (i.e. the "diffusion coefficient" is bounded away from zero) where the solution is bounded away from zero.

(2) *Maximum principle.* The solution is bounded from above and below by its initial data. This is true regardless of whether or not $m = 1$ (linear heat equation) or $m > 1$.

(3) *Well-posed distribution solutions* for smooth nonnegative initial data.

(4) *Finite speed of propagation* of the support of the solution.

The last property is special to *degenerate* diffusion ($m > 1$). The heat equation ($m = 1$) has infinite speed of propagation of the support.

Consider now a fourth order analogue of (1.6) that arises in the thin film problem (as in (1.5))

$$h_t + \nabla \cdot (f(h) \nabla \Delta h) = 0. \tag{1.7}$$

In the context of thin film dynamics, we are interested in questions of existence, uniqueness, and finite speed of propagation of the support of weak solutions of equations of the form (1.7). The important difference from the case (1.6) is that (1.7) is fourth order as opposed to second order. It is diffusive so we have property (1), instantaneous smoothing where the solution is positive, but property (2), the maximum principle, is far from guaranteed. Indeed, if we take the nondegenerate case $f(h) = 1$, then solutions can change sign. This can be seen by simply examining the heat kernel for fourth order diffusion in $\mathbb{R}^D$ and noting that unlike the well-known second order heat kernel, the fourth order heat kernel has an oscillatory exponentially decaying tail that changes sign. It is not immediately apparent why the nonlinear problem (1.7) should have solutions that preserve their sign.

What is remarkable is that the nonlinear structure of (1.7) endows it with a positivity preserving property. In particular, one can show for one space dimension [15, 21] with $f(h) = h^n$ with $n \geq 3.5$ on a periodic spatial domain, if the initial condition h_0 is positive, then the solution $h(x,t)$ is guaranteed to stay positive. In fact, one can derive an a priori pointwise lower bound depending only on the H^1 norm and minimum value of the initial data. This form of a weak maximum principle is due to the nonlinear structure of (1.7), not the structure of the fourth order diffusion operator. It is noteworthy that other fourth order degenerate diffusion equations with different nonlinear structure have weak solutions that change sign

[12]. Despite the positivity preserving property for large n, solutions of (1.7) with smaller values of n computationally exhibit finite time singularities of the form $h \to 0$. We discuss this phenomenon in more detail in the next section. The fact that such singularities may occur makes the development of a weak solution theory, analogous to that of the porous media equation, all the more difficult. Over the past few years, mathematical machinery, largely involving energy estimates and nonlinear entropies, has been developed to address these problems. Such methods have been used to prove weak maximum principles, derive existence results for weak (nonnegative) solutions, and to prove results concerning the finite speed of propagation of the support of nonnegative solutions [13, 14].

1.2 Finite time singularities and similarity solutions of lubrication equations

The study of finite time singularities and similarity solutions of (1.7) began in the early 1990's by a group at the University of Chicago. The original project addressed the breakup of a thin neck in the Hele-Shaw cell when forced by an external pressure. The experiment, performed by Goldstein at Princeton [36], can be modeled, via lubrication theory, by a fourth order degenerate diffusion equation in one space dimension:

$$h_t + (h^n h_{xxx})_x = 0 \tag{1.8}$$

with "pressure" boundary conditions $h(\pm 1) = 1$, $h_{xx}(\pm 1) = -p$, and $n = 1$. Here h denotes the thickness of the fluid neck, in the direction parallel to the plates (resulting from a 2D model). Carefully resolved numerical computations revealed that this problem yields both finite and infinite time singularities of the form $h \to 0$, for various initial data [38].

What is unusual about the structure of these singularities is that, while they locally have self-similar structure of the form

$$h(x,t) \sim \tau(t) H\left(\frac{x - a(t)}{l(t)}\right), \tag{1.9}$$

the time dependences $\tau(t)$, $l(t)$, and $a(t)$ are not completely determined by dimensional analysis of the PDE. Such anomalous "second type" scaling [7] arises in problems such as the focusing solution of the porous media equation [4], where one solves a nonlinear eigenvalue problem to determine the profile H and the relationship between scales $\tau(t)$ and $l(t)$. However, in the case of the lubrication singularities, the scaling relations are typically found by asymptotic matching involving the boundary conditions or matching to an intermediate length-scale [21].

The original work on the Hele-Shaw problem was subsequently generalized to the case of variable nonlinearity (other values of n) with other boundary conditions [21, 3] and also to different fourth order degenerate PDEs [18]. The similarity solutions observed were found using careful numerical computations involving adaptive mesh refinement near the singularity. Scaling was then observed for many orders of magnitude in $h_{\min}$, the minimum value of the solution. All of these papers make the following observations:

- All observed finite time singularities of the form $h \to 0$ as $t \to t_c$ involve second type scaling behavior in which the similarity solution described in (1.9) satisfies the quasi-static equation $(f(h)h_{xxx})_x = 0$.

- Several kinds of similarity solutions can occur for different initial conditions for the same equation. The far-field dynamics of the structure may cause the similarity solution to destabilize, at arbitrarily small length-scales, from one type of similarity solution to another [3].

As an example, consider the equation (1.8) with $n = 1/2$, periodic boundary conditions on $[-1, 1]$, and initial condition

$$h_0(x) = 0.8 - \cos(\pi x) + 0.25\cos(2\pi x). \tag{1.10}$$

The solution develops a finite time "pinching" singularity in which $h \to 0$ with a simultaneous blow-up in the fourth derivative. The third derivative forms a step function (see Fig. 2). Near the pinch-point, the solution has a leading order asymptotic form

$$h(x,t) \approx c(t_c - t) + \frac{p(x - x_c)^2}{2}, \tag{1.11}$$

where t_c is the time of pinch-off, x_c is the pinch-point, and the constant p is the curvature of the interface at the time of and location of the singularity. The blow-up in higher derivatives can be seen only in terms of higher order than (1.11). The next term satisfies

$$h_{xxx} \approx \frac{cx}{\sqrt{c(t_c - t) + p(x - x_c)^2/2}}$$

so that the local curvature h_{xx} has the form

$$p + \frac{c}{p}\sqrt{c(t_c - t) + p(x - x_c)^2/2}.$$

This scaling structure of the singularity was confirmed via numerical simulations, using a self-similar adaptive mesh refinement code, that resolved the singularity over thirty orders of magnitude in $h(x_c, t) = h_{\min}$. The solution can be continued after the initial pinch-off singularity. We discuss the continuation of this particular example later in this paper (see Figs. 3–5).

The fact that we can rigorously prove that finite time singularities cannot occur for $n \geq 3.5$ yet they are observed for certain $n < 2$ suggests the existence of a *critical exponent*. This is a value $n_* > 0$ for which solutions to (1.8) stay positive whenever $n > n_*$ and where finite time singularities are possible for $n \leq n_*$. To date, no such n_* has been established, but numerical simulations suggest that $1 < n_* < 3.5$. It is also not clear if n_* depends on boundary conditions. Since singularities appear to be localized, it is reasonable to expect that n_* might be independent of boundary conditions. It is an interesting open question whether this behavior persists in higher space dimensions ($D > 1$).

1.3 Entropies, positivity, and weak solutions

In the previous section, we discussed the open problem of finding a "critical exponent" associated with strictly positive initial data: equations with exponent larger than n_* have solutions that remain strictly positive, while ones with exponents less than n_* allow for the possibility of finite time singularities. In this section we introduce another critical exponent n_{crit} which separates powers for which the solution has increasing support from powers for which the

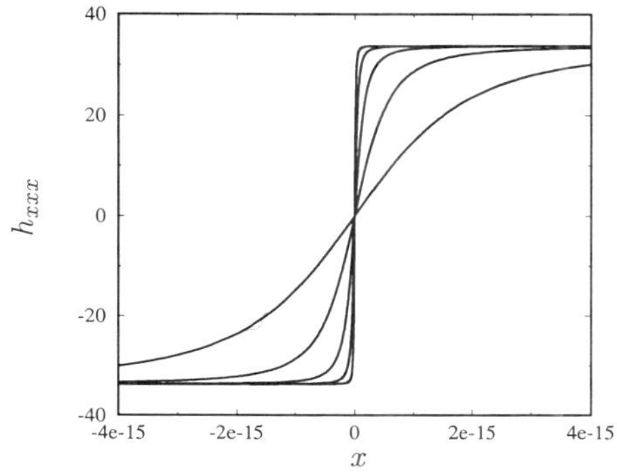

Figure 2: Onset of initial singularity in solution of (1.8) with $n = 1/2$ and initial data (1.10). Formation of a jump discontinuity in the third derivative, shown as solid lines for successive times close to singularity. The solution can be continued after the initial pinch-off singularity.

support of the solution cannot increase. For the porous media equation (1.6), the support of the solution is always monotonically increasing, regardless of the size of $m > 1$. For the 1D lubrication equation (1.8), the support of the solution cannot increase whenever $n \geq 4$ [11]. On the other hand, there are solutions with support that eventually increase to fill the entire domain for all $n < 3$ [16]. It has been conjectured that the critical exponent for increasing support is $n_{\text{crit}} = 3$, but this remains to be rigorously proved.

The exact value of the critical exponent has important ramifications for the moving contact line problem. If indeed $n_{\text{crit}} = 3$ it tells us that there are no solutions of (1.5) with $b = 0$ (no-slip) that describe a moving contact line. Such a result would not be surprising. It is consistent with the fact that such a scenario introduces a multi-valued velocity field that is known to produce infinities in the physics (hence the introduction of the slip terms that give $n < 3$).

Examples of known exact solutions support the conjecture that $n_{\text{crit}} = 3$ in both one and two space dimensions. One class of examples are the "source type solutions", solutions that start as a delta function at the origin and spread out in a self-similar way while conserving their mass. Starov [90] looked for such solutions to (1.7) with $f(h) = h^3$, in two space dimensions. He attempted to find a similarity solution to describe spreading drops under the influence of surface tension. What he discovered was that the resulting ODE for the similarity solution did not have any solutions of compact support. That is, there are no similarity solutions describing spreading drops with no-slip on the liquid/solid interface. Bernis and collaborators [16, 47] showed that in all space dimensions, source type solutions exist ($f(h) = h^n$) for $n < 3$, and they do not exist for $n \geq 3$. It is interesting that this critical exponent cannot be predicted from dimensional analysis of the equation; it is determined by properties of the ODE for the shape of the similarity solution. Traveling wave (i.e. advancing front) solutions [30] also change behavior at the critical exponent of $n = 3$.

We now discuss some of the key ideas used to prove results about the PDE (1.7). A seminal paper in the mathematical development of the theory of weak solutions for lubrication-type equations is the work [15] by Bernis and Friedman. There they showed that, in addition to conservation of mass,

$$\frac{\partial}{\partial t}\int h(x)\,dx = 0, \tag{1.12}$$

and surface tension energy dissipation,

$$\frac{\partial}{\partial t}\int |\nabla h|^2 = -\int f(h)|\nabla\Delta h|^2\,dx, \tag{1.13}$$

equations of the type (1.7) possess a nonlinear *entropy* dissipation of the form

$$\frac{\partial}{\partial t}\int G(h) = -\int |\Delta h|^2\,dx. \tag{1.14}$$

Here $G(h)$ is a convex function satisfying $G''(h) = 1/f(h)$. For the case $f(h) = h$ (e.g. Hele-Shaw) the entropy $\int G(h)$ is of the form $\int h\log h$; hence the name "entropy" was born [26] to describe this object. Using the entropy (1.14), Bernis and Friedman proved that in one space dimension, the critical exponent n_*, above which singularity formation is forbidden, satisfies $n_* < 4$. The proof uses the fact that (1.13) and conservation of mass imply an a priori bound on the $C^{1/2}$ norm of the solution while (1.14) insures a bound on $\int h^{2-n}$. This gives an a priori pointwise lower bound for the solution. It turns out that the integral $\int G(h)\,dx$ above is not the only dissipative entropy. In 1D we also have a family of entropies, satisfying

$$\frac{\partial}{\partial t}\int G^s(h) \leq 0, \qquad (G^s)'' = \frac{h^s}{f(h)}, \tag{1.15}$$

where $1/2 \leq s \leq 1$. Taking $s \to 1/2$ gives the upper bound $n_* \leq 3.5$ proved in [21].

The generalized entropies play an important role in the development of a weak existence theory for nonnegative solutions from nonnegative initial data, and for proving results on finite speed of propagation of the support. Uniqueness of weak solutions still remains an open problem. In the next section, we see how entropies have also been used recently to design numerical methods for solving these equations.

2 Positivity preserving schemes

The nonlinear structure of the PDE (1.7) presents a challenge in the design of efficient and accurate numerical methods. The degeneracy of (1.7) as $h \to 0$ can be the cause of a number of numerical instabilities. To compute a positive approximation of the solution, we may need computationally expensive local mesh refinement near the minimum of the solution in order to avoid premature or "false" singularities [21, 18].

For the computation of nonnegative weak solutions, a non-negativity preserving finite element method is proposed in [8]. Non-negativity of the solution is imposed as a constraint, so we have to solve a variational problem involving a Lagrange multiplier at every time step to advance the solution. Another non-negativity preserving scheme, that has some common elements to the work discussed here, is introduced by Grün and Rumpf [56]. The method presented here is from a joint paper by L. Zhornitskaya with the first author [102].

This section presents results from [102], illustrating that it is possible to design a finite difference (in space) scheme to satisfy discrete analogues of properties (1.12)–(1.14) above. Such a scheme preserves positivity of the solution (whenever $n \geq 2$) and has solutions that exist for all time, regardless of the size of the grid. This method improves upon previous methods (e.g. [21]) that require mesh refinement in order to avoid a premature "numerical" singularity. We show that it is possible to generalize the positivity preserving property to finite element methods on arbitrary element spaces (including those involving nonuniform grids).

2.1 The solution of the difference equations

We consider a family of continuous-time, discrete-space finite difference schemes for the one dimensional lubrication equation (1.7) with periodic boundary conditions and strictly positive initial data $h_0(x) \in H^1(S^1)$.

Notation Divide the periodic domain S^1 into N equally spaced regions of length Δx and introduce the following discrete analogues of the space derivatives:[4]

$$y_{x,i} = \frac{y_{i+1} - y_i}{\Delta x}, \qquad y_{\bar{x},i} = \frac{y_i - y_{i-1}}{\Delta x},$$
$$y_{\bar{x}x,i} = \frac{y_{\bar{x},i+1} - y_{\bar{x},i}}{\Delta x}, \qquad y_{\bar{x}x\bar{x},i} = \frac{y_{\bar{x}x,i} - y_{\bar{x}x,i-1}}{\Delta x}.$$

The discrete $H_{1,\Delta x}$ norm is

$$\|y\|_{1,\Delta x} = \left(\sum_i (y_{\bar{x},i}^2 + y_i^2)\Delta x \right)^{1/2}$$

which is equivalent to $((\sum_i y_{\bar{x},i}^2 \Delta x) + \bar{y}^2)^{1/2}$, where $\bar{y}$ denotes the mean $\sum_i y_i \Delta x$. The numerical scheme

$$\begin{aligned} &y_{i,t} + (a(y_{i-1}, y_i) y_{\bar{x}x\bar{x},i})_x = 0, \\ &i = 0, 1, \ldots, N-1, \\ &y_i(0) = h_0(x_i), \end{aligned} \tag{2.1}$$

which we can view as a coupled system of ordinary differential equations for the y_i, is a continuous-time discrete-space approximation of the PDE (1.7), provided that the nonlinear coefficient $a(s_1, s_2)$ satisfies the following definition:

2.1 Definition (*Generic finite difference scheme*)

$$a(s,s) = f(s), \tag{a}$$
$$a(s_1, s_2) = a(s_2, s_1), \tag{b}$$
$$a(s_1, s_2) \in C^4\left((0,\infty) \times (0,\infty)\right) \cap C\left([0,\infty) \times [0,\infty)\right), \tag{c}$$
$$\forall \delta > 0, \ \text{there exists } \gamma > 0 \text{ such that } s_1, s_2 > \delta \implies a(s_1, s_2) \geq \gamma > 0. \tag{d}$$

[4] We append the i on to the x and $\bar{x}$ subscripts to make it clear that we are computing a finite difference as opposed to a spatial derivative. Throughout the rest of the paper, we use ∂_x, ∇, etc. to denote space derivatives, and subscripts to denote finite differences in space.

Conditions (a)–(c) guarantee the scheme (2.1) is second order consistent with a positive smooth solution of (1.7). Condition (c) does not require $a(s_1, s_2)$ to be differentiable at the origin. Condition (d) says that $a(s_1, s_2)$ is positive whenever both its arguments are positive, but may become degenerate if at least one of its arguments approaches zero. This matches the property of the analytical solution (1.7) since $f(h) \sim h^n$ is positive whenever h is positive, but $f(h) \to 0$ as $h \to 0$. Both $a(s_1, s_2) = f(1/2(s_1 + s_2))$ and $a(s_1, s_2) = 1/2(f(s_1) + f(s_2))$ are examples of allowed discretizations. The former discretization is used, for example, in [21].

The following properties of the numerical scheme are analogues of the continuous case:

- Discrete conservation of mass:

$$\sum_{i=0}^{N-1} y_i(t)\Delta x = \sum_{i=0}^{N-1} y_i(0)\Delta x. \tag{2.2}$$

- Discrete energy dissipation:

$$\frac{1}{2}\sum_i (y_{\bar{x},i}(t))^2\Delta x + \int_0^t \sum_i a(y_{i-1}(s), y_i(s))(y_{\bar{x}x\bar{x},i}(s))^2\Delta x\, ds = \frac{1}{2}\sum_i (y_{\bar{x},i}(0))^2\Delta x. \tag{2.3}$$

Standard ODE arguments allow us to prove the following lemma.

2.2 Lemma (Existence, uniqueness and continuation of solutions of the numerical scheme [102]) *Given the coupled ODE system* (2.1) *with strictly positive initial data* $y_i(0)$ *and diffusion coefficient* $a(s_1, s_2)$ *satisfying conditions* (a)–(d) *from Definition* 2.1, *there is a time interval* $[0, \sigma]$ *for which there exists a unique positive solution of the coupled ODE system* (2.1). *Moreover, there exists a constant* C, *depending only on the discrete* $H_{1,\Delta x}$ *norm of the initial data* $y_i(0)$, *such that* $y_i(t) \le C$ *for all* i *and all* $t \le \sigma$. *The solution can be continued for arbitrarily long times provided it remains positive.*

2.2 A positivity preserving finite difference scheme

In this section we show that if $n \ge 2$ in (1.8), then there exists a way to choose the function $a(s_1, s_2)$ in (2.1) so that, for positive initial data, the solution of (2.1) stays positive for all time. This property is a discrete analogue of the weak maximum principle for the PDE (1.8). Our goal is to choose $a(s_1, s_2)$ so that the numerical scheme (2.1) satisfies a discrete form of (1.14) and show that this is sufficient to guarantee that its solutions also preserve positivity.

2.3 Lemma [102] *Let* $G''(v) = 1/f(v)$ *be a nonlinear entropy function. If we choose* $a(s_1, s_2)$ *in* (2.1) *to satisfy*

$$a(s_1, s_2) = \begin{cases} \dfrac{s_1 - s_2}{G'(s_1) - G'(s_2)} & \text{if } s_1 \ne s_2, \\ f(s_1) & \text{if } s_1 = s_2, \end{cases} \tag{2.4}$$

then the solution of (2.1), *with positive initial data, satisfies the following discrete entropy dissipation property*

$$\sum_i G(y_i(t))\Delta x + \int_0^t \left\{ \sum_i (y_{\bar{x}x,i}(s))^2 \Delta x \right\} ds = \sum_i G(y_i(0))\Delta x. \tag{2.5}$$

This leads to the following proposition showing that the scheme results in a positive solution regardless of the grid size.

2.4 Proposition *Let $f(h) \sim h^n$, as $h \to 0$, $n \geq 2$ and*

$$a(s_1, s_2) = \begin{cases} \dfrac{s_1 - s_2}{G'(s_1) - G'(s_2)} & \text{if } s_1 \neq s_2, \\ f(s_1) & \text{if } s_1 = s_2. \end{cases} \tag{2.6}$$

Then given positive initial data $y_i(0) > 0$, for every fixed Δx there exists $\bar{\delta} > 0$ such that $\min_t \min_i y_i(t) \geq \bar{\delta}$. Moreover, if $n \geq 4$ then $\bar{\delta}$ depends only on $\|y_i\|_{1,\Delta x}$ and $\min_i y_i$. In particular, it does not depend directly on Δx.

Using the discrete entropy dissipation property (2.5), we get

$$\sum_j \frac{1}{y_j^{n-2}} \Delta x \leq \tilde{C}, \quad \text{if } n > 2,$$

$$\sum_j \ln\left(\frac{1}{y_j}\right) \Delta x \leq \tilde{C}, \quad \text{if } n = 2,$$

where $\tilde{C}$ depends on the uniform positive lower and (for $n = 2$) upper bounds of the initial condition $y_i(0)$. This implies that

$$\min_t \min_i y_i(t) \geq \left(\frac{\Delta x}{\tilde{C}}\right)^{1/(n-2)} \equiv \bar{\delta}, \qquad \text{if } n > 2,$$

$$\min_t \min_i y_i(t) \geq \exp\left(-\frac{\tilde{C}}{\Delta x}\right) \equiv \bar{\delta}, \quad \text{if } n = 2.$$

This estimate is valid for any $n \geq 2$. However, it depends badly on Δx, that is, as Δx gets smaller the lower bound gets smaller as well. Since there exists a positive lower bound on the solution of the PDE (1.8) for n large enough, we would like to obtain a lower bound independent of Δx for the discrete case as well. The following argument, similar to that for the continuous case [15], shows that this is indeed possible for $n \geq 4$.

The discrete energy dissipation property and the Sobolev Lemma imply the existence of an upper bound on the discrete $C^{1/2}_{\Delta x}$ of $y_i(t)$. Let $\delta(t) = \min_{0 \leq i \leq N-1} y_i(t)$, which occurs at $i(t)$. Then we get

$$\begin{aligned} \tilde{C} &\geq \sum_j \frac{1}{y_j^{n-2}} \Delta x \geq \sum_j \frac{\Delta x}{(\delta(t) + C|(j - i(t))\Delta x|^{1/2})^{n-2}} \\ &\geq 2\frac{1}{\delta^{n-4}} \int_0^{1/\delta^2} \frac{ds}{(1 + s^{1/2})^{n-2}}. \end{aligned}$$

Now to obtain a bound independent of Δx, we consider two cases: $\delta > 1$ and $\delta \leq 1$. In the former case we already have the bound we wanted. In the latter case we get

$$\tilde{C} \geq 2\frac{1}{\delta^{n-4}} \int_0^{1/\delta^2} \frac{ds}{(1 + s^{1/2})^{n-2}} \geq 2\frac{1}{\delta^{n-4}} \int_0^1 \frac{ds}{(1 + s^{1/2})^{n-2}} = \frac{C'}{\delta^{n-4}}, \quad \text{if } n > 4,$$

$$\tilde{C} \geq 2\int_0^{1/\delta^2} \frac{ds}{(1 + s^{1/2})^2} \geq C' \ln\left(\frac{1}{\delta^2}\right), \quad \text{if } n = 4,$$

which again implies the existence of a lower bound independent of Δx.

So, with the choice of spatial discretization (2.6) the finite difference scheme (2.1) is positivity preserving for all $n \geq 2$, and the lower bound is independent of Δx provided $n \geq 4$.

Note that the above results, coupled with Lemma 2.2, prove global existence of positive solutions of the scheme (2.1) with the special choice of $a(s_1, s_2)$ in (2.6). We state this as a corollary.

2.5 Corollary *Let $f(h) \sim h^n$ with $n \geq 2$. The finite difference scheme* (2.1)*, with the special choice of $a(s_1, s_2)$ satisfying* (2.6)*, and positive initial data $y_i(0) = h_0(x_i) > 0$ has a global in time positive solution $y_i(t)$ regardless of the size of the mesh Δx.*

2.3 Consistency and convergence

To measure consistency, we introduce the local truncation error $\tau_i(t)$, defined as the result of substituting the solution of the PDE (1.8) into (2.1):

$$h_{i,t} + (a(h_{i-1}, h_i)h_{\bar{x}x\bar{x},i})_x = \tau_i(t). \tag{2.7}$$

Since a positive solution of (1.7) is infinitely differentiable [15], we can use Taylor series expansions to examine the consistency of the numerical scheme.

2.6 Lemma [102] *Let $a(s_1, s_2)$ satisfy the conditions* (a)–(d) *of Definition* 2.1 *and let $h(x,t)$ be a smooth positive solution of* (1.7)*. Then the local truncation error $\tau_i(t)$ in* (2.7) *is $O((\Delta x)^2)$ uniformly in t.*

The error depends on the size of the higher derivatives of h, which are known empirically [21, 18] to become unbounded when a positive solution approaches zero.

If the grid size is too large, a generic scheme may fail to have a solution for all time due to numerical singularities. In the case of a generic scheme existence of a solution is guaranteed only for a sufficiently small grid size.

2.7 Theorem [102] *Let $h(x,t)$ be a smooth solution of* (1.7) *such that $h(x,t) \geq \delta > 0$ for some $\delta > 0$ and $t \in [0,T]$. Then for Δx sufficiently small there exists a unique solution $y_i(t)$ of* (2.1) *for all $t \in [0,T]$. Moreover, there exists a constant $C > 0$ such that*

$$\sup_{t\in[0,T]} \| y_i(t) - h(x_i,t) \|_{1,\Delta x} \leq C\Delta x^2.$$

This convergence result includes the positivity preserving scheme which has solutions even on very coarse grids.

2.4 Modified entropy dissipating scheme

In Corollary 2.5 we showed that if $n \geq 2$ then for all Δx there exists a unique positive solution of the entropy dissipating scheme (EDS) for all t. However, for $n < 2$ the EDS scheme (2.1), (2.6) may not be positivity preserving. In examples such as the Hele-Shaw cell ((1.8) with $n = 1$), we may wish to use a positivity preserving scheme to approximate a positive solution. We now present a modification of the EDS that yields a positivity preserving scheme for all

$n > 0$. To accomplish this, we use the following regularization of the PDE (1.7), introduced in [15]:

$$h_{\epsilon t} + \partial_x(f_\epsilon(h_\epsilon)\partial_x^3 h_\epsilon) = 0, \qquad f_\epsilon(h_\epsilon) = \frac{h_\epsilon^4 f(h_\epsilon)}{\epsilon f(h_\epsilon) + h_\epsilon^4},$$

and the corresponding numerical scheme

$$\begin{aligned} &y_{\epsilon,i,t} + (a_\epsilon(y_{\epsilon,i-1}, y_{\epsilon,i})y_{\epsilon\bar{x}x\bar{x},i})_x = 0, \\ &i = 0, 1, \ldots, N-1, \\ &y_{\epsilon,i}(0) = h_0(x_i), \end{aligned} \tag{2.8}$$

$$a_\epsilon(s_1, s_2) = \begin{cases} \dfrac{s_1 - s_2}{G'_\epsilon(s_1) - G'_\epsilon(s_2)} & \text{if } s_1 \neq s_2, \\ f_\epsilon(s_1) & \text{if } s_1 = s_2. \end{cases} \tag{2.9}$$

Note that numerical method (2.8)–(2.9) depends on two independent parameters, namely Δx and ϵ. However we could choose $\epsilon = \epsilon(\Delta x)$ to be a function of the grid size.

Since $f_\epsilon(h_\epsilon) \sim h_\epsilon^4/\epsilon$ as $h_\epsilon \to 0$ we know that for all $\epsilon > 0$ and $\Delta x > 0$ the solution of (2.8)–(2.9) is positive. Moreover, the following theorem shows that the solution of the modified EDS converges to the solution of the original (nonregularized) PDE as $\epsilon \to 0$ and $\Delta x \to 0$ as long as the solution of the latter stays positive. The proof relies on the fact that for all $\bar{\delta} > 0$, $m, n \in \mathbb{N}$, there exists a constant $C(m, n, \bar{\delta})$ such that

$$\left\| \frac{\partial^{n+m}}{\partial s_1^n \partial s_2^m} a_\epsilon(s_1, s_2) - \frac{\partial^{n+m}}{\partial s_1^n \partial s_2^m} a(s_1, s_2) \right\|_{L^\infty([\bar{\delta}, M]^2)} \leq C(m, n, \bar{\delta})\epsilon. \tag{2.10}$$

2.8 Theorem *Let $h(x,t)$ be a smooth solution of (1.8) such that $h(x,t) \geq \delta > 0$ for some $\delta > 0$ and $t \in [0,T]$. Then for all Δx and ϵ there exists a unique solution $y_{\epsilon,i}(t)$ of the modified EDS (2.8)–(2.9) for all $t \in [0,T]$. Moreover, there exists a constant $C > 0$ independent of Δx and ϵ such that*

$$\sup_{t\in[0,T]} \| y_{\epsilon,i}(t) - h(x_i, t) \|_{1,\Delta x} \leq C(\Delta x^2 + \epsilon).$$

We can then choose $\epsilon \sim (\Delta x)^2$ to make the scheme (2.8)–(2.9) second order in Δx.

2.5 Finite difference schemes in 2D

In two space dimensions the lubrication equation with periodic boundary conditions is

$$\begin{aligned} h_t + &\partial_x(f(h)\partial_x(\partial_x^2 h + \partial_y^2 h)) + \\ &\partial_y(f(h)\partial_y(\partial_x^2 h + \partial_y^2 h)) = 0, \quad x \in S^1, \ y \in S^1. \end{aligned} \tag{2.11}$$

Let $z_{ij}(t)$ be a solution of the following finite difference scheme

$$\begin{aligned} z_{i,j,t} + &(a(z_{i-1,j}, z_{i,j})(z_{\bar{x}x,ij} + z_{\bar{y}y,ij})_{\bar{x}})_x + \\ &(a(z_{i,j-1}, z_{i,j})(z_{\bar{x}x,ij} + z_{\bar{y}y,ij})_{\bar{y}})_y = 0, \end{aligned} \tag{2.12}$$

where $a(s_1, s_2)$ is an approximation of $f(h)$ satisfying Definition 2.1.

As in the one-dimensional case, consider the finite difference scheme (2.12) as a coupled system of ODEs. Given positive initial data, there exists, locally in time, a unique solution of (2.12); the proof follows from the same arguments as in Lemma 2.2. Moreover, following the argument there, we can uniquely continue the solution in time provided it stays bounded away from zero and infinity. Therefore, a unique solution exists globally whenever we can prove the existence of a priori upper and lower bounds.

A generic finite difference scheme (2.12) has the following properties:

- Discrete conservation of mass:

$$\begin{aligned}\sum_{i=0}^{N-1}\sum_{j=0}^{M-1} z_{ij,t}\Delta x\Delta y &+ \sum_{i=0}^{N-1}\sum_{j=0}^{M-1}(a(z_{i-1,j},z_{i,j})(z_{\bar{x}x,ij}+z_{\bar{y}y,ij})_{\bar{x}})_x\Delta x\Delta y\\ &+\sum_{i=0}^{N-1}\sum_{j=0}^{M-1}(a(z_{i,j-1},z_{i,j})(z_{\bar{x}x,ij}+z_{\bar{y}y,ij})_{\bar{y}})_y\Delta x\Delta y = 0\end{aligned}$$

  or equivalently

$$\sum_{i=0}^{N-1}\sum_{j=0}^{M-1} z_{ij}(t)\Delta x\Delta y = \sum_{i=0}^{N-1}\sum_{j=0}^{M-1} z_{ij}(0)\Delta x\Delta y. \tag{2.13}$$

- Discrete energy dissipation.

Multiplication of (2.12)) by $(z_{\bar{x}x,ij}+z_{\bar{y}y,ij})\Delta x\Delta y$, summation over i,j and integration in time gives

$$\begin{aligned}\frac{1}{2}\sum_{i,j}\{(z_{\bar{x},ij}(t))^2 &+ (z_{\bar{y},ij}(t))^2\}\Delta x\Delta y\\ &+\int_0^t\left\{\sum_{i,j}a(z_{i-1,j}(s),z_{ij}(s))(z_{\bar{x}x,ij}(s)+z_{\bar{y}y,ij}(s))_{\bar{x}}^2\Delta x\Delta y\right\}ds\\ &+\int_0^t\left\{\sum_{i,j}a(z_{i,j-1}(s),z_{ij}(s))(z_{\bar{x}x,ij}(s)+z_{\bar{y}y,ij}(s))_{\bar{y}}^2\Delta x\Delta y\right\}ds\\ &=\frac{1}{2}\sum_{i,j}\{(z_{\bar{x},ij}(0))^2+(z_{\bar{y},ij}(0))^2\}\Delta x\Delta y.\end{aligned} \tag{2.14}$$

The above mass conservation and energy dissipation combine to give the following estimate:

2.9 Lemma *There exists a constant C such that for any z_{ij} a positive solution of* (2.12), *on the time interval $0 \le t \le T$, with initial data $z_{ij}(0) = h_0(x_i, y_j) > 0$, we have the a priori upper bound*

$$z_{ij}(t) \le \frac{C\|z_{ij}(0)\|_{1,\Delta x,\Delta y}}{\sqrt{\Delta x\Delta y}}, \tag{2.15}$$

where $\|\cdot\|_{1,\Delta x,\Delta y}$ denotes the discrete H^1 norm in 2D.

We now construct a special entropy dissipating scheme that gives an a priori pointwise lower bound. The following lemma shows that the same discretization of the diffusion coefficient that gives dissipation of the nonlinear entropy in 1D also works in 2D.

2.10 Lemma *Let $G''(v) = 1/f(v)$ be a nonlinear entropy function. Let $f(h) \sim h^n$ as $h \to 0$. Then if we choose*

$$a(s_1, s_2) = \begin{cases} \dfrac{s_1 - s_2}{G'(s_1) - G'(s_2)}, & \text{if } s_1 \neq s_2, \\ f(s_1), & \text{if } s_1 = s_2, \end{cases} \tag{2.16}$$

then any solution of (2.12) *with positive initial data $z_{ij}(0) = h_0(x_i, y_j) > 0$ satisfies the following discrete entropy dissipation property*

$$\begin{aligned} &\sum_{i,j} G(z_{ij}(t))\Delta x \Delta y + \int_0^t \left\{ \sum_{i,j} (z_{\bar{x}x,ij}(s) + z_{\bar{y}y,ij}(s))^2 \Delta x \Delta y \right\} ds \\ &= \sum_{i,j} G(z_{ij}(0))\Delta x \Delta y. \end{aligned} \tag{2.17}$$

This property implies that the solution $z_{ij}(t)$ has the following pointwise lower bound

$$\min_t \min_{ij} z_{ij}(t) \geq \left(\frac{\Delta x \Delta y}{\tilde{C}} \right)^{1/(n-2)} \equiv \bar{\delta}, \quad \text{if } n > 2, \tag{2.18}$$

$$\min_t \min_{ij} z_{ij}(t) \geq \exp\left(-\frac{\tilde{C}}{\Delta x \Delta y} \right) \equiv \bar{\delta}, \quad \text{if } n = 2. \tag{2.19}$$

In 2D we do not have a uniform lower bound independent of the grid size for large values of n as we did in the 1D case. This is due to the fact that the 1D argument used the a priori $C^{1/2}$ bound, from Sobolev embedding, which we do not know for the 2D case. The fact that we can derive an a priori pointwise lower bound implies the following corollary:

2.11 Corollary (Global existence of solutions of the 2D entropy dissipating scheme) *Consider $f(h) \sim h^n$ as $h \to 0$, for $n \geq 2$. Then for all Δx, Δy, the entropy dissipating scheme* (2.12) *with $a(s_1, s_2)$ satisfying* (2.16) *and positive initial data $z_{ij}(0) = h_0(x_i, y_j) > 0$ has a unique positive global in time solution.*

In general, the generic scheme (2.12) may only have local solutions in time. The issue is that the solution may lose positivity at some finite time and can not be continued after that time. In addition we have the following convergence result.

2.12 Theorem (Convergence) *Let $h(x, y, t)$ be a smooth solution of* (2.11) *such that $C \geq h(x, y, t) \geq \delta > 0$ for some $C > \delta > 0$ and $t \in [0, T]$. Let $z_{ij}(t)$ be a solution of the numerical scheme* (2.12)*) with $(\Delta x)^3 \ll \Delta y \ll (\Delta x)^{1/3}$. Then for $\Delta x, \Delta y$ sufficiently small the solution $z_{ij}(t)$ can be continued for all $t \in [0, T]$, and there exists a constant $C > 0$ such that we have the H^1 convergence*

$$\sup_{t \in [0,T]} \| z_{ij}(t) - h(x_i, y_j, t) \|_{1, \Delta x, \Delta y} \leq C(\Delta x^2 + \Delta y^2)$$

and the pointwise convergence

$$\sup_{t\in[0,T]}\sup_{i,j}|z_{ij}(t)-h(x_i,y_j,t)|^2\leq C\frac{\left((\Delta x)^2+(\Delta y)^2\right)^2}{\Delta x\Delta y}.$$

2.6 Positivity preserving finite element methods

A natural framework is that of finite elements. We begin by showing that the 1D EDS ((2.1) with (2.6)) finite difference scheme is equivalent to a finite element approximation in which a nonlinear function of the solution is represented in the element basis.

2.13 Example (*1D EDS*) First introduce the pressure $p=-\partial_x^2 h$. The original PDE

$$h_t+\partial_x(f(h)\partial_x^3 h)=0 \tag{2.20}$$

can be rewritten as

$$h_t-\partial_x(f(h)\partial_x p)=0, \qquad p=-\partial_x^2 h. \tag{2.21}$$

Let (η_1,η_2) denote the standard inner product on $L^2(S^1)$, $G(y)$ be a function satisfying $G''(y)=1/f(y)$, $T^{\Delta x}$ be a space of piecewise linear periodic functions on the spatial grid of size Δx. We introduce the interpolation operator $\pi^{\Delta x}:C(S^1)\to T^{\Delta x}$ such that $(\pi^{\Delta x}\eta)(x_j)=\eta(x_j)$ for all j. Define a discrete inner product on $C(S^1)$ by

$$(\eta_1,\eta_2)^{\Delta x}\stackrel{\text{def}}{=}\int_{S^1}\pi^{\Delta x}(\eta_1(x)\eta_2(x))\,dx\equiv\sum_j\eta_1(x_j)\eta_2(x_j)\Delta x.$$

The EDS is equivalent to the following finite element approximation of (2.21):
Find $y,w:G'(y)\in T^{\Delta x}, w\in T^{\Delta x}$ such that

$$(y_t,\chi)^{\Delta x}+(f(y)\partial_x w,\partial_x\chi)=0 \quad \forall\chi\in T^{\Delta x}, \tag{2.22}$$

$$(\partial_x y,\partial_x\chi)=(w,\chi)^{\Delta x} \quad \forall\chi\in T^{\Delta x}. \tag{2.23}$$

Consider the standard basis $\{\chi_i(x)\}_{i=1}^N$ for $T^{\Delta x}$:

$$\chi_i(x)=\begin{cases}\dfrac{x-x_{i-1}}{\Delta x} & \text{if } x_{i-1}\leq x\leq x_i,\\[2mm] \dfrac{x_{i+1}-x}{\Delta x} & \text{if } x_i\leq x\leq x_{i+1}.\end{cases}$$

Substitution in (2.22)–(2.23) gives $(y_t,\chi_i)^{\Delta x}=y_{i,t}\Delta x$ and

$$y_{i,t}+\frac{1}{\Delta x}\left(y_{\bar{x}x\bar{x},i+1}\frac{y_{i+1}-y_i}{G'(y_{i+1})-G'(y_i)}-y_{\bar{x}x\bar{x},i}\frac{y_i-y_{i-1}}{G'(y_i)-G'(y_{i-1})}\right)=0, \tag{2.24}$$

which indeed is the EDS (2.1) with (2.4).

The same finite element approximation on a non-uniform mesh produces an analogous scheme (2.24) with nonuniform differences. A variant of this scheme is used in [27] to compute finite time singularities in a long-wave unstable generalization of (2.20).

The finite element representation (2.22)–(2.23) has similar structure to that introduced in [8]. The main difference is that the method in [8] takes the solution y to be in the subspace $T^{\Delta x}$ spanned by the element basis, resulting in a method that requires a Lagrange multiplier to insure nonnegativity of the solution in cases where it might otherwise become negative. Here, we take the nonlinear function $G'(y) \in T^{\Delta x}$, resulting in a scheme that preserves positivity. Our choice of element representation has one consequence: we need to ensure that $\nabla y \in L^2$ in order to make sense of the inner product on the left hand side of (2.23).

The finite element approach above generalizes to positivity preserving schemes in higher dimensions $(S^1)^d$ and to more complicated finite element subspaces of $H^1((S^1)^d)$. Consider the following form of the lubrication equation in d space dimensions:

$$h_t - \nabla \cdot (f(h)\nabla p) = 0, \quad p + \Delta h = 0, \quad f(h) \sim h^n, \quad h \to 0. \tag{2.25}$$

Now let (η_1, η_2) denote an inner product in $L^2((S^1)^d)$, $T^{\Delta\vec{x}}$ be a finite dimensional subspace of $H^1((S^1)^d)$, $\pi^{I_1} : C((S^1)^d) \to T^{\Delta\vec{x}}$ be an interpolation operator and

$$(\eta_1, \eta_2)^{I_1} \stackrel{\text{def}}{=} \int_{(S^1)^d} \pi^{I_1}(\eta_1(x)\eta_2(x))\, dx,$$

be an associated inner product on $C((S^1)^d)$. Given any d-dimensional vectors $\xi_1, \xi_2 \in C((S^1)^d, R^d)$ let $(\xi_1, \xi_2)^{I_2}$ denote a numerical integration rule replacing an $L^2((S^1)^d, R^d)$ inner product.

Consider the following general finite element approximation of (2.25).

Find $z, w : G'(z) \in T^{\Delta\vec{x}}, w \in T^{\Delta\vec{x}}$ such that

$$(z_t, \chi)^{I_1} + (f(z)\nabla w, \nabla\chi)^{I_2} = 0, \qquad \forall\chi \in T^{\Delta\vec{x}}, \tag{2.26}$$

$$(\nabla z, \nabla\chi)^{I_2} = (w, \chi)^{I_1}, \quad \forall\chi \in T^{\Delta\vec{x}}. \tag{2.27}$$

Taking $\chi = G'(z)$ in (2.26) and $\chi = w$ in (2.27), we obtain

$$\left(z_t, G'(z)\right)^{I_1} + (\nabla w, \nabla z)^{I_2} = 0,$$
$$(\nabla z, \nabla w)^{I_2} = (w, w)^{I_1},$$

or

$$\frac{d}{dt}\int_{(S^1)^d} \pi^{I_1}(G(z))dx = -(w, w)^{I_1} \leq 0.$$

This yields the following a priori bound:

$$\int_{(S^1)^d} \pi^{I_1}(G(z(x,t)))\, dx \leq C.$$

Now by using an explicit form of the interpolation operator π^{I_1} we can rewrite the left hand side of the above inequality as $\sum_i a_i G(z_i(t))$ with $a_i > 0$ for all i. As before, this gives a positive (dependent on $\Delta\vec{x}$) lower bound on $z_i(t)$ for $n \geq 2$.

We have to ensure that $\nabla z \in L^2((S^1)^d, R^d)$ to use the finite element method (2.26)–(2.27). This is guaranteed as long as all $z_i(t)$ are bounded from above. In the special cases of the above example and the next example below, additional structure of the scheme allows us to prove an a priori bound on the discrete H^1 norm of the solution z, independent of the grid size. In general, however, discrete energy dissipation (analogous to (2.3)) may not occur for (2.26)–(2.27). Note that since $G'(z) \in T^{\Delta \vec{x}}$, in general $\nabla G'(z) = \frac{1}{f(z)} \nabla z \in L^2((S^1)^d)$. Thus a sufficient condition for $\nabla z \in L^2((S^1)^d, R^d)$ is that $f(z)$ is bounded. This is also a sufficient condition for $f(z)\nabla w \in L^2((S^1)^d, R^d)$ which is needed to make sense of the nonlinear term in (2.26).

Now let us consider some further examples of the general finite element method (2.26)–(2.27).

2.14 Example In two space dimensions, let $\vec{x} = (x, y)$ and $T^{\Delta \vec{x}}$ be a space of piecewise bilinear functions on the rectangles of size $\Delta x \times \Delta y$. Let $\chi_{ij}(x, y)$ be a basis for this space such that $\chi_{ij}(x, y)$ is equal to 1 at node (x_i, y_j) and 0 at all other nodes. As before, let π^{I_1} : $C((S^1)2) \to T^{\Delta \vec{x}}$ be the interpolation operator such that for any continuous function $\eta(x, y)$ $(\pi^{I_1}\eta)(x_i, y_j) = \eta(x_i, y_j)$ for all i, j. For any 2-dimensional vectors $\xi_1, \xi_2 \in C((S^1)^2, \mathbb{R}^2)$ let

$$(\xi_1, \xi_2)^{I_2} = \sum_{i,j} \int_{x_i}^{x_{i+1}} \frac{1}{2}(\xi_1^1(x, y_{j+1})\xi_2^1(x, y_{j+1})\xi_1^1(x, y_j)\xi_2^1(x, y_j))\Delta y\, dx$$
$$+ \sum_{i,j} \int_{y_j}^{y_{j+1}} \frac{1}{2}(\xi_1^2(x_{i+1}, y)\xi_2^2(x_{i+1}, y) + \xi_1^2(x_i, y)\xi_2^2(x_i, y))\Delta x\, dy.$$

On the product of first components this numerical integration rule performs integration exactly in the first variable, but uses the trapezoidal rule instead of integration in the second variable. Similarly, on the product of second components it performs integration exactly in the second variable, but uses the trapezoidal rule instead of integration in the first variable. We now show that this choice gives us the $2D$ positivity preserving scheme (2.12) with (2.16) from Section 2.5. Taking $\chi = \chi_{ij}$ in (2.26) gives $(z_t, \chi_{ij})^{I_1} = z_{ij,t}\Delta x \Delta y$ and

$$z_{ij,t} + (a(z_{i-1,j}, z_{i,j})(z_{\bar{x}x,ij} + z_{\bar{y}y,ij})_{\bar{x}})_x + (a(z_{i,j-1}, z_{i,j})(z_{\bar{x}x,ij} + z_{\bar{y}y,ij})_{\bar{y}})_y = 0,$$

which is the positivity preserving $2D$ scheme ((2.12) and (2.16)) that we introduced in Section 2.5.

In previous sections on finite differences we encountered both of the examples considered above. Now we introduce a new scheme that arises from a different choice of inner products in (2.26)–(2.27).

2.15 Example In $1D$ consider the same scheme as (2.22)–(2.23) except with trapezoidal rule instead of exact integration for computation of $(f(y)\partial_x w, \partial_x \chi)$ and $(\partial_x y, \partial_x \chi)$. Then $(z_t, \chi_i)^{I_1} = z_{i,t}\Delta x$ and we obtain the scheme

$$z_{i,t} - (a(z_{i-1}, z_i) w_{\bar{x},i})_x = 0$$

with

$$a(z_{i-1}, z_i) = \frac{f(z_{i-1}) + f(z_i)}{2}$$

and

$$w_i = -(\{G'(z_i)\}_{\bar{x}} a(z_{i-1}, z_i))_x.$$

This shows that we can use a discretization of the diffusion coefficient of the form $a(z_{i-1}, z_i) = (f(z_{i-1}) + f(z_i))/2$ and still have a positivity preserving finite difference scheme as long as we also change the definition of the numerical second derivative.

2.7 Computational example

We now present a computational example, from [102], that illustrates the effectiveness of using a positivity preserving (EDS) scheme over a generic one. Consider the example discussed earlier with initial condition (1.10) and $n = 1/2$. We discussed the formation of a finite time singularity that exhibited a discontinuiuty in third derivative as $h \to 0$. From [15, 11, 26] we know that the solution can be continued in time as the limit of a sequence of positive approximations. That is, the regularization

$$h_{\epsilon t} + (f_\epsilon(h_\epsilon) h_{\epsilon xxx})_x = 0, \; f_\epsilon(h_\epsilon) = \frac{h_\epsilon^4 f(h_\epsilon)}{\epsilon f(h_\epsilon) + h_\epsilon^4}. \tag{2.28}$$

Since $f_\epsilon(h_\epsilon) \sim h_\epsilon^4/\epsilon$ as $h_\epsilon \to 0$ we know that for all $\epsilon > 0$ the analytical solution of the regularized problem is positive. In [17], a nonnegative weak solution is computed numerically by taking successively smaller values of ϵ in (2.28). That paper used a scheme of the type (2.1) with $a(s_1, s_2) = f((s_1 + s_2)/2)$. A fine grid is required in order to resolve the spatial structure and keep the numerical solution positive in order to continue the computation. Here we show that an entropy dissipating scheme does a much better job at computing this problem without requiring excessive spatial resolution.

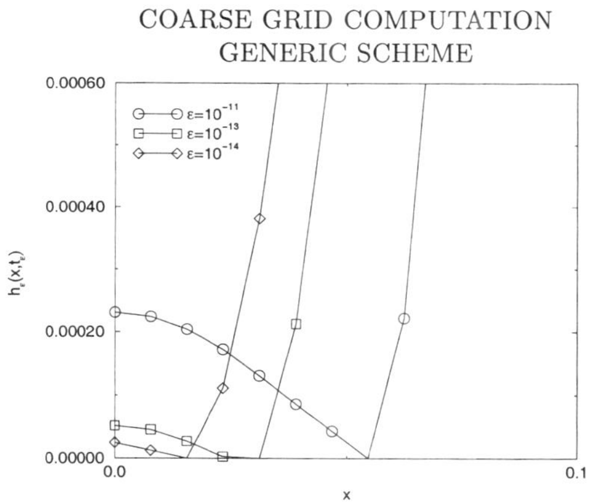

Figure 3: Failed attempt to compute, using a generic difference scheme on a coarse grid, the positive approximations of the weak continuation after the initial singularity described in Fig. 2. Final times $t \approx 0.00086$, 0.00076 and 0.00074 correspond to $\epsilon = 10^{-11}, 10^{-13}$, and 10^{-14}. 128 grid points on $[0, 1]$. $\log_{10}(\min \Delta t) = -14$. We could not continue computing beyond these times since the numerical solution becomes negative.

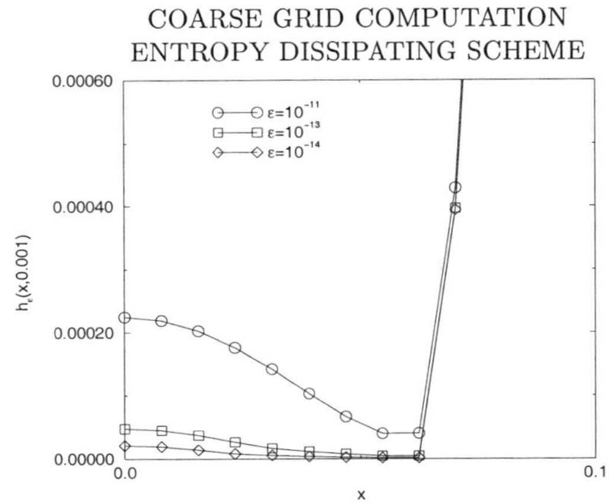

Figure 4: Successful attempt to compute, using an entropy dissipating scheme on a course grid, the numerical solution at fixed time $t = 0.001, \epsilon = 10^{-11}, 10^{-13}$, and 10^{-14}. 128 grid points on [0,1]. $\log_{10}(\min \Delta t) = -6.6$ for $\epsilon = 10^{-11}$, -7.2 for $\epsilon = 10^{-13}$ and -7.4 for $\epsilon = 10^{-14}$.

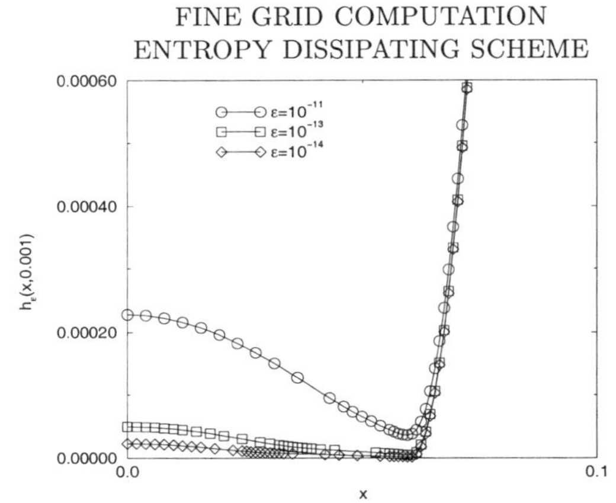

Figure 5: Comparison of computation in Fig. 4 with numerical solution on a fine grid. Same fixed time $t = 0.001, \epsilon = 10^{-11}, \epsilon = 10^{-13}$ and $\epsilon = 10^{-14}$. 1024 grid points on $[0, 1]$. $\log_{10}(\min \Delta t) = -6.7$ for $\epsilon = 10^{-11}$, -7.0 for $\epsilon = 10^{-13}$ and -7.3 for $\epsilon = 10^{-14}$.

Fig. 3 shows the computational results obtained by the generic scheme for three values of the regularization parameter $\epsilon = 10^{-11}$, 10^{-13} and 10^{-14} with uniform grid of 128 points on $[0, 1]$. In all runs we attempted to reach a final time of $t = 10^{-3}$. However, the generic scheme developed a singularity earlier, which made us unable to compute the solution at the time prescribed. Fig. 4 shows the results of the entropy dissipating scheme for the same input. In this case we successfully computed the numerical solution up to $t = 10^{-3}$. Note that in both cases we used the same purely implicit method for time integration choosing the time step Δt small enough to ensure that the discrete time system shows the same behavior as a continuous time one. Fig. 5 shows the results obtained by the entropy dissipating scheme on a much finer grid, a uniform grid of 1024 points on $[0, 1]$. Note that even though the graphs look much smoother now, they show very good agreement with those shown in Fig. 4.

3 Driven films and undercompressive shocks

In many industrial and biological applications thin films are driven by external body forces such as gravity [60, 91] or centrifugal forces [48, 71]. In other words, the external forces instill a convective flow within the film which causes it to spread across a substrate. Under these driving forces the front of the spreading film can become unstable and finger. Moreover, surface stresses, such as that produced by a thermal gradient, can produce very similar fingering patterns [35]. A common feature in these works is the observation that the base state before the instability has, near the contact line, a thick "bump" (frequently termed a capillary ridge) that is responsible for the linear instability.

Some recent driven film experiments, however, studying the problem of a thin film driven by a thermal gradient with an opposing gravitational force have yielded some surprising results. It was noticed that the front developed what appeared to be a pronounced capillary ridge (sometimes 2.5 times the thickness of the bulk film) with no apparent fingering.

In this section we consider this problem using a combination of analysis and two-dimensional numerical simulations. There are strong similarities between this problem and many other driven flow problems and the majority of the analysis contained herein is easily adaptable to their study.

3.1 An experiment

We have in mind a particular experimental problem, in which both surface stresses and body forces are present, leading to two different kinds of convective motion. However, many of the ideas presented here apply to simpler geometries.

We consider an experiment designed and studied by Cazabat's group in Paris [23, 34, 35, 83, 82]. A liquid film is driven from a macroscopic reservoir onto a tilted plate by a surface tension gradient, induced by a constant surface temperature gradient externally imposed along the plate. The liquid is a nonvolatile polydimethylsiloxane (PDMS) which completely wets the substrate, an oxidized silicon wafer. The experiments are prepared with a 7-Å-thick monolayer of PDMS on the surface. Note that this sets a height scale for the precursor layer in the model of the next section.

To increase the film thickness the plate is tilted at an angle from the vertical. Capillarity (and therefore the curvature of the meniscus between the reservoir and the film) govern the film thickness in this range, allowing for control of the film by a geometrical parameter [46], as compared to increasing the thermal gradient, which would ultimately lead to nonlinear behavior of the fluid. The experimental results are quite unexpected: with increased thickness of the flat part of the film the bump becomes more pronounced, but the contact line becomes stable. As we discuss in more detail below, the stable "bump" is actually the combination of two shocks, the leading one being undercompressive and the trailing one being compressive.

3.2 The mathematical model

We consider a mathematical model based on a lubrication approximation (see [24, 23] and references therein):

$$\frac{\partial u}{\partial t} + \frac{\partial}{\partial x}\left(\frac{\tau u^2}{2\eta} - \frac{\rho g u^3 \sin\alpha}{3\eta}\right) = -\nabla\cdot\left[\left(\frac{\gamma u^3 \nabla\Delta u}{3\eta}\right) + \left(\frac{\rho g u^3 \cos\alpha}{3\eta}\nabla u\right)\right], \tag{3.1}$$

where the film height u is a function of position along the incline x and of the transverse variable y as well as of time, t. The constants τ, η, ρ, g, γ and α respectively represent the temperature gradient, viscosity, density, gravity, surface tension and angle of inclination from the horizontal.

The left hand side of (3.1) resembles a scalar hyperbolic equation featuring a non-convex flux in the x-direction; we discuss hyperbolic conservation laws in detail below. The first term in the flux corresponds to flow driven by the Marangoni stress, while the second represents flow driven by the component of gravity tangential to the substrate. This is coupled to the second and fourth order diffusive regularizations on the right hand side. Physically, the second order term represents the effects of gravity normal to the substrate whereas the fourth order diffusive term arises from surface tension attempting to reduce gradients in the film profile.

Typically, a second order diffusive regularization on its own restricts the existence of shocks to those of purely compressive form. However, the additional fourth order regularization provides an opportunity for shocks of undercompressive form to also appear in the solution of (3.1). We discuss these details further in terms of a phase space analysis below.

Application of the dimensional rescaling

$$u = H\hat{u}, \qquad x = l\hat{x}, \qquad t = T\hat{t} \tag{3.2}$$

to (3.1) yields the relevant scales

$$H = \frac{3\tau}{2\rho g \sin\alpha}, \qquad l = \left(\frac{3\gamma\tau}{2\rho^2 g^2 \sin^2\alpha}\right)^{1/3}, \qquad T = \frac{2\eta}{\tau^2}\left(\frac{4}{9}\tau\gamma\rho g\sin\alpha\right)^{1/3}. \tag{3.3}$$

We arrive at the non-dimensional problem (the hats have been dropped for ease of notation)

$$\frac{\partial u}{\partial t} + \frac{\partial}{\partial x}[F(u)] = -\nabla\cdot(u^3\nabla\Delta u) + D\nabla\cdot(u^3\nabla u), \tag{3.4}$$

where

$$D = \left(\frac{9\tau^2}{4\gamma\rho g}\right)^{1/3}\frac{\cot\alpha}{(\sin\alpha)^{1/3}} \tag{3.5}$$

and

$$F(u) = u^2 - u^3. \tag{3.6}$$

For much of the experimental work done to date, $D \ll 1$, and thus we we focus on the reduced equation

$$\frac{\partial u}{\partial t} + \frac{\partial}{\partial x}[F(u)] = -\nabla \cdot (u^3 \nabla \Delta u). \tag{3.7}$$

A detailed investigation of the role of the second order diffusive term in (3.4) in terms of the size of D can be found in [74]. We also consider the problem in a reference frame moving with speed s, and so for the rest of this section we adjust $f(u)$ from (3.6) to be

$$F(u) = u^2 - u^3 - su. \tag{3.8}$$

To relieve the no-slip paradox that occurs at a moving contact line (as discussed in Section 3.1), we assume the presence of a flat precursor film of thickness u_+ ahead of the film (this being consistent with the precursor layer discussed in the experimental details). Behind the advancing front is a flat film with thickness u_- controlled by the meniscus [46]. Thus the equations (3.7)–(3.8) are coupled to the far-field conditions

$$u \to u_- \text{ as } x \to -\infty, \qquad u \to u_+ \text{ as } x \to +\infty. \tag{3.9}$$

We note that equation (3.7) is nearly identical in form to (1.5), the only difference being the addition of the convective term due to the external driving mechanisms. It is this term that allows shocks or jump discontinuities to form in the solution of (3.7).

It is useful here to review the role of conservation laws in general. A more comprehensive discussion can be found in [70]. Conservation laws are first order, hyperbolic equations of the form

$$u_t + [F(u)]_x = g(u, t, x), \tag{3.10}$$

where the equation is called homogeneous if $g \equiv 0$. When the flux $F(u)$ is a nonlinear function of u it is well known that shocks may develop from smooth initial data. Before the shock forms the solution is classical, and the method of characteristic gives the unique solution. After the shock forms, the solution is no longer classical, but still makes sense as a weak solution involving integration of the PDE against smooth test functions. This machinery allows us to interpret spatial jumps in the solution of a first order equation and yields a relationship, called the Rankine-Hugoniot jump condition, between the shock speed and the left and right states of the shock. Where it is smooth, the weak solution corresponds to a classical solution. Uniqueness of the weak solution requires an additional condition that arises from the physics of the problem. For (3.10), the admissible solution is taken to be the one that satisfies the Lax entropy condition, namely that

$$F(u_-) > s > F(u_+), \tag{3.11}$$

where s is the shock speed and the subscripts on u represent the left and right states on either side of the shock, consistent with (3.9). The admissible solution is sometimes termed a Lax shock and is also called compressive due to its form in characteristic space. The Lax entropy condition, however, is only a local condition and for a globally admissible solution we must also satisfy the Oleinik chord condition (the Lax entropy condition can be shown to arise out of a simple perturbation analysis of the Oleinik chord condition). We note that for fluxes with at most one inflection point, the Lax entropy condition alone is sufficient to prove admissibility, but this is not true in general. The Oleinik chord condition, which states that on the graph of the flux function, the chord connecting a left and right state of a shock cannot cross the flux function, is the correct condition for more general fluxes. One can observe this to be the case by analyzing the limiting traveling wave solution of the viscously regularized conservation law

$$u_t + [F(u)]_x = \varepsilon \Delta u \quad \text{as} \quad \varepsilon \to 0. \tag{3.12}$$

It turns out that the choice of a second order regularization is rather subtle, and taking a higher order regularization leads to some surprising alternatives. It has recently been shown that problems of the form

$$u_t + (F(u))_x = -\nabla \cdot (u^n \nabla \Delta u) \tag{3.13}$$

allow undercompressive shocks to form and we must reformulate our notion of admissibility. Undercompressive and other non-classical shocks have been well studied in systems of conservation laws [1, 31, 61, 88, 89, 99]. For the scalar law case, they were thought to be a purely mathematical curiosity and were not believed to have any experimental basis. The experiment we discuss here is perhaps the first example of a physical undercompressive shock for a scalar law problem. Of particular interest is the stability of these undercompressive solutions and their numerical modeling. Numerical schemes are typically designed to seek out compressive shocks and ignore any undercompressive effects. The creation of codes which capture the undercompressive behavior has therefore become an area of growing interest. Additionally, in numerical simulations and physical experiments it is found that the creation of an undercompressive shock is typically coupled with the appearance of a standard compressive shock and it is important to know how these multiple shock structures are affected by small perturbations.

We now return to the problem at hand. For each value of u_+ (i.e. a particular precursor thickness) there are a number of values of u_- for which transitions/bifurcations in the solution structure occur. We assume that $u_+ \ll 1$ (for our purposes, the analysis for any $u_+ < 1/3$ is equivalent; the second derivative of the flux function changes sign at $u_+ = 1/3$ and it is impossible to generate compressive shocks with $u_- > u_+$ when $u_+ > 1/3$).

We begin by reviewing the possible scenarios for long time behavior of the one dimensional problem:

(*I*) For $u_+ < u_- < u_1$ (u_1 dependent on u_+) there exists a unique capillary Lax shock connecting u_- to u_+.

(*II*) For $u_1 < u_- < u_2$ (u_2 dependent on u_+) there are multiple capillary Lax shock profiles connecting u_- to u_+. There is a critical value of u_-, denoted here by u_*, for which an infinite number of capillary Lax shocks exist.

(*III*) For $u_- > u_2$ there are no capillary Lax shocks connecting u_- to u_+. There is, however, a critical value of $u_- = u_{UC}$ (dependent on u_+) for which a single undercompressive capillary shock exists.

For a single shock profile (compressive or undercompressive) the unperturbed base state is given by the one-dimensional traveling wave solution of the evolution equation (3.7) [24] (see also [60, 64] for comparison). Imposing the solution $u = u(x)$ (the traveling wave solution becomes stationary in the moving reference frame if s is chosen suitably in (3.8), cf. (3.16)) yields

$$[F(u)]_x + (u^3 u_{xxx})_x = 0. \tag{3.14}$$

Equ. (3.14) is immediately integrable, the constant of integration being specified by the far-field conditions (3.9). Application of these conditions yields

$$\begin{aligned} F(u_+) &= u_+^2 - u_+^3 - su_+ = k, \\ F(u_-) &= u_-^2 - u_-^3 - su_- = k, \end{aligned} \tag{3.15}$$

which can be rearranged to give the Rankine-Hugoniot condition for the speed of the shock

$$s = \frac{u_+^2 - u_+^3 - (u_-^2 - u_-^3)}{(u_+ - u_-)}. \tag{3.16}$$

We choose this value of s in (3.8) for both numerical and analytical convenience.

3.3 Traveling waves with fourth order diffusion

The admissibility of non-classical shocks with higher order diffusion is related to the admissibility of traveling waves for this problem [24]. These solutions satisfy the third order ordinary differential equation (cf. (3.14))

$$F(u) - F(u_+) = -u^3 u'''. \tag{3.17}$$

(In integrating the equation once, we have assumed $u'''(x) \to 0$ and $u(x) \to u_+$ as $x \to \infty$.) Equ. (3.17) has two parameters u_+ and s, which appear in F in (3.8). The traveling reference frame with s appearing in (3.8) results in stationary waves.

Writing (3.17) as

$$u''' = g(u; u_+, s), \tag{3.18}$$

where

$$g(u; u_+, s) = -u^{-3}(F(u) - F(u_+)), \tag{3.19}$$

we see that the equilibrium $u = u_-$ (zeroes of $g(.; u_+, s)$) is given by the Rankine-Hugoniot condition (3.16) for shocks. Moreover, the linearized ordinary differential equation $\tilde{u}''' = \frac{\partial g}{\partial u}(u_-; u_+, s)\tilde{u}$ has three eigenvalues, namely the three cube roots of $\frac{\partial g}{\partial u}(u_-; u_+, s)$. Note further that $\frac{\partial g}{\partial u}(u_-; u_+, s) = -\frac{1}{u_-^3}(F'(u_-))$, so that the sign of $\frac{\partial g}{\partial u}(u_-; u_+, s)$ is related to whether characteristics at u_-, traveling with speed $F'(u_-)$, are faster or slower than the speed s of the traveling wave.

To further understand equilibria, and trajectories joining different equilibria, we write Eq. (3.18) as a first order system:

$$\begin{aligned} u' &= v, \\ v' &= w, \\ w' &= g(u; u_+, s). \end{aligned} \tag{3.20}$$

For the specific equilibrium $(u, v, w) = (u_-, 0, 0)$, with $g(u_-; u_+, s) = 0$, the three cube roots of $\frac{\partial g}{\partial u}(u_-; u_+, s)$ give rise to the following structure of invariant manifolds containing the equilibrium.

(i) If $F'(u) < s$, then $\frac{\partial g}{\partial u}(u; u_+, s) > 0$, so that $(u, 0, 0)$ has a one-dimensional unstable manifold and a two-dimensional stable manifold on which solutions spiral into the equilibrium due to the complex conjugate pair of eigenvalues with negative real part.

(ii) If $F'(u) > s$, then $\frac{\partial g}{\partial u}(u; u_+, s) < 0$, so that $(u, 0, 0)$ has a one-dimensional stable manifold and a two-dimensional unstable manifold on which solutions spiral away from the equilibrium due to the complex conjugate pair of eigenvalues with positive real part.

Now equilibria of (3.20) correspond to intersections of the line with zero slope through $(u_+, F(u_+))$ with the graph of F. Alternatively, we can fix $u_- \neq u_+$ and define s through the Rankine-Hugoniot condition (3.16). Specifically, for the flux function $F(u) = u^2 - u^3 - su$,

$$s = u_- + u_+ - u_-^2 - u_- u_+ - u_+^2. \tag{3.21}$$

We now consider the case of a weak Lax shock, specifically $0 < u_+ < u_- < 1/3$, with $u_- - u_+$ small, in detail; a similar analysis holds for the other cases. Then $(u_-, 0, 0)$ has a two-dimensional unstable manifold and $(u_+, 0, 0)$ has a two-dimensional stable manifold. If they intersect transversally (away from the equilibria), then the curve of intersection is a trajectory for (3.20) from $(u_-, 0, 0)$ to $(u_+, 0, 0)$. We refer to the corresponding solution of the partial differential equation (3.7) as a *capillary shock profile* from u_- to u_+. The existence of capillary shock profiles for weak Lax shocks away from the inflection point $u = 1/3$ is provided by the analysis of Kopell and Howard [67].

3.4 Structure of phase space

We next investigate the appearance of capillary shock profiles for different values of u_-, and hence different s, by numerically exploring the structure of the phase space. In order to do so, we determine the unstable and stable manifolds of the three equilibria

$$B = (u_+, 0, 0), \quad M = (u_-, 0, 0), \quad \text{and} \quad T = (1 - u_- - u_+, 0, 0)$$

(with $u_+ < u_- < 1 - u_- - u_+$).

Certain trajectories are visualized in a perspective plot of the three-dimensional phase space (see Fig. 6(a) for $u_- = 0.33$). Additionally, we show the intersection of the stable manifold $W^s(B)$ of B, the unstable manifold $W^u(M)$ of M, and the unstable manifold $W^u(T)$ of T with the plane

$$P = \{(u, u', u''); \; u = (2u_- + u_+)/3\}$$

in Fig. 6(b). We refer to the plane P as a Poincaré section. Since $W^s(B)$ and $W^u(M)$ are surfaces, their intersections with P appear as two curves, whereas the one-dimensional manifold $W^u(T)$ appears in the Poincaré section as a point. Capillary profiles for Lax shocks correspond to intersections of $W^u(M)$ and $W^s(B)$. A capillary profile for an undercompressive shock appears when $W^s(B)$ contains the curve $W^u(T)$.

As an example of the use of Poincaré sections in the study of multi-dimensional differential equations we study the case II from Section 3.2, $u_1 < u_- < u_2$, corresponding to multiple capillary shock profiles; a full description of the entire phase space can be found in [24].

For each value of $u_+ < 1/3$ there is a range of values for $u_- > u_1$, for which there exists more than one orbit connecting M to B. This range of u_- extends up to a finite upper bound u_2,. For $u_+ = 0.1$, we find numerically

$$u_2 = 0.3479.$$

The number of connections can be computed numerically by examining the number of intersection points of the curves $W^u(M) \cap P$ and $W^s(B) \cap P$ on the Poincaré section.

For $u_- = 0.33$, the Poincaré section, shown in Fig. 6(b), indicates three intersections of $W^s(B) \cap P$ and $W^u(M) \cap P$. Fig. 6(a) shows the corresponding orbits in three dimensional phase space; on the scale of the graph, two of the orbits nearly coincide with the largest separation near T.

There is a special value of u_-, which we denote by u_*, for which $W^U(T)$ intersects $W^S(B)$. The existence of this connection was proved rigorously by the first author and Shearer in [28]. At this special value, we observe in the phase portrait, infinitely many turns in $W^u(M) \cap P$, and consequently a countably infinite number of intersections of $W^u(M) \cap P$ with $W^s(B) \cap P$.To understand this, we first remark that, in the neighborhood of T, the connection from M to T is governed by the two complex eigenvalues of the linearization of (3.20) around this equilibrium. These eigenvalues have negative real part, so that the trajectory approaches T through a spiral with an infinite number of turns. A neighboring trajectory, contained in $W^u(M)$ but not in $W^s(T)$, will initially stay close to the orbit connecting M to T, and undergo some turns, until it finally is repulsed from the vicinity of T along one of the two branches of $W^u(T)$. The infinite number of intersections of $W^u(M) \cap P$ with $W^s(B) \cap P$ when $u = u_*$ corresponds to an infinite number of different orbits connecting M to B. Each orbit corresponds to a different traveling wave solution of (3.7).

Every solution connects the same two far-field constant states and therefore (by the Rankine-Hugoniot condition (3.16)) has the same speed. Note that Fig. 6(b) resolves only a few turns of the spirals of $W^u(M) \cap P$. In numerical trials we could resolve a total of six turns and five intersection points, in $W^u(M) \cap W^s(B) \cap P$. The accumulation point of the intersection points in the Poincaré section corresponds to the special orbit connecting T to B. This orbit yields a special traveling wave solution of (3.7). This is an undercompressive capillary shock. Note also that u_* and u_{UC} are related by $u_{UC} = 1 - u_+ - u_*$.

The analysis of traveling waves can be carried out for the more general problem resulting from (3.4) with $D > 0$. In [28] it was rigorously proved that for any $u_+ < 1/3$ and D sufficiently small there exists a special shock speed s^* for which a heteroclinic orbit from T to B exists. This orbit corresponds to an undercompressive traveling wave connecting u_+ to u_{UC}. However, for D large enough, no such connection exists and undercompressive shocks cannot arise in the solution. The phase space computations in [24] were extended to the case of positive D in (3.5) by Münch [74] and to the case of traveling waves satisfying the Navier slip condition (in lieu of having a precursor film, so that there is a true contact line) in [32]. The full problem with van der Waals interactions (wetting as as opposed to dewetting, discussed in the next section) was recently considered by Golovin et al.[51].

3.5 Two-dimensional numerical simulations

When the fingering instability arises the dynamics are no longer one-dimensional and we must study the full two-dimensional numerics to understand the phenomena. We solve (3.7)

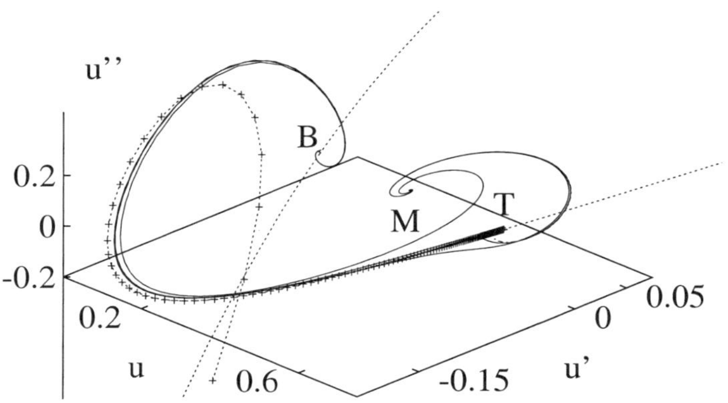

(a) Perspective view

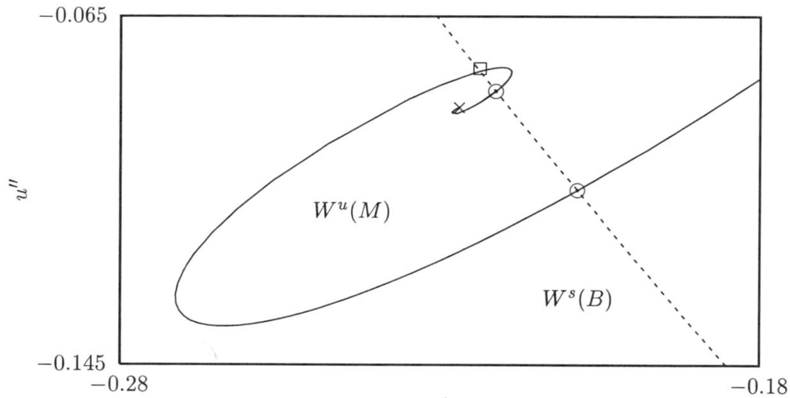

(b) Poincaré section P

Figure 6: Phase portrait for $u_- = 0.33$: Solid lines in (a) represent the three connections from M to B, which appear as three intersection points $W^u(M) \cap W^s(B) \cap P$ in (b). Circles and boxes indicate whether they correspond to stable or unstable traveling wave solutions of the PDE (3.7), respectively. For better visibility, one branch of $W^u(T)$ has been emphasized with '+'. We note that this diagram is plotted with the parameter s set to zero in (3.8).

using the ADI scheme described below. Such methods have a long history of application to second order and linear fourth order problems. More recently they have also been used in other driven thin film problems (see, for example, [84, 42]). The method applied here is an extension of those investigated in [10].

There is an inherent fear in using an ADI scheme for higher order parabolic problems as there are no convergence or existence results available. However, as illustrated in [10], schemes for fourth order problems appear to perform equally as well as other possible schemes and have a distinct speed advantage requiring only $O(N)$ computations at each time step. It is also noteworthy that starting from one-dimensional initial conditions (i.e. with no variation in the transverse y-direction) our scheme computes solutions with cross-section that matches the one-dimensional solutions found in [24]. We feel these comments justify the use of the method here.

It has been previously shown in [98] that it is impossible to attain $O((\Delta t)^2)$ accuracy with a two level ADI scheme. Equ. (3.7) is therefore approximated by a three level scheme which is formally $O((\Delta t)^2)$ accurate. We take

$$\begin{aligned} \mathcal{L}_x u^* &= \frac{1}{\gamma}[\alpha u^n + \beta u^{n-1}] - \frac{1}{\gamma}\Delta t_1[\partial_x(\tilde{f}^{n+1}\partial_{yyx}) + \partial_y(\tilde{f}^{n+1}\partial_{xxy})]\tilde{u}^{n+1} \\ &\quad - \frac{1}{\gamma}\Delta t_1 \mathcal{D}_y \tilde{u}^{n+1}, \end{aligned} \tag{3.22}$$

$$\mathcal{L}_y u^{n+1} = u^* + \frac{1}{\gamma}\Delta t_1 \mathcal{D}_y \tilde{u}^{n+1}, \tag{3.23}$$

where

$$\mathcal{L}_x = \mathcal{I} + \frac{1}{\gamma}\Delta t_1 \mathcal{P}_x + \frac{1}{\gamma}\Delta t_1 \mathcal{D}_x, \qquad \mathcal{L}_y = \mathcal{I} + \frac{1}{\gamma}\Delta t_1 \mathcal{D}_y,$$

$$\mathcal{D}_x = \partial_x(\tilde{f}^{n+1}\partial_{xxx}), \qquad \mathcal{D}_y = \partial_y(\tilde{f}^{n+1}\partial_{yyy}), \qquad \mathcal{P}_x u^* = \partial_x(u^{*2} - u^{*3} - Uu^*),$$

$$\tilde{u}^{n+1} = \sigma u^n + \rho u^{n-1}, \qquad \tilde{f}^{n+1} = f(\tilde{u}^{n+1}),$$

and the constants α, β, γ, ρ and σ are defined as

$$\alpha = \frac{\Delta t_1 + \Delta t_2}{\Delta t_2}, \qquad \beta = -\frac{(\Delta t_1)^2}{\Delta t_2(\Delta t_1 + \Delta t_2)}, \qquad \gamma = \frac{\Delta t_2 + 2\Delta t_1}{\Delta t_2 + \Delta t_1},$$

$$\rho = -\frac{\Delta t_1}{\Delta t_2}, \qquad \sigma = \frac{\Delta t_1 + \Delta t_2}{\Delta t_2}.$$

Here u^* denotes an intermediate value for the numerical scheme. Note that the three time levels are contained within the definition of $\tilde{u}^{n+1}$ (and therefore $\tilde{f}^{n+1}$) which is a linear approximation for u^{n+1} while the effects of including an adaptive time step are seen in the choice of α, β, γ, ρ and σ. Most importantly, by using the $\tilde{f}^{n+1}$ approximation for f^{n+1}, Equ. (3.23)) is linear and the only nonlinearity in (3.22) comes from $\mathcal{P}_x$. Although we have made a further approximation to gain linearity in one of the equations this will not introduce errors larger than $O((\Delta t)^2)$ at any time step, and we do not believe that the accumulated error over several time steps will be any larger than that already introduced by our approximation of (3.7) by (3.22)–(3.23). If this did become an issue, it is a relatively simple matter to introduce an iteration procedure at each time step that updates our approximation $\tilde{f}^{n+1}$ until the difference between successive approximations is below a certain preset tolerance (implying that we are close to the true value of f^{n+1}); see [10] for a discussion of iteration methodology for these problems.

We now turn our consideration to the choice of suitable discretizations for the spatial derivatives in (3.22)–(3.23), including those inherent in $\mathcal{D}_x$, $\mathcal{D}_y$ and $\mathcal{P}_x$. The discretizations adopted here are cell centered, second order accurate, thereby making the whole scheme $O((\Delta t)^2 + (\Delta x)^2)$ accurate. The first order derivatives with respect to x are given by

$$\partial_x p_{i,j} = \frac{p_{i+1/2,j} - p_{i-1/2,j}}{\Delta x} + O((\Delta x)^2)$$

and similarly for $\partial_y p_{i,j}$. The cell centered third derivative takes the form

$$\partial_{xxx} u_{i+1/2,j} = \frac{u_{i+2,j} - 3u_{i+1,j} + 3u_{i,j} - u_{i-1,j}}{(\Delta x)^3} + O((\Delta x)^2)$$

and similarly for $\partial_{yyy} u_{i,j+1/2}$. The mixed derivatives are also given by cell centered differences, namely

$$\partial_{xyy} u_{i+1/2,j} = \frac{u_{i+1,j+1} - u_{i,j+1} - 2u_{i+1,j} + 2u_{i,j} + u_{i+1,j-1} - u_{i,j-1}}{\Delta x (\Delta y)^2} + O((\Delta x)^2) + O((\Delta y)^2)$$

and similarly for $\partial_{yxx} u_{i,j+1/2}$.

Finally, we must select a discretization for the nonlinear diffusion in the fourth order term. We take the standard averaged approximation

$$f^{n+1}_{i+1/2,j} = \frac{1}{2}(f^{n+1}_{i,j} + f^{n+1}_{i+1,j}) + O((\Delta x)^2).$$

At each time step we solve the nonlinear equation (3.22) using Newton iteration and then solve the linear equation (3.23) using a banded matrix solver with LU decomposition. Obviously, the approximations we make here are by no means unique. For a more detailed discussion of ADI methods for high order parabolic problems see [10].

Fig. 7 shows a two-dimensional simulation of a separating double shock structure. Transverse sinusoidal perturbations are added in front of the leading shock (left). These bumps impact the undercompressive wave (middle) but do not destabilize it. However, their effect results in perturbations traveling along characteristics from the undercompressive wave to the trailing Lax wave (right). In contrast, a perturbation applied to a compressive wave, resulting from single step initial data, produces the characteristic fingering instability shown in Fig. 8.

3.6 Linear stability analysis

To understand the transition to fingering we must perform a linear stability analysis. We begin by invoking the solution ansatz

$$u(x, y, t) = \hat{u}(x, t) + \varepsilon h(x, t)\cos(ky) \quad \varepsilon \ll 1, \tag{3.24}$$

and substitution of (3.24) into (3.7) leads to a linear evolution equation for h at $O(\epsilon)$:

$$h_t = -[hF'(\hat{u})]_x - [\partial_x(\hat{u}^3\partial_x^3 h + 3h\hat{u}^2\partial_x^3\hat{u})] - k^2[\partial_x(\hat{u}^3\partial_x h) + \hat{u}^3(\partial_x^2 h)] - k^4\hat{u}^3 h = \mathcal{L}_k h \tag{3.25}$$

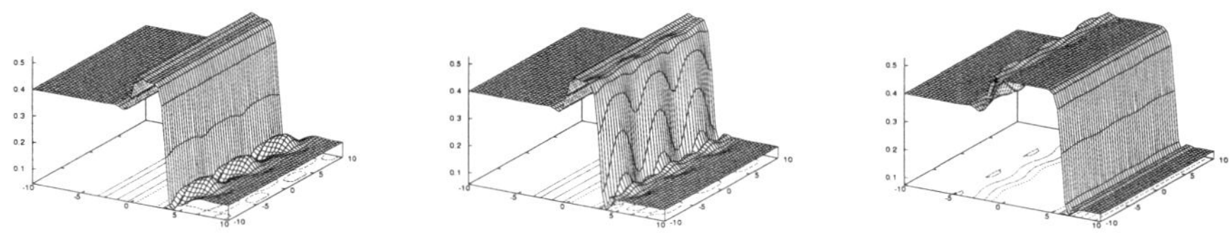

Figure 7: Surface and contour plots showing the effect of transverse perturbations on the double shock structure of an undercompressive-Lax wave pair for the thermal-gravity flux (3.8).

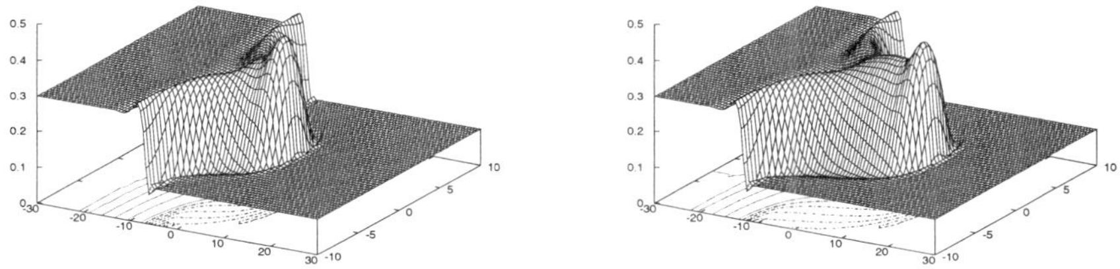

Figure 8: Surface and contour plots showing the fingering instability for a compressive wave.

where k denotes the y wavenumber. The symmetry in y leads to only even powers of k in (3.25).

For self-adjoint problems this linear stability analysis is sufficient to completely describe the behavior of solutions. However, the problem considered here is not self-adjoint and the large time growth rates tell only part of the whole story. There happen to be transient timescales on which growing modes can excite nonlinearities of the original equation [20].

For simplicity of analysis we begin by considering the case of a single, well-established shock (either compressive or undercompressive), in the form of a traveling wave $\hat{u}(x,t) = \hat{u}(x)$. Note that we choose the reference frame to travel at the same speed as the wave, which is equivalent to setting s in (3.8) so that the speed of the shock is zero. Thus the traveling wave becomes a stationary solution $\hat{u}$. Looking for solutions to (3.25) of the separable form $h(x,t;k) = e^{\beta(k)t}\psi(x,k)$ leads to the relationship

$$\mathcal{L}_k \psi = \beta \psi, \tag{3.26}$$

so that β acts as an eigenvalue with ψ as the corresponding eigenfunction. We are primarily interested in small k corresponding to long wavelength perturbations; large k perturbations are always smoothed rapidly by surface tension effects. In the small k limit we can expand β and ψ as power series in even powers of k due to the symmetry in y. Whether solutions grow or decay for large time is thereby dependent on the sign of β_1, which can be found from the solvability condition

$$\beta_1 \int_{-\infty}^{\infty} \hat{u}_x \pi_0 \, dx = \int_{-\infty}^{\infty} \pi_{0x} \hat{u}^3 \hat{u}_{xx} \, dx + \int_{-\infty}^{\infty} \pi_0 [F(\hat{u}) - F(u_+)] \, dx, \tag{3.27}$$

where $\pi_0(x)$ is the left eigenfunction that satisfies the formal adjoint of $\mathcal{L}_0$, namely $\mathcal{L}_0^* \pi_0 = 0$, where

$$\mathcal{L}_0^* = -\frac{d^3}{dx^3}\left(\hat{u}^3 \frac{d}{dx}\right) + \left(F'(\hat{u}) + 3\hat{u}^2 \frac{d^3\hat{u}}{dx^3}\right)\frac{d}{dx}. \tag{3.28}$$

3.6.1 Compressive shocks

The work of [25] showed (by application of spectral theory) that for compressive shocks there exists a unique left eigenfunction, π_0, given by the constant function $\pi_0 = 1/(u_+ - u_-)$. The fact that π_0 is constant validates the stability analysis considered in [20] and [63]. The solvability condition (3.27) then simplifies to

$$\beta_1 = \frac{1}{u_+ - u_-} \int_{-\infty}^{\infty} [F(\hat{u}) - F(u_+)] \, dx, \tag{3.29}$$

where the integral can be calculated numerically once $\hat{u}(x)$ is known. As $u_+ - u_-$ is a fixed quantity, the stability of compressive shocks therefore comes down to the sign of the integral in (3.29) which must change at a critical wavenumber k_c; for $k < k_c$ the compressive shock is unstable whereas for $k > k_c$ the shock is stable.

3.6.2 Undercompressive shocks

Analysis of the evolution equation (3.7) illustrates that double shock structures can form from simple step initial data [24]. In fact, as mentioned previously, we only expect to have an undercompressive shock existing independently for the choice of $u_- = u_{UC}$. For $u_- \neq u_{UC}$ we must necessarily have either a compressive shock or an undercompressive/compressive shock pair. The double shock structures seen in experiment [23] consist of a leading undercompressive shock followed by a trailing compressive shock. The undercompressive shock is a jump from the precursor film thickness u_+ to a large height u_{UC}, while the compressive shock is a jump from u_{UC} down to the flat film height u_-, which is fixed by the curvature of the meniscus in the experiment.

It was shown in [25] that the unique left eigenfunction for undercompressive shocks is not constant and takes no simple analytic form. A far-field and numerical study led to the understanding that for a range of $k > 0$, the right eigenfunction grows exponentially as $x \to -\infty$, and this is the mechanism by which perturbations are swept away from behind the undercompressive front. The fact that perturbations do not remain localized to an undercompressive shock has important implications for these shock pairs. From the linear stability analysis above, we see that it is possible for the perturbation spreading from the undercompressive shock to reach the compressive shock which may then destabilize, leading to a fingering instability down the plate toward the fluid bath. This scenario is consistent with the numerical simulation shown in Fig. 7. The important consequence of this analysis is that undercompressive shocks are stable to all transverse perturbations; it is these special stability properties which provide the mechanism for producing stable moving contact lines.

In contrast, for compressive shocks there exists a band of long wavelength perturbations capable of destabilizing the shock and triggering a fingering instability.

4 Dewetting films

We now turn to the problem of dewetting films. The competition of attractive van der Waals forces and short range repulsive forces, like Born repulsion, can produce complex instabilities in layers of thin liquid films on solid substrates. Experimental studies have shown this behavior for polymer films [66, 80, 86, 87, 100], liquid crystal films [58, 92, 93], liquid metals [29, 58] and evaporating films [41, 76]. During these dewetting processes, large droplets are formed which are connected by ultra-thin films. In analogy to spinodal decomposition in phase separation for binary alloys described by the Cahn-Hilliard equation, this evolution is sometimes called spinodal dewetting [72, 73, 76].

We consider a model and results from [22]. The lubrication approximation for dewetting films under the influence of both attractive van der Waals forces and short range repulsive forces has a general form [37, 78]

$$\frac{\partial h}{\partial t} - \nabla \cdot (f(h)\nabla p) = 0, \tag{4.1}$$

where the pressure p in the thin film is given by

$$p = P(h) - \Delta h. \tag{4.2}$$

The explicit form of the nonlinear mobility f again depends on the boundary condition at the liquid-solid interface, as in (1.5). The forms

$$f(h) = h^3, \quad \text{and} \quad f(h) = h^3 + \beta h^b, \tag{4.3}$$

with $\beta > 0$, $b \in (0,3)$, correspond to *no-slip* and various classes of *slip* boundary conditions, respectively.

The term $P(h)$ encompasses forces exerted on the film, such as a hydrostatic body force and disjoining/conjoining intermolecular forces due to van der Waals interactions and Born repulsion [43]. For dewetting, we consider functions

$$P(h) = \frac{1}{h^n} - \frac{\epsilon^{m-n}}{h^m} = \frac{1}{h^n}\left(1 - \left[\frac{\epsilon}{h}\right]^{m-n}\right), \quad 0 < n < m, \tag{4.4}$$

where ϵ is a small positive parameter [42]. The standard 6–12 Lennard-Jones potential [62, 72, 73, 78] corresponds to the exponents $(n, m) = (3, 9)$ in this model. The h^{-n} term describes the aforementioned long-range attractive force, while the second term in (4.4) models the short range stabilizing effect of Born repulsion. Oron et al. [76, 77] consider (4.4) with $(n, m) = (3, 4)$ to describe a thin film on a layered solid substrate. Our results show that the solutions of these models have the same qualitative structure for a large set of (n, m) values.

Early studies of van der Waals driven instabilities of thin films, by Williams and Davis [33, 95] and de Gennes [37], considered this problem without repulsive pressure terms, i.e. $P(u)$ given by equation (4.4) with $\epsilon = 0$. In general, classical solutions of this problem can cease to exist in finite time due to singularities which occur if the film ruptures, $u \to 0$. In this context, for $n = 3$, self-similar finite-time rupture solutions were studied by Zhang and Lister [101], and Witelski and Bernoff [96, 97].

In contrast, for $\epsilon > 0$, the problem is globally well-posed, and solutions exist for all times [22, 54, 55, 56]. Numerical simulations of thin films with generalized van der Waals disjoining pressure show complex pattern formation during which films evolve to a meta-stable state composed of a collection of droplets connected by a thin film of thickness approximately ϵ. A comparison with physical experiments shows striking similarities to the characteristic features of dewetting including the formation of capillary ridges at the edges of growing holes (cf. [56] and [85] for drying dewetting paint).

We show that the parameter ϵ in (4.4) gives a minimum thickness for non-trivial thin film equilibrium states. Viewed as a regularization, it is natural to consider the limit of these "post-rupture" solutions as $\epsilon \to 0$. We show via numerical simulations and asymptotic analysis of the equilibria, that *concentrations* [44] occur in this model. That is, this limit makes sense only as a measure; for fixed times after rupture numerical simulations show the $\epsilon \to 0$ limit concentrates mass as a δ-distribution.

Appropriate boundary conditions for equation (4.1) corresponding to the physical problem of a fluid confined in an impermeable solid container are

$$\mathbf{n} \cdot \nabla p = 0, \quad \mathbf{n} \cdot \nabla h = 0, \quad \text{on} \quad \partial\Omega. \tag{4.5}$$

The latter condition specifies a 90° contact angle between the fluid surface and vertical boundaries at $\partial\Omega$. This condition makes a uniform fluid layer, $h(x, y) = \bar{h}$, an exact solution of the full problem and is physically equivalent to neglecting any positive or negative meniscus due to wetting or non-wetting properties of the container holding the fluid. By the former condition, mass is conserved, as is the average film thickness $\bar{h}$ which is given by

$$\bar{h} = \frac{1}{\mathcal{A}} \int_\Omega h \, dA, \tag{4.6}$$

where $\mathcal{A} = \text{area}(\Omega)$. We use $\bar{h}$ as a bifurcation parameter for the set of equilibrium solutions of (4.1). An energy, or Liapunov integral, for (4.1) is given by

$$\mathcal{E} = \int_\Omega \frac{1}{2} |\nabla h|^2 + Q(h) \, dA, \tag{4.7}$$

where Q is the primitive of P,

$$Q(h) = -\int_h^\infty P(v) \, dv. \tag{4.8}$$

This energy is monotone decreasing, with the rate of dissipation given by

$$\frac{d\mathcal{E}}{dt} = -\int_\Omega f(h) |\nabla p|^2 \, dA \leq 0. \tag{4.9}$$

In one dimension the local minimizers of the energy correspond to states with a constant uniform pressure, $p = \bar{p}$, and they solve the semilinear elliptic equation

$$\bar{p} = P(h) - \Delta h \quad \text{on} \quad \Omega \tag{4.10}$$

with Neumann boundary conditions. Solutions satisfy the compatibility condition

$$\bar{p} = \frac{1}{\mathcal{A}} \int_{\Omega} P(h)\, dA. \tag{4.11}$$

We consider a two-dimensional thin film resulting in a one-dimensional PDE for the film thickness h. Ω is given as an interval $(0, L)$ and h solves the equation

$$\frac{\partial h}{\partial t} + \frac{\partial}{\partial x}\left(f(h)\frac{\partial}{\partial x}\left[\frac{\partial^2 h}{\partial x^2} - P(h)\right]\right) = 0. \tag{4.12}$$

4.1 Minimizers

In this section we discuss the existence theory for the one-dimensional problem and prove existence of a smooth global minimizer of the Liapunov functional (4.7). For a pressure of the form (4.4), we have the following theorem for positive initial data:

4.1 Theorem *Consider initial data satisfying*

$$h_0 > 0, \quad h_0 \in H^1(\Omega), \quad \textit{and} \quad \frac{1}{2}\int_{\Omega} |\partial_x h_0|^2\, dx + \int_{\Omega} Q(h_0)\, dx < \infty, \tag{4.13}$$

and assume that $0 < n < m$, $m \geq 3$ *in* (4.4). *Then a unique positive smooth solution of* (4.1)–(4.4) *exists for all* $t > 0$.

Proof Following arguments from previous papers [15, 27] it suffices to derive a priori pointwise upper and lower bounds for the solution. We derive an a priori bound on the H^1 norm and show this implies pointwise bounds. Then uniform parabolicity implies the solution is completely smooth. Similar arguments are presented in [15] for the case $P(h) = 0$ and in [27] for a destabilizing non-singular $P(h)$.

First we note that the Liapunov functional implies that for any time $T > 0$,

$$\frac{1}{2}\int_{\Omega} |\partial_x h(T)|^2\, dx \leq \frac{1}{2}\int_{\Omega} |\partial_x h_0|^2\, dx + \int_{\Omega} Q(h_0)\, dx - \int_{\Omega} Q(h(T))\, dx.$$

The initial data is in H^1 and positive. Since $m > n$, $-Q(h)$ has an a priori upper bound independent of h, implying an a priori bound for $\int |\partial_x h|^2\, dx$ at the later time T. In one space dimension this immediately implies an a priori pointwise upper bound on h and also on the $C^{1/2}$ norm, at time T.

To show that h has a pointwise a priori lower bound at time T, note that the above implies $\int Q(h(T))\, dx \leq C$. Let $h_{\min} = \min_{\Omega} h$ be attained at x_0. By Hölder continuity, $h(x) \leq u_{\min} + C_H|x - x_0|^{1/2}$, where C_H is the Hölder constant. Hence:

$$C > \int Q(h)dx \geq C(\epsilon)\int (h_{\min} + C_H|x - x_0|^{1/2})^{-m+1}\, dx + O(1) \geq C(\epsilon)\eta(h_{\min}) + O(1)$$

where $\eta(h_{\min}) = -\log h_{\min}$ for $m = 3$ and $\eta(h_{\min}) = h_{\min}^{3-m}$ for $m > 3$. Thus the solution is forbidden to go below a positive threshold for $m \geq 3$. Note that all the bounds depend strongly on ϵ as it nears zero. In particular, these results are not true for the case $\epsilon = 0$. □

The long-time behavior of solutions of the PDE is related to minimizers of the Liapunov functional (4.7). Define $Q(h)$ to be ∞ for $h \leq 0$; we can then prove the following results concerning the global minimizer.

4.2 Theorem [22] *Let Ω be a bounded domain of class $C^{0,1}$, let $\bar{h} > 0$ and assume $0 < n < m$ in (4.4). Then a global minimizer of $\mathcal{E}(\cdot)$ (defined in (4.7)) exists in the class $V := \{h \in H^1(\Omega) : \int_\Omega h\,dx = \bar{h}\}$. In particular, $h > 0$ almost everywhere in Ω.*

One can go on to prove that the global minimizer satisfies the same positivity result as in Theorem 4.2. Moreover the minimizer satisfies the steady state ODE:

4.3 Theorem [22] *Let $m \geq 3$ and $0 < n < m$ in (4.4). Then a minimizer h as constructed in Theorem 4.2 is a solution of the ODE on $\Omega = \{x | 0 \leq x \leq L\}$,*

$$P(h) - h_{xx} = \bar{p},$$

$$h_x(0) = h_x(L) = 0, \qquad \frac{1}{L}\int_0^L h\,dx = \bar{h}.$$

Moreover, h is smooth.

4.2 One-dimensional steady-state solutions

From Theorem 4.3 we know that all minimizers of $\mathcal{E}$ are smooth steady state solutions on $0 \leq x \leq L$, satisfying

$$P(h) - h_{xx} = \bar{p}, \tag{4.14}$$

$$h_x(0) = h_x(L) = 0, \qquad \bar{p} = \frac{1}{L}\int_0^L P(h)\,dx. \tag{4.15}$$

In this section we examine the number and structure of steady state solutions.

While aspects of this problem parallel the Cahn-Hilliard and constrained Allen-Cahn equations [5, 6, 40, 53, 75], the nonlinear function $P(h)$ in (4.14) has a different form than considered in most previous works and leads to a different structure of the bifurcation diagram.

Equ. (4.14) can be written as an autonomous phase plane system,

$$h_x = v, \qquad v_x = P(h) - \bar{p}, \tag{4.16}$$

where $\bar{p}$ is a control parameter. Observe that every Neumann solution of (4.14)–(4.15) on $x \in (0, L)$ can be extended by reflection symmetry to yield a periodic solution of (4.16) on $x \in (0, 2L)$. Hence, a necessary condition for the existence of Neumann solutions is that (4.16) admits periodic solutions. Moreover, the symmetry of (4.16) guarantees that periodic solutions on $[0, 2L]$ can be shifted to yield Neumann solutions on $[0, L]$.

Let $P_{\max}$ be the maximum of $P(h)$ on $h > 0$. Observe that for any $P(h)$ of the form (4.4) and $\bar{p} \in (0, P_{\max})$, (4.16) has a unique elliptic center point at h_c satisfying $P(h_c) - \bar{p} = 0$ with $P'(h_c) < 0$. Note that for $\bar{p} < 0$ there are no elliptic equilibria, hence no periodic orbits and no Neumann solutions of (4.14). In addition, for $0 \leq \bar{p} < P_{\max}$, (4.16) has a hyperbolic saddle point at $h_s \sim \epsilon + \epsilon^{n+1}\bar{p}/(m-n)$ as $\epsilon \to 0$ with $h_c > h_s$. The stable and unstable manifolds of the saddle point form a homoclinic orbit that encloses h_c and all of the periodic solutions of (4.16). Periodic solutions surrounding the center h_c have a minimum greater than the saddle point value h_s. Therefore every non-trivial equilibrium solution is bounded from below by h_s, and thus $h(x) > \epsilon$.

Equ. (4.15) has the trivial uniform constant solution $h = \bar{h}$ for any positive value $\bar{h} > 0$; the corresponding average pressure is $\bar{p} = P(\bar{h})$. Branches of non-trivial solutions can bifurcate from the constant solutions only at singular points $\bar{h}_*$ where the linear operator $\mathcal{L} = P'(\bar{h}_*) - \partial_{xx}$ has a non-trivial nullspace [65]; hence the bifurcation points are given by

$$P'(\bar{h}_k) = -k^2\pi^2/L^2, \quad k = 1, 2, \ldots, K. \tag{4.17}$$

For $\epsilon > 0$, $P'(h)$ is bounded from below for $h > 0$, therefore there are only a finite number of branches, $k = 1, 2, \ldots, K$. For each k, there are two roots, $\bar{h}_k^+ > \bar{h}_k^-$, with $P''(\bar{h}_k^+) > 0$ and $P''(\bar{h}_k^-) < 0$. It is straightforward to show that as $\epsilon \to 0$, $K = O(\epsilon^{-(n+1)/2})$ by noting that the minimum scales like $P'(h^*) = O(\epsilon^{-(n+1)})$, where the critical point scales as $h^* = O(\epsilon)$. For $\epsilon = 0$ there are an infinite number of upper solutions $\bar{h}_k^+$ while the lower solutions $\bar{h}_k^-$ disappear.

We consider an interval of fixed length L and examine the bifurcating branches of solutions as the total mass, $L\bar{h} = \int_0^L h(x)dx$, varies. For $\epsilon > 0$, for each k there is a branch of non-trivial steady state solutions that bifurcates from $\bar{h}_k^+$ and reconnects at $\bar{h}_k^-$. This can be proved rigorously for m and n satisfying the following conditions.

4.4 Definition We call a pair of positive exponents (n, m) in (4.4) *admissible* if they satisfy $m > n$ and either (i) or (ii):

(i) $n > 1/2$ and $3n^2 - 10mn + 3m^2 - m - n + 2 > 0$;

(ii) $3n^2 - 14mn + 3m^2 - 2m - 2n + 7 < 0$.

The following theorem is proved in [22].

4.5 Theorem *Consider $P(h)$ as given in* (4.4) *and admissible (n, m) in the sense of Definition* 4.4. *Then, for each positive integer $k \leq K$ there exists a non-trivial branch of solutions to* (4.14), (4.15). *Parametrized by $\bar{p} \in (P(\bar{h}_k^+), P(\bar{h}_k^-))$, each branch of non-trivial solutions bifurcates from $\bar{h}_k^+$ and reconnects to the constant state branch at $\bar{h}_k^-$. Furthermore, all solutions of* (4.14), (4.15) *are either trivial constant solutions or lie on one of these bifurcating branches.*

Fig. 9 shows the bifurcation diagram for the set of equilibria of (4.14), (4.15). At each $\bar{h}$, the diagram gives the maximum and minimum values of each solution $h(x)$. Specifying two out of three of the values $h_{\min}$, $h_{\max}$, $\bar{h}$, specifies an equilibrium solution $h(x)$, unique up to the reflection symmetry, $x \to L - x$.

The local structure of the bifurcation diagram can computed analytically using perturbation methods. This results in an amplitude equation that predicts either subcritical or supercritical pitchfork bifurcations at each of the points on the constant solution curve where an eigenvalue changes sign. For details, the reader is referred to [22].

4.3 Stability of the equilibrium solutions

In this section we consider the stability of equilibria with respect to mass-preserving perturbations. We begin with the linear analysis for uniform constant equilibrium solutions and go on to make more global statements for the entire set of equilibrium solutions.

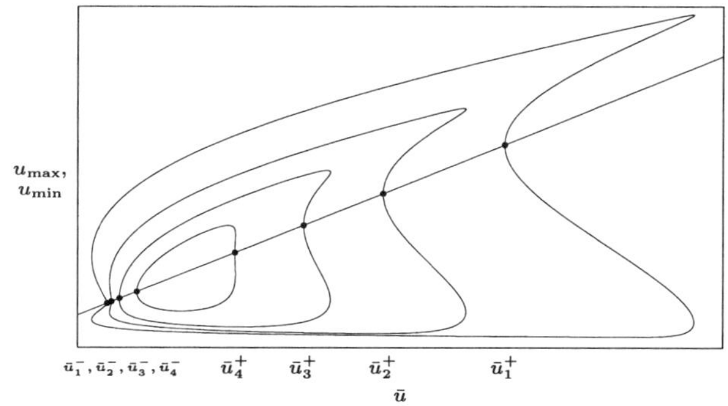

Figure 9: Bifurcation diagram of equilibria for (4.4) with $(n, m) = (3, 4)$ and $\epsilon = 0.14$, yielding $K = 4$ branches.

If the initial data for the PDE (4.12), with $f(h), P(h)$ given by (4.3), (4.4), is near an equilibrium $\underline{h}(x)$, then it is convenient to write the solution in the form $h(x,t) = \underline{h}(x) + \tilde{h}(x,t)$, where $\tilde{h}$ is a zero-mean perturbation. Linearizing (4.12) yields the evolution equation for $\tilde{h}$,

$$\frac{\partial \tilde{h}}{\partial t} = \mathcal{L}\tilde{h} \equiv \partial_x(f(\underline{h})\partial_x[P'(\underline{h}) - \partial_{xx}])\tilde{h}, \tag{4.18}$$

with the corresponding linearized Neumann boundary conditions (4.5),

$$\partial_{xxx}\tilde{h} = 0, \quad \partial_x \tilde{h} = 0, \quad \text{at} \quad x = 0, 1. \tag{4.19}$$

The choice $\tilde{h}(x,t) = \hat{h}(x)e^{\lambda t}$ in (4.18) yields the associated Neumann eigenvalue problem, $\mathcal{L}\hat{h} = \lambda\hat{h}$.

We note that the linear stability of the uniform constant solutions $\underline{h}(x) = \bar{h}$ can be computed directly [97]. For these states

$$\lambda_\ell = -f(\bar{h})\ell^2\pi^2(P'(\bar{h}) - \ell^2\pi^2), \quad \ell = 1, 2, 3 \ldots, \tag{4.20}$$

with corresponding eigenmodes $A\cos(\pi\ell x)$.

The operator (4.18) is the composition of two second-order operators, $\mathcal{L}\hat{h} = \bar{\mathcal{M}}\bar{\mathcal{L}}\hat{h}$, where the mobility operator, $\bar{\mathcal{M}}$, and nonlocal operator, $\bar{\mathcal{L}}$, are respectively

$$\bar{\mathcal{M}}v \equiv \partial_x(f(\underline{h})\partial_x v), \qquad \bar{\mathcal{L}}w \equiv P'(\underline{h})w - w_{xx} - \frac{1}{L}\int_0^L P'(\underline{h})w\,dx, \tag{4.21}$$

for zero-mean $w(x)$ satisfying Neumann boundary conditions. We note that $\mathcal{L}$ is self-adjoint with respect to a weighted H^{-1} norm using an inner product defined by the mobility. For every zero-mean function $w(x)$, there exists a unique zero-mean potential, W, such that $\bar{\mathcal{M}}W = w$. The weighted H^{-1} inner-product of two zero-mean functions w and v is then $\langle w, v\rangle \equiv \int f(\underline{h})\partial_x W\,\partial_x V\,dx$. This norm can be used to describe the thin film evolution as a gradient flow in H^{-1}. As the operator $\mathcal{L}$ is self-adjoint and the domain is compact, the spectrum is real and discrete [45, 69]. We note that related results on the properties of the linearized spectrum can be obtained from analysis of a Rayleigh quotient, as considered by Bates and Fife for the Cahn-Hilliard equation [9] and by Laugesen and Pugh [68, 69] for equations of the type (1.5).

In the case of constant mobility $f(h) = c$, Bai et al. [5] and Grinfeld and Novick-Cohen [53] showed that the properties of the spectrum of (4.18) can be obtained from analysis of the second order problem for $\bar{\mathcal{L}}$. Since the equilibria are smooth and bounded away from zero, the problem with non-constant mobility $f(u)$ can be treated equivalently; the operator $\bar{\mathcal{M}}$ is invertible and hence $\mathcal{L}\hat{h} = \lambda\hat{h}$ can be written as $\bar{\mathcal{L}}\hat{h} = \lambda\bar{\mathcal{M}}^{-1}\hat{h}$. These results directly extend to (4.18) using the homotopy

$$\bar{\mathcal{L}}\hat{h} = \lambda[\alpha + (1-\alpha)\bar{\mathcal{M}}^{-1}]\hat{h}, \tag{4.22}$$

where $\alpha = 0$ gives the fourth order operator $\mathcal{L}$ and $\alpha = 1$ gives the second order operator $\bar{\mathcal{L}}$. Lemma 1.2 in [53] and Theorem 3.1 in [5] show that the number of positive and negative eigenvalues of (4.22) is independent of $0 \leq \alpha \leq 1$. Furthermore, Grinfeld and Novick-Cohen's Theorem 5.1 [53] for excluding $\lambda = 0$ from the spectrum of $\mathcal{L}$ directly extends to our problem and can be stated as

4.6 Theorem *Consider the linearized equation* (4.18) *with* $f(h), P(h)$ *given by* (4.3), (4.4). *Then, away from the bifurcation points* $\bar{h}_k^{\pm}$ *on the line of constant solutions, and apart from turning points,* $\lambda = 0$ *is not in the spectrum of* (4.18) *on the branches of non-trivial solutions. No secondary bifurcations are possible.*

For the Neumann problem for (1.5), Laugesen and Pugh have proved that equilibrium solutions with interior critical points are unstable (see Theorem 4 [69]). We now combine these results to prove the existence of a family of stable non-trivial solutions for (4.12).

4.7 Theorem *For problem* (4.12) *with* $P(h)$ *in* (4.4) *satisfying Theorem* 4.5 *and* $f(h)$ *given by* (4.3), *there exists an interval* $I = (\bar{h}_1^{\min}, \bar{h}_1^{\max})$ *containing* $(\bar{h}_1^-, \bar{h}_1^+)$ *such that for each* $\bar{h} \in I$, *there exists a stable non-trivial steady state solution. These stable monotone solutions lie on the first branch of solutions,* $k = 1$.

Fig. 9 with $\epsilon = 0.14$ shows a $k = 1$ branch with a very simple structure. There is precisely one turning point for each subcritical bifurcation. We conjecture that this is generically true for this problem with $P(h)$ of the form (4.4). In this case, we can determine the number of positive eigenvalues for all solutions along the first bifurcating branch. The first branch has two unstable portions, one near each of the bifurcation points. By Theorem 4.7, on the portion of the non-trivial branch between the turning points, all of the solutions must be stable. In the next section we study the asymptotic structure of this branch in the limit as $\epsilon \to 0$.

4.8 Corollary *For* $k = 2, \ldots, K$, *a subset of solutions along the* k^{th} *branch of non-trivial solutions (connecting* h_k^- *to* h_k^+*) is stable with respect to perturbations with the same k-fold symmetry.*

This result follows directly from Theorem 4.7 by noting that k^{th} bifurcating branch for the problem with domain size L is the first bifurcating branch for the problem with domain size L/k. Instability of these reduced-domain-stable solutions to perturbations on the full domain is called *coarsening* [9].

4.4 Asymptotic behavior of equilibrium solutions

We now study the structure of the equilibrium solutions of (4.14), (4.15) in the limit as $\epsilon \to 0$. This branch has subcritical pitchfork bifurcations at $\bar{h}_1^+$ and $\bar{h}_1^-$. Turning points occur on this branch at $\bar{h}_1^{\max}$ and $\bar{h}_1^{\min}$, respectively, describing the maximum and minimum average film thicknesses for non-trivial solutions. As described above, the turning points separate families of stable and unstable solutions on this branch. The unstable solutions have a finite amplitude structure, with $h_{\min} = O(1)$ as $\epsilon \to 0$, while the stable solutions are localized droplets connected by an ultra-thin film with thickness $h \sim \epsilon$ (see Fig. 10) and approach the homoclinic solution of (4.16). These stable droplet solutions concentrate mass as a δ-distribution in the limit of vanishing ϵ. Moreover, using matched asymptotic analysis [22], one can obtain an estimate for the turning point $\bar{h}_1^{\max}$ as a function of ϵ.

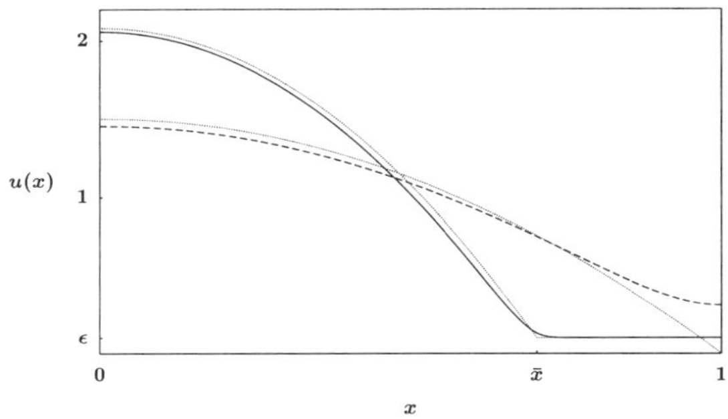

Figure 10: The stable (solid) and unstable (dashed) equilibrium solutions, and the leading-order asymptotic approximations (dotted), for $\epsilon = 0.1$ and $(n, m) = (3, 4)$ with $\bar{h} = 1$, and the leading-order asymptotic approximations (dotted).

As $\epsilon \to 0$, for each allowable fixed value of $\bar{h} \in (\bar{h}_1^{\min}, \bar{h}_1^{\max})$, we get the solution

$$h(x) \sim \begin{cases} \dfrac{3\bar{h}}{2\bar{x}^3}(\bar{x}^2 - x^2) & 0 \le x < \bar{x} = (\frac{9}{2}\bar{h}^2/|Q(\epsilon)|)^{1/4}, \\ \epsilon & \bar{x} < x \le 1. \end{cases} \tag{4.23}$$

From the form of Q, for $\epsilon \to 0$, $Q(\epsilon) = O(\epsilon^{-(n-1)})$ and we obtain

$$\bar{x} \sim \sqrt{\bar{h}}O(\epsilon^{(n-1)/4}) \to 0, \qquad \bar{p} \sim O(\epsilon^{-3(n-1)/4})/\sqrt{\bar{h}} \to \infty. \tag{4.24}$$

Hence for $\epsilon \to 0$, (4.23) is a delta sequence, which converges to a Dirac δ-distribution with mass $\bar{h}$, $\bar{h}\delta(x)$. As ϵ decreases, (4.23) approaches a parabolic profile with fixed area and support that vanishes as $\epsilon \to 0$. An estimate of the maximum mass $\bar{h}_1^{\max}$ of the first branch of the bifurcation diagram is given by the intersection of the two asymptotic families of solutions. This occurs when $h_{\max} \sim 3\bar{h}/2$ and (4.24) are set equal to yield

$$\bar{h} \le \bar{h}_1^{\max} \sim \frac{2}{9}\sqrt{2|Q(\epsilon)|}. \tag{4.25}$$

4.5 Discussion

In our study, for finite $\epsilon > 0$, the conjoining pressure term in (4.4) regularizes the $\epsilon = 0$ finite-time rupture problem. For $\epsilon > 0$ the evolution is well-posed through topological transitions, as in diffusive interface models for Hele-Shaw flows. And, for $\epsilon \to 0$, we will present numerical simulations of the dynamics of (4.12) that show the occurrence of concentrations.

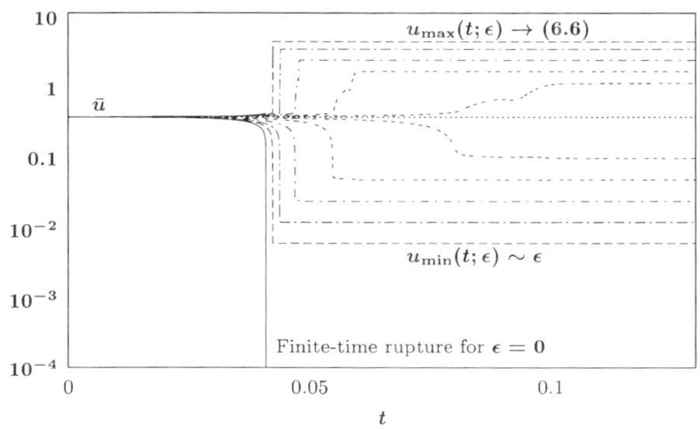

Figure 11: A sequence of dynamic simulations in the $\epsilon \to 0$ limit. Numerical results indicate convergence to the equilibrium rapidly after rupture for $\epsilon \to 0$.

To examine the formation of mass concentrations as $\epsilon \to 0$ after rupture, we consider a sequence of numerical solutions $h^\epsilon(x,t)$ of the initial value problem for (4.12) with $L = 1$, $f(h) = h^3$, and $P(h) = h^{-3} - \epsilon h^{-4}$, with the monotone decreasing initial data $h_0(x) = 0.4 + 0.002\cos(\pi x)$. Fig. 11 shows the maximum and minimum of the solution as a function of time for the following values of ϵ: 0.1, 0.05, 0.025, 0.0125, 0.00625, and $\epsilon = 0$. The $\epsilon = 0$ problem exhibits finite-time rupture at time $T_R \approx 0.041$. For $\epsilon > 0$, the solutions do not rupture, but each "de-wets" shortly after T_R; that is, the minimum of each solutions becomes small, $h_{\min}(t) = O(\epsilon)$. Thereafter, the $\epsilon > 0$ solutions rapidly converge to the unique non-trivial stable steady state solution for this problem. The graphs of the minimum and maximum of the solutions $h^\epsilon(x,t)$ shown in Fig. 11 give one indicator of the convergence to the steady state. We find that for smaller values of ϵ the convergence to steady state happens on shorter time scales. Since the steady states converge to a Dirac δ-distribution, the computations suggest that for each time after the rupture time T_R, the time-dependent solution sequence converges to a δ-distribution, i.e. for all $t \geq 0$,

$$\lim_{\epsilon\to 0} h^\epsilon(x,t) = \begin{cases} h^0(x,t), & t \leq T_R, \\ \bar{h}\delta(x), & t > T_R. \end{cases} \tag{4.26}$$

Thus *concentrations* occur in this solution sequence for $\epsilon \to 0$. In fact the only estimate, independent of ϵ, that is known for the PDE is the L^1 norm, that is the conservation of mass. This simulation of $h^\epsilon(x,t)$ suggests that sharper analytical control is not possible. Finally, it is clear from this simulation that there is a *droplet formation time scale*, T_D, i.e. the time it takes for the solution to converge to a parabolic droplet after the onset of dewetting at T_R. This preliminary numerical study suggests that $T_D \to 0$ as $\epsilon \to 0$.

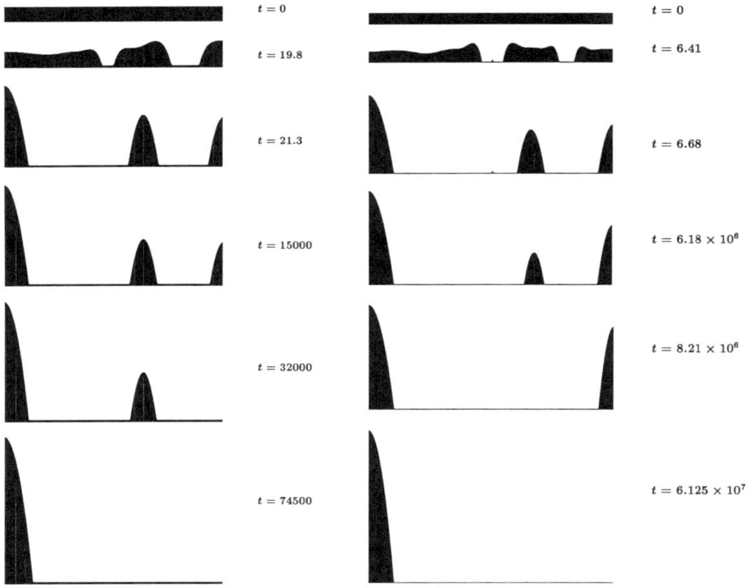

Figure 12: The long time dynamics for (4.12) with $n = 3$, $m = 4$ for $\epsilon = 0.1$ (left) and $\epsilon = 0.01$ (right). Note that the time-scale for the coarsening process increases as $\epsilon \to 0$ and that the final state converges to the global minimizer.

To briefly study the influence of the limit $\epsilon \to 0$ on the dynamics of topological transitions (described in [78] as "morphological phase separation"), we look at simulations illustrating long-time coarsening dynamics in (4.12). We use the algorithm in [55] to compute the solution of (4.12) with $f(h) = h^3$, $P(h) = h^{-3} - \epsilon h^{-4}$, $L = 20$, and the initial data $h_0(x) = 1 + 0.005\sin(6(x-13)^2)$, for $\epsilon = 0.1$ and $\epsilon = 0.01$. Unlike the simple monotone initial data used in Fig. 11, this new initial data has the potential to form complicated patterns of droplets that may depend sensitively on the structure of $h_0(x)$. As shown in Fig. 12, these simulations illustrate the dynamics of dewetting (or formation of "dry spots") and coarsening starting from small perturbations of a flat thin film. Similar results have been observed in other studies for fixed values of ϵ [73, 76, 78]; our interest here is the dependence of the dynamics on ϵ, in the limit that $\epsilon \to 0$. Fig. 12 shows the results for $\epsilon = 0.1$ and $\epsilon = 0.01$. In both situations, after an initial period of dewetting dynamics, the solutions converge to a series of droplets, at arbitrary positions in the domain, generally with different masses, all connected by an ultra-thin film. Each of these droplets is close to an equilibrium solution, and in fact they persist for very long times. However, from Section 4.3 we know that there cannot be any stable equilibria with interior extrema [69], hence these states are unstable. These states are called meta-stable, and their long-time dynamics has been studied for Cahn-Hilliard equations [81, 94]. The long-time dynamics are characterized by coarsening instabilities that ultimately lead to a final stable droplet which is the global minimizer. The time needed for the coarsening process depends on ϵ, with smaller ϵ requiring more time to evolve to the global minimizer. Thus there is a *coarsening time scale*, T_C, that describes the time it takes for the number of droplets in the solution to decrease. As suggested by Fig. 12, this time scale increases, $T_C \to \infty$, as $\epsilon \to 0$, a dramatically different dependence on ϵ than the droplet formation time scale, $T_D \to 0$. A model for the coarsening dynamics, in terms of droplet motion and mass transfer, is developed in [49].

Acknowledgements The first author wishes to thank the American Mathematical Society for permission to excerpt parts of Section 1 from the review article [19].

This work is supported by NSF grant DMS–0074049 and ONR grant N000140110290.

References

[1] R. Abeyaratne and J. Knowles, Implications of viscosity and strain gradient effects for the kinetics of propagating phase boundaries in solids, *SIAM J. Appl. Math.* **51** (1991), 1205–1221.

[2] D. J. Acheson, *Elementary Fluid Dynamics*, Oxford University Press, Oxford, 1990.

[3] R. Almgren, A. L. Bertozzi, and M. P. Brenner, Stable and unstable singularities in the unforced Hele-Shaw cell, *Phys. Fluids* **8** (6) (1996), 1356–1370.

[4] D. G. Aronson and J. Graveleau, A self-similiar solution to the focusing problem for the porous medium equation, *European J. Appl. Math.* **4** (1993), 65.

[5] F. S. Bai, C. M. Elliott, A. Gardiner, A. Spence, and A. M. Stuart, The viscous Cahn-Hilliard equation. I. Computations, *Nonlinearity* **8** (2) (1995), 131–160.

[6] F. S. Bai, A. Spence, and A. M. Stuart, Numerical computations of coarsening in the one-dimensional Cahn-Hilliard model of phase separation, *Phys. D* **78** (3–4) (1994), 155–165.

[7] G. I. Barenblatt, *Scaling, Self-similarity, and Intermediate Asymptotics*, Cambridge University Press, Cambridge, 1996.

[8] J. W. Barrett, J. F. Blowey, and H. Garcke, Finite element approximation of a fourth order degenerate parabolic equation, *Numer. Math.* **80** (4) (1998), 525–556.

[9] P. W. Bates and P. C. Fife, Spectral comparison principles for the Cahn-Hilliard and phase-field equations, and time scales for coarsening, *Phys. D* **43** (2–3) (1990), 335–348.

[10] R. M. Beam and R. F. Warming, Alternating direction implicit methods for parabolic equations with a mixed derivative, *SIAM J. Sci. Statist. Comput.* **1** (1980), 131–159.

[11] E. Beretta, M. Berstch, and R. D. Passo, Nonnegative solutions of a fourth order nonlinear degenerate parabolic equation, *Arch. Rational Mech. Anal.* **129** (1995), 175–200.

[12] F. Bernis, Change in sign of the solutions to some parabolic problems, in: *Nonlinear Analysis and Applications (Arlington, Tex., 1986)* (V. Lakshmikantham, ed.), Lecture Notes in Pure and Appl. Math. **109**, Dekker, New York, 1987, 75–82.

[13] F. Bernis, Finite speed of propagation and continuity of the interface for slow viscous flows, *Adv. Differential Equations* **1** (3) (1996), 337–368.

[14] F. Bernis, Finite speed of propagation for thin viscous flows when $2 \leq n < 3$, *C. R. Acad. Sci. Paris Sér. I Math.* **322** (1996), 1169–1174.

[15] F. Bernis and A. Friedman, Higher order nonlinear degenerate parabolic equations, *J. Differential Equations* **83** (1990), 179–206.

[16] F. Bernis, L. A. Peletier, and S. M. Williams, Source type solutions of a fourth order nonlinear degenerate parabolic equation, *Nonlinear Anal.* **18** (3) (1992), 217–234.

[17] A. L. Bertozzi, Loss and gain of regularity in a lubrication equation for thin viscous films, in: *Free Boundary Problems: Theory and Applications* (J. I. Díaz et al., eds.), Pitman Res. Notes Math. Ser. **323**, Longman, Harlow, 1995, 72–85.

[18] A. L. Bertozzi, Symmetric singularity formation in lubrication-type equations for interface motion, *SIAM J. Appl. Math.* **56** (3) (1996), 681–714.

[19] A. L. Bertozzi, The mathematics of moving contact lines in thin liquid films, *Notices Amer. Math. Soc.* **45** (6) (1998), 689–697.

[20] A. L. Bertozzi and M. P. Brenner, Linear stability and transient growth in driven contact lines, *Phys. Fluids* **9** (3) (1997), 530–539.

[21] A. L. Bertozzi, M. P. Brenner, T. F. Dupont, and L. P. Kadanoff, Singularities and similarities in interface flow, in: *Trends and Perspectives in Applied Mathematics* (L. Sirovich, ed.), Appl. Math. Sci. **100**, Springer-Verlag, New York, 1994, 155–208.

[22] A. L. Bertozzi, G. Grün, and T. P. Witelski, Dewetting films: bifurcations and concentrations, *Nonlinearity* **14** (6) (2001), 1569–1592.

[23] A. L. Bertozzi, A. Münch, X. Fanton, and A.-M. Cazabat, Contact line stability and 'undercompressive shocks' in driven thin film flow, *Phys. Rev. Lett.* **81** (23) (1998), 5169–5172.

[24] A. L. Bertozzi, A. Münch, and M. Shearer, Undercompressive shocks in thin film flows, *Phys. D* **134** (4) (1999), 431–464.

[25] A. L. Bertozzi, A. Münch, M. Shearer, and K. Zumbrun, Stability of compressive and undercompressive thin film travelling waves, *European J. Appl. Math.* **12** (3) (2001), 253–291.

[26] A. L. Bertozzi and M. C. Pugh, The lubrication approximation for thin viscous films: regularity and long time behavior of weak solutions, *Comm. Pure Appl. Math.* **49** (2) (1996), 85–123.

[27] A. L. Bertozzi and M. C. Pugh, Long-wave instabilities and saturation in thin film equations, *Comm. Pure Appl. Math.* **51** (6) (1998), 625–661.

[28] A. L. Bertozzi and M. Shearer, Existence of undercompressive travelling waves in thin film equations, *SIAM J. Math. Anal.* **32** (1) (2000), 194–213.

[29] J. Bischof, D. Scherer, S. Herminghaus, and P. Leiderer, Dewetting modes of thin metallic films: Nucleation of holes and spinodal dewetting, *Phys. Rev. Lett.* **77** (8) (1996), 1536–1539.

[30] S. Boatto, L. Kadanoff, and P. Olla, Travelling wave solutions to thin film equations, *Phys. Rev. E* **48** (1993), 4423.

[31] A. Bourlioux and A. J. Majda, Theoretical and numerical structure of unstable detonations, *Philos. Trans. Roy. Soc. London Ser. A* **350** (1995), 29–69.

[32] R. Buckingham, M. Shearer, and A. L. Bertozzi, Thin films traveling waves and the navier slip condition, preprint, Duke University, 2002.

[33] J. P. Burelbach, S. G. Bankoff, and S. H. Davis, Nonlinear stability of evaporating/condensing liquid films, *J. Fluid Mech.* **195** (1988), 463–494.

[34] P. Carles and A.-M. Cazabat, The thickness of surface-tension-gradient-driven spreading films, *J. Colloid Interface Sci.* **157** (1993), 196–201.

[35] A.-M. Cazabat, F. Heslot, S. M. Troian, and P. Carles, Finger instability of this spreading films driven by temperature gradients, *Nature* **346** (6287) (1990), 824–826.

[36] P. Constantin, T. F. Dupont, R. E. Goldstein, L. P. Kadanoff, M. J. Shelley, and S.-M. Zhou, Droplet breakup in a model of the Hele-Shaw cell, *Phys. Rev. E* **47** (6) (1993), 4169–4181.

[37] P. G. de Gennes, Wetting: statics and dynamics, *Rev. Modern Phys.* **57** (1985), 827–880.

[38] T. F. Dupont, R. E. Goldstein, L. P. Kadanoff, and S.-M. Zhou, Finite-time singularity formation in Hele Shaw systems, *Phys. Rev. E* **47** (6) (1993), 4182–4196.

[39] E. B. Dussan V and S. Davis, On the motion of a fluid-fluid interface along a solid surface, *J. Fluid Mech.* **65** (1974), 71–95.

[40] J. C. Eilbeck, J. E. Furter, and M. Grinfeld, On a stationary state characterization of transition from spinodal decomposition to nucleation behaviour in the Cahn-Hilliard model of phase separation, *Phys. Lett. A* **135** (4–5) (1989), 272–275.

[41] M. Elbaum, S. G. Lipsom, and J. F. Wettlaufer, Evaporation pre-empts complete wetting, *Europhys. Lett.* **29** (6) (1995), 457–462.

[42] M. H. Eres, L. W. Schwartz, and R. V. Roy, Fingering phenomena for driven coating films, *Phys. Fluids* **12** (6) (2000), 1278–1295.

[43] T. Erneux and D. Gallez, Can repulsive forces lead to stable patterns in thin liquid films?, *Phys. Fluids* **9** (4) (1997), 1194–1196.

[44] L. C. Evans, *Weak Convergence Methods for Nonlinear Partial Differential Equations*, CBMS Regional Conf. Ser. in Math. **74**, Amer. Math. Soc., Providence, RI, 1990.

[45] L. C. Evans, *Partial Differential Equations*, Grad. Stud. Math. **19**, Amer. Math. Soc., Providence, RI, 1998.

[46] X. Fanton, A.-M. Cazabat, and D. Quéré, Thickness and shape of films driven by a Marangoni flow, *Langmuir* **12** (24) (1996), 5875–5880.

[47] R. Ferreira and F. Bernis, Source-type solutions to thin-film equations in higher dimensions, *European J. Appl. Math.* **8** (5) (1997), 507–524.

[48] N. Fraysse and G. M. Homsy, An experimental study of rivulet instabilities in centrifugal spin coating of viscous Newtonian and non-Newtonian fluids, *Phys. Fluids* **6** (4) (1994), 1491–1504.

[49] K. Glasner, A diffuse interface approach to Hele-Shaw flow, preprint, Duke University, 2001.

[50] S. Goldstein, ed., *Modern Developments in Fluid Dynamics*, vol. 2, Dover Publ., New York, 1965.

[51] A. A. Golovin, B. Y. Rubinstein, and L. M. Pismen, Effect of van der Waals interaction on the fingering instability of thermally driven thin wetting films, *Langmuir* **17** (2001), 3930–3936.

[52] H. P. Greenspan, On the motion of a small viscous droplet that wets a surface, *J. Fluid Mech.* **84** (1978), 125–143.

[53] M. Grinfeld and A. Novick-Cohen, The viscous Cahn-Hilliard equation: Morse decomposition and structure of the global attractor, *Trans. Amer. Math. Soc.* **351** (6) (1999), 2375–2406.

[54] G. Grün, On the convergence of entropy consistent numerical schemes for lubrication-type equations in multiple space dimensions, to appear in *Math. Comp.*

[55] G. Grün, On the numerical simulation of wetting phenomena, in: *Proc. 15th GAMM-Seminar Kiel, Numerical Methods of Composite Materials* (W. Hackbusch and S. Sauter, eds.), Vieweg-Verlag, Braunschweig, to appear.

[56] G. Grün and M. Rumpf, Nonnegativity preserving convergent schemes for the thin film equation, *Numer. Math.* **87** (1) (2000), 113–152.

[57] P. J. Haley and M. J. Miksis, The effect of the contact line on droplet spreading, *J. Fluid Mech.* **223** (1991), 57–81.

[58] S. Herminghaus, K. Jacobs, K. Mecke, J. Bischof, A. Fery, M. Ibn-Elhaj, and S. Schlagowski, Spinodal dewetting in liquid crystal and liquid metal films, *Science* **282** (5390) (1998), 916–919.

[59] C. Huh and L. E. Scriven, Hydrodynamic model of steady movement of a solid/liquid/fluid contact line, *J. Colloid Interface Sci.* **35** (1971), 85–101.

[60] H. Huppert, Flow and instability of a viscous current down a slope, *Nature* **300** (1982), 427–429.

[61] E. Isaacson, D. Marchesin, and B. Plohr, Transitional waves for conservation laws, *SIAM J. Math. Anal.* **21** (1990), 837–866.

[62] J. N. Israelachvili, *Intermolecular and Surface Forces*, 2nd ed., Academic Press, New York, 1992.

[63] D. E. Kataoka and S. M. Troian, A theoretical study of instabilities at the advancing front of thermally driven coating films, *J. Colloid Interface Sci.* **192** (1997), 350–362.

[64] D. E. Kataoka and S. M. Troian, Stabilizing the advancing front of thermally driven climbing films, *J. Colloid Interface Sci.* **203** (1998), 335–344.

[65] J. B. Keller and S. Antman, eds., *Bifurcation Theory and Nonlinear Eigenvalue Problems*, Benjamin, New York-Amsterdam, 1969.

[66] R. Konnur, K. Kargupta, and A. Sharma, Instability and morphology of thin liquid films on chemically heterogeneous substrates, *Phys. Rev. Lett.* **84** (5) (2000), 931–934.

[67] N. Kopell and L. N. Howard, Bifurcations and trajectories joining critical points, *Adv. Math.* **18** (1975), 306–358.

[68] R. S. Laugesen and M. C. Pugh, Energy levels of steady states for thin film type equations, `http://xyz.lanl.gov/abs/math.AP/0003208`, to appear in *J. Differential Equations*.

[69] R. S. Laugesen and M. C. Pugh, Linear stability of steady states for thin film and Cahn-Hilliard type equations, *Arch. Rational Mech. Anal.* **154** (1) (2000), 3–51.

[70] P. D. Lax, *Hyperbolic Systems of Conservation Laws and the Mathematical Theory of Shock Waves*, CBMS-NSF Regional Conf. Ser. in Appl. Math. **11**, SIAM, Philadelphia, PA, 1973.

[71] F. Melo, J. F. Joanny, and S. Fauve, Fingering intability of spinning drops, *Phys. Rev. Lett.* **63** (18) (1989), 1958–1961.

[72] V. S. Mitlin, Dewetting of a solid surface: analogy with spinodal decomposition, *J. Colloid Interface Sci.* **156** (1993), 491–497.

[73] V. S. Mitlin and N. V. Petviashvili, Nonlinear dynamics of dewetting: kinetically stable structures, *Phys. Lett. A* **192** (1994), 323–326.

[74] A. Münch, Shock transitions in Marangoni-gravity driven thin film flow, *Nonlinearity* **13** (2000), 731–746.

[75] A. Novick-Cohen and L. A. Peletier, Steady states of the one-dimensional Cahn-Hilliard equation, *Proc. Roy. Soc. Edinburgh Sect. A* **123** (6) (1993), 1071–1098.

[76] A. Oron and S. G. Bankoff, Dewetting of a heated surface by an evaporating liquid film under conjoining/disjoing pressures, *J. Colloid Interface Sci.* **218** (1999), 152–166.

[77] A. Oron and S. G. Bankoff, Dynamics of a condensing liquid film under conjoining/disjoining pressures, *Phys. Fluids* **13** (5) (2001), 1107–1117.

[78] A. Oron, S. H. Davis, and S. G. Bankoff, Long-scale evolution of thin liquid films, *Rev. Modern Phys.* **69** (3) (1997), 931–980.

[79] L. A. Peletier, The porous media equation, in: *Applications of Nonlinear Analysis in the Physical Sciences* (H. Amann et al., eds.), Pitman, New York, 1981, 229–241.

[80] G. Reiter, Dewetting of thin polymer films, *Phys. Rev. Lett.* **68** (1) (1992), 75–78.

[81] L. G. Reyna and M. J. Ward, Metastable internal layer dynamics for the viscous Cahn-Hilliard equation, *Methods Appl. Anal.* **2** (3) (1995), 285–306.

[82] M. Schneemilch and A.-M. Cazabat, Shock separation in wetting films driven by thermal gradients, *Langmuir* **16** (25) (2000), 9850–9856.

[83] M. Schneemilch and A.-M. Cazabat, Wetting films in thermal gradients, *Langmuir* **16** (23) (2000), 8796–8801.

[84] L. W. Schwartz, Viscous flows down an inclined plane: Instability and finger formation, *Phys. Fluids A* **1** (3) (1989), 443–445.

[85] L. W. Schwartz, R. V. Roy, R. R. Eley, and S. Petrash, Dewetting patterns in a drying liquid film, *J. Colloid Interface Sci.* **234** (2) (2001), 363–374.

[86] A. Sharma and R. Khanna, Pattern formation in unstable thin liquid films, *Phys. Rev. Lett.* **81** (16) (1998), 3463–3466.

[87] A. Sharma and R. Khanna, Pattern formation in unstable thin liquid films under the influence of antagonistic short- and long-range forces, *J. Chem. Phys.* **110** (10) (1999), 4929–4936.

[88] M. Shearer, D. G. Schaeffer, D. Marchesin, and P. Paes-Leme, Solution of the Riemann problem for a prototype 2×2 system of non-strictly hyperbolic conservation laws, *Arch. Rational Mech. Anal.* **97** (1987), 299–320.

[89] M. Slemrod, Admissibility criteria for propagating phase boundaries in a van der Waals fluid, *Arch. Rational Mech. Anal.* **81** (1983), 301–315.

[90] V. M. Starov, Spreading of droplets of nonvolatile liquids over a flat solid, *J. Colloid Interface Sci. USSR* **45** (1983), 1154.

[91] S. M. Troian, E. Herbolzheimer, S. A. Safran, and J. F. Joanny, Fingering instabilities of driven spreading films, *Europhys. Lett.* **10** (1) (1989), 25–30.

[92] M. P. Valignat, S. Villette, J. Li, R. Barberi, R. Bartolino, E. Dubois-Violette, and A.-M. Cazabat, Wetting and anchoring of a nematic liquid crystal on a rough surface, *Phys. Rev. Lett.* **77** (10) (1996), 1994–1997.

[93] F. Vandenbrouck, M. P. Valignat, and A.-M. Cazabat, Thin nematic films: metastability and spinodal dewetting, *Phys. Rev. Lett.* **82** (13) (1999), 2693–2696.

[94] M. J. Ward, Metastable patterns, layer collapses, and coarsening for a one-dimensional Ginzburg-Landau equation, *Stud. Appl. Math.* **91** (1) (1994), 51–93.

[95] M. B. Williams and S. H. Davis, Nonlinear theory of film rupture, *J. Colloid Interface Sci.* **90** (1982), 220–228.

[96] T. P. Witelski and A. J. Bernoff, Stability of self-similar solutions for van der Waals driven thin film rupture, *Phys. Fluids* **11** (9) (1999), 2443–2445.

[97] T. P. Witelski and A. J. Bernoff, Dynamics of three-dimensional thin film rupture, *Phys. D* **147** (2000), 155–176.

[98] T. P. Witelski and M. Bowen, ADI methods for high order parabolic equations, preprint, Duke University, 2001.

[99] C. C. Wu, New theory of MHD shock waves, in: *Viscous Profiles and Numerical Methods for Shock Waves* (M. Shearer, ed.), SIAM, Philadelphia, PA, 1991.

[100] R. Xie, A. Karim, J. F. Douglas, C. C. Han, and R. A. Weiss, Spinodal dewetting of thin polymer films, *Phys. Rev. Lett.* **81** (6) (1998), 1251–1254.

[101] W. W. Zhang and J. R. Lister, Similarity solutions for van der Waals rupture of a thin film on a solid substrate, *Phys. Fluids* **11** (9) (1999), 2454–2462.

[102] L. Zhornitskaya and A. L. Bertozzi, Positivity-preserving numerical schemes for lubrication-type equations, *SIAM J. Numer. Anal.* **37** (2) (2000), 523–555 (electronic).

Numerical turbulent combustion: an asymptotic view via an idealized test-case

Anne BOURLIOUX
Département de mathématiques et de statistique
Université de Montréal
C.P. 6128, succ. Centre-ville
Montréal, Qué., H3C 3J7
Canada

Abstract

One key issue in numerical turbulent combustion is to account for the interaction of the turbulent flow with the flame structure, which, for many practical simulations, occurs on scales much too small for direct resolution. A precise formulation of this reactive turbulence closure problem is introduced via an idealized test-case consisting of a multiple-scale advection-reaction-diffusion equation. Both the premixed and nonpremixed flame regimes are described. Practical ad-hoc strategies for modeling turbulent flames typically attempt to capitalize on the scale separation between the flame thickness and the turbulent scales: a rigorous asymptotic approach is used here to formulate mathematically sound models of this type.

1 Introduction

1.1 The idealized test-case

The idealized test-case describes the evolution of the reactive scalar T which can be viewed as either the temperature field or a reactant mass fraction. The scalar obeys an advection-diffusion-reaction equation of the form:

$$\frac{\partial T}{\partial t} + (\overline{\mathbf{v}}(\mathbf{x}, t) + \mathbf{v}'(\mathbf{x}/\epsilon^\alpha, t/\epsilon^\alpha)) \cdot \nabla T = \epsilon \Delta T + \frac{1}{\epsilon} f(T) \tag{1.1}$$

with $\epsilon > 0$ and $0 < \alpha < 1$ given. The nonlinear reaction rate function $f(T)$ must also be given; examples corresponding respectively to premixed and nonpremixed flames are described in the next section. The scaling in the equation is such that, in the absence of turbulent advection, the combination of slow diffusion (order ϵ, with ϵ small) and fast reaction (reaction rate $f(T)/\epsilon$) leads to a very thin flame whose width scales like ϵ.

The synthetic velocity field on the left hand side must also be specified a priori. It has two components, both assumed to be incompressible: a large scale flow $\overline{\mathbf{v}}$ and a perturbation $\mathbf{v}'$ that acts at the scale ϵ^α intermediate between the flame thickness ϵ and the large scale (normalized here to be of order 1). An example of perturbation flow is given by the Childress-Soward flow $(u, v) = (-\partial F/\partial y', \partial F/\partial x')$ with the stream function

$$F(x', y') = \sin(\pi x'/2)\sin(\pi y'/2) + \delta\cos(\pi x'/2)\cos(\pi y'/2),$$

A. Bourlioux and M.J. Gander (eds.), Modern Methods in Scientific Computing and Applications, 81–102.

where $(x', y') = (x/\epsilon^\alpha, y/\epsilon^\alpha)$.

When $\delta = 0$, the flow is a simple shear tilted at 45 degrees; when $\delta = 1$, the flow corresponds to a periodic array of vortices. In later sections, we will use the intermediate value $\delta = 0.5$, whose streamlines are shown in Fig. 1: the flow is seen to consist of an interesting mix of shears and eddies.

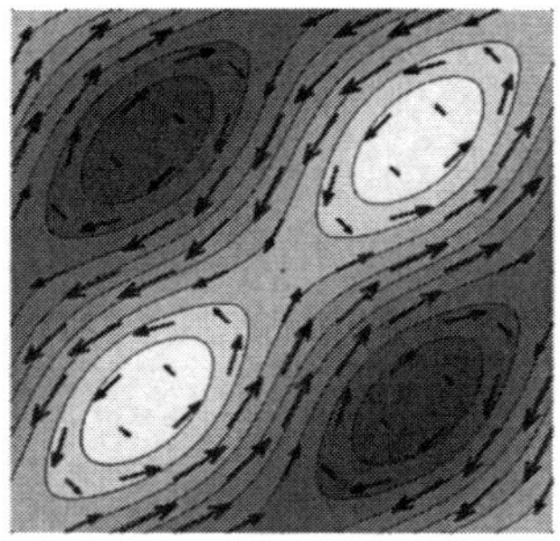

Figure 1: Streamlines for the Childress-Soward flow with $\delta = 0.5$.

1.2 Premixed and nonpremixed regimes

Practical flames belong to one of two types: premixed or nonpremixed. In this section, we describe how either type can be described by Equ. (1.1) depending on the specific choice for the nonlinear reaction rate $f(T)$ and for the boundary conditions. We ignore for now the effect of the perturbations due to small-scale advection, hence focusing on laminar, essentially one-dimensional flames—turbulent effects will be discussed at length in later sections.

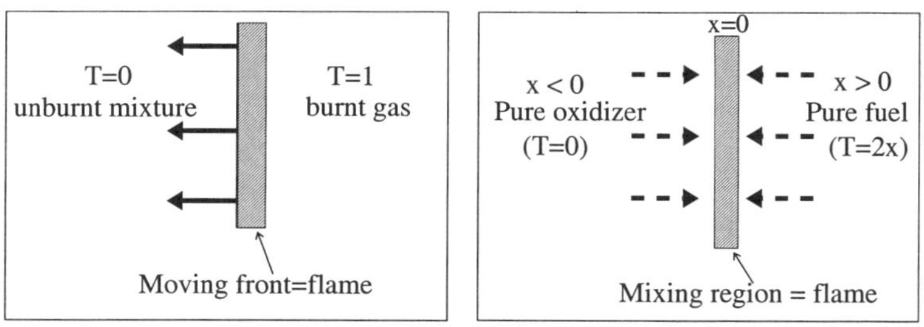

Figure 2: Typical set-ups for premixed (left) and nonpremixed (right) laminar flames.

A practical example of a premixed flame can be found in a car engine: fuel and oxidizer are mixed in the injection nozzle before entering the combustion chamber. A spark initiates the combustion process, the flame started around the spark propagates as a thin front through the chamber. The propagation is typically in the turbulent regime—this is a good thing, as turbulence enhances the speed of propagation; the laminar speed would actually be too small for engines to be useful! Ignoring such important turbulent effects for now, a typical set-up for an idealized version of a laminar premixed flame is shown in Fig. 2 (left). The flame is represented as a front initially located at $x = 0$ and moving to the left. One should view T as the temperature, with the normalization such that $T = 1$ corresponds to the hot, burnt gas temperature, and $T = 0$ corresponds to the cold, unburnt mixture of fuel and oxidizer. A typical choice for the reaction rate is given by the so-called KPP rate, with corresponding left and right states:

$$f_{\text{premixed}} = T(1 - T) \quad \text{with} \quad T_{\text{left}} = 0 \ (x \leq 0), \quad T_{\text{right}} = 1 \ (x \geq 0).$$

A simple practical example of a nonpremixed flame is the candle. The radiating heat from the flame causes the wax to melt and to move up the wick by capillarity. Upon reaching the top of the wick, the liquid will vaporize because of the direct heating from the flame. There will be an interface between the region rich in fuel (the vaporized wax, "inside" the flame) and the region rich in oxidizer (the air, "outside" the flame). As soon as the two mix, due to molecular diffusion in the laminar case, also due to turbulent mixing in the turbulent case, reaction proceeds very rapidly, allowing the process to self-maintain as long as there is wax (and air). An idealized set-up to describe this type of dynamics in the case of a laminar nonpremixed flame is shown in Fig. 2 (right). The thin reaction zone is located around $x = 0$, with typical reaction rate and left and right states given by:

$$f_{\text{nonpremixed}} = T(2x - T) \quad \text{with} \quad T_{\text{left}} = 0, \quad T_{\text{right}} = 2x.$$

For this case, one should view T as the fuel mass fraction. In Section 3, $2x - T$ will be shown to actually represent the oxidizer mass fraction. Therefore, the left state T_{left} corresponds to pure oxidizer, the right state T_{right} to pure fuel.

A straightforward mathematical analysis of the two set-ups leads to the following conclusions [8]:

(1) In both cases, the left state T_{left} and the right state T_{right} are equilibrium points of the given reaction rate $f(T_{\text{left}}) = f(T_{\text{right}}) = 0$. The flame can therefore be described as the interface between two distinct equilibrium points.

(2) In the case of the nonpremixed flame, both equilibrium points are stable. In the case of the premixed flame however, only the right state is stable, the left state is unstable and will convert to the right state at the slightest perturbation.

As a result, the two reaction rates describe regimes with very different dynamics representative of the two distinct types of practical flames described earlier.

Nonpremixed flames To the left of the nonpremixed flame, there is only oxidizer, so that the reaction cannot proceed because of the lack of fuel, and the reverse to the right of the flame. Because of diffusion, fuel and oxidizer mix at a very slow rate across the interface and react very rapidly once mixed. The reaction zone is confined to a very thin zone that does not propagate but is fed reactants by diffusion from either side. The key issue in understanding the rate at which the combustion proceeds is to get a good description of the mixing of the two ingredients. Turbulence always enhances mixing, so in this case, it will also enhance the combustion rate in the turbulent case compared to the laminar case. How to formulate a turbulent model based on this key idea is described in detail in Section 3.

Premixed flames To the left of the premixed flame, one finds the fresh mixture with the right stoichiometry for reaction, except that the gas mixture is too cold. As soon as enough heat reaches this unstable cold side by diffusion, the reaction starts, irreversibly, producing more heat. The flame interface is now a front propagating into the unburnt side and the key issue is to predict the speed of propagation of that front. Turbulence will wrinkle the front, which results in an enhanced burning speed as a result of the increase in burning area: predicting quantitatively this enhancement is the modeling challenge studied in Section 4.

1.3 Key simplifying assumptions

Typically, numerical turbulent combustion is modeled mathematically using the compressible reactive Navier-Stokes equations, a coupled system of nonlinear partial differential equations that describe conservation of mass, momentum, energy, chemical species. The idealized model here corresponds to a single species mass fraction equation, or to the energy conservation equation written in terms of temperature, under the following assumptions:

(1) The heat release is weak: the flow can therefore be considered as incompressible, mass conservation is satisfied by imposing that the velocity field is divergence-free, there is no feedback from the flame on the flow.

(2) Combustion can be described by a single, irreversible reaction. Realistic models can include dozens, sometimes hundreds of chemical reactions. Moreover, we assume that the molecular properties (conductivity, diffusivity) are constant and the same for temperature and all chemical species.

(3) The flow field is specified a priori and not a solution of Navier-Stokes equations.

The motivation for the simplification is that the full system of equations is extremely complex, for all practical purposes intractable analytically, and such that numerical simulations are very costly and hard to validate. Instead, with the simple model:

- Rigorous asymptotic results are available to provide useful information regarding the effective behavior at large scales, for example: scaling of subgrid models, upper bounds on effective transport coefficients, etc.;
- Equ. (1.1) is orders of magnitude easier to solve numerically than the full reactive Navier-Stokes equations so that cheap, extensive, reliable numerical databases can be generated.

This overall strategy of scaling the problem from the Navier-Stokes equations down to a single scalar equation has been used very successfully in the past in the case of passive scalar turbulent advection—see [9] for a self-contained introduction to that topic. Many of the passive scalar results will actually turn out to be directly relevant to modeling nonpremixed flames—see Section 3.

2 The closure problem

2.1 Direct numerical simulations

One could try to solve Equ. (1.1) for T by *direct numerical simulation*. The partial differential equation needs to be discretized by finite differences (or finite elements, finite volumes, or any other type of discretization), and the solution T will be known only at discrete mesh points (x_i, y_j) and discrete times t^n. For the nonpremixed cases to be discussed in Section 3, steady-state solutions will be sought, so that one will need to solve a coupled system of nonlinear algebraic equations. This can be handled by using a quasi-Newton procedure coupled with the GMRES iterative method for non-symmetric problems. For the premixed cases in Section 4, unsteady solutions will be computed, with typically the advection terms discretized explicitly in time while the reaction and diffusion terms are discretized implicitly. One convenient strategy is to split the sub-processes: because of the periodic boundary conditions in y (see the set-up in Section 4), the diffusion operator can be dealt with by Fourier transform, while the reactive term can be solved analytically.

This type of discretization is very standard and presents no particular technical difficulty. The challenge is that the key for its success at computing an accurate solution for T is that all the problem length scales must be resolved by the discrete mesh—they include the large scale, of order 1; the intermediate scale from the velocity perturbation, of order ϵ^α; the small scale associated with the flame thickness, of order ϵ. The regimes of practical interest correspond to $\epsilon \to 0$: the computational cost for resolving simultaneously the large scales and the fine scales becomes prohibitive. This is true up to a point for the simple scalar equation (1.1), but definitely so when dealing with the full reactive Navier-Stokes equations.

2.2 The asymptotic view and the practical view

The closure problem can now be introduced. It stems from the following two observations:

- A directly resolved solution for T as described above is typically too expansive for practical purposes.
- In any case, one is usually not interested in the fine details of the solution for T, only in its large scale behavior.

One strategy to obtain those large scale features of T would be to simply ignore the fine scale behavior, so that one would not need to use a very fine mesh to capture the small turbulent and reactive scales. However, this is known to be incorrect. Turbulent mixing by the intermediate scale velocity interacting with the small scale flame structure always leads to an overall enhancement of the effective burning rate that determines in some essential way the large scale behavior.

There are two related points of view as far as characterizing that enhancement, and more generally addressing the issue of computing the solution for T at large scales in some efficient way:

(1) **Mathematical statement of the problem:** what is the solution for T in the limit of $\epsilon, \epsilon^\alpha \to 0$, i.e. what is the effective behavior of the thin flame front at large scales, taking into account the effects of its distortion at the intermediate scale by the velocity perturbation $\mathbf{v}'$?

(2) **Practical statement of the problem:** what is an efficient procedure to compute numerically the solution for T at large scales when both the very small flame thickness ϵ and the intermediate scale ϵ^α associated with the turbulent flow $\mathbf{v}'$ are smaller than the discretization grid size one can afford?

The practical formulation of the problem in (2) is the classical closure problem that one must tackle in large eddy simulations: the challenge is to account properly for the subgrid effects that contribute to the solution at large scales. Why this is a challenge can be understood by considering the following strategy: one way of defining the large scale solution $\overline{T}$ is to view it as the average of T over the intermediate scale periodic box:

$$\overline{T}(x,y) = \frac{1}{{\Delta_\alpha}^2} \int_{x-\Delta_\alpha/2}^{x+\Delta_\alpha/2} \int_{y-\Delta_\alpha/2}^{y+\Delta_\alpha/2} T(x',y')\, dx'\, dy',$$

where $\Delta_\alpha = \epsilon^\alpha$ the size of the intermediate velocity periodic box. With that definition, $T = \overline{T} + T'$ with T' the small scale fluctuations of T, of zero mean. As $\overline{T}$ is the averaged solution of Equ. (1.1), one could hope to obtain it directly as the solution of the equation obtained by averaging Equ. (1.1). This gives:

$$\frac{\partial \overline{T}}{\partial t} + \overline{\mathbf{v}}(\mathbf{x},t) \cdot \nabla \overline{T} = \epsilon \Delta \overline{T} + T_1 + T_2, \tag{2.1}$$

where the terms T_1 and T_2 are the results of averaging nonlinear terms in the fluctuations:

- $T_1 = -\overline{v' \cdot \nabla T'}$: this is the usual turbulent mixing term resulting from averaging the turbulent advection of the scalar fluctuations. In the non-reactive case, this term can be simply modeled in closed form (i.e. only as a function of the resolved quantities $\overline{T}$, etc.): an introduction to the rigorous theory for this term can be found in [9]. The model consists in expressing T_1 as diffusive term with $T_1 = \kappa_T \Delta \overline{T}$, where κ_T is the turbulent diffusivity for which formulas, bounds, etc. can be established.
- $T_2 = \overline{f(\overline{T} + T')}$: this is the nonlinear term specific to the reactive case. A very bad model for most nonlinearities would be to simply ignore the fluctuations by setting $T_2 = f(\overline{T})$. This ignores entirely the combustion enhancement by turbulence and is wrong.

How to model effectively T_2 in terms of the resolved variables is the key closure issue in turbulent combustion and still an essentially open question for practical problems. Ad-hoc strategies frequently resorted to in practice consist in attempts at exploiting the separation of scale between the flame thickness and the turbulent scales. Pioneering work regarding this topic from the engineering point-of-view can be found in [10]. In the following sections, strategies based on the rigorous results from the mathematical statement (1) above will be described to offer a systematic way of approaching the very difficult problem as in the practical statement (2). This is done for nonpremixed flames in the next section, where it will be shown that the ad-hoc modeling strategies actually coincide with the rigorous mathematical prediction. The premixed flame case is discussed in Section 4: for that case, the mathematical strategy can differ significantly from the ad-hoc approach.

3 Nonpremixed flames

3.1 Specific set-up for nonpremixed flames

The starting point in deriving the model equations for nonpremixed flames is to write down the conservation equations for the fuel and oxidizer mass fractions, respectively Y_1 and Y_2.

Those equations describe their advection by the velocity field, diffusion, and reaction:

$$\frac{\partial Y_1}{\partial t} + \mathrm{Pe}\,\mathbf{v}\cdot\nabla Y_1 = \Delta Y_1 - \mathrm{Da}\,Y_1 Y_2, \tag{3.1}$$

$$\frac{\partial Y_2}{\partial t} + \mathrm{Pe}\,\mathbf{v}\cdot\nabla Y_2 = \Delta Y_2 - \mathrm{Da}\,Y_1 Y_2. \tag{3.2}$$

Here, $\mathbf{v}$ is the Childress-Soward flow with $\delta = 0.5$ whose streamlines are shown in Fig. 1. The flow field is scaled such that its average kinetic energy over one periodic domain is one. The two equations above correspond to Equ. (1.1), with slightly different reference scales: the new reference length scale is the perturbation flow $\mathbf{v}$ period P (its value was $P = \epsilon^\alpha$ before) and the time scale has been rescaled such that the molecular diffusivity D is now one (it was $D = \epsilon$ before). This non-dimensionalisation leads to the introduction of two classical non-dimensional numbers.

- *Peclet number* Pe: It is defined as $\mathrm{Pe} = P\lambda/D$, where λ is some constant with dimension of velocity which measures the magnitude of the velocity field, and where D is the diffusivity of both scalars. The Peclet number measures the effect of advection compared to diffusion. High turbulence intensities correspond to large Peclet numbers.

- *Damköhler number* Da: It is defined as $\mathrm{Da} = kP^2/D$ with k the reaction source term constant (its value was given by $1/\epsilon$ in Equ. (1.1)). It measures the relative effect of reaction compared to diffusion. Large Damköhler numbers mean that reaction is much faster than diffusion, which corresponds to the flamelet regime with flame thickness very small.

The reaction term corresponds to a one-step irreversible reaction $Y_1 + Y_2 \to Y_p$ (with Y_p the mass fraction of the product) at the finite reaction rate $\omega = kY_1Y_2$. One of the two reactive equations can be replaced by the simpler advection-diffusion (no reaction) equation if one introduces the passive scalar $Z = (Y_1 - Y_2)/2$. Taking the difference between Equs. (3.1) and (3.2) results in the following evolution equation for the passive scalar Z:

$$\frac{\partial Z}{\partial t} + \mathrm{Pe}\,\mathbf{v}\cdot\nabla Z = \Delta Z. \tag{3.3}$$

The system will be solved for $Y_i(x, y), Z(x, y)$ in a rectangular domain with $-L_G/2 \leq x \leq L_G/2$ and $0 \leq y \leq P$, where L_G represents the large scale of the system, taken to be much larger than P (in the present calculations, $L_G/P = 10$). Precise boundary conditions for all three scalars are stated next with a full explanation later, the set-up is illustrated in Fig. 3. The passive scalar $Z(x, y)$ is assumed to be the sum of a mean value $Z_G = x/L_G$, corresponding to a mean gradient in x with slope $1/L_G$, and of a bi-periodic perturbation Z_P of mean zero and period $P = 1$:

$$Z(x, y) = x/L_G + Z_P(x, y) \tag{3.4}$$

This is achieved by imposing the following boundary conditions:

$$Z(x, y = 0) = Z(x, y = 1) \quad \text{for} \quad -L_G/2 \leq x \leq L_G/2, \tag{3.5}$$

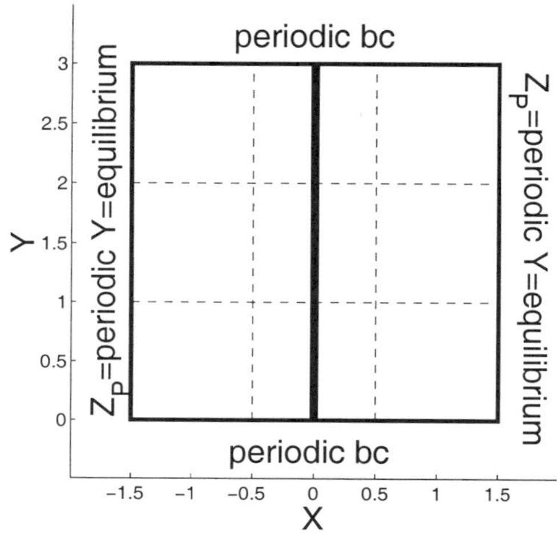

Figure 3: Nonpremixed set-up.

$$Z(-L_G/2, y) = Z(L_G/2, y) - 1 \quad \text{for} \quad 0 \le y \le 1. \tag{3.6}$$

The reactive scalars Y_1 and Y_2 are also periodic in y and satisfy the following boundary conditions in x at x_L and x_R, assuming that the size of the domain is much larger than the thickness of the flame:

$$Y_i(x_A, y) = Y_{i,\text{eq}}(x_A, y), \tag{3.7}$$

where the equilibrium values $Y_{i,\text{eq}}$ are given by:

$$\begin{aligned} Y_{1,\text{eq}}(x,y) &= Z(x,y) + |Z(x,y)| \\ Y_{2,\text{eq}}(x,y) &= -Z(x,y) + |Z(x,y)| \end{aligned} \tag{3.8}$$

as discussed in the following section.

Examples of direct simulation results for this set-up are given in Fig. 4 for the passive scalar Z and in Fig. 5 for the reaction rate, computed a posteriori as $Y_1Y_2 = Y_1(Y_1 - 2Z)$. Fig. 4 corresponds to a large Peclet number (Pe $= 200$): the advection by the "turbulent" flow is strong and the iso-levels of Z are very distorted.

The top plots in Fig. 5 show the reaction rate corresponding to the same value of Pe for two different values of Da. The left plot corresponds to a large value of Da, selected such that the laminar flame thickness is $P/8$ (see next section); indeed, even in the turbulent case, the reaction rate is nonzero only in a thin zone at the interface between the fuel-rich side and the oxidizer-rich side (the interface corresponds to $Z = 0$). The low value of the Damköhler number used in the computation corresponding to the plot on the right corresponds to a laminar flame thickness equal to P. In that case, as expected, the turbulent reaction zone occupies a much wider area; models based on thin flames will perform poorly

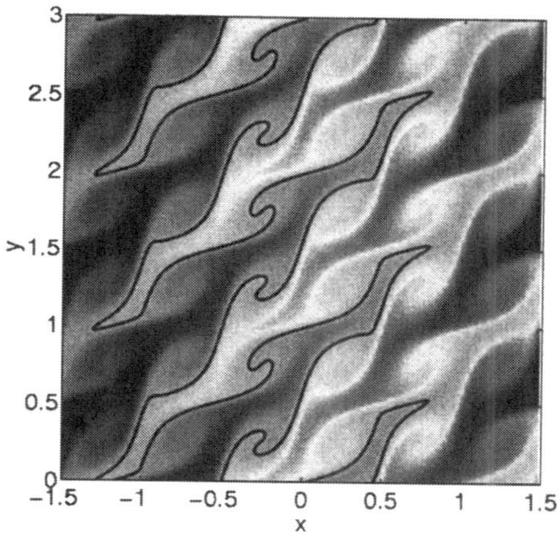

Figure 4: Nonpremixed flame—direct simulation result: passive scalar field with zero level shown as solid black line.

for such cases. The two bottom plots correspond to the same values for the Damköhler numbers, but at much lower Peclet numbers. Because the turbulence intensity is much lower, the flame is hardly distorted; such test-cases should be much less stringent than the ones with corresponding Damköhler numbers much higher turbulent intensities. The direct numerical simulation results were obtained using the numerical strategy briefly outlined in the introduction section. They will be used below to validate the rigorous asymptotic modeling strategy.

3.2 Theoretical predictions for the laminar case

The flamelet modeling strategy is based on exploiting the well understood behavior of the flame front in the laminar case and adapting it to the case where the flame front is actually distorted by turbulence, assuming that its internal structure is not affected in a major way. It is therefore necessary to first summarize what is known about the laminar case. This will incidentally help clarify a number of assumptions at the core of the problem formulation as stated earlier.

The laminar case is studied by setting $\mathbf{v} = 0$ in Equs. (3.1), (3.2), (3.3). It is very easy to check that in that case the solution for the passive scalar is trivial: $Z = x/L_G$, there is no perturbation to the steady state mean gradient profile. This explains in particular the expression for the reaction rate that was used to describe the stability of the equilibrium states in the introduction section.

This set-up corresponds to the steady one-dimensional laminar flame problem introduced in the classical paper by Williams [12] with the asymptotic results summarized next.

The length scale P is not relevant in the absence of advection, so one uses instead L_G as a

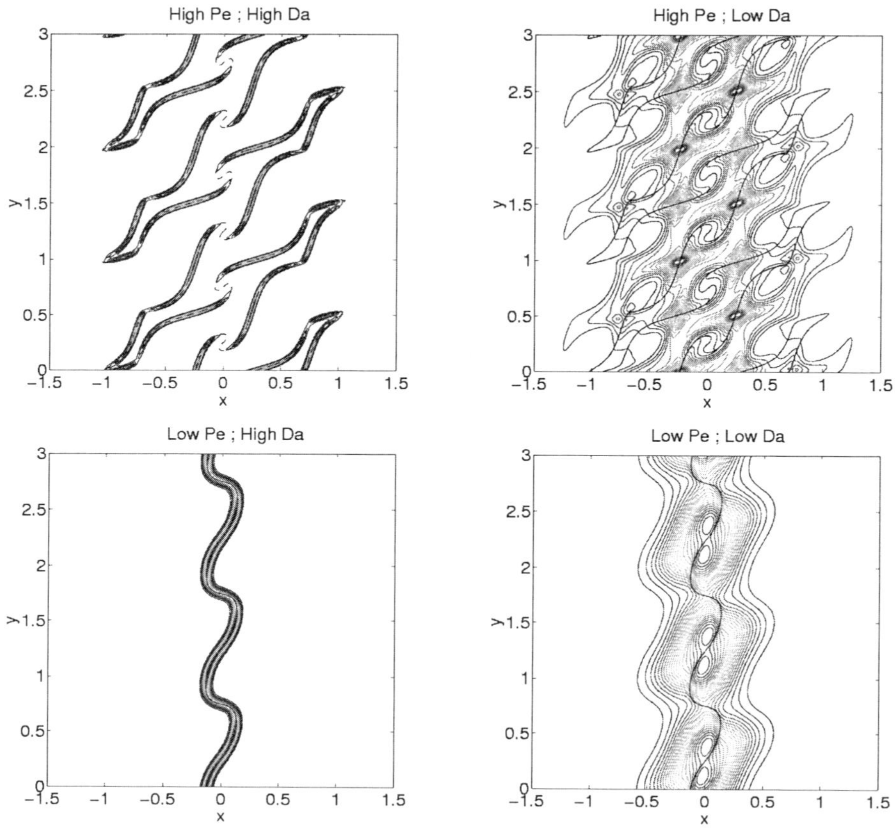

Figure 5: Nonpremixed flame—direct simulation result: reaction rate for different combinations of Pe and Da numbers.

reference length scale and defines the corresponding Damköhler group $\alpha = kL_G^2/D$ (therefore $\alpha = \mathrm{Da}(L_G/P)^2$ with Da the Damköhler number used in the previous section). Williams' theory describes the limit behavior of Y_1, Y_2 in the limit of very large Damköhler numbers.

3.3 Order 0 asymptotics in the limit $\mathrm{Da} \to \infty$

The lowest order term in the asymptotic expansion corresponds to the limit of α, $\mathrm{Da} \to \infty$. It is clear from Equs. (3.1), (3.2) that in that limit, Y_1 and Y_2 can never coexist to keep the reactive source term finite. Practically, it means that as soon as Y_1 and Y_2 mix through diffusion across the stoichiometric level $Z = 0$, they react—the flame is simply the zero-thickness interface $Z = 0$. To the right of this level, there is only fuel ($Y_{1,\mathrm{eq}} = 2Z$ and $Y_{2,\mathrm{eq}} = 0$), to the left, there is only oxidizer ($Y_{2,\mathrm{eq}} = -2Z$ and $Y_{1,\mathrm{eq}} = 0$) so that one recovers the expressions in Equ. 3.8.

3.4 Order 1 asymptotics for large but finite Da

For large but finite values of α (equivalently, Da), the departure $dY_1(Z) = Y_1 - Y_{1,\mathrm{eq}}$ can be expected to be small and confined to a small area. This is confirmed by a systematic asymptotic expansion, as summarized next.

(1) *Step 1: change to Z-coordinates.* With $Z = x/L_G$, it is trivial to study $Y_1(Z)$ instead of $Y_1(x)$. In the new coordinate system, the equation for Y_1 is given by:

$$\frac{d^2Y_1}{dZ^2} = \alpha Y_1(2Z - Y_1),$$

where $\alpha = \mathrm{Da}\, L_G^2$.

(2) *Step 2: rescale Z and Y_1.* The rescaled variables are:

$$Z_{\mathrm{norm}} = \alpha^{1/3} Z, \qquad Y_{1,\mathrm{norm}} = \alpha^{1/3} Y_1.$$

This leads to the following parameter-free equation:

$$\frac{d^2Y_{1,\mathrm{norm}}}{dZ_{\mathrm{norm}}^2} = Y_{1,\mathrm{norm}}(2Z_{\mathrm{norm}} - Y_{1,\mathrm{norm}}).$$

There is no approximation up to this point. The approximation stems from the fact that this equation is solved with the equilibrium conditions for $Y_{1,\mathrm{norm}}$ applied in the limit $Z_{\mathrm{norm}} \to \pm\infty$ while for real cases, they should be applied at a finite Z_{norm} that would depend on the size of the domain (in real space, not Z-space) and on α. Nevertheless, asymptotically, with $\alpha = \mathrm{Da}\, L_G^2$ very large, imposing the boundary conditions at $Z_{\mathrm{norm}} \to \infty$ instead of a very large finite value should be a good approximation.

The consequences of this formulation are the following:

- The equation for $Y_{1,\mathrm{norm}}$ is now independent of α. It can be solved once for all for $Y_{1,\mathrm{norm}}(Z_{\mathrm{norm}})$. Alternatively, it is convenient to work in terms of the first order asymptotic correction $dY_{1,\mathrm{norm}} = Y_{1,\mathrm{norm}} - Y_{1,\mathrm{norm,eq}}$. The result of the numerical integration is shown in Fig. 6. This can be viewed as a "flamelet library", i.e. a pre-computed solution for a thin flame structure.

- To compute the non-normalized correction $dY_1(Z)$ at a specific (large) value of α, one simply needs to normalize the value for Z, read the correction $dY_{1,\mathrm{norm}}$ from the library, and de-normalize that value to obtain dY_1.

- As a direct result of the normalization, to first order in the asymptotic expansion, the correction dY_1 scales like $\alpha^{-1/3}$ and the reaction zone thickness scales like $L_G\alpha^{-1/3}$.

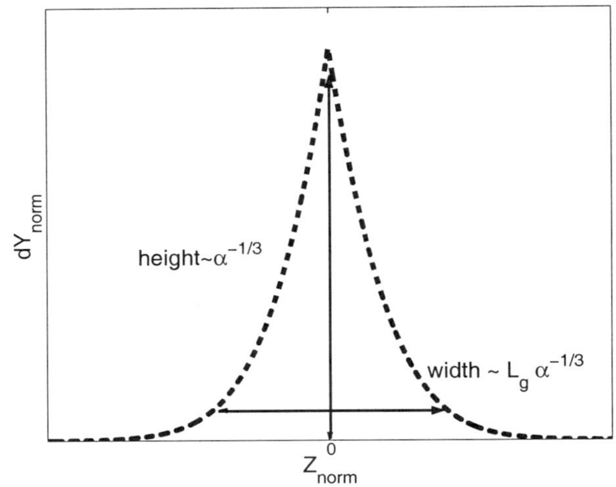

Figure 6: Nonpremixed laminar flamelet.

3.5 Turbulent flamelet approach

The turbulent flamelet approach exploits the asymptotic knowledge summarized in the previous section as a way to formulate a closure strategy for the reactive term. It is based on the two following key assumptions:

- **Flamelet model assumption I.** A detailed knowledge of the statistics of the passive scalar fluctuations is available. In particular, one knows the detailed PDF (probability density function) of Z. For the type of idealized problem discussed here, this is the case; for more general cases, it is true up to a point, see [9].
- **Flamelet model assumption II.** It is possible to express the reactive variable Y_1 as a function of the passive scalar Z (as was done above in the laminar case).

When both assumptions are satisfied, then it is a rigorous result that the large scale solution for Y_1 defined as $\overline{Y_1}$ can be obtained as follows:

$$\overline{Y_1} = \frac{1}{P^2}\iint Y_1(x,y)\,dx\,dy = \int Y_1(Z)\mathrm{PDF}(Z)\,dZ.$$

Similarly, for any function of Y_1 (for example, the reaction rate) whose large scale behavior needs to be computed:

$$\overline{f(Y_1)} = \frac{1}{P^2}\iint f(Y_1(x,y))\,dx\,dy = \int f(Y_1(Z))\mathrm{PDF}(Z)\,dZ.$$

In other words, closure for the reactive scalar is not achieved at the level of the evolution equation for the scalar, but directly at the level of the scalar! Here are two strategies to implement such closure, corresponding to the two asymptotic models.

3.5.1 Equilibrium model

The equilibrium model is very simple: if the Damköhler number is sufficiently large, the reactive scalars can be approximated by their equilibrium values $Y_{i,\mathrm{eq}}(Z)$ (see Equ. (3.8)) with a small error of order $\alpha^{-1/3}$, where $\alpha = \mathrm{Da}\, L_G^2$.

3.5.2 Finite rate flamelet

The key ingredient in computing the next order correction to the asymptotic expansion is the number $\alpha = \mathrm{Da}(L_G/P)^2 = kL_G^2/D$. It can be generalized to the turbulent case by rewriting it as $\alpha = k/(D|\nabla Z|^2)$. The denominator is called the dissipation. Using the pre-computed library in the turbulent case is now trivial: compute a local value for $\alpha(x, y)$ based on the local value of the large scale dissipation (i.e. the norm of the large scale gradient of the passive scalar) and proceed as in the laminar case. The resulting approximation should agree asymptotically with the exact value for the reactive scalar with an error of the order $\alpha^{-2/3} \sim \mathrm{Da}^{-2/3}$.

3.6 Validation

Based on rigorous mathematical results for Equs. (3.1), (3.2) (see the contribution to this volume by P. Souganidis), one expects convergence of the flamelet model, at any Peclet number, as the Damköhler number $\mathrm{Da} \to \infty$, even at large Peclet numbers. This result is verified numerically by comparing the predictions based on Assumption II of the model with the results from the direct simulations. In the direct simulations, pair values for (Z, Y_1) have been computed at each point in the domain without any modeling approximation. To validate the flamelet procedure, one follows either of the two models to predict the value of Y_1 in terms of the local value of Z and, for the finite rate model, of the local value of the dissipation, which can be easily computed by finite difference, and then compare that value $Y_{1,\mathrm{model}}$ with the value Y_1 computed directly. The plots in Fig. 7 show the L_1-norm of the difference between the two. Each subplot corresponds to a different Peclet number. The x-axis corresponds to $\mathrm{Da}^{-1/3}$, errors are expected to converge to 0 as one moves towards $\mathrm{Da}^{-1/3} \to 0$. The top line corresponds to the equilibrium model, the bottom line to the finite rate flamelet model. Here are some comments regarding those plots:

- As predicted by the theory, the norm of the error decreases at fixed Peclet number as $\mathrm{Da} \to \infty$.
- As expected from the derivation of the two models, the norm of the error with the equilibrium model (first term of the asymptotic expansion) is always larger than the norm of the error with the finite rate model (first two terms in the asymptotic expansion).
- As expected intuitively, for instance by inspection of Fig. 5, the errors are larger at higher Peclet numbers.

The plots in Fig. 7 are logarithmic plots. In all cases, the convergence appears to be linear. The slope of those curves will give the order of convergence of the models. The theoretical predictions are as follows:

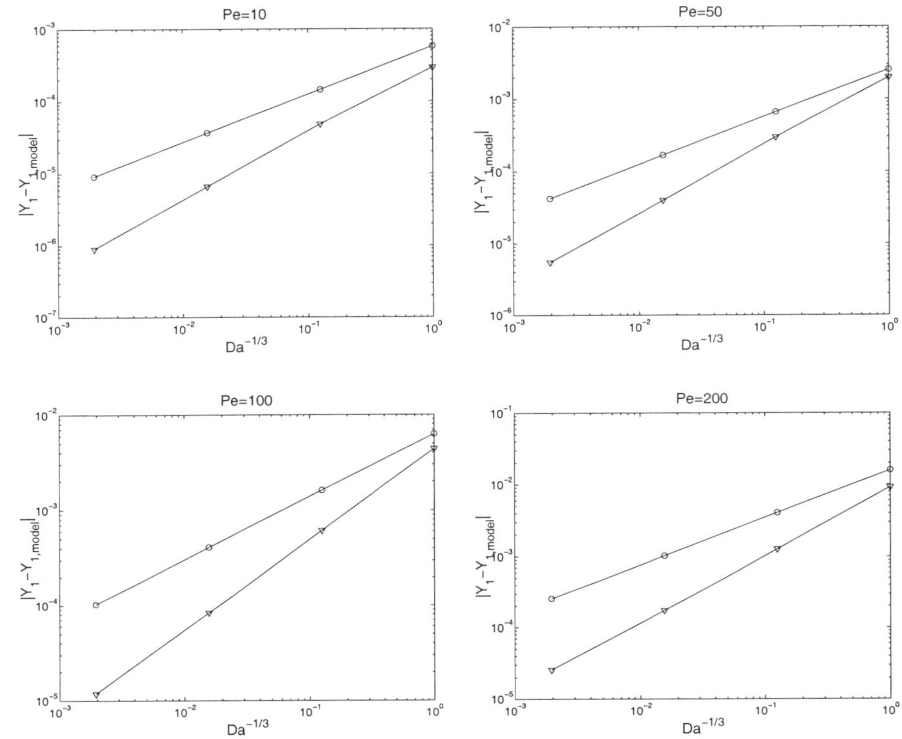

Figure 7: L_1-norms of the errors between directly computed and modeled values for Y_1, as a function of $\mathrm{Da}^{-1/3}$. Triangles correspond to the finite rate model, circles to the equilibrium model.

- In the equilibrium model, the local error on Y_1 is on the order of $\alpha^{-1/3}$; this error is confined to the flame thickness whose size also scales like $\alpha^{-1/3}$. Integrating over one periodic box will therefore result in an error on the order of $(\alpha^{-1/3})^2$, corresponding to a slope equal to 2 in the equilibrium curves in Fig. 7.
- A similar argument for the finite rate model leads to a prediction for the error to scale like $(\alpha^{-1/3})^3$, corresponding to a slope equal to 3 in the finite rate curves in Fig. 7.

The slopes from Fig. 7 were processed and are reported in Fig. 8. The agreement with the theoretical predictions is remarkable: the values are close to the theoretical predictions and are remarkably insensitive to the Peclet number (turbulence intensity). This confirms unambiguously the soundness of the approach. The strategy described here can also be used to study other practical issues related to the implementation of the models [1]. It also motivates further study of the statistical behavior of the passive scalar, in particular of its PDF and its dissipation [2].

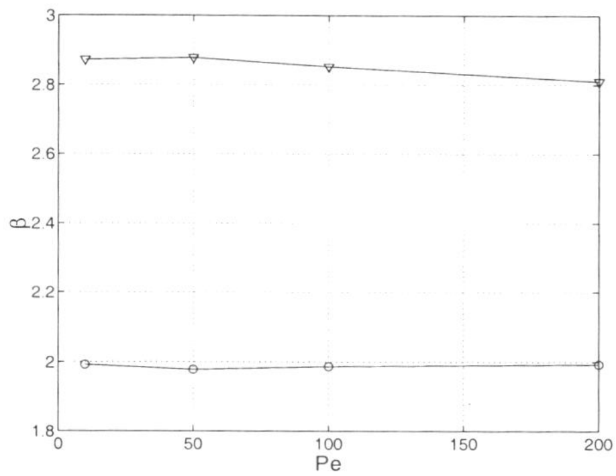

Figure 8: Scaling exponents as a function of Peclet numbers extracted from Fig. 7. Triangles: finite rate model; circles: equilibrium model.

4 Premixed flames

4.1 Specific set-up for premixed flames

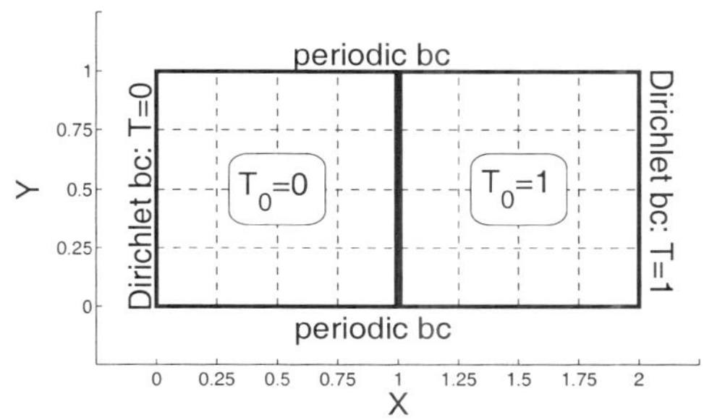

Figure 9: On the left: set-up for the premixed flame test-case with $\epsilon^\alpha = 1/4$. Each ϵ^α square tile corresponds to the periodic box with the streamlines from Fig. 1.

The idealized set-up for a premixed flame is illustrated in Fig. 9. The reaction rate is assumed to be $f(T) = T(1-T)$. Equ. (1.1) is solved in a two-dimensional domain with periodic boundary conditions in y. The initial conditions correspond to unburnt gas on the left side of the domain ($T = 0$) and burnt gas on the right ($T = 1$), so that overall the flame front is moving from right to left. The computational domain is re-centered around the front at every time-step. The large scale velocity field is a simple horizontal shear $\overline{\mathbf{v}} = (\sin(2\pi y), 0)$ and the intermediate scale flow field is obtained from the Childress-Soward stream function with $\delta = 0.5$. Direct numerical simulation results with $\epsilon = 1/16$ and $\epsilon^\alpha = 1/4$ (taking $\alpha = 1/2$) are shown in Fig. 10. The plot on the left shows the reaction rate: it is non-

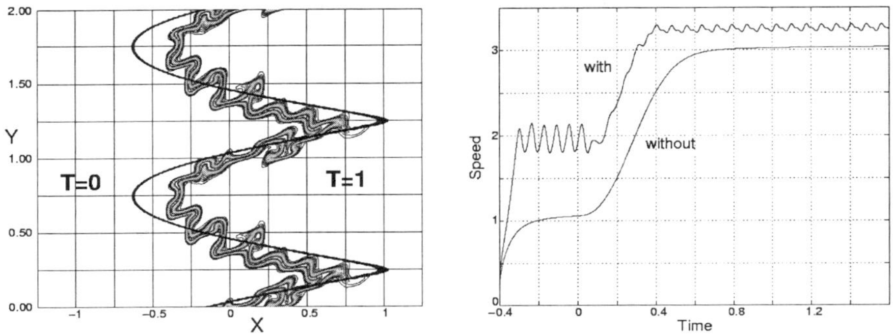

Figure 10: Premixed flame: direct numerical simulation results for the front shape and effective speed with and without the small scale velocity fluctuations.

zero only in a thin zone that is wrinkled by the velocity perturbation. The laminar flame thickness is expected to be equal to ϵ; one sees from Fig. 10 that it remains roughly the case even though the solution is now turbulent. Also shown as a solid black line is the shape of the front when there is no velocity perturbation, only the large scale shear: the overall shape is clearly affected by the turbulent wrinkling. On the right is the plot of the effective speed of the flame as a function of time, with and without the ϵ^α-velocity perturbation. The initial data at $t = -0.5$ correspond to a flat flame. At that time, only the small scale flow is turned on (when there is one) and the solution settles into a speed of propagation as a result of self-propagation through burning and wrinkling by the small scale flow. Then at time $t = 0$, the large scale shear flow is turned on and the overall speed of the flame converges to a final propagation speed. Again, one observes that the overall speed is different with and without the velocity fluctuations, with the wrinkling in the turbulent case resulting as expected in a propagation speed increase. The objective of the model is to be able to capture both the overall front shape and effective speed without resolving in detail the small length scales associated with the front wrinkling, and certainly not resolving the even smaller length scale that corresponds to the flame thickness.

4.2 Asymptotic predictions

The rigorous asymptotic theory to predict the behavior of the solution of Equ. 1.1 with the present set-up in the limit of $\epsilon \to 0$ was developed in [7], see also Souganidis' contribution in this volume.

- In the limit of $\epsilon \to 0$, the reaction zone tends to an infinitely thin front moving normal to itself with an enhanced burning speed $S_T(x, y, t) > S_L$ (with $S_L = 1$ the laminar burning speed) as well as due to advection by the large scale flow.

- The enhanced speed $S_T(x, y, t)$ is obtained via the minimization over a scalar of the eigenvalue of a nonlinear Hamilton-Jacobi partial differential equation, the so-called "cell-problem". This is a challenging computational problem, see below.

- In the present case, $S_T(x, y, t)$ depends to leading order only on the following local parameters: the local large scale flow intensity $|\overline{v}(x, y)|$ and the local inclination angle $\theta(x, y, t)$ of the front. (Given the present set-up, only those two parameters control the speed at which the front is advected across a small periodic box, the faster the front is swept along, the smaller the burning speed enhancement.)

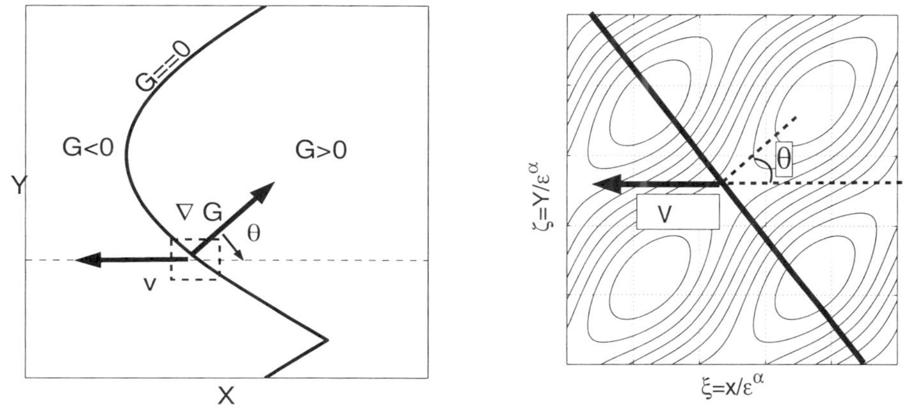

Figure 11: Left: illustration for the level-set tracking procedure for the asymptotic front. The zero-level curve, $G = 0$, indicates the location of the front. Right: close-up of the auxiliary cell-problem corresponding to the small box in the plot on the left.

The predictions are illustrated qualitatively in Fig. 11. The left plot shows the representation of the front as a zero-thickness interface, computed via a level-set method (see explanation below). The right plot shows the local cell-problem. It corresponds to zooming in on a small periodic box along the front, shown as a small rectangle in the left plot. The approximation is that, locally, the large scale front is represented as a straight line, even though it is not exactly straight at large scales.

4.2.1 The auxiliary cell-problem

Here is a qualitative discussion of the auxiliary cell-problem that needs to be solved as part of the prediction for the local effective burning speed of the flame S_T. The objective of the discussion is to provide some intuition on how this theory describes the mechanism by which turbulence leads to an enhancement of the propagation speed of the flame $S_e = S_T - S_L$ as a result of the mixing taking place near the cold side boundary.

The first step is to change variable in order to focus on the pre-heat zone near $T = 0$:

$$Z^\epsilon = \epsilon^{-1} \log T.$$

This clearly emphasizes the exponential decay of T near zero. The equation for the new variable is such that the diffusion operator splits into two terms, a small (order ϵ) diffusion term for Z^ϵ which is negligible in the limit of $\epsilon \to 0$ and a new nonlinear term (order 1) which has the same order of magnitude as the advection term: it is the balance between this nonlinear term and the advection term that controls the turbulent mixing mechanism in the pre-heat zone:

$$\frac{\partial Z^\epsilon}{\partial t} - |\nabla Z^\epsilon|^2 + (\overline{\mathbf{v}}(x,y) + \mathbf{v}(x/\epsilon^\alpha, y/\epsilon^\alpha)) \cdot \nabla Z^\epsilon = \frac{f(T)}{T} + \epsilon \Delta Z^\epsilon.$$

When ϵ goes to zero the solution Z^ϵ converges uniformly on compact subsets to the viscosity solution of the variational inequality of Hamilton-Jacobi type:

$$\max(Z_t + H(\nabla Z, x, y), Z) = 0 \tag{4.1}$$

with

$$\left.\begin{array}{ll} Z = 0 & \text{when } T = 1 \text{ (burnt side)} \\ Z = -\infty & \text{when } T = 0 \text{ (unburnt side)} \end{array}\right\}$$

and H is the effective Hamiltonian, computed as the eigenvalue of the so-called cell-problem

$$-|\mathbf{p} + \mathbf{D}w(\xi,\eta)|^2 + (\overline{\mathbf{v}}(x,y) + \mathbf{v}(\xi,\eta)) \cdot (\mathbf{p} + \mathbf{D}w(\xi,\eta)) = -H(\mathbf{p}, x, y), \tag{4.2}$$

where $\mathbf{D}$ represents the gradient with respect to the variables $(\xi,\eta) = (x/\epsilon^\alpha, y/\epsilon^\alpha)$. The theory predicts that there exists a unique value for the eigenvalue H such that Equ. (4.2) can be solved for w, the associated eigenfunction, which must be bi-periodic with zero mean on one periodic box.

Under some additional technical assumptions, the asymptotic flame front obeys a Huygens principle and evolves according to the following geometric equation:

$$G_t = F(\nabla G), \tag{4.3}$$

where the zero-level set $\{G(x,t) = 0\}$ represents the flame interface, so that on any point on the interface, the gradient ∇G is aligned with the unit normal to the front $\mathbf{n}$. F represents the overall speed of the front (advection and combustion). It must be computed according to the following minimization problem:

$$F(\mathbf{n}) = S_T(\mathbf{n}) + \overline{\mathbf{v}} \cdot \mathbf{n} = \min_{r>0} \frac{H(r\mathbf{n}) + f'(0)}{r}, \tag{4.4}$$

where H is the eigenvalue in Equ. (4.2) with $\mathbf{p} = -r\mathbf{n}$.

4.2.2 Preliminary step: generating the asymptotic flamelet library

As for the nonpremixed flame, an efficient procedure to implement the asymptotic model is to pre-compute once and for all the solution to the flamelet model. In the case of the premixed flame, this amounts to pre-computing the speed S_T via the cell-problem and minimization steps, for all expected values of the two parameters $\bar{\mathbf{v}}$ and θ. Unlike the case of nonpremixed flames, this is a very challenging, computationally intensive step. The minimization in Equ. (4.4) is a standard one-dimensional minimization problem for which efficient numerical procedures can be found. On the other hand, the cell problem in Equ. (4.2) is far less standard. Except for one-dimensional flows (simple shear layers) and a particular family of rotating flows [3, 4, 7], analytical solutions to the cell-problem are unknown and the problem must be solved numerically. The challenge is that it is both nonlinear (in particular, the eigenfunction is not expected to be smooth) and it is an eigenvalue problem. In [6], a new numerical scheme was introduced and validated specifically for the purpose of computing the flamelet library in an efficient and robust fashion. The method consists in differentiating Equ. (4.2) with respect to the small scale variables (ξ, η) in order to eliminate the eigenvalue from the computation. After differentiation, one gets the following system of conservation laws for the eigenfunction gradient $\mathbf{D}w$:

$$\mathbf{D}\{-|\mathbf{p}+\mathbf{D}w(\xi,\eta)|^2 + (\overline{\mathbf{v}}(x,y)+\mathbf{v}(\xi,\eta))\cdot(\mathbf{p}+\mathbf{D}w(\xi,\eta))\} = 0. \tag{4.5}$$

Differentiation has effectively removed the eigenvalue problem, which can be recovered a posteriori by integration of the cell-problem over a small periodic box:

$$H(\mathbf{p}, x, y) = \langle |\mathbf{D}w|^2\rangle + |\mathbf{p}|^2 - \overline{\mathbf{v}}(x,y)\cdot\mathbf{p}.$$

The system of equations (4.5) is still a complex nonlinear system of equations; the solution for $\mathbf{D}w$ is obtained as the steady state solution of the following system of nonlinear conservation laws:

$$\begin{aligned}\frac{\partial u_1}{\partial s} + \frac{\partial}{\partial \xi}\left\{-|\mathbf{p}+\mathbf{u}(\xi,\eta)|^2 + (\overline{\mathbf{v}}(x,y)+\mathbf{v}(\xi,\eta))\cdot(\mathbf{p}+\mathbf{u}(\xi,\eta))\right\} = 0\\ \frac{\partial u_2}{\partial s} + \frac{\partial}{\partial \eta}\left\{-|\mathbf{p}+\mathbf{u}(\xi,\eta)|^2 + (\overline{\mathbf{v}}(x,y)+\mathbf{v}(\xi,\eta))\cdot(\mathbf{p}+\mathbf{u}(\xi,\eta))\right\} = 0\end{aligned} \tag{4.6}$$

with $\mathbf{u} = (u_1, u_2)$ the new unknown representing $\mathbf{D}w$, and with s a pseudo-time marching variable. Because the system of equations (4.6) is not strictly hyperbolic, convergence to steady state cannot be guaranteed when solving it with classical methods. The method in [6] circumvents that difficulty by enforcing explicitly that the discrete vector solution to (4.6) is a discrete gradient—see [6] for implementation details and validation.

4.2.3 Implementation at large scales via a level-set formulation

A computationally efficient strategy to implement the theoretical predictions relies on the level-set representation for the asymptotic front as a numerical procedure to represent a propagating front obeying a Huygens principle. The front interface is embedded in a function G defined in all space, with its zero-level $\{G = 0\}$ representing the asymptotic flame front and G positive on the burnt side (for example) and negative on the unburnt side (see Fig. 11). Equ. (4.1) can be used in the entire domain with a non-constant $\overline{\mathbf{v}}$:

$$G_t(x,y,t) + \overline{\mathbf{v}}(x,y,t)\cdot\nabla G(x,y,t) = S_T(x,y,t)|\nabla G|, \tag{4.7}$$

the effective turbulent speed $S_T(x,y,t) = S_T(\nabla G(x,y,t), \overline{\mathbf{v}}(x,y)) = S_T(\theta(x,y,t), \overline{\mathbf{v}}(x,y))$ being obtained from the pre-computed library at each grid point and each time-step according to Equ. (4.4).

The numerical method to solve Equ. (4.7) is a standard level-set method [11]. It is based on upwind finite differencing that guarantees the convergence to the correct limit solution in the viscosity sense. The method is very efficient at capturing the cusps developed by the flame front; these cusps correspond to singularities in the first order derivative of G and are typical of viscosity solutions of Hamilton-Jacobi equations.

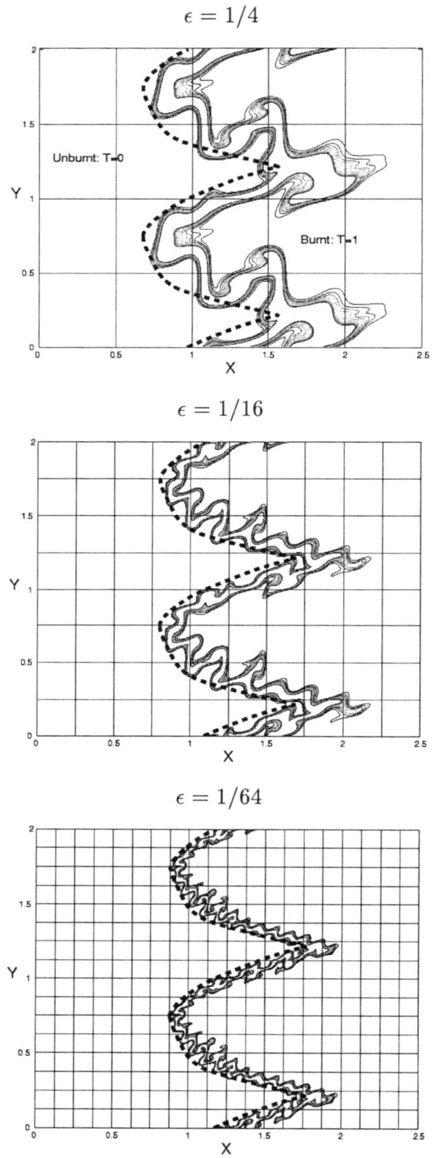

Figure 12: Premixed flame fronts from direct simulations with $\epsilon = 1/4, 1/16, 1/64$ along with the asymptotic large scale front (dashed line). The small squares within each plot indicate the small periodic boxes of size ϵ^α.

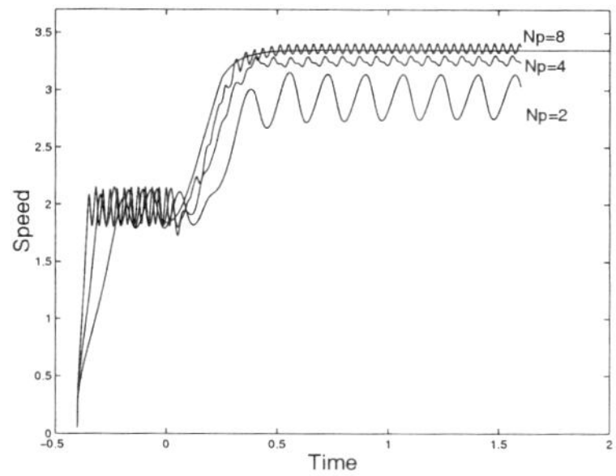

Figure 13: Effective burning speed from direct simulations for $\epsilon = 1/4$, $\epsilon = 1/16$, $\epsilon = 1/32$ corresponding to the number of small periodic boxes $N_p = 2, 4, 8$. The smooth solid line is the asymptotic prediction.

4.3 Validation of the asymptotic model

The asymptotic large scale strategy is validated by comparison with the results obtained via direct numerical simulations of Equ. 1.1 for the values of $\epsilon = 1/4, 1/16, 1/64$. In Fig. 12, the resolved flames from the direct simulations are shown by plotting the reaction rate contours. Also shown is the prediction for the asymptotic large scale front (dashed line). The convergence of the direct simulation results to the asymptotic curve is excellent. Examining more closely the plot for the intermediate test-case with $\epsilon = 1/16$, one sees that the asymptotic curve corresponds precisely to the cold side envelope of the wrinkled front. This is the proper generalization of the well-known fact that the propagation of a KPP-premixed flame depends essentially on the behavior of the solution on the cold side of the flame: the generalized cold side of a wrinkled flame is its cold side envelope as far as predicting its propagation speed.

Very good convergence of the effective propagation speed is also observed in Fig. 13. The direct numerical simulation predictions show small scale oscillations corresponding to the unsteady crossing by the front of each small periodic box, those fluctuations are off course averaged out in the asymptotic limit.

As for nonpremixed flames, the idealized test-case can also be used to validate more specific practical implementations. For example, a discussion that compares the rigorous theory with a popular ad-hoc model known as the G-equation can be found in [4]. Another example of application can be found in [5], where one examines the possibility of replacing the detailed flamelet library by an effective parameterization.

5 Conclusion

The use of rigorous asymptotic results to formulate closure for turbulent flames has been illustrated for both premixed and nonpremixed combustion via an idealized model. Asymptotics give a formal procedure to exploit the scale separation between the large scale flow and the smaller turbulent and reactive scales. Convergence of the strategies has been validated by comparing the predictions with direct numerical simulations. The asymptotic strategies rely on pre-computed solutions, "turbulent flamelet libraries". In the case of nonpremixed flames, the flamelet library is trivial to compute. In the case of premixed flames, it requires sophisticated numerical tools for an auxiliary nonlinear eigenvalue cell-problem. The idealized model allows one to approach the challenging turbulent reactive closure problem with a combination of rigorous asymptotic tools and very reliable and affordable numerical simulations, so that validation of modeling strategies can be performed very systematically.

References

[1] A. Bourlioux and A. J. Madja, An elementary model for the validation of flamelet approximations in non-premixed turbulent combustion, *Combustion Theory Modelling* **4** (2) (2000), 189–210.

[2] A. Bourlioux and A. J. Madja, Elementary models with probability distribution function intermittency for passive scalars with a mean gradient, *Phys. Fluids* **14** (2) (2002), 881–897.

[3] P. Embid, A. J. Majda, and P. E. Souganidis, Effective geometric front dynamics for premixed turbulent combustion with separated velocity scales, *Combustion Sci. Technol.* **103** (1–6) (1994), 85–115.

[4] P. Embid, A. J. Majda, and P. E. Souganidis, Comparison of turbulent flame speeds from complete averaging and the G-equation, *Phys. Fluids* **7** (8) (1995), 2052–2060.

[5] B. Khouider, A. Bourlioux, and A. J. Majda, Parametrizing the burning speed enhancement by small-scale periodic flows: I. Unsteady shears, flame residence time and bending, *Combustion Theory Modelling* **5** (3) (2001), 295–318.

[6] B. Khouider and A. Bourlioux, Computing the effective Hamiltonian in the Majda-Souganidis model of turbulent premixed flames, to appear in *SIAM J. Numer. Anal.*

[7] A. J. Majda and P. Souganidis, Large scale front dynamics for turbulent reaction-diffusion equations with separated velocity scales, *Nonlinearity* **7** (1) (1994), 1–30.

[8] A. J. Majda and P. Souganidis, The effect of turbulence on mixing in prototype reaction-diffusion systems, *Comm. Pure Appl. Math.* **53** (10) (2000), 1284–1304.

[9] A. J. Majda and P. R. Kramer, Simplified models for turbulent diffusion: Theory, numerical modelling, and physical phenomena, *Phys. Rep.* **314** (4–5) (1999), 238–574.

[10] N. Peters, Laminar diffusion flamelet models in non-premixed models in turbulent combustion, *Progr. Energy Combust. Sci.* **10** (1984), 319.

[11] J. Sethian, *Level Set Methods and Fast Marching Methods, Evolving Interfaces in Computational Geometry, Fluid Mechanics, Computer Vision, and Materials Science*, Cambridge University Press, 1999.

[12] F. Williams, Theory of combustion in laminar flows, *Annual Rev. Fluid Mech.* **3** (1971), 171–188.

Multigrid methods: from geometrical to algebraic versions

Gundolf HAASE
and
Ulrich LANGER
Institute of Computational Mathematics
Johannes Kepler University of Linz
A-4040 Linz
Austria

Notes taken by Stefan REITZINGER

Abstract

Nowadays the multigrid technique is one of the most efficient methods for solving a large class of problems including elliptic boundary value problems for partial differential equations (PDEs) or systems of PDEs. Our lecture notes start with a motivation for the multigrid idea and a brief review of the multigrid history, its situation and the perspectives. The next part of our paper is devoted to the algorithmical aspects of the construction of multigrid methods including also algebraic approaches and parallelization techniques. Algebraic multigrid methods are now quite popular in practical applications because they can be included in conventional finite element packages without changing the data structure of the package. In the theoretical part of our lecture, we present some general approaches to the convergence and efficiency analysis of multigrid methods. Finally, we discuss implementation issues and present some numerical results obtained from the application of our algebraic multigrid package PEBBLES to large-scale problems in medicine and engineering.

1 Motivation of the multigrid idea

Let us consider linear systems

$$K_h \underline{u}_h = \underline{f}_h \tag{1.1}$$

of algebraic equations arising from a Finite Element (FE), Finite Difference (FD), or a Finite Volume (FV) discretization of some formally self-adjoint, second order, scalar, elliptic partial differential equation in a bounded computational domain $\Omega \subset \mathbb{R}^d$ $(d = 1, 2, 3)$, where $K_h \in \mathbb{R}^{N_h \times N_h}$, $\underline{f}_h \in \mathbb{R}^{N_h}$, and $\underline{u}_h \in \mathbb{R}^{N_h}$ denote the system matrix (stiffness matrix), the right-hand-side vector (load vector), and the solution vector of the unknown nodal parameters, respectively. The number N_h of unknowns is related to the usual discretization parameter h by the relation $N_h = O(h^{-d})$. The stiffness matrix K_h is supposed to be sparse, symmetric

A. Bourlioux and M.J. Gander (eds.), Modern Methods in Scientific Computing and Applications, 103–153.

and positive definite. The spectral condition number $\kappa(K_h)$ of the stiffness matrix behaves typically like $O(h^{-2})$ as the discretization parameter tends to zero. Practically this means that the stiffness matrix has a large condition number on a fine grid.

The structure and the bad conditioning of the stiffness matrix K_h considerably affect the efficiency of classical direct and iterative solvers. In the case of direct solvers (e.g. Gaussian elimination, Cholesky factorization), the memory demand M and number Q of arithmetical operations asymptotically grow as $O(h^{-2d+1})$ and $O(h^{-3d+1})$, respectively. The bad conditioning of K_h results in the loss of about $\lg \kappa(K_h)$ valid digits independently of d. In the one-dimensional case ($d = 1$), the direct methods are asymptotically optimal with respect to the growth of M and Q. The superior proportional growth of M and Q in 2D ($d = 2$) and 3D ($d = 3$) is due to the so-called "fill-in" during the elimination, or factorization step. That is the reason why the direct methods are in general only efficient for small systems related to coarser grids. In contrast to the direct methods, the memory demand of the classical iterative methods such as the Jacobi and the Gauss-Seidel methods is proportional to the number of unknowns N_h. This fact is due to the sparsity structure of the stiffness matrix K_h. The number of nonzero elements per row, or column, of K_h is basically constant for a reasonable mesh refinement procedure. Hence, the total Number of Nonzero Elements (NNE) in the stiffness matrix K_h is proportional to the number of unknowns $N_h = O(h^{-d})$. Therefore, the matrix-by-vector multiplication of the type $K_h * \underline{u}_h$ that is the basic operation in every iterative process can be carried out with $O(h^{-d})$ arithmetical operations. So, the memory demand and the costs per iteration step are asymptotically optimal. However, the classical iterative methods suffer from the bad conditioning of the stiffness matrix that leads to a dramatical growth in the number of iterations as h tends to zero. More precisely, the number of iterations $I(\varepsilon)$ that is necessary to reduce the initial iteration error by the factor $\varepsilon \in (0,1)$ behaves like $O(\kappa(K_h)\ln(\varepsilon^{-1})) = O(h^{-2}\ln(\varepsilon^{-1}))$. Summing up, we observe that the classical iterative methods require asymptotically $Q(\varepsilon) = O(h^{-d-2}\ln(\varepsilon^{-1}))$ arithmetical operations in order to reduce the initial iteration error by the factor ε [3, 59].

Let us have a closer look at the convergence properties of the classical iterative methods for solving finite element equations. For the simplicity of the analysis, we consider the simplest finite (piecewise linear) element discretization of the simplest second order boundary value problem: Find some function $u(\cdot)$ defined on the interval $[0,1]$ such that the differential equation

$$-u''(x) = f(x), \quad x \in (0,1) \tag{1.2}$$

and the homogeneous Dirichlet boundary condition

$$u(0) = u(1) = 0 \tag{1.3}$$

are fulfilled, where the right-hand side f in (1.2) is a given function on $(0,1)$. The boundary value problem (1.2)–(1.3) is obviously the one-dimensional analog of the well-known Dirichlet problem for the Poisson equation that plays an important role in many of practical applications. The starting point for the finite element discretization of the boundary value problem (1.2)–(1.3) is its variational formulation. Multiplying the differential equation (1.2) by some test function v that vanishes at the Dirichlet boundary $x = 0$ and $x = 1$, integrating over our computational domain $\Omega = (0,1)$, and finally integrating by parts, we arrive at the variational formulation: Find some function $u \in V_0 = H_0^1(0,1)$ such that

$$\int_0^1 u'(x)v'(x)\,dx = \int_0^1 f(x)v(x)\,dx \quad \forall v \in V_0, \tag{1.4}$$

where the Sobolev space $H_0^1(0,1)$ contains all quadratically integrable function (L_2-functions) possessing quadratically integrable weak derivatives and vanishing at $x = 0$ and $x = 1$. Now the finite element approximation u_h to our solution u is searched in the form of the piecewise linear, continuous function

$$u_h(x) = \sum_{j=1}^{n-1} u_j p_j(x) \tag{1.5}$$

with unknown coefficients (nodal values) u_j and given ansatz functions $p_j(x)$ corresponding to the nodes $x_j = jh$, where $j = 1, 2, \ldots, N_h$, with $N_h = n-1$ and $h = 1/n$ (see Fig. 1). Inserting the ansatz (1.5) into the variational formulation (1.4) and testing with all basis

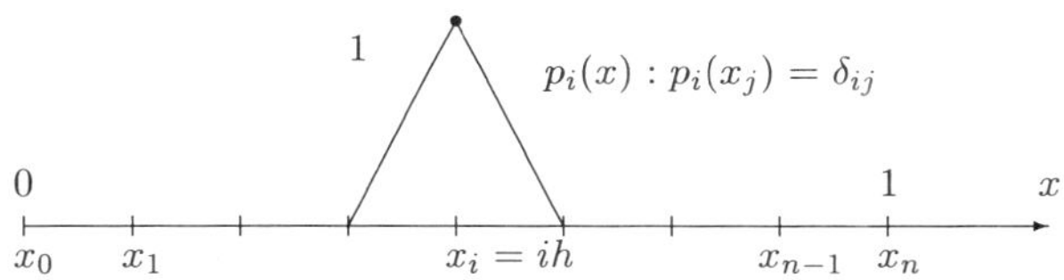

Figure 1: Piecewise linear ansatz functions (linear elements).

functions $v = p_i$ ($i = \overline{1, n-1}$), we obtain the finite element equations

$$\sum_{j=1}^{n-1} u_j \int_0^1 p_j'(x)p_i'(x)\,dx = \int_0^1 f(x)p_i(x)\,dx, \quad \forall i = \overline{1, n-1} \tag{1.6}$$

that are equivalent to a system of the form (1.1) with the tridiagonal stiffness matrix

$$K_h = K_h^{FE} = \frac{1}{h}\begin{bmatrix} 2 & -1 & & & \\ -1 & 2 & -1 & & \\ & \ddots & \ddots & \ddots & \\ & & -1 & 2 & -1 \\ & & & -1 & 2 \end{bmatrix}_{N_h \times N_h}, \tag{1.7}$$

the vector $\underline{u}_h = [u_j]_{j=\overline{1,n-1}} \in \mathbb{R}^{N_h}$ of the unknown nodal values and the load vector $\underline{f}_h = [f_i]_{i=\overline{1,n-1}} \in \mathbb{R}^{N_h}$, where

$$f_i = \int_0^1 f(x)p_i(x)\,dx, \qquad i = \overline{1, N_h}, \qquad N_h = n-1. \tag{1.8}$$

We note that the standard finite difference discretization, where the second-order derivative is replaced by the second difference quotient, results in a similar system with the system matrix $K_h^{FD} = h^{-1}K_h^{FE}$ and the coefficients $f_i = f(x_i)$ of the right-hand side $\underline{f}_h$.

Now we solve system (1.1) with the system matrix (1.7) and the right-hand side (1.8) by means of the ω-Jacobi method

$$D\frac{\underline{u}^{k+1}-\underline{u}^k}{\omega}+K\underline{u}^k=\underline{f},\quad k=0,1,\dots, \tag{1.9}$$

with $D=\operatorname{diag}(K)=[2/h\ \delta_{ij}]_{i,j=\overline{1,n-1}}$ and the given initial guess $\underline{u}^0\in\mathbb{R}^{N=n-1}$. Here and in the following we omit the discretization subscript h in the matrix and vector notations. For $\omega=1$, the ω-Jacobi method coincides with the classical Jacobi iteration. Later on we will use the relaxation parameter ω for controlling the smoothing properties of the ω-Jacobi iteration. The iteration error

$$\underline{z}^k=\underline{u}-\underline{u}^k \tag{1.10}$$

obviously satisfies the error iteration scheme

$$\underline{z}^{k+1}=\left(I-\omega\frac{h}{2}K\right)\underline{z}^k=S(\omega)\underline{z}^k=S^{k+1}(\omega)\underline{z}^0 \tag{1.11}$$

with the iteration operator (error transition operator) $S(\omega)=I-\omega D^{-1}K$. In order to make a Fourier analysis of the propagation of the iteration error, we need the eigenvalues and the eigenvectors of the stiffness matrix K. It is easy to verify that the eigenvalues λ_l and the eigenvectors $\underline{\varphi}_l$ of the stiffness matrix K are given by the formulas

$$\lambda_l=\frac{4}{h}\sin^2\left(\frac{l\pi}{2n}\right)\quad\text{and}\quad\underline{\varphi}_l=\left[\sqrt{2}\sin\left(l\pi\frac{i}{n}\right)\right]_{i=\overline{1,n-1}}, \tag{1.12}$$

respectively, with $l=1,2,\dots,n-1$. The scaling of the eigenvectors $\{\underline{\varphi}_l\}$ is chosen in such a way that they are orthonormal with respect to the discrete L_2-scalar product, i.e.

$$(\underline{\varphi}_l,\underline{\varphi}_j)_h=\sum_{i=1}^{n-1}h\sqrt{2}\sin\left(l\pi\frac{i}{n}\right)\sqrt{2}\sin\left(j\pi\frac{i}{n}\right)=\delta_{lj}. \tag{1.13}$$

The eigenvalues λ_l obviously satisfy the inequalities

$$8h<\lambda_1=\frac{4}{h}\sin^2\frac{\pi}{2n}<\lambda_2<\dots<\lambda_{n-1}=\frac{4}{h}\sin^2\frac{n-1}{n}\frac{\pi}{2}<\frac{4}{h}, \tag{1.14}$$

from which we obtain the spectral condition number estimate

$$\kappa(K):=\frac{\lambda_{n-1}}{\lambda_1}=\frac{\sin^2\frac{n-1}{n}\frac{\pi}{2}}{\sin^2\frac{\pi}{2n}}\le\frac{1}{2}h^{-2}, \tag{1.15}$$

i.e. the spectral condition number behaves like $O(h^{-2})$, as mentioned above. Inserting the Fourier decomposition of the initial iteration error

$$\underline{z}^0=\sum_{l=1}^{n-1}\alpha_l\underline{\varphi}_l \tag{1.16}$$

into the error iteration scheme (1.11), we obtain the Fourier decomposition

$$\underline{z}^{k+1} = \left(I - \omega\frac{h}{2}K\right)^{k+1} \underline{z}^0 = \sum_{l=1}^{n-1} \alpha_l \left(1 - \omega\frac{h}{2}\lambda_l\right)^{k+1} \underline{\varphi}_l \tag{1.17}$$

of the iteration error after $k+1$ iterations, where the coefficients

$$\alpha_l = (\underline{z}^0, \underline{\varphi}_l)_h = \sum_{i=1}^{n-1} h z_i^0 \underline{\varphi}_l(i) = \sqrt{2} \sum_{i=1}^{n-1} h z_i^0 \sin l\pi \frac{i}{h} \tag{1.18}$$

denote the Fourier coefficients of the initial iteration error $\underline{z}^0$. Using Parseval's equality

$$\|\underline{z}^0\|_h^2 = (\underline{z}^0, \underline{z}^0)_h = \sum_{l=1}^{n-1} \alpha_l^2, \tag{1.19}$$

we finally arrive at the iteration error estimate

$$\begin{aligned} \|\underline{z}^{k+1}\|_h^2 = (\underline{z}^{k+1}, \underline{z}^{k+1})_h &= \sum_{l=1}^{n-1} \alpha_l^2 \left(1 - \omega\frac{h}{2}\lambda_l\right)^{2(k+1)} \\ &\le \left[\max_{l=\overline{1,n-1}} \left|1 - \omega\frac{h}{2}\lambda_l\right|\right]^{2(k+1)} \sum_{l=1}^{n-1} \alpha_l^2 \\ &= \left[\max_{l=\overline{1,n-1}} \left|1 - \omega\frac{h}{2}\lambda_l\right|\right]^{2(k+1)} \|\underline{z}^0\|_h^2. \end{aligned} \tag{1.20}$$

Therefore, the spectral radius

$$\rho(h,\omega) = \rho(S) = \max\left\{\left|1 - \omega\frac{h}{2}\lambda_l\right| : l = \overline{1, n-1}\right\} \tag{1.21}$$

of the iteration operator S describes the error damping per iteration. The error damping factor $\rho(h,\omega)$ is often called *convergence factor*, or *convergence rate*. The analysis of the convergence factor $\rho(h,\omega)$, which is illustrated in Fig. 2, shows us that the optimal (minimal) rate

$$\rho_{\text{opt}} = \rho(h,1) = 1 - 2\sin^2\frac{\pi}{2n} = \cos \pi h = 1 - O(h^2) \approx 1 - \frac{\pi^2}{2}h^2 \tag{1.22}$$

is attained at $\omega = \omega_{\text{opt}} := 1$, i.e. for the classical Jacobi method. The ω-Jacobi method (1.9) converges if and only if $\omega \in (0, 4/(h\lambda_{n-1}))$, where the upper bound $4/(h\lambda_{n-1})$ tends to 1. From Fig. 2 we can already observe an important property of the ω-Jacobi method, namely a fast reduction of the high frequencies can be obtained via underrelaxation, e.g. for $\omega \approx 1/2$. The analysis of the reduction factors

$$\left(1 - \omega\frac{h}{2}\lambda_l\right) = 1 - 2\omega\sin^2\frac{l\pi}{2n} = 1 - \omega(1 - \cos(l\pi h)) \tag{1.23}$$

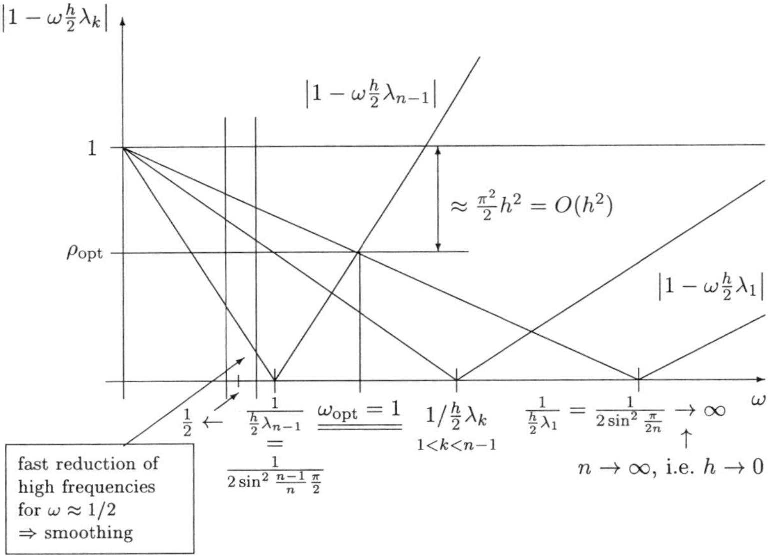

Figure 2: Reduction factors.

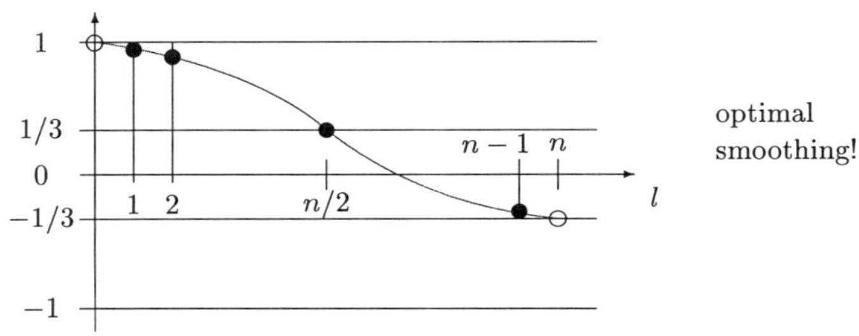

Figure 3: $\omega = \frac{2}{3}$: $1 - \frac{2}{3}\frac{h}{2}\lambda_l = 1 - \frac{4}{3}\sin^2\frac{l\pi}{2n} = \frac{1}{3} + \frac{2}{3}\cos\frac{l\pi}{n}$.

for different values of ω in dependence of $l = \overline{1, n-1}$ (frequencies) shows that the best smoothing is attained at $\omega \approx 2/3$ (see Fig. 3). Indeed, let us define the *smoothing factor*

$$\begin{aligned}\mu(h,\omega) = \mu(S) &= \max\left\{\left|1 - \omega\frac{h}{2}\lambda_l\right| : \frac{n}{2} \le l \le n-1\right\} \\ &= \max\{|1-\omega|,\ |1-\omega(1+\cos\pi h)|\}\end{aligned} \tag{1.24}$$

and the *asymptotic smoothing factor*

$$\mu^*(\omega) = \sup_{h\le 1/4} \mu(h,\omega) = \max\{|1-\omega|,\ |1-2\omega|\} \tag{1.25}$$

as measures for the reduction of the high frequencies in the error belonging to the indices $l = n/2, \ldots, n-1$. Here and below we assume for simplicity that n is even and greater or equal than 4. The optimal (minimal) values

$$\mu(h, \omega_h^*) = \min_{0 \le \omega \le 1} \mu(h, \omega) = \frac{\cos \pi h}{2 + \cos \pi h} \tag{1.26}$$

and

$$\mu^*(\omega^*) = \min_{0 \le \omega \le 1} \mu^*(\omega) = \frac{1}{3} \tag{1.27}$$

are attained at $\omega_h^* = 2/(2 + \cos \pi h)$ and $\omega^* = 2/3$, respectively. Therefore, the underrelaxed ω-Jacobi method is able to reduce the high frequency part in the iteration error dramatically after only a few iterations whereas there is almost no reduction of the very low frequencies in the iteration error. For instance, in the case of $h = 10^{-3}$ and $\omega = 2/3$, the high frequency part is already reduced by the factor $0.7 \cdot 10^{-4}$ after 10 iterations, whereas the overall error (i.e. the low frequency part) is only reduced by the factor 0.999967! However, as illustrated in the left-hand part of Fig. 4, the low frequencies ($l = 1, \ldots, (n/2) - 1$) in the iteration error can be well approximated on the coarser grid with the grid size $2h$, whereas the high frequencies ($l = n/2, \ldots, n-1$) in the iteration error cannot be represented on the coarser grid at all (see the right-hand part of Fig. 4).

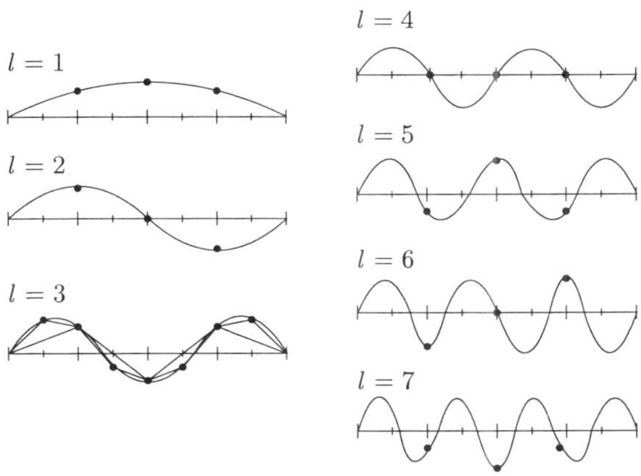

Figure 4: Error reduction via coarse grid correction ($n = 8$).

The combination of the *smoothing property* of the underrelaxed ω-Jacobi method with the *approximation property* of the coarser grid allows us to kill the high frequency part as well as the low frequency part very efficiently. The successive use of *smoothing* and *coarse grid correction* results in a twogrid iteration process called *twogrid method* (TGM) that is sketched below and that is algorithmically described in Section 3.1 more precisely:

Start: $k = 0; \underline{u}_h^0$ given initial guess, e.g. $\underline{u}_h^0 = 0$.

Step 1: Smooth, e.g. by the ω-Jacobi-method ($\omega = 2/3$):
$\underline{u}_h^{k,\nu} \longleftarrow \underline{u}_h^{k,0} := \underline{u}_h^k$ (ν smoothing steps)

Step 2: Compute the defect:
$\underline{d}_h^k = \underline{f}_h - K_h \underline{u}_h^{k,\nu} \quad [= K_h \underline{u}_h - K_h \underline{u}_h^{k,\nu} = K_h \underline{z}_h^{k,\nu}]$

Step 3: Coarse-Grid-Correction (CGC):
$\underline{d}_h^k \xrightarrow[\text{restriction}]{\text{projection}} \underline{d}_{2h}^k \equiv \underline{d}_{2h}$ | e.g. by pointwise injection

$\downarrow$

$K_{2h} \underline{w}_{2h} = \underline{d}_{2h}$

$\downarrow$

$\underline{w}_h \xleftarrow[\text{prolongation}]{\text{interpolation}} \underline{w}_{2h}$ | e.g. by linear interpolation

Step 4: CGC of the "smoothed" approximation:
$\underline{u}_h^{k+1} = \underline{u}_h^{k,\nu} + \underline{w}_h$,
$k := k + 1$, goto step 1.

In the twogrid method, the coarse grid defect system

$$K_{2h} \underline{w}_{2h} = \underline{d}_{2h} \tag{1.28}$$

is solved by some direct method. However, in many practical applications, the direct solution of the coarse grid systems (1.28) is too expensive because the coarse grid systems are simply too large. In this case, one can repeat steps 1–4 for the coarse grid systems on the grid "$2h$" γ times with the initial guess $\underline{w}_{2h}^0 = 0$, where γ is usually equal to 1 (V-cycle) or to 2 (W-cycle). Now we have to solve coarse grid systems on the grid "$4h$". Again, repeating the steps 1–4 γ times, and taking into account coarser and coarser grids up to some coarsest grid where the direct solver can be used efficiently, we end up with the *multigrid method* (MGM) that is described in Section 3.2 in detail.

2 History—situation—perspectives

We observed in Section 1 that the combination of the following algorithmic ingredients should result in efficient multigrid algorithms:

(1) smoothing of rough error parts via classical iterative methods;

(2) approximation of smooth errors on a coarser grid;

(3) recurrent call of (1) and (2) on a sequence of coarser and coarser grids;

(4) nested iteration for producing good initial guesses.

In his pioneering paper [28] from 1961, R. P. Fedorenko proposed the first twogrid method for solving the finite difference equations arising from the five-point approximation of the

Poisson equation in a rectangular domain. Already the first numerical results presented in this paper showed the great potential of this method. In his second paper [29] from 1964, he gave a first convergence analysis and proposed to combine (1) and (2) with (3) that results in a complete multigrid cycle. In 1966, N. S. Bachwalov provided the first rigorous convergence proof on the basis of the sum splitting technique (see also Section 3.1) and showed that the nested iteration technique with the multigrid method as nested iteration process allows us to produce approximate solutions that differ from the exact solution of the boundary value problem in the order of the discretization error with optimal complexity [4]. In the 70s, these techniques were generalized to variational finite difference equations and general finite element equations by G. P. Astrachanzev [2] and by V. G. Korneev [53], respectively.

Also in the 70s, A. Brandt in Israel [16, 17] and W. Hackbusch in Germany [37] reinvented the multigrid method independently of the Russian school, generalized it to new classes of problems, and developed the theory. In particular, W. Hackbusch proposed the so-called product splitting that reduces the twogrid rate estimate to the proof of the approximation property and the smoothing property [39]. At the end of the 70s and at the beginning of the 80s there was a real boom in research and in publication of research results on multigrid methods. The proceedings [39] of the first European Conference on Multigrid Methods, which was held at Cologne-Porz in 1981, reflect the state-of-the-art in the research on multigrid methods, give the first interesting applications and present a multigrid bibliography with more than 200 entries. This conference at Cologne-Porz created the series of the European Conferences on Multigrid Methods the proceedings of which give a good reflection of the research results during the last 20 years [39, 40, 41, 43, 42, 26]. The first monograph on multigrid methods was published by W. Hackbusch in 1985 [38]. In this monograph, the reader finds a complete overview of the most important results on multigrid and related methods obtained till 1985.

The theoretical convergence analysis is one thing. An efficient implementation of the multigrid method for certain classes of problems is another game. Let us first have a look at the practical convergence behavior of the basic multigrid cycles (V-cycle, generalized V-cycle, W-cycle) for solving the discrete Poisson equation that was already studied by R.P. Fedorenko in his very first paper [28]. Here we consider the Poisson equation $-\Delta u = 1$ in the unit square $\Omega := (0,1) \times (0,1)$ under homogeneous Dirichlet conditions $u = 0$ on the boundary $\partial\Omega$ of the unit square, and use the simplest finite element discretization with linear triangular elements on a uniform $n \times n$ triangular grid. This finite element discretization results in a linear system of the form (1.1) with $N_h = (n-1)^2$ finite element equations and the discretization size $h = 1/(n-1)$. In this case, the finite element stiffness matrix coincides with the well-known five-point finite difference approximation of the Laplacian up to some scaling factor h^{-1}. Table 1 shows the number $I(\varepsilon)$ of iterations that are needed in order to reduce the initial defect $\underline{d}_h^0 := \underline{f}_h - K_h \underline{u}_h^0 = \underline{f}_h$ (zero initial guess) by the factor $\varepsilon := 10^{-4}$ in the Euclidean norm, the convergence rate η and the CPU time in seconds for the basic multigrid cycles (V-cycle, generalized V-cycle[1], W-cycle). The cycles are built by the following components: ν_1 lexicographically forward Gauss-Seidel pre-smoothing sweeps, ν_2 lexicographically backward Gauss-Seidel post-smoothing sweeps, linear finite element interpolation as prolongation, and the transposed operation as restriction. In the upper part of Table 1 we study the dependency of the multigrid convergence on the levels for the case $(\nu_1, \nu_2) = (1,0)$ (only one pre-smoothing

[1] See Section 3.2 for the definition of the generalized V-cycle.

step!). The last two rows show the results for the cases $(\nu_1, \nu_2) = (1,1)$ and $(\nu_1, \nu_2) = (2,1)$ on the finest grid ($l = 12$, corresponding to 16.744.464 unknowns) that we considered. The calculations were carried out by the multigrid C code MGTEST on a PENTIUM IV PC with 800 MHz clock rate. The multigrid test code MGTEST can be downloaded.[2]

Table 1: Convergence behavior of different multigrid cycles.

		V-cycle			gener. V-cycle			W-cycle		
l	n	I	η	CPU [sec]	I	η	CPU [sec]	I	η	CPU [sec]
6	64	15	0.521	0.01	11	0.429	0.01	10	0.365	0.01
7	128	16	0.538	0.08	12	0.431	0.07	10	0.366	0.06
8	256	16	0.559	0.39	12	0.438	0.40	10	0.366	0.37
9	512	17	0.571	1.73	12	0.445	1.75	10	0.366	1.64
10	1024	18	0.582	7.44	12	0.454	5.73	10	0.367	5.52
11	2048	18	0.599	29.76	12	0.463	23.68	10	0.367	23.22
12	4096	19	0.607	130.89	13	0.465	107.66	10	0.367	105.36
12	4096	12	0.285	73.74	5	0.135	54.73	5	0.135	65.74
12	4096	6	0.210	65.08	4	0.062	54.17	4	0.062	62.98

The classical convergence analysis techniques mentioned above do not cover the practically important case of the V-cycle. Another drawback of the classical techniques consists in the assumption of "sufficiently many" smoothing steps. This contradicts the practical experience which is clearly reflected in Table 1. The V-cycle converges almost uniformly even if only 1 smoothing sweep is used per cycle! Therefore, new convergence theories that cover these interesting cases at least for some classes of problems under additional assumptions have been developed by D. Braess and W. Hackbusch [12] and others [14, 15, 74]. A successful commercial use of multigrid methods for solving hard practical problems depends not only on the level-independent convergence behavior that is typical for multigrid methods, but also on their robustness against bad parameters (e.g. geometrical parameters, large coefficient jumps, singularities, anisotropies, etc.) arising often in many practical applications. The construction of efficient and at the same time robust multigrid algorithms for problems with bad parameters is a hot topic in the current research activities (see, e.g., [65]).

The multigrid boom at the beginning of the 80s initiated many interesting research directions and many new applications of multigrid and related methods. Algebraic multigrid (AMG) methods are one of the most important issues at least from a practical point of view. In contrast to the geometrical multigrid method, its algebraic version recovers algebraically the ingredients for a complete multigrid algorithm from the fine grid data only (see Section 4). The first serious approach to AMG was made in 1982 by A. Brandt, S. McCormick and J.W. Ruge [21]. This approach was afterwards improved in several papers [20, 18, 63, 64]. There was a revival of the research activities in AMG methods and their analysis in the mid-nineties [19, 71, 9, 72, 30, 52]. Since then a lot of new ideas have been published (see, e.g., [23, 47, 44] and the review paper [68] by K. Stüben). There is an urgent need for efficient

[2] http://www.numa.uni-linz.ac.at/Staff/haase/Lectures/Montreal01/index.html

implementation of robust AMG methods that can be used in commercial codes. It is much easier to use an AMG code in an existing software package as solver than redesign this package in such a way that hierarchical data structures for implementing geometrical multigrid solvers are available. In Section 4, we give an introduction to AMG and discuss the issues connected with an efficient implementation of some AMG methods.

Multigrid methods can exploit new computer architectures, especially parallel computers with distributed memory very well (see, e.g. [27]). In principle, there are two approaches to the parallelization of multigrid methods. Firstly, multigrid methods can be used as local solvers in domain decomposition preconditioners (see e.g. [35]). Secondly, multigrid methods can be directly parallelized with slight modifications of the multigrid algorithm on the basis of data partitioning techniques [6, 31, 34, 54, 45]. In Section 5, we will discuss the direct approach to the parallelization of multigrid methods via some data partitioning technique that is borrowed from the nonoverlapping domain decomposition method. Parallel multigrid methods implemented on professional parallel computers will allow us to solve really large-scale problems. Teraflop parallel computers such as the ASCII-white can handle 10^9 complexity problems in seconds. On the other side, parallel multigrid application software will also be superior to conventional software on low cost PC clusters due to the asymptotical optimality and efficiency of parallel multigrid algorithms.

During the last 20 years the development of multigrid software has been mainly pushed by research groups working at universities or at similar research institutions. R. Bank was certainly one of the pioneers in developing general purpose finite element multigrid codes. His code PLTMG was used by many researchers [5]. Similar research codes have been also developed for different classes of applications by other groups. Here we mention only the parallel multigrid package UG that is now widely used in research as well as in commercial applications [7] and our parallel geometrical and algebraic multigrid packages FEPP [55] and PEBBELS [60].

Since the pioneering monograph [38] by W. Hackbusch, many further books have been published wherein the reader can find interesting additional information on the theory and application of multigrid methods in survey form, e.g. the books by W. L. Briggs, V. E. Henson and S. F. McCormick [24], D. Braess [10], J. H. Bramble [13], S. C. Brenner and L. R. Scott [22], S. F. McCormick [58], S. Reitzinger [60], U. Rüde [62], V. V. Shaidurov [66], S. Vandewalle [70], P. Wesseling [73], U. Trottenberg, C. Oosterlee and A. Schüller [69] with a contribution by K. Stüben [67]. The most recent information about publications on multigrid method, related conferences, multigrid software, etc., can be found on the multigrid home page `http://www.mgnet.org` on the internet.

3 Multigrid algorithms and their analysis

3.1 The Twogrid Method

At the end of Section 1, we sketched our first twogrid algorithm for solving system (1.1). Algorithm 1 now gives a complete algorithmic description of one step of a twogrid method for solving (1.1), where the subscripts $h = h_q$ and $H = h_{q-1}$ indicate the quantities belonging to the fine grid and the coarse grid, respectively (c.f. also Section 3.2). Let us emphasize here that the use of different fonts for the vectors ($\underline{\mathfrak{u}}_h, \underline{\mathfrak{w}}_h, \ldots$ and $\underline{\mathsf{f}}_h, \underline{\mathsf{d}}_h, \ldots$ for the so-called accumulated types and distributed types, respectively) and different arrows in the assignments (the arrows $\Leftarrow$ and $\leftarrow$ indicate assignments with and without communication, respectively) is due to the parallelization of Algorithm 1 and is explained in Section 5. The same is true for the remaining algorithms in this section. For the serial version of algorithms discussed in this section, this differentiation is not relevant! We note that the standard coarsening is

Algorithm 1 Twogrid Method: $\underline{\mathfrak{u}}_h \Leftarrow \mathrm{TGM}(\mathsf{K}_h, \underline{\mathfrak{u}}_h, \underline{\mathfrak{f}}_h, h)$

for $i = 1$ step 1 until ν_1 **do**
  $\underline{\mathfrak{u}}_h \Leftarrow \mathsf{G}_h^{\text{pre}}(\underline{\mathfrak{u}}_h)$ {presmoothing}
end for
$\underline{\mathsf{d}}_h \leftarrow \underline{\mathsf{f}}_h - \mathsf{K}_h \cdot \underline{\mathfrak{u}}_h$ {defect calculation}
$\underline{\mathsf{d}}_H \leftarrow I_h^H \cdot \underline{\mathsf{d}}_h$ {restriction}
$\underline{\mathfrak{w}}_H \Leftarrow K_H^{-1} \cdot \underline{\mathsf{d}}_H$ {direct coarse grid solver}
$\underline{\mathfrak{w}}_h \leftarrow I_H^h \cdot \underline{\mathfrak{w}}_H$ {prolongation}
$\underline{\mathfrak{u}}_h \leftarrow \underline{\mathfrak{u}}_h + \underline{\mathfrak{w}}_h$ {coarse grid correction}
for $i = 1$ step 1 until ν_2 **do**
  $\underline{\mathfrak{u}}_h \Leftarrow \mathsf{G}_h^{\text{post}}(\underline{\mathfrak{u}}_h)$ {postsmoothing}
end for

connected with the doubling of the discretization parameter, i.e. $h_{q-1} = 2h_q$. We assume that the smoothing procedures $G_h^{\text{pre/post}}(\cdot)$ are some kind of regular, affine-linear fixed point iteration that can be rewritten in the form

$$G_h(\underline{u}_h) := S_h \underline{u}_h + T_h \underline{f}_h, \tag{3.1}$$

where S_h and T_h denote the iteration operator and the affine-linear part, respectively. For consistency reasons, the affine-linear part must have the form $T_h = (I_h - S_h)K_h^{-1}$, where I_h again denotes the identity. Primarily, we think of the underrelaxed ω-Jacobi method, the Gauss-Seidel method, or block-versions of these relaxation methods as smoothers. The numbers ν_1 and ν_2 of pre- and post-smoothing sweeps are usually small, e.g., for standard problems (Poisson-like problems), $(\nu_1, \nu_2) = (1,0), (1,1)$, or $(2,1)$ are quite sufficient (c.f. Table 1 in Section 2). The restriction operator $I_h^H : R^{N_h} \to R^{N_H}$ maps a fine grid vector onto some coarse grid vector, whereas the prolongation operator $I_H^h : R^{N_H} \to R^{N_h}$ maps a coarse grid vector onto some fine grid vector. In the standard FE case where the prolongation operator is naturally defined by the FE interpolation and the restriction operator is simply the transposed prolongation operator, i.e. $I_h^H = (I_H^h)^T$, the coarse grid stiffness matrix K_H fulfills the so-called Galerkin relation

$$K_H = I_h^H K_h I_H^h. \tag{3.2}$$

On the other hand, the Galerkin relation (3.2) can be used to derive the coarse grid matrix K_H from the fine grid stiffness matrix K_h. By the way, the Galerkin projection (3.2) is the standard technique for deriving the coarse grid matrices in the Algebraic Multigrid Method discussed in Section 4.

It is easy to see from Algorithm 1 that the twogrid method itself can be rewritten in the form of an affine-linear fixed point iteration

$$\underline{u}_h^{k+1} = M_h^H \underline{u}_h^k + A_h^H \underline{f}_h, \quad k = 0, 1, \ldots, \tag{3.3}$$

with the twogrid iteration operator M_h^H and its affine-linear part

$$A_h^H = (I_h - M_h^H)K_h^{-1}, \tag{3.4}$$

where the superscript k again denotes the iteration index. Setting $\underline{f}_h$ formally to 0, we can directly derive the representation

$$M_h^H = (S_h^{\text{post}})^{\nu_2} C_h^H (S_h^{\text{pre}})^{\nu_1}, \tag{3.5}$$

of the twogrid iteration operator M_h^H from Algorithm 1, the representation (3.1) of the smoothing procedure and the twogrid iteration scheme (3.3), with the coarse-grid-correction operator

$$C_h^H = I_h - I_H^h K_H^{-1} I_h^H K_h. \tag{3.6}$$

The twogrid iteration operator M_h^H transforms the old iteration error $\underline{z}_h^k := \underline{u}_h - \underline{u}_h^k$ into the new iteration error $\underline{z}_h^{k+1}$, i.e.,

$$\underline{z}_h^{k+1} := \underline{u}_h - \underline{u}_h^{k+1} = M_h^H \underline{z}_h^k. \tag{3.7}$$

It is well known from the theory of iterative methods that the twogrid iteration converges if and only if the spectral radius $\rho(M_h^H)$ of the twogrid iteration operator M_h^H is strictly smaller than 1. In Section 1, we observed that the spectral radius of the ω-Jacobi method tends to 1 if h tends to 0. This behavior is typical for all classical iterative methods. In contrast to this, the spectral radius $\rho(M_h^H)$ of the twogrid iteration operator M_h^H remains well separated from 1 if h tends to 0. In practice, one is interested in convergence rate estimates with respect to some norm $\|\cdot\|$ rather than in spectral radius estimates. From (3.7), we immediately derive the norm estimate

$$\|\underline{z}_h^{k+1}\| \leq \|M_h^H\| \, \|\underline{z}_h^k\|, \tag{3.8}$$

where we assume that the operator (matrix) norm corresponds to the vector norm in which we are interested. The straightforward estimation

$$\|M_h^H\| \leq \|S_h^{\text{post}}\|^{\nu_2} \, \|C_h^H\| \, \|S_h^{\text{pre}}\|^{\nu_1} \tag{3.9}$$

of the twogrid iteration operator M_h^H fails because $\|S_h^{\text{pre/post}}\| = 1 - O(h^2)$ (c.f. ω-Jacobi smoother in Section 1) and $\|C_h^H\| \geq 1$ (exercise!). In order to obtain an h-independent bound less than 1, one has to take into account our observation from Section 1, namely that the smoothers $S_h^{\text{pre/post}}$ kill the high frequency part in the iteration error, whereas the coarse-grid-correction C_h^H reduces only the low frequency part.

Let us briefly discuss two classical proof techniques that make heavy use of this observation. It is enough to consider only the presmoothing case $\nu_2 = 0$. Indeed, if K_H satisfies the Galerkin relation (3.2), then the coarse-grid-correction C_h^H is a projection, i.e $C_h^H = C_h^H \cdot C_h^H$. Therefore, the estimation of $\|M_h^H\|$ can obviously be reduced to separate estimations of the presmoothing and the postsmoothing cases.

The *first technique* (sum splitting) was mainly developed by the Russian school [4, 2, 53, 56]. Inserting the identity I_h in form of the sum $I_h = P_h^{\text{low}} \oplus P_h^{\text{high}}$ of two projections into M_h^H between C_h^H and S_h^{pre}, we obtain the estimate

$$\|M_h^H\| \leq \|C_h^H P_h^{\text{low}}\| \, \|S_h^{\text{pre}}\|^{\nu_1} + \|C_h^H\| \, \|P_h^{\text{high}} (S_h^{\text{pre}})^{\nu_1}\|, \tag{3.10}$$

where P_h^{low} and P_h^{high} are the projections onto the subspaces spanned by the low frequency eigenvectors and the high frequency eigenvectors, respectively. For sufficiently many presmoothing steps ν_1 and under some additional regularity assumptions, the right-hand side of estimate (3.10) can be bounded by some $\sigma_* < 1$ that is independent on h. This is due to the *approximation property* that allows us to make the term $\|C_h^H P_h^{\text{low}}\|$ small and to the *smoothing property* that allows us to make the term $\|P_h^{\text{high}}(S_h^{\text{pre}})^{\nu_1}\|$ arbitrary small for sufficiently many presmoothing steps ν_1.

The *second technique* (product splitting) was proposed by W. Hackbusch [39, 38]. Now the identity I_h is inserted into M_h^H between C_h^H and S_h^{pre} in the form of the product $I_h = K_h^{-1}K_h$. If we are able to show the so-called *approximation property*

$$\|C_h^H K_h^{-1}\| \le ch^\delta \tag{3.11}$$

and the so-called *smoothing property*

$$\|K_h(S_h^{\text{pre}})^{\nu_1}\| \le \eta(\nu_1)h^{-\delta}, \tag{3.12}$$

then we immediately arrive at the rate estimate

$$\|M_h^H\| \le \|C_h^H K_h^{-1}\|\,\|K_h(S_h^{\text{pre}})^{\nu_1}\| \le c\eta(\nu_1) =: \sigma_*, \tag{3.13}$$

with positive, h-independent constants c and δ and with some non-negative function $\eta(\nu_1)$ that tends to 0 for $\nu_1 \to \infty$. Therefore, the bound $\sigma_* = c\eta(\nu_1)$ in (3.13) can again be made less than 1 for sufficiently many presmoothing steps ν_1. In order to show the approximation property (3.11), we usually need some regularity results for the boundary value problem under consideration. In [11], this analysis is extended to nonconforming FE discretization. The drawback of these classical proof techniques consists in the fact that sufficiently many smoothing steps are required for convergence. This strong requirement contradicts the practical experience where often one smoothing step gives satisfactory results (c.f. also Section 2). Therefore, new proof techniques are needed. Such techniques were developed in the 80s and 90s [12, 38, 14, 15, 13, 74].

3.2 The Multigrid Method

The direct solution of the coarse grid systems in the twogrid Algorithm 1 is too expensive in many practical applications. So, it is reasonable to apply the same technique to the coarse grid systems taking into account coarser and coarser grids (c.f. Section 1). The right-hand side of the coarse grid systems is some residual (defect). Thus, the initial guess for the iterative solution of the coarse grid systems should be 0. One step of a complete *multigrid method* (MGM), or more precisely, of an l-grid method, is described by Algorithm 2, where the subscript q indicates all quantities belonging to the grid level characterized by the discretization parameter h_q, and the level index q is running from the the coarsest grid $q = 1$ to the finest grid $q = l$. As in the twogrid case, K_q, $G_q^{\text{pre/post}}(\cdot)$, I_q^{q-1} and I_{q-1}^q denote the stiffness matrices, the smoothing operators, the restrictions and the prolongations, respectively. We refer to Sections 3.1 and 5 for explanations of the meaning of different fonts for the vectors and of the use of different arrows. We recall that this differentiation is only aligned with the parallelization of the algorithms.

Algorithm 2 Multigrid Method: $\underline{\mathsf{u}}_q \Leftarrow \text{MGM}(\mathsf{K}_q, \underline{\mathsf{u}}_q, \underline{\mathsf{f}}_q, q)$

```
if q == 1 := COARSESTLEVEL then
  u_q ⇐ SOLVE(K_q · u_q = f_q)                  {solution on the coarsest grid}
else
  for i = 1 step 1 until ν1(q) do
    u_q ⇐ G_q^pre(u_q)                           {presmoothing}
  end for
  d_q ← f_q − K_q · u_q                          {defect calculation}
  d_{q−1} ← I_q^{q−1} · d_q                      {restriction}
  w_{q−1} ← 0                                    {coarse grid initial guess}
  for j = 1 step 1 until γ(q − 1) do
    w_{q−1} ⇐ MGM(K_{q−1}, w_{q−1}, d_{q−1}, q − 1)   {recurrent call}
  end for
  w_q ← I_{q−1}^q · w_{q−1}                      {prolongation}
  u_q ← u_q + w_q                                {coarse grid correction}
  for i = 1 step 1 until ν2(q) do
    u_q ⇐ G_q^post(u_q)                          {postsmoothing}
  end for
end if
```

Of course, the procedure $\text{MGM}(\mathsf{K}_q, \underline{\mathsf{u}}_q, \underline{\mathsf{f}}_q, q)$ is applied to the fine grid system (c.f. (1.1) with $h = h_l$)

$$K_l \underline{u}_l = \underline{f}_l, \tag{3.14}$$

in whose the solution we are interested. The iteration with procedure MGM(·) will be terminated if some accuracy test is attained. The algorithmic description of the multigrid solution of (3.14) is given in Algorithm 3.

Algorithm 3 Multigrid solver for the fine grid system $\mathsf{K}_l \underline{\mathsf{u}}_l = \underline{\mathsf{f}}_l$.

```
u_l ← u_l^0                                      {define the initial guess}
while ||f_l − K_l u_l|| > ε||f_l − K_l u_l^0|| do
  u_l ⇐ MGM(K_l, u_l, f_l, l)                    {multigrid iteration}
end while
```

The integers $\gamma(q)$ $(q = 2, \ldots, l-1)$ control the multigrid cycling regime. If $\gamma(q) = 1$ for all $q = 2, \ldots, l-1$, then the cycling regime is called *V-cycle*. In the case that $\gamma(q) = 2$ for all $q = 2, \ldots, l-1$, we call the cycling regime *W-cycle*. In Fig. 5 we illustrate these cycles by so-called pictograms that explain also the names V-cycle and W-cycle. We speak about an *alternating cycle* if $\gamma(q)$ changes between 1 and 2 from one level to the next. If $\gamma(q) = 2$ on coarser grid and 1 on finer grids, then such a cycling regime is called *composed cycle*. Usually the numbers $\nu_1(q)$ and $\nu_2(q)$ of pre- and postsmoothing steps are small and constant for all levels $q = 2, \ldots, l$ (see Section 3.1). However, it may be reasonable to increase the numbers of pre- and postsmoothing steps on the coarser grids. If the numbers of pre- and postsmoothing steps are increased from one grid to the next coarser grid, e.g. $\nu_1(q) = 2\nu_1(q+1)$ and $\nu_2(q) = 2\nu_2(q+1)$ for all $q = 2, \ldots, l-1$, in the V-cycle, then such a cycle is called a

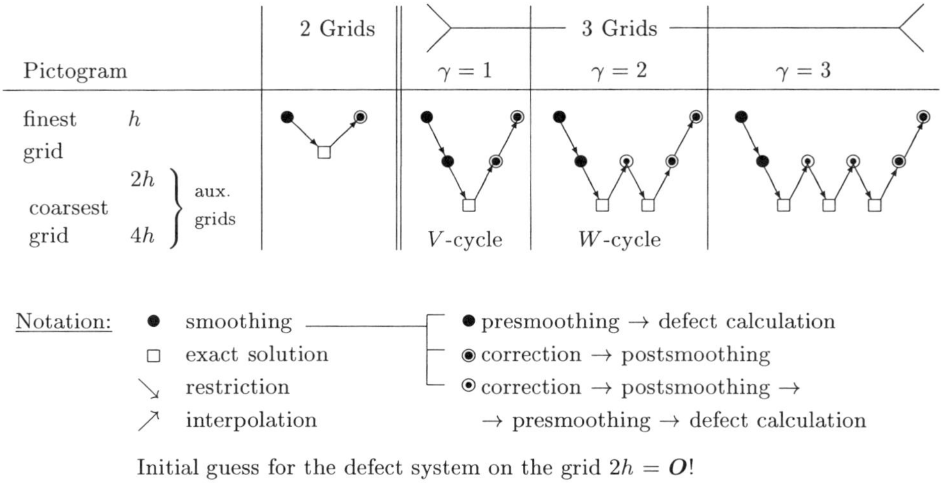

Figure 5: Pictograms.

generalized V-cycle. The system matrices K_q on the coarser grids $q = 1, \ldots, l-1$ can be also defined by Galerkin projection (3.2).

At a first glance it seems that the multigrid algorithm requires much more memory M than the corresponding single grid method. But this is not true if the right coarsening strategy is used. Let us assume that the standard coarsening is used where the discretization parameter is doubled from one grid to the next coarser grid, i.e. $h_q = 2h_{q+1}$ for all $q = 2, \ldots, l-1$. In this case, we can assume without loss of generality that the numbers of unknowns N_q fulfill the relations

$$N_q \leq N_{q+1}/\tau \quad \forall q = 1, \ldots, l-1, \tag{3.15}$$

with the enlargement factor $\tau = 2^d$. Taking into account the sparsity of the stiffness matrices, we observe from Algorithm 2 that the memory M_q needed on level q is only proportional to the number N_q of unknowns on level q, i.e. there is a q-independent positive constant c such that

$$M_q \leq cN_q \quad \forall q = 1, \ldots, l. \tag{3.16}$$

Inequalities (3.15) and (3.16) immediately yield the estimate

$$M \leq \sum_{q=1}^{l} cN_q \leq cN_l \sum_{q=1}^{l} \left(\frac{1}{2^d}\right)^{q-1} = c\frac{1-(1/2^d)^l}{1-1/2^d} N_l \leq c\frac{2^d}{2^d-1} N_l \tag{3.17}$$

of the total memory M required by the multigrid method for solving (3.14) that gives the final estimates

$$M \leq \begin{cases} c\frac{4}{3}N_l, & \text{for } d=2 \\ c\frac{8}{7}N_l, & \text{for } d=3 \end{cases} \tag{3.18}$$

for the 2D and 3D cases, respectively. Therefore, the multigrid method does not need much more memory than the corresponding single grid method.

In analogy to the twogrid method (3.3), the multigrid method described by Algorithm 3 can also be rewritten in the form of an affine-linear fixed point iteration

$$\underline{u}_l^{k+1} = M_l \underline{u}_l^k + (I_l - M_l) K_l^{-1} \underline{f}_l, \quad k = 0, 1, \ldots, \tag{3.19}$$

with the *multigrid iteration operator* M_l and the iteration index k, provided that the smoothing procedures $G_q^{\text{pre/post}}(\cdot)$ itself are affine-linear fixed point iterations of the form

$$G_q(\underline{u}_q) := S_q \underline{u}_q + (I_q - S_q) K_q^{-1} \underline{f}_q, \tag{3.20}$$

where S_q denotes the iteration operator of the smoothing procedure and q runs from 2 to l. Replacing the exact solution $\underline{w}_{l-1} = K_{l-1}^{-1} \underline{d}_{l-1}$ of the coarse grid system by the approximate solution

$$\underline{w}_{l-1}^{\gamma(l-1)} = (I_{l-1} - M_{l-1}) K_{l-1}^{-1} \underline{d}_{l-1} \tag{3.21}$$

obtained by $\gamma(l-1)$ $(l-1)$-grid cycles with the zero initial guess $\underline{w}_{l-1}^0 = 0$, we arrive at the recurrent definition of the multigrid iteration operator M_l given in the following lemma.

3.1 Lemma *The multigrid iteration operator M_l is defined by the recurrence relations*

$$M_q = (S_q^{\text{post}})^{\nu_2(q)} (I_q - I_{q-1}^q (I_{q-1} - M_{q-1}^{\gamma(q-1)}) K_{q-1}^{-1} I_q^{q-1} K_q)(S_q^{\text{pre}})^{\nu_1(q)}, \tag{3.22}$$

where $q = 3, 4, \ldots, l$. If the systems on the coarsest grid with the level index $q = 1$ are solved by a direct method (which is assumed throughout this paper), then M_2 is nothing else than the twogrid iteration operator M_2^1 defined by relations (3.5)–(3.6) with $h = h_2$ and $H = h_1$.

A closer look at (3.22) shows that M_q can be represented in the form

$$M_q = M_q^{q-1} + A_{q-1}^q M_{q-1}^{\gamma(q-1)} B_q^{q-1} \tag{3.23}$$

of a perturbed twogrid iteration operator M_q^{q-1}, where

$$A_{q-1}^q = (S_q^{\text{post}})^{\nu_2(q)} I_{q-1}^q \quad \text{and} \quad B_q^{q-1} = K_{q-1}^{-1} I_q^{q-1} K_q (S_q^{\text{pre}})^{\nu_1(q)}. \tag{3.24}$$

The following lemma provides recurrence relations for the rates that are the basis for uniform rate estimates under appropriate cycling regimes.

3.2 Lemma *Let us assume that*

$$\|M_q^{q-1}\| \le \sigma_* \quad \text{and} \quad \|A_{q-1}^q\| \cdot \|B_q^{q-1}\| \le c \tag{3.25}$$

for all $q = 2, 3, \ldots, l$ $(l = 3, 4, \ldots)$, with level-independent positive constants $\sigma_* < 1$ (twogrid rate, see Section 3.1) and c. Then

$$\|M_l\| \le \eta_l, \tag{3.26}$$

where η_l is recurrently defined by the relation

$$\eta_q = \sigma_* + c(\eta_{q-1})^{\gamma(q-1)}, \quad q = 2, 3, \ldots, l \quad (l = 3, 4, \ldots), \tag{3.27}$$

with $\eta_2 = \sigma_*$.

The proof of Lemma 3.2 follows immediately from the assumptions (3.25) and simple norm estimates applied to the representation (3.23) of the q-grid iteration operators. □

If the (h_q, h_{q-1})-TGM converges with a sufficiently small rate σ_* for all q and if the cycling regime is appropriately chosen, then the recurrence (3.27) of Lemma 3.2 will yield level-independent convergence rate estimates for the MGM (3.19). Indeed, in the case of the W-cycle, i.e. $\gamma(q) = \gamma = 2$ for all q, the recurrence (3.27) gives the uniform rate estimate

$$\|M_l\| \le \eta_* := \frac{1}{2c}(1 - \sqrt{1 - 4c\sigma_*}) \le 2\sigma_*, \tag{3.28}$$

provided that $\sigma_* \le 1/(4c)$. Similar rate estimates can be derived for the alternating cycle, the composed cycle and for the generalized V-cycle (exercise). However, in the case of the pure V-cycle, i.e. $\gamma(q) = \gamma = 1$ for all q, the fixed point equation $\eta = \sigma_* + c\eta$ corresponding to the recurrence (3.27) has a positive solution $\eta_* = \sigma_*/(1-c)$ if and only if $c < 1$. Unfortunately, in many practical applications, $c \ge 1$. Therefore, this proof technique does not provide a rate estimate for the V-cycle even if the twogrid rate is small. This contradicts all practical experience where reasonable convergence rates have been observed even for the V-cycle with only one smoothing step (see also Table 1 in Section 2). For some special classes of problems, this case was covered by new convergence theories [12, 38, 14, 15, 13, 74].

Let us finally count the arithmetical operations that are needed for one step of the multigrid iteration (3.19), i.e. for one call of the multigrid procedure $\mathrm{MGM}(\mathsf{K}_l, \underline{\mathsf{u}}_l, \underline{\mathsf{f}}_l, l)$. For simplicity, we assume that $\gamma(q) = \gamma$ for all $q = 2, \dots, l-1$, and that the numbers of pre- and post-smoothing steps are equal $\nu = \nu_1 + \nu_2$ at all levels. Furthermore, we suppose that the number of unknowns is growing by the factor τ from level to level. More precisely, inequalities (3.15) are valid with some enlargement factor $\tau > 1$ (the standard enlargement factor is $\tau = 2^d$). Now it is clear that at each level $q = 2, \dots, l$ the presmoothing, the defect calculation, the restriction, the prolongation, the correction and the postsmoothing need not more than $c_2 N_q$ arithmetical operations, where c_2 is some level-independent positive constant. Summing up over all levels, we obtain the following estimate for the number of arithmetical operations $Q(M_l)$ required by one call of the multigrid procedure at the finest level:

$$\begin{aligned} Q(M_l) &= c_2 N_l + \gamma\left\{c_2\frac{N_l}{\tau} + \gamma\left\{c_2\frac{N_l}{\tau^2} + \dots + \gamma\left\{c_2\frac{N_l}{\tau^{l-2}} + c_1\frac{N_l}{\tau^{l-1}}\right\}\cdots\right\}\right\} \\ &\le \max\left\{c_2, \frac{c_3}{\gamma}\right\}\sum_{j=0}^{l-1}\left(\frac{\gamma}{\tau}\right)^j N_l, \end{aligned} \tag{3.29}$$

where $c_1 N_1$ is the bound on the number of arithmetical operations needed on the coarsest grid. Calculating the sum for the possible choices of γ and τ, we arrive at the estimates

$$Q(M_l) \le \begin{cases} O(N_l) & \text{for } 1 \le \gamma < \tau, \\ O(N_l \ln(N_l)) & \text{for } \gamma = \tau, \\ O(N_l^{1+\ln(\beta)}) & \text{for } \gamma > \tau. \end{cases} \tag{3.30}$$

Summarizing the results on the convergence and on the operation count for multigrid methods, we can finally prove the following theorem:

3.3 Theorem *Let us assume that*

(1) $1 \le \gamma < \tau$ *and*

(2) $\|M_l\| \le \eta_* < 1$,

where η_* *is supposed to be independent of* h_l. *Then the multigrid method* (3.19) *algorithmically described by Algorithms* 3 *and* 2 *needs*

$$Q(\varepsilon) = O(\ln(\varepsilon^{-1})N_l) \tag{3.31}$$

arithmetical operations in order to reduce the initial error $\|\underline{u}_l - \underline{u}_l^0\|$ *by the factor* $\varepsilon := 10^{-s} \in (0,1)$.

The proof follows from the fact that $Q(\varepsilon) = I(\varepsilon)Q(M_l)$ is nothing but the product of the number of iterations $I(\varepsilon)$ and the costs of arithmetical operations per iteration step $Q(M_l)$. Relation (3.30) provides the estimate $Q(M_l)$ for $1 \le \gamma < \tau$. The number of iterations $I(\varepsilon)$ results from the condition $\eta_*^{I(\varepsilon)} \le \varepsilon$ that gives $I(\varepsilon) \ge \ln(\varepsilon^{-1})/\ln(\eta_*^{-1})$. □

Therefore, multigrid methods are asymptotically optimal with respect to the operation count (Theorem 3.3) and memory demand (estimates (3.18)). These properties make the multigrid methods so attractive for the solution of large-scale problems.

The multigrid iteration scheme (3.19) can be rewritten in the canonical form of a preconditioned Richardson iteration

$$C_l \frac{\underline{u}_l^{k+1} - \underline{u}_l^k}{\omega} + K_l \underline{u}_l^k = \underline{f}_l, \quad k = 0, 1, \ldots, \tag{3.32}$$

with the relaxation parameter $\omega = 1$ and the so-called multigrid preconditioner

$$C_l = K_l(I_l - M_l)^{-1}. \tag{3.33}$$

On one hand, this observation, together with well-known convergence results for the preconditioned Richardson method, can be used for the convergence analysis of multigrid methods itself [14, 15]. On the other hand, we can accelerate the convergence of the basic multigrid algorithm by an appropriate choice of the relaxation parameter ω. In particular, if the stiffness matrix K_l and the multigrid preconditioner C_l are symmetric and positive definite (spd), then other acceleration techniques such as the gradient technique, the Chebyshev acceleration, or the Conjugate Gradient (CG) technique can be used. Since the CG method is the most popular acceleration technique for spd systems in practice, we will present a detailed description of the CG method with spd multigrid preconditioners of the form (3.33) in the next subsection.

3.3 The Multigrid Preconditioned CG-Method

In hard practical applications, geometrical and, especially, algebraic multigrid methods are commonly used to construct very efficient preconditioners C_l that are then used as preconditioner in another iteration process such as a Krylov subspace iteration. In this subsection we assume that the stiffness matrix K_l is spd and in that case, the Preconditioned Conjugate Gradient (PCG) method is certainly the most widely used basic iteration scheme [3, 59].

Algorithm 4 Multigrid preconditioned CG: $\underline{\mathsf{u}}_l \Leftarrow \mathrm{MGPCG}(\mathsf{K}_l, \underline{\mathsf{u}}_l, \underline{\mathsf{f}}_l)$

{Initialization step:}
$\underline{\mathsf{u}}_l \leftarrow \underline{\mathsf{u}}_l^0$ {define the initial guess}
$\underline{\mathsf{d}}_l \leftarrow \underline{\mathsf{f}}_l - \mathsf{K}_l \cdot \underline{\mathsf{u}}_l$ {defect calculation}
$\underline{\mathfrak{w}}_l \Leftarrow \mathsf{C}_l^{-1} \cdot \underline{\mathsf{d}}_l$ {multigrid preconditioning}
$\delta^0 \Leftarrow (\underline{\mathfrak{w}}_l, \underline{\mathsf{d}}_l)$
$\delta \leftarrow \delta^0$
$\underline{\mathfrak{s}}_l \leftarrow \underline{\mathfrak{w}}_l$ {search direction}
{Iteration:}
while $\delta > \varepsilon^2\, \delta^0$ **do**
  $\alpha \Leftarrow \delta / (\mathsf{K}_l \underline{\mathfrak{s}}_l, \underline{\mathfrak{s}}_l)$
  $\underline{\mathsf{u}}_l \leftarrow \underline{\mathsf{u}}_l + \alpha \underline{\mathfrak{s}}_l$ {update of the iterate}
  $\underline{\mathsf{d}}_l \leftarrow \underline{\mathsf{d}}_l + \alpha \mathsf{K}_l \underline{\mathfrak{s}}_l$ {update of the defect}
  $\underline{\mathfrak{w}}_l \Leftarrow \mathsf{C}_l^{-1} \cdot \underline{\mathsf{d}}_l$ {multigrid preconditioning}
  $\hat{\delta} \Leftarrow (\underline{\mathfrak{w}}_l, \underline{\mathsf{d}}_l)$
  $\beta \leftarrow \hat{\delta} / \delta$
  $\delta \leftarrow \hat{\delta}$
  $\underline{\mathfrak{s}}_l \leftarrow \underline{\mathfrak{w}}_l + \beta \underline{\mathfrak{s}}_l$ {update of the search direction}
end while

Algorithm 4 gives a detailed algorithmic description of the PCG method. We recall again that different notations for the vectors and different arrows are only relevant for the parallel version of Algorithm 4 (see also Section 5). Before discussing the construction of the preconditioner C_l via multigrid techniques in detail, we shall recall the well-known rigorous convergence estimate for the PCG method [3, 59]. In order to reduce the initial iteration error in the K_l-energy norm $\|\cdot\|_{K_l}^2 := (K_l \cdot, \cdot)$ by some factor $\varepsilon := 10^{-s} \in (0,1)$, i.e.

$$\|\underline{u}_l - \underline{u}_l^{I(\varepsilon)}\|_{K_l} \le \varepsilon \|\underline{u}_l - \underline{u}_l^0\|_{K_l}, \tag{3.34}$$

we need at most

$$I(\varepsilon) = \mathrm{entire}\left(\frac{\ln(\varepsilon^{-1} + \sqrt{\varepsilon^{-2} - 1})}{\ln(\varrho^{-1})} + 1 \right) \tag{3.35}$$

PCG iterations provided that the preconditioner C_l is spd and spectrally equivalent to K_l, i.e. there exist positive constants γ_1 and γ_2 such that

$$\gamma_1 (C_l \underline{v}_l, \underline{v}_l) \le (K_l \underline{v}_l, \underline{v}_l) \le \gamma_2 (C_l \underline{v}_l, \underline{v}_l) \quad \forall \underline{v}_l \in R^{N_l}, \tag{3.36}$$

where $\varrho = (1 - \sqrt{\gamma_1/\gamma_2})/(1 + \sqrt{\gamma_1/\gamma_2})$. Of course, the accuracy test in the K_l-energy norm is not practicable because the solution $\underline{u}_l$ of system (3.14) is unknown. We prefer to use the $K_l C_l^{-1} K_l$-energy norm test

$$(\underline{w}_l^{I(\varepsilon)}, \underline{d}_l^{I(\varepsilon)}) = \|\underline{u}_l - \underline{u}_l^{I(\varepsilon)}\|^2_{K_l C_l^{-1} K_l} \le \varepsilon^2 \|\underline{u}_l - \underline{u}_l^0\|^2_{K_l C_l^{-1} K_l} \tag{3.37}$$

rather than the defect norm test. Indeed, if C_l is close to K_l as is the case for a multigrid preconditioner, then $K_l C_l^{-1} K_l$ is close to K_l as well.

The multigrid preconditioning step

$$C_l^{-1} \cdot \underline{d}_l := (I_l - (M_l)^j) K_l^{-1} \cdot \underline{d}_l \tag{3.38}$$

in Algorithm 4 means nothing else but the application of j (in practice: $j = 1$, or 2) multigrid cycles to the defect system

$$K_l \underline{w}_l = \underline{d}_l \tag{3.39}$$

with the zero initial guess $\underline{w}_l^0 = 0$.

The parameters and the components of the multigrid algorithm defining the multigrid preconditioning step (3.38) must be chosen in such a way that the multigrid preconditioner

$$C_l = K_l (I_l - (M_l)^j)^{-1} \tag{3.40}$$

is spd and satisfies the spectral equivalence inequalities (3.36) with spectral equivalence constants γ_1 and γ_2 such that the relative spectral condition number $\kappa(C_l^{-1} K_l) \leq \gamma_2 / \gamma_1$ is as small as possible.

The next two lemmas provide precise conditions ensuring the symmetry of the multigrid preconditioner (3.40).

3.4 Lemma *Let us assume that the multigrid iteration operator M_l defined by the recursion* (3.22) *is self-adjoint in the K_l-energy scalar product, i.e.,*

$$(M_l \underline{u}_l, \underline{v}_l)_{K_l} = (\underline{u}_l, M_l \underline{v}_l)_{K_l} \quad \forall \underline{u}_l, \underline{v}_l \in R^{N_l}, \tag{3.41}$$

briefly, $M_l = M_l^*$. Then the multigrid preconditioner (3.40) *is symmetric.*

Proof $M_l = M_l^*$ is obviously equivalent to the relation $K_l^{-1} M_l^T = M_l K_l^{-1}$, where M_l^T denotes the transpose of M_l. Using this relation, we immediately see that C_l^{-1} is symmetric. Indeed,

$$\begin{aligned}(C_l^{-1})^T &= ((I_l - (M_l)^j) K_l^{-1})^T = K_l^{-1} - K_l^{-1} M_l^T \cdots M_l^T \\ &= K_l^{-1} - M_l \cdots M_l K_l^{-1} = C_l^{-1}.\end{aligned}$$

Therefore, C_l is symmetric too. □

3.5 Lemma *Let us assume that*

$$I_q^{q-1} = (I_{q-1}^q)^T \quad \text{and} \quad \bar{S}_q^{\text{post}} = (\bar{S}_q^{\text{pre}})^* \quad \forall q = 2, \ldots, l, \tag{3.42}$$

where $\bar{S}_q^{\text{pre}} = (S_q^{\text{pre}})^{\nu_1(q)}$ and $\bar{S}_q^{\text{post}} = (S_q^{\text{post}})^{\nu_2(q)}$ for all $q = 2, \ldots, l$. Then the multigrid iteration operator M_l is self-adjoint in the K_l-energy scalar product.

Proof A simple calculation shows that under the assumptions (3.42) all twogrid iteration operators are self-adjoint in the K_q-energy scalar product, i.e. $M_q^{q-1} = (M_q^{q-1})^*$ for all $q = 2, \ldots, l$. Using representation (3.23) of the multigrid iteration operator M_q, we can easily proof its self-adjointness by induction (see [50, 49] for details). □

Now we can prove our main theorem on multigrid preconditioners.

3.6 Theorem *Let us suppose that the symmetry conditions* (3.42) *of Lemma* 3.5 *are fulfilled. Furthermore, we assume that all restrictions* I_q^{q-1} $(q = 2, \ldots, l)$ *have full rank and that all stiffness matrices fulfil the so-called Galerkin condition*

$$K_{q-1} = I_q^{q-1} K_q I_{q-1}^q \quad \forall q = 1, \ldots, l. \tag{3.43}$$

Finally, we suppose that there is a rate estimate for the multigrid method involved, i.e. there exists some positive constant $\eta = \eta_* \in (0, 1)$ *such that*

$$\varrho(M_l) = \|M_l\|_{K_l} := \sup_{\underline{v}_l \in R^{N_l}} \frac{\|M_l \underline{v}_l\|_{K_l}}{\|\underline{v}_l\|_{K_l}} (=) \leq \eta < 1, \tag{3.44}$$

where $\varrho(M_l)$ *denotes the spectral radius of* M_l. *Then the multigrid preconditioner* (3.40) *is spd and satisfies the spectral equivalence inequalities* (3.36) *with the spectral equivalence* $\gamma_1 = 1 - \eta^j$ *and* $\gamma_2 = 1$.

Proof The symmetry of the multigrid preconditioner C_l directly results from Lemmas 3.4 and 3.5. The positive definiteness would follow from the spectral equivalence inequalities (3.36). Thus, it remains to prove the spectral equivalence inequalities (3.36) with the spectral constants given in Theorem 3.6. Let us suppose for the moment that the multigrid iteration operator M_l is non-negative in the K_l-scalar product, i.e.

$$(M_l \underline{v}_l, \underline{v}_l)_{K_l} = (K_l M_l \underline{v}_l, \underline{v}_l) \geq 0 \quad \forall \underline{v}_l \in R^{N_l}. \tag{3.45}$$

Representation (3.40) and inequality (3.45) yield the estimates

$$\begin{aligned} (C_l^{-1}\underline{u}_l, \underline{u}_l) &= (K_l^{-1}\underline{u}_l, \underline{u}_l) - ((M_l)^j K_l^{-1}\underline{u}_l, \underline{u}_l) \\ &= (K_l^{-1}\underline{u}_l, \underline{u}_l) - ((M_l)^j \underline{v}_l, \underline{v}_l)_{K_l} \\ &\leq 1 \cdot (K_l^{-1}\underline{u}_l, \underline{u}_l) \quad \forall \underline{u}_l \in R^{N_l}, \end{aligned} \tag{3.46}$$

with the substitution $\underline{u}_l = K_l \underline{v}_l$. On the other side, we get the inequalities

$$\begin{aligned} (C_l^{-1}\underline{u}_l, \underline{u}_l) &\leq (K_l^{-1}\underline{u}_l, \underline{u}_l) - \|M_l\|_{K_l}^j (\underline{v}_l, \underline{v}_l)_{K_l} \\ &\leq (1 - \eta^j) \cdot (K_l^{-1}\underline{u}_l, \underline{u}_l) \quad \forall \underline{u}_l \in R^{N_l}. \end{aligned} \tag{3.47}$$

Inequalities (3.46) and (3.47) imply $\gamma_2 = 1$ and $\gamma_1 = 1 - \eta^j$, respectively. Of course, it remains to show inequality (3.45). Again, this inequality is proved in two steps. In a first step, inequality (3.45) is shown for all twogrid iteration operators M_q^{q-1} instead of M_l. Then, using the representation (3.23) of the multigrid iteration operator M_q, we can easily prove their non-negativeness by induction (see [50, 49] for details). □

The second part of the assumptions (3.42) ensuring the symmetry of the multigrid preconditioner imposes a strong condition on the choice and the arrangement of the pre- and post-smoothing procedures. The following lemma represents a practical proposal for the right choice of the pre- and post-smoothing iteration operators S_q^{pre} and S_q^{post} in the pre- and post-smoothing iteration processes (3.20).

3.7 Lemma *Assume that pre- and post-smoothing iteration operators have the form*

$$S_q^{\text{pre}} := I_q - \omega_q B_q^{-1} K_q \quad \textit{and} \quad S_q^{\text{post}} := I_q - \omega_q (B_q^{-1})^T K_q \tag{3.48}$$

with regular $N_q \times N_q$ *matrices* B_q *and appropriately chosen parameters* ω_q. *If the number* $\nu_1(q)$ *of pre-smoothing sweeps is equal to the number* $\nu_2(q)$ *of post-smoothing sweeps, then* $\bar{S}_q^{\text{post}} = (\bar{S}_q^{\text{pre}})^*$ *for all* $q = 2, \ldots, l$.

The proof is left to the reader (see also [50, 49] for the proof). □

The ω-Jacobi as well as the Gauss-Seidel method fit into the framework of Lemma 3.7. Indeed, the ω-Jacobi method corresponds to $B_q = D_q = \text{diag}(K_q)$. In the Gauss-Seidel case, the pre-smoothing and post-smoothing sweeps have to be arranged in a symmetrical way, e.g. lexicographically forward Gauss-Seidel sweeps in the pre-smoothing steps correspond to lexicographically backward Gauss-Seidel sweeps in the post-smoothing steps.

3.4 The Full Multigrid Method

The choice of the initial guess in Algorithm 3 is certainly important for the efficiency of multigrid methods in practical applications. If the boundary value problem under consideration allows a coarse discretization, then we can solve the corresponding coarse grid system

$$K_1 \underline{u}_1 = \underline{f}_1 \tag{3.49}$$

by means of some direct method and we can use the obtained coarse grid solution $\underline{u}_1$ for the adaptive mesh refinement and for constructing some initial guess $\underline{u}_2^0$ for the next finer mesh by interpolation. Then we improve this interpolated coarse grid solution by calling the multigrid (twogrid) procedure MGM($\mathsf{K}_2, \underline{\mathfrak{u}}_2, \underline{\mathfrak{f}}_2, 2$) several times. The continuation of this *nested iteration* process leads to the so-called *Full Multigrid Method* (FMGM) that is formally described by Algorithm 5.

Algorithm 5 Full multigrid method: $\underline{\mathfrak{v}}_q \Leftarrow$ FMGM($\mathsf{K}_q, \underline{\mathfrak{f}}_q, q = 1, l$)

```
v_1 ⇐ SOLVE (K_1 · u_1 = f_1)                {solution on the coarsest grid}
for q = 2 step 1 until l do
  v_q ← I^q_{q-1} · v_{q-1}                   {prolongation}
  for k = 1 step 1 until k_q do
    v_q ⇐ MGM(K_q, v_q, f_q, q)               {nested multigrid iteration}
  end for
end for
```

We mentioned in Section 2 that N. S. Bachvalow already proposed this nested iteration approach with the multigrid iteration as nested iteration and showed for his problem that under some assumptions the number of arithmetical operations for computing some FMG approximate $\underline{v}_l^{k_l}$ that differ from the exact solution in the order of the discretization error is directly proportional to the number N_l of unknowns on the finest grid. In this sense, the FMGM is asymptotically optimal. Indeed, this result is stated in Theorem 3.8 under quite general assumptions.

3.8 Theorem *Let us assume that there are positive and level-independent constants* $\eta_* \in (0,1)$ *(multigrid rate),* $c_A(u)$ *(approximation constant),* c_I *(interpolation constant) and* $\alpha = \sqrt[d]{\tau}$ *(enlargement constant) such that the assumptions*

(A1) $\|M_q\| \le \eta_* < 1$ *(see Section 3.2);*

(A2) $\|I_{q-1}^q \underline{u}_{q-1} - \underline{u}_q\| \le c_A(u) h_q^p$*;*

(A3) $\|I_{q-1}^q\| := \sup_{\underline{v}_{q-1} \in \mathbb{R}^{N_{q-1}}} \frac{\|I_{q-1}^q \underline{v}_{q-1}\|_{\mathbb{R}^{N_q}}}{\|\underline{v}_{q-1}\|_{\mathbb{R}^{N_{q-1}}}} \le c_I$*;*

(A4) $h_{q-1}/h_q \le \alpha$ *and* $1 \le \gamma < \tau$

hold for $q = 2,3,\ldots,l$ *and for all* $l = 2,3,\ldots,$ *where* $\underline{u}_{q-1}$ *and* $\underline{u}_q$ *are the exact solution of our finite element equations (1.1) for* $h = h_{q-1}$ *and* $h = h_q$*, respectively. The notation* $c_A(u)$ *indicates nothing else but the dependence of* $c_A(u)$ *on the solution* u *of the underlying boundary value problem. If the number* $i := k_q$ *of nested iteration satisfies the condition*

$$\eta_*^i < 1/c_1 \tag{3.50}$$

for all $q = 2,3,\ldots,l$*, then the estimate*

$$\|\underline{u}_q - \underline{v}_q^i\| \le c_2(\eta_*) c_A(u) h_q^p \tag{3.51}$$

holds for $q = l$*, where* $c_1 = c_I \alpha^p$*,* $c_2(\eta_*) = \eta_*^i/(1 - c_1 \eta_*^i)$*,* $\underline{u}_l$ *again denotes the exact solution of (1.1) at the finest level* l*, and* $\underline{v}_l^i$ *is the FMG approximate produced by Algorithm 5. The number of arithmetical operations that are needed to produce the FMG approximate* $\underline{v}_l^i$ *is proportional to the number* N_l *of unknowns on the finest grid corresponding to the discretization parameter* $h = h_l$*.*

Proof The crucial estimate (3.51) is proved by induction. Indeed, for $q = 1$, estimate (3.51) is trivially fulfilled since we assumed that the coarse grid system (3.49) is solved by some direct method. Let us suppose that estimate (3.51) is valid for $q-1$. We now show that (3.51) also holds for q. Indeed, using assumptions (A1)–(A4), we arrive at the estimates

$$\begin{aligned}
\|\underline{u}_q - \underline{v}_q^i\| &\le \eta_*^i \|\underline{u}_q - \underline{v}_q^0\| = \eta_*^i \|\underline{u}_q - I_{q-1}^q \underline{v}_{q-1}^i\| \\
&= \eta_*^i \|\underline{u}_q - I_{q-1}^q \underline{u}_{q-1} + I_{q-1}^q \underline{u}_{q-1} - I_{q-1}^q \underline{v}_{q-1}^i\| \\
&\le \eta_*^i \{\|\underline{u}_q - I_{q-1}^q \underline{u}_{q-1}\| + \|I_{q-1}^q\| \, \|\underline{u}_{q-1} - \underline{v}_{q-1}^i\|\} \\
&\le \eta_*^i \{c_A(u) h_q^p + c_I c_2(\eta_*) c_A(u) h_{q-1}^p\} \\
&= \eta_*^i (1 + c_I c_2(\eta_*) (h_{q-1}/h_q)^p) c_A(u) h_q^p \\
&\le \eta_*^i (1 + c_I c_2(\eta_*) (\alpha)^p) c_A(u) h_q^p = c_2(\eta_*) c_A(u) h_q^p
\end{aligned}$$

which finally prove estimate (3.51). The cost estimate that is stated in the theorem follows from Theorem 3.3 and a simple counting of the arithmetical operations needed by Algorithm 5. □

It is even sufficient to solve the coarse grid system (3.49) only approximately, e.g. via some algebraic multigrid method (see Section 4). The only thing that we have to ensure is the accuracy prescribed by (3.51) for $q = 1$. Such a *hybrid multigrid method* can certainly be very efficient in those practical applications where the the coarse grid system (3.49) is already quite large.

In order to illustrate the practical relevance of estimate (3.51), we suppose that $c_I = 1$, $\alpha = 2$, $p = 1$ and $\eta_* = 0.1$. We notice that these constants are typical for linear finite elements and standard refinement. For this setting of the involved constant, we observe that only one ($i = 1$) nested multigrid iteration is sufficient for producing FMGM iterates that differ from the exact solution in the order of the discretization error.

Of course, instead of the multigrid method MGM($\mathsf{K}_q, \underline{\mathfrak{v}}_q, \underline{\mathsf{f}}_q, q$), we can also use the multigrid preconditioned CG method MGPCG($\mathsf{K}_q, \underline{\mathfrak{u}}_q, \underline{\mathsf{f}}_q$) described by Algorithm 4 as nested iteration in Algorithm 5. In contrast to this MGPCG nested iteration process, P. Deuflhard proposed some cascadic nested iteration process that uses a decreasing number of unpreconditioned (!) CG iterations on the finer grids [25] (see [8] for the analysis). Nowadays, this nested iteration process is called *CASCADE algorithm*.

4 Algebraic multigrid methods

4.1 Basic idea of Algebraic Multigrid

Let us assume that only the fine grid information is available, i.e,

- the matrix $K_h = K_\ell$ and the right hand side $f_h = f_\ell$;
- the mesh $\tau_h = \tau_\ell$ and the node set ω_ℓ.

Then we want to construct the following coarse grid components from that given information:

- matrices K_q and right hand sides f_q, $q = \overline{1, \ell - 1}$;
- coarse mesh τ_q and coarse nodes set ω_q, $q = \overline{1, \ell - 1}$;
- intergrid transfer operators I_{q-1}^q, I_q^{q-1}, $q = \overline{2, \ell}$.

If we had all of the above components, then we could define appropriate smoothing operators S_q and we could apply the standard multigrid algorithm MGM on the sequence of matrices $\{K_q\}_{q=1}^{\ell}$. The pure (= classical) algebraic multigrid method (AMG) uses only the matrix information stored in K_h.

In order to derive an AMG algorithm, we start with an ideal twogrid approach. This idea cannot be realized in 2D or 3D but its approximation will lead us to good interpolation operators I_{q-1}^q. The coarse operators are derived immediately afterwards and an introduction into coarsening strategies finishes this section. The fine grid index h will be omitted for matrices in the remainder of Section 4.

4.2 An exact twogrid method

Assumption: There exists already a classification of nodes into N_C coarse grid nodes ("C") and N_F fine grid nodes ("F") with index sets ω_C and ω_F such that $\omega_\ell = \omega_C \cup \omega_F$ and $\omega_C \cap \omega_F = \emptyset$.

This induces the block structure of the fine grid problem (with $\underline{u}_C \in \mathbb{R}^{N_C}$ and $\underline{u}_F \in \mathbb{R}^{N_F}$)

$$K \cdot \underline{u} = \begin{bmatrix} K_{CC} & K_{CF} \\ K_{FC} & K_{FF} \end{bmatrix} \cdot \begin{bmatrix} \underline{u}_C \\ \underline{u}_F \end{bmatrix} = \begin{bmatrix} \underline{f}_C \\ \underline{f}_F \end{bmatrix} = \underline{f} \; . \tag{4.1}$$

By introducing the transformation matrix $T = \begin{bmatrix} I_C & \mathbf{O} \\ K_{FF}^{-1} K_{FC} & I_F \end{bmatrix}$ we can express K from (4.1) as a triple product

$$K = T^T \cdot \begin{bmatrix} \widehat{S}_C & \mathbf{O} \\ \mathbf{O} & K_{FF} \end{bmatrix} \cdot T \tag{4.2}$$

with the Schur complement matrix $\widehat{S}_C = K_{CC} - K_{CF} K_{FF}^{-1} K_{FC}$. Rearranging (4.2) leads to

$$\begin{aligned} \widehat{K} &:= \begin{bmatrix} \widehat{S}_C & \mathbf{O} \\ \mathbf{O} & K_{FF} \end{bmatrix} = T^{-T} \cdot K \cdot T^{-1} \\ &= \begin{bmatrix} I_C & -K_{CF} K_{FF}^{-1} \\ \mathbf{O} & I_F \end{bmatrix} \cdot \begin{bmatrix} K_{CC} & K_{CF} \\ K_{FC} & K_{FF} \end{bmatrix} \cdot \begin{bmatrix} I_C & \mathbf{O} \\ -K_{FF}^{-1} K_{FC} & I_F \end{bmatrix} \end{aligned} \tag{4.3}$$

and we extract $\widehat{R} = \begin{bmatrix} I_C & -K_{CF} K_{FF}^{-1} \end{bmatrix}$ and $\widehat{P} = \begin{bmatrix} I_C \\ -K_{FF}^{-1} K_{FC} \end{bmatrix}$ such that we get the obvious relation

$$\widehat{S}_C = \widehat{R} \cdot K \cdot \widehat{P}. \tag{4.4}$$

If we interpret $\widehat{S}_C$ as a coarse grid matrix K_H and $\widehat{P}$ as (exact) interpolation I_H^h, respectively $\widehat{R}$ as restriction I_h^H, then (4.4) represents the Galerkin approach (3.2) , and we can define the twogrid Algorithm 1 with these components. This twogrid method with exact interpolation and restriction acts as a direct solver.

Example 1D: (see (1.2) and (1.3) in Section 1: $-u''(x) = f(x)$, $x \in (0,1)$; $u(0) = u(1) = 0$). An FEM discretization with linear elements and $N = 7$, $h = 1/8$ leads to components:

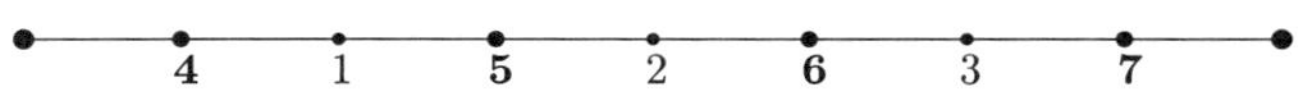

$$K = \frac{1}{h} \left[\begin{array}{ccc|cccc} 2 & & & -1 & -1 & & \\ & 2 & & & -1 & -1 & \\ & & 2 & & & -1 & -1 \\ \hline -1 & & & 2 & & & \\ -1 & -1 & & & 2 & & \\ & -1 & -1 & & & 2 & \\ & & -1 & & & & 2 \end{array} \right] \begin{array}{l} \\ \mathbf{C} \\ \\ \\ \mathbf{F} \\ \\ \end{array} = \begin{bmatrix} K_{CC} & K_{CF} \\ K_{FC} & K_{FF} \end{bmatrix},$$

$$\qquad\qquad \underbrace{\qquad\qquad}_{\mathbf{C}} \quad \underbrace{\qquad\qquad\qquad}_{\mathbf{F}}$$

$$\widehat{P} = \frac{1}{2} \begin{bmatrix} 2 & & \\ & 2 & \\ & & 2 \\ 1 & & \\ 1 & 1 & \\ & 1 & 1 \\ & & 1 \end{bmatrix} = -K_{FF}^{-1} K_{FC} \quad (\widehat{R} \equiv \widehat{P}^T), \text{ i.e., linear prolongation (!)},$$

$$\widehat{S}_C = \widehat{R} \cdot K \cdot \widehat{P} = \frac{1}{(2h)} \begin{bmatrix} 2 & -1 & 0 \\ -1 & 2 & -1 \\ 0 & -1 & 2 \end{bmatrix}.$$

The coarse grid matrix $\widehat{S}_C$ is identical with a matrix K_H resulting from the FEM discretization of the coarse mesh. The exact twogrid method with the components above works perfectly in 1D since K_{FF} is a diagonal matrix and therefore easy to invert.

Example 2D: (see also Section 2: $-\Delta u = f$ in $\Omega = (0,1) \times (0,1)$, $u = 0$ on $\partial\Omega$). An FEM discretization with linear triangular elements and $N = 49$, $h = 1/8$ leads to a system matrix K with the stencil $K = \begin{bmatrix} & -1 & \\ -1 & 4 & -1 \\ & -1 & \end{bmatrix}$. Now, K_{FF} is no longer a diagonal matrix:

$\Longrightarrow$ Inverting K_{FF} for calculation of interpolation operator is as expensive as solving the original problem

$\Longrightarrow$ Exact twogrid is not applicable in 2D (and 3D)!

Observation: The structure of the entries in $\widehat{P}$ ($\widehat{R}$) and $\widehat{S}_C$ calculated via (4.3)–(4.4) coincides rather well with the structure of a bilinear prolongation/restriction and a coarse matrix resulting from a 5-point stencil discretization, respectively a 9-point stencil discretization. Therefore, there is a chance to approximate $\widehat{P}$ with something cheaper.

4.3 Prolongation

Idea: Approximation of $-K_{FF}^{-1}K_{FC}$ by an appropriate operator B_{FC}. This approximation has to be good enough, otherwise the coarse operator derived via the Galerkin approach does no longer reflect the characteristics of the fine grid operator, e.g., null space and changing coefficients.

4.3.1 Matrix weighted prolongation in 1D

Let us have a look at the simple 1D example:

$$-(a(x)u'(x))' = 0 \quad \forall x \in (0, 1/2) \cup (1/2, 1) \tag{4.5}$$

with interface conditions

$$u(1/2 - 0) = u(1/2 + 0)$$
$$a(x)u'(x)|_{x=1/2-0} = a(x)u'(x)|_{x=1/2+0}$$

with boundary conditions

$$u(0) = 1, \qquad u(1) = 0,$$

where

$$a(x) = \begin{cases} a_1 = 10 & x \in [0, 1/2] \\ a_2 = 1 & x \in (1/2, 1] \end{cases}.$$

The exact solution

$$u(x) = \begin{cases} -\frac{2}{11}x + 1 & x \in [0, 1/2] \\ -\frac{20}{11}(x - 1) & x \in [1/2, 1] \end{cases}$$

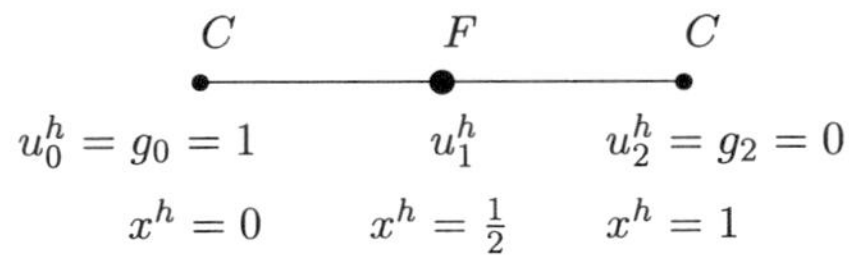

Figure 6: A simple 1D grid.

is nothing but the harmonic extension of the boundary data into the domain $(0, 1)$ with respect to the differential operator (4.5).

The FEM discretization in Fig. 6 with linear elements and the grid nodes

$$\tau_h = \{x_0^h = 0, x_1^h = 1/2, x_2^h = 1\}$$

defines an FE basis $\Phi^h = \{\varphi_0^h(x), \varphi_1^h(x), \varphi_2^h(x)\}$ and results, after a partial modification with respect to the Dirichlet boundary conditions, in the system

$$K \cdot \underline{u}^h = \frac{1}{2} \begin{bmatrix} 2 & 0 & 0 \\ -a_1 & a_1 + a_2 & a_2 \\ 0 & 0 & 2 \end{bmatrix} \cdot \begin{bmatrix} u_0^h \\ u_1^h \\ u_2^h \end{bmatrix} = \begin{bmatrix} 1 \\ 0 \\ 0 \end{bmatrix} = \underline{f}^h \tag{4.6}$$

with the solution

$$u_0^h = 1, \quad u_2^h = 0 \quad \text{and} \quad u_1^h = \frac{a_1 u_0 + a_2 u_2}{a_1 + a_2} = -\frac{K_{10} u_0 + K_{12} u_2}{K_{11}} = \frac{10}{11}.$$

The FE solution $u^h(x) = \Phi^h \cdot \underline{u}$ coincides with $u(x)$ in this example. The classical FE basis $\Phi^H = \{\varphi_0^H(x), \varphi_1^H(x)\}$ of the coarse grid $\tau_H = \{x_0^H = 0, x_1^H = 1\}$ can be derived from the fine basis Φ^h by using the linear interpolation $I_h^H = \begin{bmatrix} 1 & 0 \\ \frac{1}{2} & \frac{1}{2} \\ 0 & 1 \end{bmatrix}$ and the relation $\Phi^H = \Phi^h \cdot I_h^H$, i.e., $\varphi_0^H = \varphi_0^h + 1/2\varphi_1^h$ and $\varphi_1^H = 1/2\varphi_1^h + \varphi_2^h$. The related coarse grid solution $u^H(x) = \Phi^H \cdot \left[\begin{smallmatrix} 1 \\ 0 \end{smallmatrix}\right]$ is a really poor approximation of the fine grid solution $u^h(x)$. This approximation can be improved by choosing a non-standard coarse basis $\widehat{\Phi}^H$ derived from $\widehat{\Phi}^h$ with the exact interpolation $\widehat{P} = \begin{bmatrix} I_C \\ -K_{FF}^{-1} K_{FC} \end{bmatrix} = \begin{bmatrix} 1 & 0 \\ \frac{a_1}{a_1+a_2} & \frac{a_2}{a_1+a_2} \\ 0 & 1 \end{bmatrix}$ (matrix K from (4.6)). Therefore, the new coarse basis is

$$\widehat{\Phi}^H = \Phi^h \cdot \widehat{P}, \quad \text{i.e.,} \quad \varphi_0^H = \varphi_0^h + 10/11\varphi_1^h \quad \text{and} \quad \varphi_1^H = 1/11\varphi_1^h + \varphi_2^h,$$

and the coarse FE solution $\widehat{u}^H(x) = \widehat{\Phi}^H \cdot \begin{bmatrix} 1 \\ 0 \end{bmatrix}$ is again identical with the fine grid solution u^h. The coarse function $\widehat{u}^H(x)$ is the solution of (4.5), i.e., $\widehat{u}^H(x)$ is the harmonic extension, and the matrix $\widehat{P}$ is the discrete extension operator from coarse grid nodes to fine grid nodes. If we write $\widehat{P} = \begin{bmatrix} 1 & 0 \\ \frac{K_{10}}{K_{10}+K_{12}} & \frac{K_{12}}{K_{10}+K_{12}} \\ 0 & 1 \end{bmatrix}$ then $\widehat{P}$ is the matrix weighted linear interpolation, i.e.,

$$u_1^h = \frac{1}{K_{10} + K_{12}} (K_{10} u_0 + K_{12} u_2). \tag{4.7}$$

4.3.2 Matrix weighted prolongation for a 5-point stencil

The 2D- or 3D-approximation of $\widehat{P}$ has to be a local approximation of the discrete extension operator from coarse space $\mathbb{R}^{N_C}$ to fine space $\mathbb{R}^{N_C+N_F}$.

Let us consider a prolongation for a 5-point stencil on an equidistant tensor product mesh.

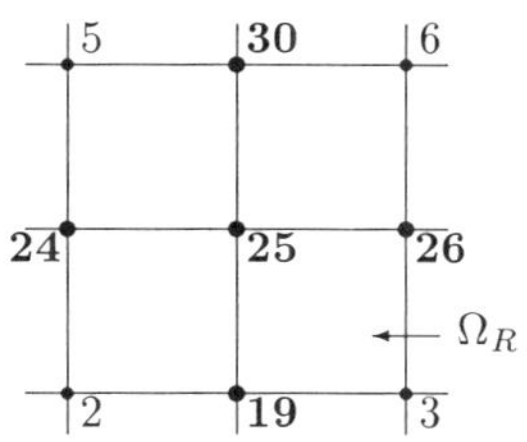

$\{2,3,5,6\} \subset \omega_C$
$\{19,24,25,26,30\} \subset \omega_F$

Figure 7: A simple 2D grid.

(1) $u_{19}, u_{24}, u_{26}, u_{30}$ are calculated with matrix weighted interpolation (4.7);

(2) u_{25}: we determine the discrete harmonic extension in Ω_R, i.e., we multiply the 25th row of the matrix K with the vector of unknowns $\underline{u}$ and solve $0 = \sum_{i=1}^{N} K_{25,i} \cdot u_i$, and get $u_{25} = -\{K_{25,19}u_{19} + K_{25,24}u_{24} + K_{25,26}u_{26} + K_{25,30}u_{30}\}/K_{25,25}$;

$\vdots$

substitute $u_{19}, \ldots, u_{30}$ with result from item (1).

Further algebraic transformations lead to the matrix weighted bilinear interpolation

$$
\begin{aligned}
u_{25} = \frac{-1}{K_{25,25}} \Bigg\{ & \left(\frac{K_{25,19}K_{19,2}}{K_{19,2}+K_{s,3}} + \frac{K_{25,24}K_{24,2}}{K_{24,2}+K_{24,5}} \right) u_2 + \left(\frac{K_{25,19}K_{19,3}}{K_{19,2}+K_{19,3}} + \frac{K_{25,26}K_{26,3}}{K_{26,3}+K_{26,6}} \right) u_3 \\
& + \left(\frac{K_{25,24}K_{24,5}}{K_{24,2}+K_{24,5}} + \frac{K_{25,30}K_{30,5}}{K_{30,5}+K_{30,6}} \right) u_5 + \left(\frac{K_{25,26}K_{26,6}}{K_{26,3}+K_{26,6}} + \frac{K_{25,30}K_{30,6}}{K_{30,5}+K_{30,6}} \right) u_6 \Bigg\}
\end{aligned}
$$

4.3.3 Prolongation for general meshes

Let us illustrate the graph of an arbitrary FE matrix as a mesh, and mark the fine and coarse nodes (ω_C, ω_F) as in Fig. 8. There we have:

- Set of coarse neighbors: $C^i = \{\text{nodes} \in \omega_C, \text{ connected with } \underbrace{\text{node } i}_{\in \omega_F}\}$,
  e.g., $C^6 = \{1,2,3\}$, $C^5 = \{1,2,\ldots\}$, $C^4 = \{1,3,\ldots\}$.
  Set of fine neighbors: $F^i = \{\text{nodes} \in \omega_F, \text{ connected with node } i \in \omega_F\}$,
  e.g., $F^6 = \{4,5\}$.

- We know that the interpolation weights $\alpha_{jj} = 1 \ \forall j \in \omega_C$, and we have to determine:
  $\alpha_{ij} \ \forall i \in \omega_F \ \forall j \in C^i$;
  $\alpha_{ij} = 0$ otherwise.

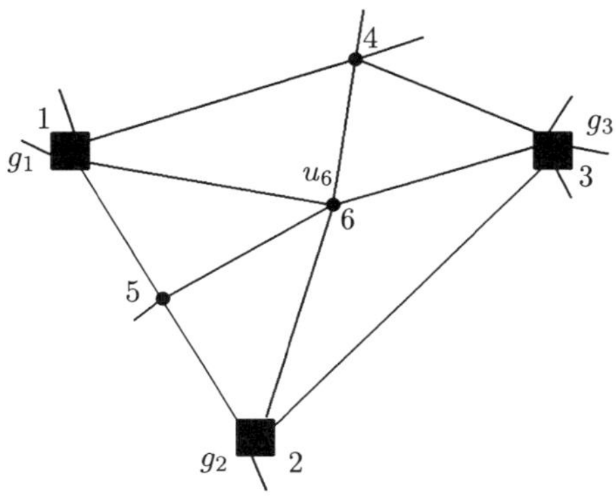

$\{1,2,3\} \subset \omega_C$
$\{4,5,6\} \subset \omega_F$

Figure 8: Neighborhood of node 6.

Now, we will determine the integration weights α_{ij}:

Variant 0:
Linear interpolation (4.7) between (coarse) father and (fine) son nodes.
Unstructured grid has no unique father-son relations anymore.
Therefore, node 6 may use all 3 coarse nodes for interpolation, i.e. $j \in C^6$.

Variant 1 (improved):

$$\begin{aligned}
\text{Interpolation in } i = 6 \quad &\Longleftrightarrow \quad \text{harmonic extension inside the polygon} \\
&\qquad\quad 1 \to 5 \to 2 \to 3 \to 4 \to 1 \\
&\Longleftrightarrow \quad \sum_{j \in F^i \cup C^i \cup \{i\}} K_{ij} u_j = 0 + \text{B.C. } u_j = g_j \quad \forall j \in F^i \cup C^i \\
&\Longleftrightarrow \quad K_{66} u_6 = - \underbrace{\sum_{j=1}^{3} K_{6j} g_j}_{\substack{j \in C^j \\ g_j \text{ known}}} - \underbrace{\sum_{j=4}^{5} K_{6j} g_j}_{\substack{j \in F^j \\ g_j \text{ unknown!}}} . \qquad (*)
\end{aligned}$$

Now, linear interpolation (4.7) for calculation of unknowns g_j, $j \in F^i$:

$$g_5 = \frac{K_{51} g_1 + K_{52} g_2}{K_{51} + K_{52} [+ \underbrace{K_{53}}_{=0}]} [= \widehat{c}_{51} g_1 + \widehat{c}_{52} g_2] \quad \text{and} \quad g_4 = \frac{K_{41} g_1 + K_{43} g_3}{K_{41} + K_{43}} .$$

$\overset{(*)}{\Longrightarrow}$

$$6 \in F^i$$

$$u_6 = \frac{-1}{K_{66}} \Bigg\{ \left(K_{61} + \frac{K_{65} K_{51}}{K_{51} + K_{52}} + \frac{K_{64} K_{41}}{K_{41} + K_{43}} \right) g_1 + \left(K_{62} + \frac{K_{65} K_{52}}{K_{51} + K_{52}} \right) g_2 + \left(K_{63} + \frac{K_{64} K_{43}}{K_{41} + K_{43}} \right) g_3 \Bigg\}$$

Introducing

$$\widehat{c}_{ij} := \sum_{k\in F^i} \frac{K_{ik}\cdot K_{kj}}{\sum_{l\in C^i} K_{kl}} \tag{4.8}$$

leads to

$$u_i^h = u_i = \frac{-1}{K_{ii}}\left\{\sum_{j\in C^i} (K_{ij}+\widehat{c}_{ij}) \underbrace{g_j}_{=u_j^H}\right\},$$

and finally to the interpolation weights

$$\widehat{\alpha}_{ij} = \begin{cases} 1 & i=j\in\omega_C \\ \dfrac{-(K_{ij}+\widehat{c}_{ij})}{K_{ii}} & i\in\omega_F,\ j\in C^i \\ 0 & \text{otherwise} \end{cases}. \tag{4.9}$$

Formulas (4.9) can be further improved without changing the pattern of the prolongation matrix P.

Variant 2: J. W. Ruge and K. Stüben proposed in [63, 64] an improved calculation of g_5, g_4, as follows. We investigate g_5 in the local system (Fig. 9) $K_{51}g_1 + K_{52}g_2 + \widetilde{K}_{55}g_5 + K_{56}u_6 = 0$

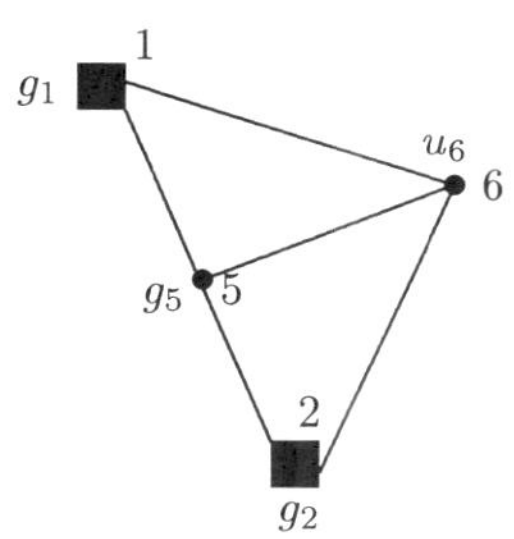

Figure 9: Restricted neighborhood of node 5.

with known values g_1, g_2, u_6 and $\widetilde{K}_{55} = -(K_{51}+K_{52}+K_{56})$ (the row sum has to be zero)
$\Longrightarrow g_5 = \dfrac{K_{51}}{K_{56}+K_{51}+K_{52}}g_1 + \dfrac{K_{52}}{K_{56}+K_{51}+K_{52}}g_2 + \dfrac{K_{56}}{K_{56}+K_{51}+K_{52}}u_6$ (also g_4).
The same methodology as in variant 1 leads to the formulas

$$c_{ij} := \sum_{k\in F^i} \frac{K_{ik}\cdot K_{kj}}{\sum_{l\in C^i} K_{kl} + K_{ki}} \tag{4.10}$$

$$\alpha_{ij} := \begin{cases} 1 & i=j\in\omega_C \\ \dfrac{-(K_{ij}+c_{ij})}{K_{ii}+c_{ii}} & i\in\omega_F,\ j\in C^i \\ 0 & \text{otherwise} \end{cases}. \tag{4.11}$$

4.4 Coarsening

4.4.1 Motivation

The pure algebraic multigrid uses only matrix information to derive the coarse system, i.e.,

known: matrix $K = K_\ell$, rigth-hand side $f = f_\ell$, find grid indices ω_ℓ

unknown: $\left.\begin{array}{l}\text{coarse grid indices } \omega_k = \omega_{k+1,C},\\ \text{transfer operators } P_k, R_k,\\ \text{matrices } K_k,\\ \text{rigth-hand side } f_k,\end{array}\right\} k = \overline{1, \ell - 1}$

known: $[\omega,]K$;

unknown: $[\omega_C, \omega_F,]P, R, K_H = RKP$

The main problem consists in finding the coarse nodes ω_C such that the main properties of the operator K are reflected in the coarse operator K_H. The following determination of the prolongation operator will be done analogously to Section 4.3.2.

Example anisotropic operator: $\varepsilon \partial^2 u/\partial x^2 + \partial^2 u/\partial y^2 = f$ in $(0,1)^2$ with boundary conditions. The unit square is discretized by an equidistant tensor product grid and the standard finite difference discretization results in the 5-point stencil

$$\frac{1}{h^2}\begin{bmatrix} & -1 & \\ -\varepsilon & 2\varepsilon + 2 & -\varepsilon \\ & -1 & \end{bmatrix},$$

i.e., if ε tends to zero then the problems in y-direction become completely independent. Therefore, a coarsening in the y-direction can be applied but no coarsening in the x-direction, see Fig. 10 for an illustration. This semi-coarsening cannot be reproduced by a simple general coarsening strategies without taking into account the operator characteristics. (Special line smoothers could solve the problem in the given example but this is not a general approach.)

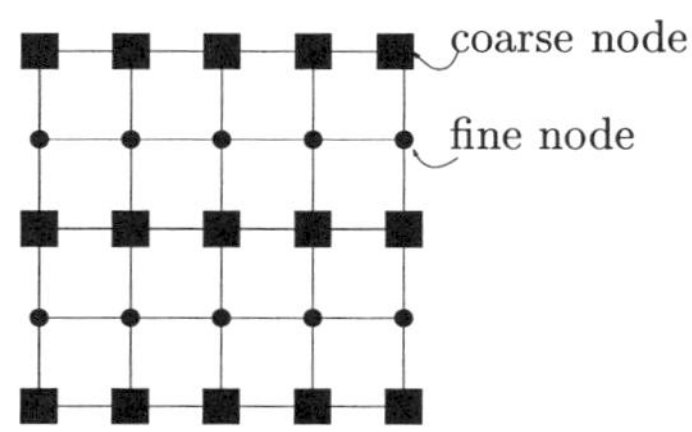

Figure 10

A closer look at the matrix entries resulting form the stencil gives us the idea that coarsening is allowed into the direction of relatively large entries. This is indeed the basic idea of the following coarsening strategy.

4.4.2 Coarsening by Ruge/Stüben

Let us define the strength of coupling to a node i with respect to a set of nodes $I \subset V = \omega$ (mostly a single node) by

$$d(i, I) := \frac{1}{\max_k\{|K_{ik}|\}} \sum_{j \in I} -K_{ij}.$$

This leads to the set of nodes whose coupling to node i is stronger than a threshold parameter α:

$$S^i := \{j \in V | d(i, \{j\}) \geq \alpha\}, \qquad \overset{\bullet}{i} \longleftarrow \overset{\bullet}{j}$$

and to the set of nodes to which node i possesses a strong coupling:

$$S^{i,T} := \{j \in V | i \in S^j\}. \qquad \overset{\bullet}{i} \longrightarrow \overset{\bullet}{j}$$

The relation $S^i = S^{i,T}$ holds for matrices K resulting from a discretization of the Laplace operator, but this is no longer true for anisotropic operators.

The coarsening algorithm by J. W. Ruge and K. Stüben [30, 63, 64] tries in phase I to find a set of coarse nodes ω_C with the property that no two points are strongly connected to each other. Phase II of the algorithm has to guarantee that each node in F is either strongly connected to a node from C, or that is coupled to a node from F with a strong coupling to a node from C.

- All parts marked sans serif in the algorithms above are heuristic, i.e., other strategies can be used.
- Parameters α, β are also heuristically. In [30], the authors chose $\alpha = 0.25$ (obvious for the 5-point stencil) and $\beta = 0.35$.
- The Ruge-Stüben algorithm produces again a regular coarse grid from a regular fine grid if the node numbering supports the heuristic part in the algorithms properly.
- If the geometry is known then we can improve the rough approximation of the geometry by initializing C with these geometric nodes.
  $\rightarrow$ start Ruge-Stüben algorithm with $C = \{\text{geometric nodes}\}$

Algorithm 6 Coarsening Phase I: Split $V = \omega$ into C, F

```
C = ∅, F = ∅
while C ∪ F ≠ V do
  pick i ∈ V \ {C ∪ F} with maximal |S^{i,T}| + |S^{i,T} ∩ F|      {heuristical}
  if |S^{i,T}| + |S^{i,T} ∩ F| = 0 then
    F := V \ C                                                      {≙ STOP}
  else
    C := C ∪ {i}, F := F ∪ {S^{i,T} \ C}
  end if
end while
```

Algorithm 7 Coarsening Phase II: Split $V = \omega$ into C, F

```
T := ∅                                              {Set of already checked nodes}
while T ⊂ F do
  pick i ∈ F \ T  T := T ∪ {i}; C̃ := ∅
  C^i = S^i ∩ C                                     {strong neighbor in C}
  F^i = S^i ∩ F                                     {strong neighbor in F}
  while F^i ≠ ∅ do
    pick j ∈ F^i; F^i := F^i \ {j}
    if d(j, C^i) ≤ β · d(i, {j}) then
      if |C̃| = 0 then
        C̃ := {j}; C^i = C^i ∪ {j}
      else
        C := C ∪ {i}; F = F \ {i}
      end if
    end if
    C := C ∪ C̃, F = F \ C̃
  end while
end while
ω_C = C, ω_F := F
```

The coarsening algorithm shown above requires a criterion for strong connections. We used the criterion $|K_{ij}| \geq \alpha \max_k\{|K_{ik}|\}$ for a node j to be strongly coupled to node i [63]. Another choice is $|K_{ij}| \geq \Theta\sqrt{K_{ii} \cdot K_{jj}}$ with a level dependent parameter Θ, proposed by P. Vanek, J. Mandel, M. Brzina [72].

4.4.3 Other coarsening strategies

A simple but useful coarsening technique is the agglomeration technique by D. Braess [9] which determines aggregates by finding patterns in the matrix graph. Each aggregate will be handled as one (virtual) coarse node. P. Vanek et al. [72] used their coarsening criterion for considering operator properties in the construction of agglomerates. This technique is fast, well suited for block systems resulting from systems of PDEs, and produces a sparse coarse matrix K_H. The agglomeration technique uses piecewise constant interpolation for intergrid transfer so that robustness with respect to varying material parameters cannot be achieved. If this interpolation is followed by (at least) one smoothing step, then robustness can be achieved [71, 72]. But each smoothing sweep in the prolongation and restriction extends the interpolation stencil by one layer of fine nodes so that the coarse matrix is getting denser and the whole algorithm will be more expensive.

Another very interesting approach is called AMGe and was introduced by the Livermore group [23]. The main idea consists in preserving non-overlapping elements and their matrices on all grids. Neighboring elements will be combined to patches, allowing one to determine the coarse nodes with respect to the quality of the interpolation. The quality of interpolation is described by the functional

$$\frac{\langle (I - PR)\underline{e}, (I - PR)\underline{e} \rangle}{\langle K\underline{e}, \underline{e} \rangle}$$

which has to be minimized for all test vectors $\underline{e} \in \mathbb{R}^N \setminus \ker(K)$ under the constraint of small arithmetic work in the multigrid algorithm. The minimization of the global functional is too expensive, and therefore the appropriate functionals on the patches are minimized. This approach has been refined for element agglomeration techniques [47], and the strategy can be applied if no element information is available [44].

4.5 Calculation of the coarse matrix

The approximation of the exact prolongation $\widehat{P} = \begin{bmatrix} I_C \\ -K_{FF}^{-1}K_{FC} \end{bmatrix}$ by an approximation $P = \begin{bmatrix} I_C \\ B_{FC} \end{bmatrix}$ in Section 4.3 changes the formulas (4.3) into

$$\begin{bmatrix} K_H & \neq \mathbf{O} \\ \neq \mathbf{O} & K_{FF} \end{bmatrix} = \begin{bmatrix} I_C & B_{FC}^T \\ \mathbf{O} & I_F \end{bmatrix} \cdot \begin{bmatrix} K_{CC} & K_{CF} \\ K_{FC} & K_{FF} \end{bmatrix} \cdot \begin{bmatrix} I_C & \mathbf{O} \\ B_{FC} & I_F \end{bmatrix} \tag{4.12}$$

with $K_H = \widehat{S}_C + T_C$ and the distortion $T_C = (B_{FC}^T + K_{CF}K_F^{-1})K_F(B_{FC} + K_F^{-1}K_{FC})$. The non-zero off-diagonal blocks of the left term in (4.12) cannot be considered in the solution process, and therefore we use $C := \begin{bmatrix} K_H & \mathbf{O} \\ \mathbf{O} & K_{FF} \end{bmatrix}$ to approximate the exact operator in (4.3). This approximation can be quantified via the condition number κ:

$$\kappa(C^{-1}\widehat{K}) = \left(\sqrt{\mu} + \sqrt{\mu + 1}\right)^2,$$

where $\mu = \varrho(\widehat{S}_C^{-1}T_C)$ denotes the spectral radius of $\widehat{S}_C^{-1}T_C$ [35]. The coarse matrix K_H is generated by the Galerkin approach (3.2)

$$K_H = R \cdot K \cdot P = K_C - K_{CF}B_{FC} - B_{FC}^T K_{FC} + B_{FC}^T K_F B_{FC}. \tag{4.13}$$

with transfer operators P, $R = P^T$ calculated via (4.11). Note that (4.13) does not contain K_F^{-1}, and that the sparsity pattern of K_H is determined by the term $B_{FC}^T K_F B_{FC}$. We can calculate the coarse matrix entries with the formulas

$$[K_H]_{ij} = \left[\sum_{k \in \mathcal{N}_i} \sum_{l \in \mathcal{N}_j} \alpha_{ki} K_{kl} \alpha_{lj} \right]_{ij}, \tag{4.14}$$

where $\mathcal{N}_i = C^i \cup F^i \cup \{i\}$ denotes the (strong) neighborhood of a node i. A straightforward implementation of (4.14) for generating the coarse system matrix results in very large computational times since many search procedures with complexity $O(N_C^2)$ are necessary. An interchange of the inner and outer loops lead to an optimal algorithm of complexity $O(N_F) = O(N_C)$, see Algorithm 8. F^i, C^i are defined as in Section 4.2.

4.6 Problem discussion

The classical AMG by J. W. Ruge and K. Stüben was originally designed for M-matrices as system matrices. An M-matrix is, roughly spoken, a weakly diagonally dominant matrix ($|K_{ii}| \geq \sum_{j \neq i} |K_{ij}|$) with positive diagonal entries and non-positive off-diagonal entries. This M-matrix property represents the discrete maximum principle.

Algorithm 8 Algorithm for generating the coarse matrix according to (4.14)

```
K_H := 0
for all k ∈ ω_F ∪ ω_C do
  Determine F^k, C^k
  for all l ∈ F^k ∪ C^k ∪ {k} do
    T := C^k ∪ C^l ∪ ({l} ∩ ω_C)
    for all i ∈ T do
      for all j ∈ T do
        K_{H,i,j} := K_{H,i,j} + α_{ki} K_{kl} α_{lj}
      end for
    end for
  end for
end for
```

If we use quadrilateral finite elements, or FE basis functions of higher polynomial degree, or the matrix results from a system of PDEs, or if we have a poor triangulation, then the M-matrix property is easily lost in the system matrix. We list a few approaches how to handle this case:

(a) Ignore small positive off-diagonal entries or add (lump) them to the diagonal. The lumping changes the discrete operator, i.e., the matrix does no longer represent the original differential operator and therefore this method gets unstable easily.

(b) Use agglomeration, see Section 4.4.3.

(c) Use AMGe techniques, see Section 4.4.3.

(d) If we have access to the element stiffness matrices $K^{(r)}$, which accumulate to the stiffness matrix K, then we can use the element preconditioning technique [36]. This technique consists in finding element matrices $C^{(r)}$ which are M-matrices that best approximate the original matrices $K^{(r)}$ in the spectral sense. The matrix C accumulated from the $C^{(r)}$ matrices will be used in the AMG. This technique results in very good results for certain classes of finite elements.

(e) Another technique [34] constructs an auxiliary matrix B by using knowledge of fine mesh coordinates and operator properties. This matrix B is an M-matrix by construction and it will be used for finding the coarse nodes. All remaining components in AMG are standard.

5 Parallelization

5.1 Data partitioning and basic parallel routines

Our data partitioning is based on the non-overlapping distribution of finite elements that form a subdomain, but all the statements and techniques in this section can be easily adapted to other data decompositions. In 2D this induces three classes of nodes: inner nodes, edge nodes

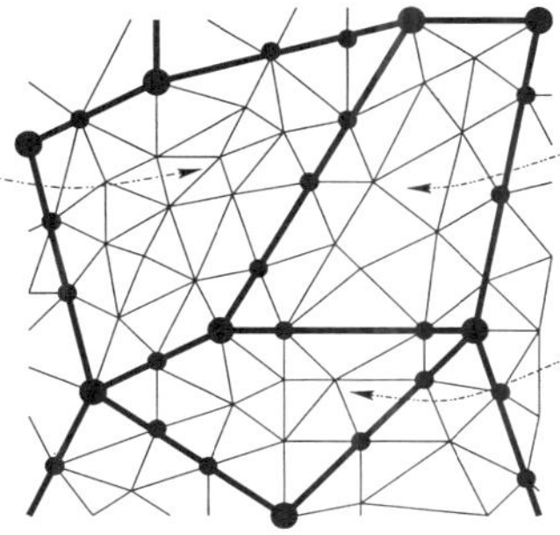

Figure 11: Nonoverlapping domain decomposition.

and vertex nodes. Face nodes are collected in an additional class in 3D. All nodes excluding the inner nodes are often called interface nodes.

The mapping of a vector $\underline{w} = (w^{[i]})_i \in \mathbb{R}^{N_h}$ in global numbering onto a local vector $\underline{w}_s \in \mathbb{R}^{N_s}$ in subdomain $\overline{\Omega}_s$ $(s = \overline{1,P})$ is represented symbolically by $\mathcal{A}_s \cdot \underline{w}$ with subdomain connectivity matrices $\mathcal{A}_s : \mathbb{R}^{N_h} \mapsto \mathbb{R}^{N_s}$. The transpose $\mathcal{A}_s^T$ of these binary matrices $\mathcal{A}_s$ maps a local vector back to the global one. The index set of all subdomains, a component $w^{[j]}$, $j \in \omega_h$ belongs to, is denoted by

$$\sigma^{[j]} := \{s : \exists i \in \omega_s : \mathcal{A}_s^{[i,j]} \neq 0\} \quad \forall j \in \omega_h \ . \tag{5.1}$$

There are (at least) two possibilities for storing those components and consequently that vector (see also Fig. 12).

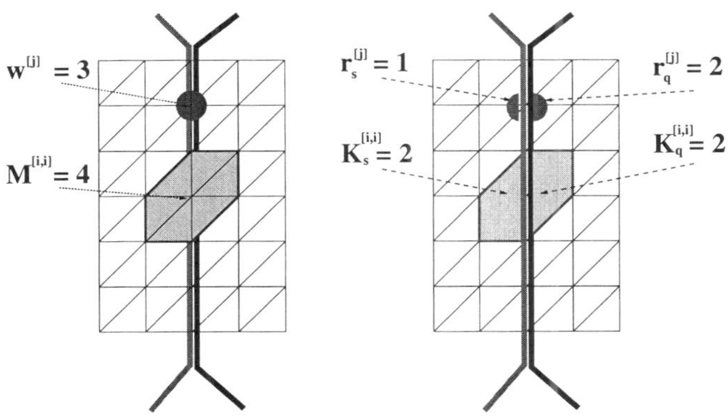

Figure 12: Illustration of accumulated and distributed vectors and matrices.

A vector $\underline{\mathfrak{w}}$ is called an *accumulated vector* if each vector component $\mathfrak{w}^{[i]}$ is stored in all subdomains Ω_s, $s \in \sigma^{[i]}$, with its full value. The local vectors $\underline{\mathfrak{w}}_s$ can be represented as

$$\underline{\mathfrak{w}}_s := \mathcal{A}_s \cdot \underline{\mathfrak{w}}. \tag{5.2}$$

A vector $\underline{\mathsf{r}}$ is called a *distributed vector* if it is decomposed into local vectors$\underline{\mathsf{r}}_s$ such that

$$\underline{\mathsf{r}} = \sum_{s=1}^{P} \mathcal{A}_s^T \cdot \underline{\mathsf{r}}_s \tag{5.3}$$

holds, i.e., all subdomains Ω_s, $s \in \sigma^{[i]}$ store only $\underline{\mathsf{r}}_s$ and possess a portion of the full vector value $\mathsf{r}^{[i]}$ which can be determined only by taking the sum in (5.3). The conversion of a distributed vector $\underline{\mathsf{v}}$ into an accumulated vector $\underline{\mathfrak{w}}$ can be done by evaluating the sum in (5.3) followed by the restriction (5.2),

$$\underline{\mathfrak{w}} \Leftarrow \underline{\mathsf{v}} \quad : \quad \underline{\mathfrak{w}}_s := \mathcal{A}_s \cdot \underline{\mathfrak{w}} = \mathcal{A}_s \cdot \sum_{s=1}^{P} \mathcal{A}_s^T \cdot \underline{\mathsf{v}}_s . \tag{5.4}$$

A matrix can also be stored in two ways. A matrix $\mathfrak{P}$ is called *accumulated* if its local restrictions $\mathfrak{P}_s$ possess the full entries of it, and we can write

$$\mathfrak{P}_s := \mathcal{A}_s \cdot \mathfrak{P} \cdot \mathcal{A}_s^T . \tag{5.5}$$

We call a matrix K *distributed* if it is the sum of locally stored matrices K_s,

$$\mathsf{K} := \sum_{s=1}^{P} \mathcal{A}_s^T \cdot \mathsf{K}_s \cdot \mathcal{A}_s . \tag{5.6}$$

That is, each subdomain $\overline{\Omega}_s$ stores only a part of its full entries. Distributed matrices K_s are automatically obtained after the FE-accumulation in the subdomains Ω_s due to our non-overlapping construction principle. The sum in (5.6) is equivalent to a global matrix accumulation which will not be carried out in the parallel case.

It is trivial that additive combinations of vectors of the same type do not require any communication. The inner product of different type vectors requires one global reduction operation of the local inner products:

$$\langle \underline{\mathfrak{w}}, \underline{\mathsf{r}} \rangle = \underline{\mathfrak{w}}^T \underline{\mathsf{r}} = \underline{\mathfrak{w}}^T \sum_{s=1}^{P} \mathcal{A}_s^T \underline{\mathsf{r}}_s = \sum_{s=1}^{P} (\mathcal{A}_s \underline{\mathfrak{w}})^T \underline{\mathsf{r}}_s = \sum_{s=1}^{P} \langle \underline{\mathfrak{w}}_s, \underline{\mathsf{r}}_s \rangle . \tag{5.7}$$

Any other combination of vector types requires more effort. The multiplication of a distributed matrix K with an accumulated vector $\underline{\mathfrak{w}}$,

$$\mathsf{K} \cdot \underline{\mathfrak{w}} = \sum_{s=1}^{P} \mathcal{A}_s^T \mathsf{K}_s \mathcal{A}_s \cdot \underline{\mathfrak{w}} = \sum_{s=1}^{P} \mathcal{A}_s^T (\mathsf{K}_s \cdot \underline{\mathfrak{w}}_s) = \sum_{s=1}^{P} \mathcal{A}_s^T \underline{\mathsf{v}}_s = \underline{\mathsf{v}}, \tag{5.8}$$

results in a distributed vector. The realization requires no communication at all because we only have to compute locally $\underline{\mathsf{v}}_s = \mathsf{K}_s \cdot \underline{\mathfrak{w}}_s$.

The situation changes if we use an accumulated matrix $\mathfrak{P}$. Here we have to ensure that the pattern of $\mathfrak{P}$ fulfills the condition

$$\forall i, j \in \omega_h : \quad \sigma^{[i]} \not\subseteq \sigma^{[j]} \Longrightarrow \mathfrak{P}^{[i,j]} = 0. \tag{5.9}$$

If the pattern condition (5.9) holds, then the operations

$$\underline{\mathfrak{w}} = \mathfrak{P} \cdot \underline{\mathfrak{w}}_H \quad \text{and} \quad \underline{\mathfrak{r}}_H = \mathfrak{P}^T \cdot \underline{\mathfrak{r}} \tag{5.10}$$

can be performed locally without any communication, i.e., $\forall s = \overline{1, P}$

$$\underline{\mathfrak{w}}_s = \mathfrak{P}_s \cdot \underline{\mathfrak{w}}_{H,s} \quad \text{and} \quad \underline{\mathfrak{r}}_{H,s} = \mathfrak{P}_s^T \cdot \underline{\mathfrak{r}}_s \ . \tag{5.11}$$

Under the assumption (5.9), the triple product $\mathsf{K}_H := \mathfrak{P}^T \cdot \mathsf{K} \cdot \mathfrak{P}$ can be performed locally without communication

$$\mathsf{K}_{H,s} = \mathfrak{P}_s^T \cdot \mathsf{K}_s \cdot \mathfrak{P}_s. \tag{5.12}$$

5.2 Parallel multigrid

The operations (5.7) and (5.8) allow already to formulate a parallel preconditioned conjugate gradient (PCG) algorithm for solving the matrix equation $\mathsf{K}\underline{\mathfrak{u}} = \underline{\mathsf{f}}$ with a preconditioner C_K, see [57] and Algorithm 4.

One possible choice for the preconditioner is $C_K^{-1} = (I - M_K)\mathsf{K}^{-1}$, with M_K being the multigrid iteration operator for K, see Section 3. The parallel multigrid iteration is presented in Algorithm 9, where $\ell = 1$ now stands for the finest grid since the number of coarser grids is unknown in the beginning! Therefore, the preconditioner is called as $\text{PMG}(\{\mathsf{K}^1, \mathfrak{P}^1\}, \underline{\mathfrak{u}}, \underline{\mathsf{f}}, 1)$. The algorithm applies the parallel version of the smoother $\text{SMOOTH}(\mathsf{K}, \underline{\mathfrak{u}}, \underline{\mathsf{f}})$ and its transpose, e.g., a block Jacobi smoother with Gauss-Seidel smoothing in blocks containing interior unknowns of the subdomains, see also [48]. Furthermore, the interpolation $\mathfrak{P}_h$ has to fulfill the pattern condition (5.9), and we take $\mathfrak{P}_h^T$ as restriction. The coarse grid system can be solved directly (either in parallel or sequentially on one processor) or again by some iterative method similar to the parallel CG. We change the notation of level index from subscript K_j to superscript K^j to emphasize that now $j = 1$ denotes the fine grid.

Algorithm 9 Parallel multigrid $\underline{\mathfrak{u}} \Leftarrow \text{PMG}(\{\mathsf{K}^j, \mathfrak{P}^j\}, \underline{\mathfrak{u}}, \underline{\mathsf{f}}, j)$

if $j == \text{COARSELEVEL}$ **then**
  $\underline{\mathfrak{u}} \Leftarrow \text{SOLVE}(\mathsf{K}^j \cdot \underline{\mathfrak{u}} = \underline{\mathsf{f}})$
else
  $\widetilde{\underline{\mathfrak{u}}} \Leftarrow \text{SMOOTH}(\mathsf{K}^j, \underline{\mathfrak{u}}, \underline{\mathsf{f}})$
  $\underline{\mathsf{d}} \leftarrow \underline{\mathsf{f}} - \mathsf{K}^j \cdot \widetilde{\underline{\mathfrak{u}}}$
  $\underline{\mathsf{d}}_H \leftarrow (\mathfrak{P}^j)^T \cdot \underline{\mathsf{d}}$
  $\underline{\mathfrak{w}}_H \leftarrow 0$
  $\underline{\mathfrak{w}}_H \Leftarrow \text{PMG}(\{\mathsf{K}^j, \mathfrak{P}^j\}, \underline{\mathfrak{w}}^H, \underline{\mathsf{d}}^H, j+1)$
  $\underline{\mathfrak{w}} \leftarrow \mathfrak{P}^j \cdot \underline{\mathfrak{w}}_H$
  $\widetilde{\underline{\mathfrak{u}}} \leftarrow \widetilde{\underline{\mathfrak{u}}} + \underline{\mathfrak{w}}$
  $\underline{\mathfrak{u}} \Leftarrow \text{SMOOTH}^T(\mathsf{K}^j, \widetilde{\underline{\mathfrak{u}}}, \underline{\mathsf{f}})$
end if

All steps in Algorithm 9 with single-lined arrows $\leftarrow$ are local on each processor using local data only. On the other hand, the use of a double-lined arrow $\Leftarrow$ indicates that communication is involved. Despite the coarse grid solver, only the smoothing sweeps require communication.

5.3 Parallel setup for AMG

In the design of our parallel AMG code we look for components which minimize the communication overhead without increasing the overall effort perceptibly. If the prolongation matrices $\mathfrak{P}^\ell$ satisfy the pattern condition (5.9), we conclude from (5.11) and (5.12) that all the operations in Algorithm 12 and Algorithm 9 involving $\mathfrak{P}^\ell$ are local. The pattern of $\mathfrak{P}^\ell$ is controlled by the strong connections $\{S^{i,T}\}^\ell$ as input parameters in the function WEIGHTS, calculating the prolongation weights according to Section 4.3. Therefore, we have to control the sets of strong connections in the coarsening step to ensure the required interpolation pattern, see also [31].

5.3.1 Communication groups and node sets

Condition (5.9) is a good criterion for checking an already given matrix. But for AMG we need to derive a constructive algorithm from that criterion for the construction of the prolongation operators $\mathfrak{P}^\ell$. Obviously, each node belongs to a communication group σ_k which is characterized by the identical set of subdomains for all group members. Now, subset relations between these communication groups as in condition (5.9) are used to derive the admissible pattern for prolongation. These subset relations will group the nodes into four disjoint sets of nodes with respect to the given communication group σ_k. We define the set of all communication groups by

$$\Sigma = \{\sigma_k\}_k := \bigcup_{i\in\omega_h} \sigma^{[i]}. \tag{5.13}$$

The elements of Σ are uniquely ordered by the rules

(1) $|\sigma_k| > |\sigma_m| \Rightarrow k < m$;

(2) $|\sigma_k| = |\sigma_m|$ and σ_k lexicographically larger than $\sigma_m \Rightarrow k < m$.

This ordering prevents communication deadlocks in the algorithm. Furthermore, we introduce the node sets for a given communication group $\sigma_k \in \Sigma$:

(a) active nodes: $\omega_a := \{i \in \omega_h : \sigma^{[i]} \equiv \sigma_k\}$;

(b) visible nodes: $\omega_v := \{i \in \omega_h : \sigma^{[i]} \subset \sigma_k\}$;

(c) grouped nodes: $\omega_g := \{i \in \omega_h : \sigma^{[i]} \supset \sigma_k\}$;

(d) invisible nodes: $\{i \in \omega_h : \sigma^{[i]} \not\subseteq \sigma_k \wedge \sigma^{[i]} \not\supseteq \sigma_k\} \equiv \omega_h \setminus (\omega_a \cup \omega_v \cup \omega_g)$.

The model example in Fig. 13 (which uses a linear FE discretization) possesses by definition (5.1) the communication groups $\{1,2,3,4\}$, $\{1,2\}$, $\{1,3\}$, $\{2,4\}$, $\{3,4\}$ and trivially $\{1\}$, $\{2\}$, $\{3\}$, $\{4\}$. Further, for subdomain 1, i.e., Ω_1, we get four groups where $\sigma_1 = \{1,2,3,4\}$, $\sigma_2 = \{1,2\}$, $\sigma_3 = \{1,3\}$ and $\sigma_4 = \{1\}$. The last one does not require any communication but simplifies the algorithms. It is obvious from Fig. 13, that $\sigma_1 = \sigma^{[25]}$, $\sigma_2 = \sigma^{[18]}$, $\sigma_3 = \sigma^{[24]}$ and $\sigma_4 = \sigma^{[1]}$. In addition we obtain for subdomain Ω_1 with respect to $\sigma_3 = \{1,3\}$ the following derived node sets $\omega_a = \{22,23,24\}$, $\omega_g = \{25\}$ and $\omega_v = \{1,2,3,8,9,10,15,16,17\}$, and the invisible nodes $\{4,11,18\}$.

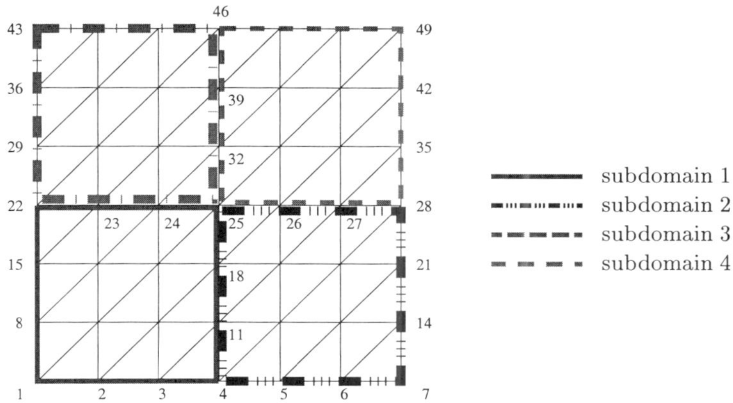

Figure 13: Non-overlapping subdomains with triangulation and row-wise numbering.

5.3.2 Parallel coarsening

The determination of strong connections as well as the calculation of interpolation weights require the accumulated matrix entries which cannot be provided by the distributed matrix K. Therefore, we have to store the stiffness matrix temporarily twice, i.e., as distributed matrix and as a temporary accumulated matrix. The latter is locally stored as $\widetilde{\mathfrak{K}}_s = \mathcal{A}_s\widetilde{\mathfrak{K}}\mathcal{A}_s^T$.

A first step to control the admissible interpolation pattern consists in changing the accumulated matrix such that (5.9) is fulfilled

$$\widetilde{\mathfrak{K}} = \sum_{s=1}^{P} \mathcal{A}_s^T \widetilde{\mathsf{K}}_s \mathcal{A}_s \qquad \text{with} \quad \widetilde{\mathsf{K}}_s^{[i,j]} = \begin{cases} \mathsf{K}_s^{[i,j]} & \text{if } \sigma^{[i]} \subseteq \sigma^{[j]} \vee \sigma^{[i]} \supseteq \sigma^{[j]} \\ 0 & \text{otherwise} \end{cases}, \tag{5.14}$$

where $\widetilde{\mathsf{K}}_s^{[i,j]}$ is introduced only for purposes of notation. This means for Fig. 13, that the resulting matrix $\widetilde{\mathfrak{K}}$ has zero entries $\widetilde{\mathfrak{K}}^{[18,26]} = \widetilde{\mathfrak{K}}^{[24,32]} = 0$, although the original entries may be $\mathfrak{K}^{[18,26]} \neq 0$, $\mathfrak{K}^{[24,32]} \neq 0$. The same holds for the transposed entries. The sets of strong connections S^i, $S^{i,T}$ can be determined by the sequential function coarse$(i, j, \widetilde{\mathfrak{K}}_s)$ (see Section 4.4) and, automatically, these sets do not contain forbidden node connections.

The second step takes advantage of the fact that our ordering of the communication groups from Section 5.3.1 already provides the right relation between the nodes due to ω_a, ω_v and ω_g. The local coarsening will be done subsequently on the active sets of all communication groups σ_k. We have to guarantee coherence of the resulting coarse and fine nodes for active sets shared by more than one subdomain. The easiest way consists in a coarsening only on the root process of communication group σ_k, broadcasting the result within σ_k, and doing the coarsening with the marked coarse nodes ω_m on the remaining processes. The two coarsening parts before and after the communication can be handled by the same routine

$$(\omega_C, \omega_F) \leftarrow \textsc{CoarseP}(\{S_h^{i,T}\}, \omega_m, \omega_a, \omega_v, \omega_C, \omega_F),$$

Algorithm 10 Coarsening $(\omega_C, \omega_F) \leftarrow \text{COARSEP}(\{S_h^{i,T}\}, \omega_m, \omega_a, \omega_v, \omega_C, \omega_F)$

```
while ω_C ∪ ω_F ⊅ ω_a do
  k ← PICK({S_h^{i,T}}, ω_m \ (ω_C ∪ ω_F), ω_F)
  if |S_h^{k,T}| + |S_h^{k,T} ∩ ω_F| = 0 then
    ω_F ← ω_F ∪ (ω_a \ ω_C)
  else
    ω_C ← ω_C ∪ {k}
    ω_F ← ω_F ∪ ((S_h^{k,T} \ ω_C) ∩ ω_v)
  end if
end while
```

see Algorithm 10, which can be derived from the sequential version. The sequential routine is contained in the parallel one via the call

$$(\omega_C, \omega_F) \leftarrow \text{COARSEP}(\{S_h^{i,T}\}, \omega_h, \omega_h, \omega_h, \omega_C, \omega_F).$$

A third step has to restrict the strong connections such that no fine node from the active set ω_a has connections to the visible nodes from ω_v. These three steps guarantee the admissible pattern of the prolongation matrix for operation (5.11). Algorithm 11 presents the routine for parallel coarsening on each subdomain Ω_s.

Algorithm 11 Parallel coarsening $(\{S_h^{i,T}\}, \omega_C, \omega_F) \leftarrow \text{PARCOARSE}(\{\mathrm{S_h^{i,T}}\}, \omega_\mathrm{h})$

```
Determine the set of communication groups Σ
ω_C ← ∅, ω_F ← ∅
for all k = 1, ..., |Σ| do
  ω_a ← {i ∈ ω_h : σ^[i] ≡ σ_k}
  ω_v ← {i ∈ ω_h : σ^[i] ⊂ σ_k}
  ω_g ← {i ∈ ω_h : σ^[i] ⊃ σ_k}
  if ROOT(σ_k) then
    (ω_C, ω_F) ← COARSEP({S_h^{i,T}}, ω_a, ω_a, ω_v, ω_C, ω_F)
    ω_m ← ω_C ∩ ω_a
  end if
  ω_m ⇐ BROADCAST(σ_k, ω_m)
  if not (ROOT(σ_k)) then
    (ω_C, ω_F) ← COARSEP({S_h^{i,T}}, ω_m, ω_a, ω_v, ω_C, ω_F)
  end if
  for all i ∈ ω_a do
    if i ∈ ω_F then
      S_h^{i,T} ← S_h^{i,T} ∩ (ω_a ∪ ω_g)
    else
      S_h^{i,T} ← S_h^{i,T} ∩ (ω_a ∪ ω_v)
    end if
  end for
end for
```

5.3.3 Coarse grid matrix and parallel setup

The computation of the interpolation matrix $\mathfrak{P}_h$ is based on the set of strong connections $\{S_h^{i,T}\}$. That is why $\mathfrak{P}_h$ fulfills the pattern condition (5.9) by construction. Therefore, we call

$$\mathfrak{P}_h \leftarrow \text{WEIGHTS}(\{S_h^{i,T}\}, \widetilde{\mathfrak{K}}, \omega_C, \omega_F)$$

on each processor without any communication. Moreover, we can apply (5.12) with the distributed fine grid matrix K_h so that the coarse matrix K_H is again distributed stored and can be calculated fully in parallel without communication, i.e., we call

$$\mathsf{K}_H \leftarrow \mathfrak{P}_h^T \mathsf{K}_h \mathfrak{P}_h$$

locally on each processor.

Collecting the subroutines of previous sections we obtain Algorithm 12 which realizes the parallel setup with the call $\text{PARSETUP}(\mathsf{K}^1, \omega^1, 1)$ on the finest level. It is to be executed on all processors. The coarsening part in Algorithm 12 terminates if the global number of coarse nodes $|\omega|$ becomes smaller than COARSEGRID, cf. the sequential part. The operators $\mathcal{A}_s^\ell$ denote the subdomain-connectivity matrices with respect to level ℓ.

Algorithm 12 Parallel Setup $\{\mathsf{K}^j, \mathfrak{P}^j\}_{j=1}^{\text{COARSELEVEL}} \leftarrow \text{PARSETUP}(\mathsf{K}_h, \omega_h, \ell)$

if $|\omega| > \text{COARSEGRID}$ **then**
  $\mathsf{K}^\ell \leftarrow \mathsf{K}_h$
  $\widetilde{\mathfrak{K}} \Leftarrow \left(\sum_{s=1}^P (\mathcal{A}_s^\ell)^T \widetilde{\mathsf{K}}_s \mathcal{A}_s^\ell\right)_{\text{pattern}}$
  $\{S^{i,T}\}^\ell \leftarrow \text{GETSTRONG}(\widetilde{\mathfrak{K}})$
  $(\{S^{i,T}\}^\ell, \omega_C, \omega_F) \Leftarrow \text{PARCOARSE}(\{S^{i,T}\}^\ell, \omega^\ell)$
  $\mathfrak{P}^\ell \leftarrow \text{WEIGHTS}(\{S^{i,T}\}^\ell, \widetilde{\mathfrak{K}}, \omega_C, \omega_F)$
  $\mathsf{K}_H \leftarrow (\mathfrak{P}^\ell)^T \mathsf{K}^\ell \mathfrak{P}^\ell$
  $\omega_H \leftarrow \omega_C$
  $\text{PARSETUP}(\mathsf{K}_H, \omega_H, \ell + 1)$
else
  $\text{COARSELEVEL} \leftarrow \ell$
end if

6 Applications

6.1 Potential equation in the human head

People with epileptic fits can only be cured by removing those parts of their brain causing the fit. The main difficulty consists in identifying the location and size of the malfunctioning brain-matter, called medical source reconstruction. In order to do this, one has to solve about 10,000 times the potential equation

$$-\operatorname{div}(k(x,y,z)\nabla u(x,y,z)) = f(x,y,z) \tag{6.1}$$

in the human head Ω with Neumann boundary conditions on $\partial\Omega$ and the source distribution $f(x,y,z)$. Several hundred thousand tetrahedral elements are necessary to reflect the main characteristics of the human head in the discrete model in Fig. 14. The conductivity $k(x, y, z)$ varies from $0.33\,[1/\Omega m]$ in all skin- and brain-elements, through $0.0042\,[1/\Omega m]$ in the skull elements, to $1.0\,[1/\Omega m]$ in the CSF, i.e., within the layer between brain and skull and within the ventricular system [1].

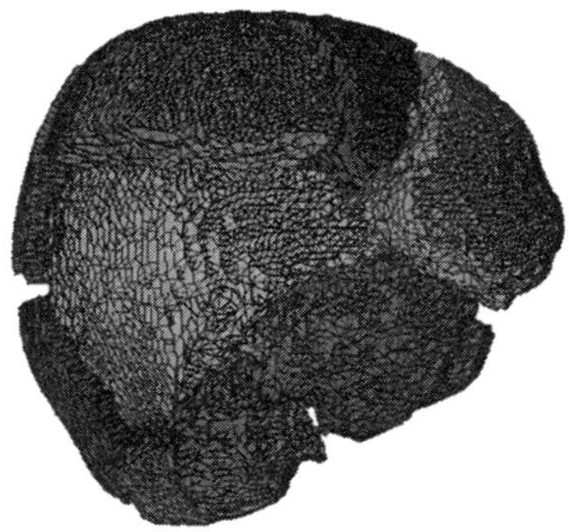

Figure 14: FE-meshes of human head, partitioned for 12 processors with METIS and visualized with PMVIS: 713733 tetrahedra and 118299 nodes.

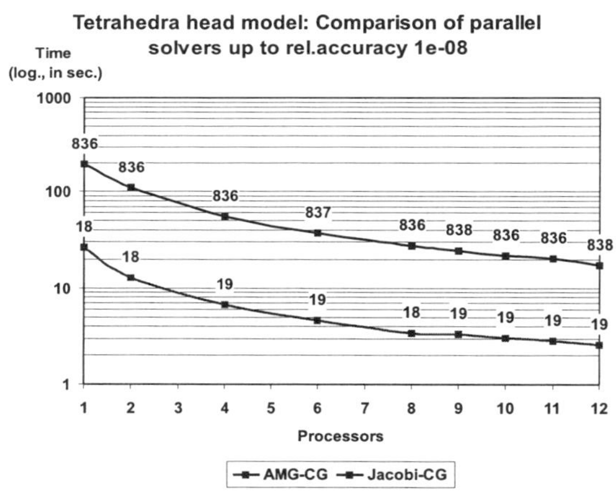

Figure 15: WCPU with respect to number of processors for the solver part. The numbers of iterations are shown with the graphs.

We ran the experiment on an SGI ORIGIN 200 with R10000, 195 MHz processors and overall 6 GB of main memory and solved the finite element equations derived from (6.1) with an AMG-preconditioned CG using one $V(1,1)$-cycle in the AMG-preconditioner. The local accumulation of the matrix on one processor took 173.4 seconds; parallelized on 12 processors a setup time of 14.89 seconds was achieved.

Fig. 15 shows the wall-clock time (WCPU) of our parallel AMG-preconditioned CG solver PEBBLES [60] compared to the standard parallel Jacobi-preconditioned CG. The number of iterations for both solvers, necessary for the required accuracy, is indicated in the graphs. The time for the setup of the preconditioner is not included, since it has to be carried out only once per head model and is thus negligible with regard to the solution of the inverse problem. To give an idea, the setup of the AMG on one processor took 29.9 seconds and parallelized on 12 processors 7.4 seconds. The 3D potential distribution was calculated on one processor within 195.8 seconds with the Jacobi-CG method, whereas the parallel AMG-CG method on 12 processors needed 2.6 seconds. This is a factor of about 75 (7.5 through multi-level preconditioning and 10 through parallelization).

6.2 Magnetostatic valve

One cutting edge in combustion engine design is the development of very fast valves for more fuel injections into the combustion chamber per cycle than can nowadays be achieved. Magnetic valves are one candidate for such superfast valves. One major part in the simulation of these components consists in solving the 3D magnetostatic Maxwell equations

$$\mathbf{curl}(\nu\,\mathbf{curl}\,\mathbf{u}) + \sigma\mathbf{u} = \mathbf{f} \tag{6.2}$$

in the computational domain Ω with appropriate boundary conditions on $\partial\Omega$, where $\sigma = 10^{-4}$, $\nu(x, y, z)$ and $\mathbf{f}(x, y, z)$ are given [51]. Fig. 16 presents the mesh of the FEM discretization

N_h	it	setup [sec.]	solver [sec.]
8714	14	0.53	2.49
65219	28	4.17	44.01
504246	63	34.49	792.49

Figure 16: Material distribution in one quarter of the valve, Numerical results.

with Nedelec-elements containing half a million unknowns, and the iteration counts for an AMG preconditioned CG with the stopping accuracy $\varepsilon = 10^{-8}$. The code PEBBLES runs on an SGI Octane, R12000 300 MHz. We mention that the discrete Maxwell equations require special AMG components [61].

6.3 C-magnet: geometric versus algebraic multigrid

Let us consider the C-magnet depicted in Fig. 17. The magnet field is generated by a current in the coil and can be described by (6.2). The coarse mesh contains 2907 unknowns and the

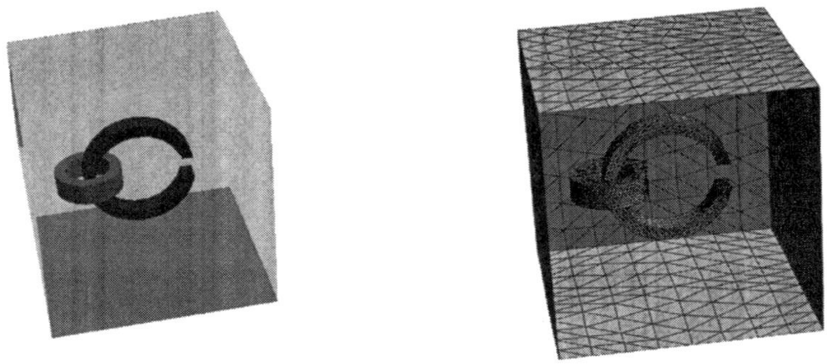

Figure 17: C-Magnet: geometry and mesh with 153408 elements and 206549 nodes.

the fine mesh 180426 unknowns after two refinement steps in each tetrahedral element. We want to compare algebraic multigrid applied to the fine grid problem with geometric multigrid (GMG) using the hierarchy of the three meshes. The hybrid smoother proposed in [46] is used, see also [32]. The required decomposition into subdomains for solving the problem on a parallel computer (SGI ORIGIN 2000, 300 MHz) has been obtained by recursive spectral bisection. In order to analyze the parallel performance we consider the following components of the algorithm: generation of the system matrix, setup and solver, see Table 2.

			Geometric MG					Algebraic MG					
	SysMat		Setup		Solver			Setup		Solver			
P	T	S	T	S	It.	T	S	T	S	It.	T	S	S(1)
1	40.3	1.0	13.1	1.0	13	99.4	1.0	44.4	1.0	60	247.3	1.0	1.0
2	19.6	2.0	10.8	1.2	13	69.2	1.4	24.4	1.8	52	127.1	1.9	1.7
4	9.7	4.1	9.1	1.4	13	31.8	3.1	12.6	3.5	52	48.9	5.0	4.3
8	4.7	8.5	8.8	1.5	13	16.2	6.1	7.6	5.8	58	28.2	8.8	8.4

Table 2: Number of processors (P), wall clock time (T) in seconds for generating the system matrix, the setup phase and the solver. Corresponding speedup (S), number of iterations (It.) and speedup with respect to one iteration (S(1)).

As expected, the generation of the system matrix shows optimal speedup since this component is free of communication. In the case of GMG, the setup phase involves the setup of the smoother and the LU-decomposition of the coarse-grid system matrix with 2907 unknowns. Since the setup is dominated by the sequential LU-decomposition, only low speedups can be observed. In the case of AMG, the setup involves the coarsening itself and the LU-decomposition of the coarse-grid system which has less than 500 unknowns. Hence, the setup shows reasonable speedups since it is dominated by the coarsening of the inner nodes. In both cases, AMG and GMG, the solver shows a similar speedup behavior. Although most of the gain is due to the additional CPU capacity, the additional cache which comes with each processor also contributes to the acceleration. We refer to [33] for more detailed information.

Acknowledgements The authors would like to thank the NATO Advanced Study Institute and the Austrian Fund for the Advancement of Scientific Research (FWF) for supporting this contribution to the summer workshop on "Modern Methods in Scientific Computing and Applications". Special thanks are given to Johanna Kienesberger and Stefan Reitzinger who helped us in preparing the lectures and these lecture notes. We are also indebted to Carsten Wolters (Max Planck Institute of Cognitive Neuroscience Leipzig, Germany) and to Manfred Kaltenbacher (Department of Sensor Technology at the University of Erlangen-Nürnberg, Germany) for providing the examples used in the numerical experiments with our AMG code PEBBELS in Sections 6.1 and 6.2, respectively.

References

[1] A. Anwander, M. Kuhn, S. Reitzinger, and C. Wolters, A parallel algebraic multigrid solver for the finite element method source localization in the human brain, *Computing and Visualization in Science*, 2002 (to appear).

[2] G. P. Astrachancev, An iterative method for solving elliptic net problems, *USSR Computational Math. and Math. Phys.* **11** (2) (1971), 171–182.

[3] O. Axelsson, *Iterative Solution Methods*, Cambridge University Press, Cambridge, 1994.

[4] N. S. Bachvalov, On the convergence of a relaxation method with natural constraints on the elliptic operator, *USSR Computational Math. and Math. Phys.* **6** (5) (1966), 101–135.

[5] R. Bank, *PLTMG: A Software Package for Solving Elliptic Partial Differential Equations*, SIAM, Philadelphia, PA, 1994.

[6] B. Bastian, *Parallele Adaptive Mehrgitterverfahren*, Teubner Skr. Numer., B.G. Teubner, Stuttgart, 1996.

[7] P. Bastian, UG version 2.0 — short manual, Preprint 92–14, IWR Heidelberg, 1992.

[8] F. A. Bornemann and P. Deuflhard, The cascadic multigrid method for elliptic problems, *Numer. Math.* **75** (1996), 135–152.

[9] D. Braess, Towards algebraic multigrid for elliptic problems of second order, *Computing* **55** (1995), 379–393.

[10] D. Braess, *Finite Elements: Theory, Fast Solvers and Applications in Solid Mechanics*, Cambridge University Press, Cambridge, 1997.

[11] D. Braess, M. Dryja, and W. Hackbusch, A multigrid method for nonconforming fe-discretisations with application to non-matching grids, *Computing* **63** (1999), 1–25.

[12] D. Braess and W. Hackbusch, A new convergence proof for the multigrid method including the V-cycle, *SIAM J. Numer. Anal.* **20** (1983), 967–975.

[13] J. H. Bramble, *Multigrid Methods*, Pitman Res. Notes Math. Ser. **294**, Longman Sci. Tech., Harlow, 1993.

[14] J. H. Bramble and J. E. Pasciak, New convergence estimates for multigrid algorithms, *Math. Comput.* **49** (180) (1987), 311–329.

[15] J. H. Bramble and J. E. Pasciak, New estimates for multilevel algorithms including the V-cycle, *Math. Comput.* **60** (1993), 447–471.

[16] A. Brandt, Multi-level adaptive techniques (MLAT) for fast numerical solution to boundary value problems, in: *Proc. 3rd Internat. Conf. on Numerical Methods in Fluid Mechanics, Paris, 1972*, Lecture Notes in Phys., Springer-Verlag, Berlin-Heidelberg-New York, 1973.

[17] A. Brandt, Multi-level adaptive solutions to boundary-value problems, *Math. Comput.* **31** (1977), 333–390.

[18] A. Brandt, Algebraic multigrid theory: The symmetric case, *Appl. Math. Comput.* **19** (1986), 23–56.

[19] A. Brandt, Multiscale scientific computation: Review 2001, in: *Multiscale and Multiresolution Methods* (T. Barth, R. Haimes, and T. Chan, eds.), Springer-Verlag, Berlin-Heidelberg-New York, 2001.

[20] A. Brandt, S. F. McCormick, and J. W. Ruge, Algebraic multigrid (AMG) for sparse matrix equations, in: *Sparsity and Its Applications* (D. J. Evans, ed.), Cambridge University Press, Cambridge, 1985, 257–284.

[21] A. Brandt, S. F. McCormick, and J. W. Ruge, Algebraic multigrid (AMG) for automatic multigrid solution with application in geodetic computations, Technical Report CO POB 1852, Inst. Comp. Studies State Univ., 1982.

[22] S. C. Brenner and L. R. Scott, *The Mathematical Theory of Finite Element Methods*, Springer-Verlag, New York, 1994.

[23] M. Brezina, A. Cleary, R. Falgout, V. Henson, J. Jones, T. Manteuffel, S. McCormick, and J. Ruge, Algebraic multigrid based on element interpolation (AMGe), *SIAM J. Sci. Comput.* **22** (5) (2000), 1570–1592.

[24] W. L. Briggs, V. E. Henson, and S. F. McCormick, *A Multigrid Tutorial*, 2nd ed., SIAM, Philadelphia, PA, 2000.

[25] P. Deuflhard, Cascadic conjugate gradient methods for elliptic partial differential equations I: Algorithm and numerical results, Preprint SC 93–23, Konrad-Zuse-Zentrum für Informationstechnik Berlin, 1993.

[26] E. Dick, K. Riemslagh, and J. Vierendeels, eds., *Multigrid Methods VI. Proceedings Sixth European Multigrid Conference*, Springer-Verlag, Berlin, 2000, 27-30.

[27] C. Douglas, G. Haase, and U. Langer, *A Tutorial on Elliptic PDEs and Parallel Solution Methods*, SIAM, 2002, in preparation.

[28] R. P. Fedorenko, A relaxation method for elliptic difference equations, *USSR Computational Math. and Math. Phys.* **1** (5) (1961), 1092–1096.

[29] R. P. Fedorenko, The speed of convergence of one iterative process, *USSR Computational Math. and Math. Phys.* **4** (3) (1964), 227–235.

[30] T. Grauschopf, M. Griebel, and H. Regler, Additive multilevel-preconditioners based on bilinear interpolation, matrix dependent geometric coarsening and algebraic-multigrid coarsening for second order elliptic PDEs, SFB–Bericht Nr. 342/02/96, Technische Universität, München, 1996.

[31] G. Haase, A parallel AMG for overlapping and non-overlapping domain decomposition, *Electron. Trans. Numer. Anal.* **10** (2000), 41–55.

[32] G. Haase, M. Kuhn, and U. Langer, Parallel multigrid 3d Maxwell solvers, *Parallel Comput.* **6** (27) (2001), 761–775.

[33] G. Haase, M. Kuhn, U. Langer, S. Reitzinger, and J. Schöberl, Parallel Maxwell solvers, in: *Scientific Computing in Electrical Engineering* (U. van Rienen, M. Günther, and D. Hecht, eds.), Springer-Verlag, Berlin-Heidelberg-New York, 2000, 71–78.

[34] G. Haase, M. Kuhn, and S. Reitzinger, Parallel AMG on distributed memory computers, *SIAM SISC* (2002), to appear.

[35] G. Haase, U. Langer, and A. Meyer, The approximate Dirichlet decomposition method. I, II, *Computing* **47** (1991), 137–167.

[36] G. Haase, U. Langer, S. Reitzinger, and J. Schöberl, Algebraic multigrid methods based on element preconditioning, *Internat. J. Computer Math.* **80** (3–4) (2001).

[37] W. Hackbusch, Implementation of the multi-grid method for solving partial differential equations, Technical Report RA 82, IBM T. J. Watson Research Centre, 1976.

[38] W. Hackbusch, *Multigrid Methods and Applications*, Springer-Verlag, Berlin, 1985.

[39] W. Hackbusch and U. Trottenberg, eds., *First European Conference on Multigrid Methods*, Lecture Notes in Math. **960**, Springer-Verlag, Berlin-Heidelberg-New York, 1982.

[40] W. Hackbusch and U. Trottenberg, eds, *Second European Conference on Multigrid Methods*, Lecture Notes in Math. **1228**, Springer-Verlag, Berlin-Heidelberg-New York, 1986.

[41] W. Hackbusch and U. Trottenberg, eds, *Third European Conference on Multigrid Methods*, Internat. Ser. Numer. Math. **98**, Birkhäuser, Basel, 1991.

[42] W. Hackbusch and G. Wittum, eds., *Multigrid Methods V. Proceedings of the Fifth European Multigrid Conference*, Springer-Verlag, Berlin, 1998.

[43] P. W. Hemker and P. Wesseling, eds., *Multigrid Methods IV. Proceedings of the Fourth European Multigrid Conference*, Birkhäuser, Basel, 1994.

[44] V. Henson and P. Vassilevski, Element-free AMGe: General algorithms for computing interpolation weights in AMG, *SIAM J. Sci. Comput.* **23** (2) (2001), 629–650.

[45] V. E. Henson and U. M. Yang, BoomerAMG: a parallel algebraic multigrid solver and preconditioner, Technical Report UCRL–JC–141495, Lawrence Livermore National Laboratory, 2000.

[46] R. Hiptmair, Multigrid methods for Maxwell's equations, *SIAM J. Numer. Anal.* **36** (1999), 204–225.

[47] J. Jones and P. Vassilevski, AMGe based on element agglomeration, *SIAM J. Sci. Comput.* **23** (1) (2001), 109–133.

[48] M. Jung, On the parallelization of multi-grid methods using a non-overlapping domain decomposition data structure, *Appl. Numer. Math.* **23** (1) (1997), 119–137.

[49] M. Jung and U. Langer, Applications of multilevel methods to practical problems, *Surveys Math. Industry* **1** (1991), 217–257.

[50] M. Jung, U. Langer, A. Meyer, W. Queck, and M. Schneider, Multigrid preconditioners and their applications, in: *Proc. 3rd Multigrid Seminar, Biesenthal, GDR, 1989* (G. Telschow, ed.), Report-Nr. R-MATH-03/89, Karl-Weierstrass-Institute of the Academy of Science of the GDR, Berlin, 1989, 11–52.

[51] M. Kaltenbacher, S. Reitzinger, and J. Schöberl, Algebraic multigrid for solving 3D nonlinear electrostatic and magnetostatic field problems, *IEEE Trans. Magnetics* **36** (4) (2000), 1561–1564.

[52] F. Kickinger, Algebraic multigrid for discrete elliptic second-order problems, in: *Multigrid Methos V. Proc. 5th European Multigrid Conf.* (W. Hackbusch, ed.), Lecture Notes in Comput. Sci. Engrg. **3**, Springer-Verlag, New York, 1998, 157–172.

[53] V. G. Korneev, *Finite Element Schemes of Higher Order of Accuracy*, Leningrad University Press, Leningrad, 1977, (Russian).

[54] A. Krechel and K. Stüben, Parallel algebraic multigrid based on subdomain blocking, *Parallel Comput.* **8** (27) (2001), 1009–1031.

[55] M. Kuhn, U. Langer, and J. Schöberl, Scientific computing tools for 3d magnetic field prpblems, in: *The Mathematics of Finite Elements and Applications* (J. R. Witheman, ed.), Elsevier, Amsterdam, 2000, 239–258.

[56] U. Langer, On the choice of iterative parameters in the relaxation method on a sequence of meshes, *USSR Computational Math. and Math. Phys.* **22** (5) (1982), 98–114.

[57] K. H. Law, A parallel finite element solution method, *Comput. & Structures* **23** (6) (1989), 845–858.

[58] S. McCormick, *Multilevel Adaptive Methods for Partial Differential Equations*, Frontiers Appl. Math. **6**, SIAM, Philadelphia, PA, 1989.

[59] G. Meurant, *Computer Solution of Large Systems*, North-Holland, Amsterdam, 1999.

[60] S. Reitzinger, *Algebraic Multigrid Methods for Large Scale Finite Element Equations*, Universitätsverlag Rudolf Trauner, Linz, 2001.

[61] S. Reitzinger and J. Schöberl, An algebraic multigrid method for finite element discretization with edge elements, *Numer. Linear Algebra Appl.* (2002), to appear.

[62] U. Rüde, *Mathematical and Computational Techniques for Multilevel Adaptive Methods*, Frontiers Appl. Math. **13**, SIAM, Philadelphia, PA, 1993.

[63] J. W. Ruge and K. Stüben, Efficient solution of finite difference and finite element equations by algebraic multigrid (AMG), in: *Multigrid Methods for Integral and Differential Equations* (D. J. Paddon and H. Holstein, eds.), Inst. Math. Appl. Conf. Ser., Clarendon Press, Oxford, 1985, 169–212.

[64] J. W. Ruge and K. Stüben, Algebraic multigrid (AMG), in: *Multigrid Methods* (S. F. McCormick, ed.), Frontiers Appl. Math. **3**, SIAM, Philadelphia, PA, 1987, 73–130.

[65] J. Schöberl, Multigrid methods for a parameter dependent problem in primal variables, *Numer. Math.* **84** (1999), 97–119.

[66] V. V. Shaidurov, *Multigrid Methods for Finite Elements*, Kluwer, Dordrecht, 1995.

[67] K. Stüben, Algebraic multigrid: An introduction with applications, in: *Multigrid* (U. Trottenberg, C. Oosterlee, and A. Schüller, eds.), Academic Press, 2000, 413–532.

[68] K. Stüben, A review of algebraic multigrid, *J. Comput. Appl. Math.* **128** (2001), 281–309.

[69] U. Trottenberg, C. Oosterlee, and A. Schüller, eds., *Multigrid*, Academic Press, 2000.

[70] St. Vandewalle, *Parallel Multigrid Waveform Relaxation for Parabolic Problems*, Teubner, Stuttgart, 1993.

[71] P. Vaněk, Acceleration of convergence of a two level algorithm by smoothing transfer operators, *Appl. Math.* **37** (1992), 265–274.

[72] P. Vaněk, J. Mandel, and M. Brezina, Algebraic multigrid based on smoothed aggregation for second and fourth order problems, *Computing* **56** (1996), 179–196.

[73] P. Wesseling, *An Introduction to Multigrid Methods*, Wiley, Chichester, 1992.

[74] H. Yserentant, Old and new convergence proofs for multigrid methods, in: *Acta Numerica*, Cambridge University Press, 1993, 285–326.

One-way operators, absorbing boundary conditions and domain decomposition for wave propagation

Laurence HALPERN

and

Adib RAHMOUNI

Département de Mathématiques, Institut Galilée
Université Paris 13
93430 Villetaneuse
France

Abstract

We present the notion of paraxial operators and applications to absorbing boundary conditions and domain decomposition for hyperbolic problems. We also describe the new concept of perfectly matched layers.

1 Introduction

We treat here the question of computing the solution of a problem in a large, possibly infinite, domain. The applications are numerous: geophysics, aeronautics, oceanography ... The tool for large problems is domain decomposition, where different problems with different scales and/or different numerical discretization methods can be used in different subdomains. Many of these problems are also posed in infinite domains. In some cases the equation in the domain is replaced by a simpler one which for instance represents the propagation in a preferred direction: this leads to the notion of paraxial equations. In any case, the computational domain has to be bounded in all directions. One has to introduce fictitious boundaries with so-called absorbing or farfield boundary conditions. There is a strong connection between these notions, as a transparent boundary operator is a paraxial operator, as we shall demonstrate here. We shall also demonstrate that the optimal transmission condition between two subdomains in domain decomposition is produced by the paraxial operator. This will provide the link between these three notions.

The guiding line here is the notion of wave propagation. In Section 2 we start with the transport equation, where we introduce the notions of finite speed of propagation, well-posedness, numerical schemes, and the necessity of adding extra boundary conditions. In Section 3 we turn to the one-dimensional wave equation, where we introduce the concepts of paraxial operators, absorbing boundary conditions and optimal transmission conditions for domain decomposition. A numerical scheme is described, and a domain decomposition procedure is applied to heterogeneous media. In Section 4 we give precise definitions for the three notions for the wave equation in two dimensions, and we highlight the links between

A. Bourlioux and M.J. Gander (eds.), Modern Methods in Scientific Computing and Applications, 155–209.

them. We give some preliminary results on domain decomposition with optimal transmission conditions in this case.

In the last five years, there has been a tremendous activity about a new notion of perfectly matched layers, which is very competitive to absorbing boundary conditions for linear constant coefficient hyperbolic models. Therefore the last section is devoted to the description of this notion, and several improvements.

All the references are to be found in the body of the text.

Contents

2 The one-dimensional transport equation

2.1 The continuous problem

Let us consider the scalar transport equation

$$\frac{\partial u}{\partial t} + a(x)\frac{\partial u}{\partial x} = f \quad \text{on} \quad \mathbb{R} \times (0,T), \tag{2.1}$$

where a is a continuous function of x with a definite sign: a is either positive or negative in the domain. We can write (2.1) as

$$\frac{d}{dt}u(x(t),t) = f(x(t),t), \tag{2.2}$$

where $x(t)$ is defined as the characteristic curve

$$\frac{dx}{dt} = a. \tag{2.3}$$

This expresses the fact that the solution propagates along the *characteristics curves*. If the velocity a is a constant, they are straight lines as described in Fig. 1.

If a is a constant, there are special solutions of this equation, the *plane waves*. They are given by

$$u(x,t) = e^{i(kx-\omega t)}.$$

Such a function is a solution of the equation if and only if ω and k are related through the *dispersion relation*

$$\omega = ak.$$

For general waves, the function $v_\varphi(k) = \omega/k$ is called the phase velocity. Here the dispersion relation is a linear function of k: the model is *non-dispersive*.

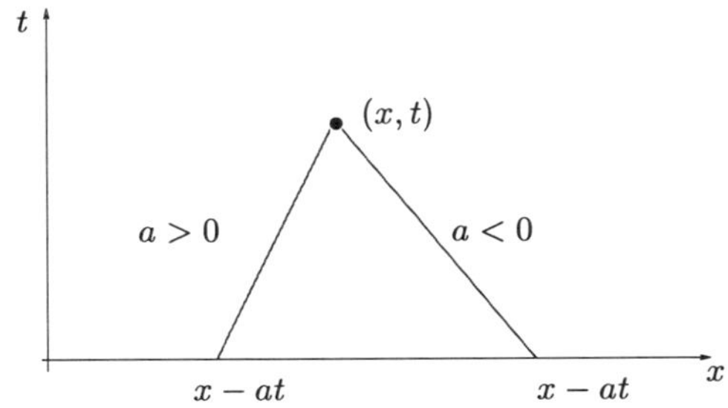

Figure 1: Characteristic lines for a constant velocity a.

2.1.1 The Cauchy problem

We introduce now the *Cauchy problem* by adding initial data

$$u(x,0) = u^{(0)}(x) \quad \text{on} \quad \mathbb{R}. \tag{2.4}$$

We define the *well-posedness* in the sense of Hadamard: a problem is well-posed if

- it has a smooth solution;
- the solution is unique;
- the solution depends continuously on the data.

The following theorem asserts that the Cauchy problem for the transport equation is well-posed:

2.1 Theorem *Assume that the initial velocity $u^{(0)}$ is in $L^2(\mathbb{R})$ and the right-hand side f is in $L^2(0,T;L^2(\mathbb{R}))$. Then problem* (2.1), (2.4) *has a unique solution u in $L^2(0,T;L^2(\mathbb{R}))$. This solution is given on the characteristic curves* (2.3) *by*

$$u(x(t),t) = u^{(0)}(x(0)) + \int_0^t f(x(s),s)\,ds. \tag{2.5}$$

In particular, if a is constant we have

$$u(x,t) = u^{(0)}(x-at) + \int_0^t f(x-a(t-s),s)\,ds.$$

Proof Using (2.2) and the theory of differential equations, a unique characteristic exists through any given point (x,t) and we obtain (2.5). □

This result shows that the solution u at a point (x,t) depends only on the values of the data along the characteristic curve. It proves that *the solution propagates with speed a*: if at time 0 the solution vanishes outside the interval $[\alpha,\beta]$, then at any time t the solution for $f := 0$ vanishes outside the interval $[\alpha + a_* t, \beta + a_* t]$ with $a_* = \sup_{x\in\mathbb{R}} a(x)$.

2.1.2 The initial boundary value problem

We shall suppose here the velocity a to be constant, and introduce the convection equation on the half-line $\mathbb{R}^+$:

$$\frac{\partial u}{\partial t} + a\frac{\partial u}{\partial x} = f \quad \text{in} \quad \mathbb{R}^+ \times (0, T), \tag{2.6}$$

with the initial data

$$u(x, 0) = u^0(x) \quad \text{in} \quad \mathbb{R}^+. \tag{2.7}$$

If a is negative, the characteristics leave the domain, the solution travels to the left, so we do not need a condition on the boundary $x = 0$. In the case where a is positive, the characteristics enter the domain, the solution travels to the right, so we need a boundary condition at $x = 0$. This is illustrated in Fig. 2.

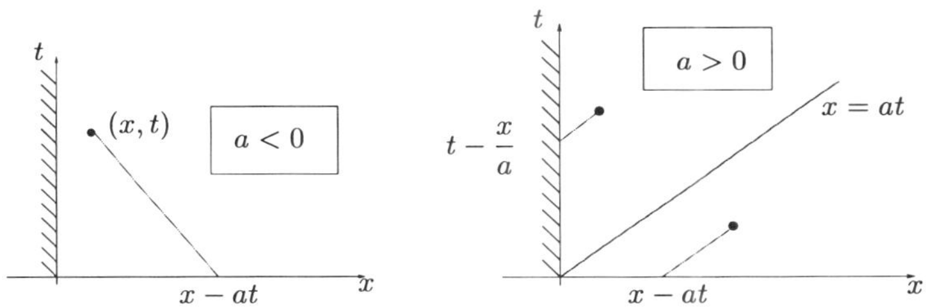

Figure 2: Propagation on the half-line.

The simplest boundary condition is to prescribe the solution on the boundary, that is,

$$u(0, t) = g(t) \quad \text{on} \quad (0, T), \tag{2.8}$$

and the related *initial boundary value problem* is well-posed:

2.2 Theorem *If a is negative, Problem* (2.6), (2.7) *admits a unique solution*

$$u(x, t) = u^0(x - at).$$

If a is positive, Problem (2.6), (2.7) *with boundary condition* (2.8) *has a unique solution*

$$u(x, t) = \begin{cases} u^0(x - at) & \textit{for } x > at, \\ g(t - x/a) & \textit{for } x < at. \end{cases}$$

The initial boundary value problem is well-posed in $L^2(0, T; L^2(\mathbb{R}_+))$. Furthermore, if the compatibility condition

$$u^0(0) = g(0)$$

is satisfied, then it is well-posed in the space of continuous functions.

2.2 The discrete algorithm

We are given a mesh in time and space, with size Δt and Δx; the mesh points in space are $x_j = j\Delta x$, and in time they are $t^n = n\Delta t$. As usual we denote by u_j^n an approximation of $u(x_j, t^n)$.

2.2.1 Examples

The simplest scheme is the upwind scheme (Fig. 3) given by

$$\frac{u_j^{n+1} - u_j^n}{\Delta t} + a\frac{u_j^n - u_{j-1}^n}{\Delta x} = 0. \tag{2.9}$$

It is of order one in time and in space. For a better accuracy in time and space, the leap-frog scheme is very classical (Fig. 4). It is of order 2 in time and space, but it is centered, which can alter the dispersion properties,

$$\frac{u_j^{n+1} - u_j^{n-1}}{2\Delta t} + a\frac{u_{j+1}^n - u_{j-1}^n}{2\Delta x} = 0. \tag{2.10}$$

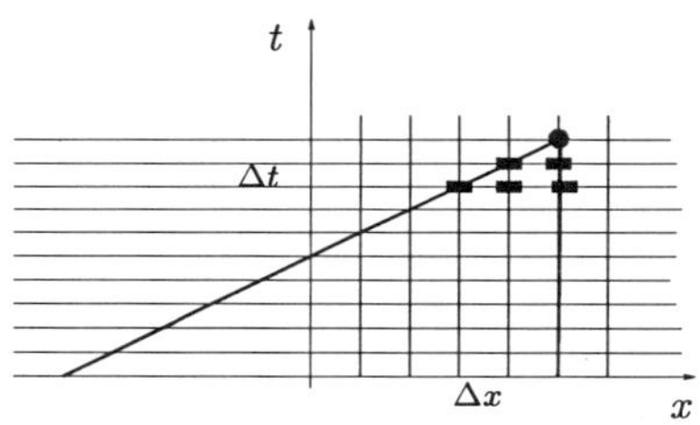

Figure 3: Upwind scheme (2.9).

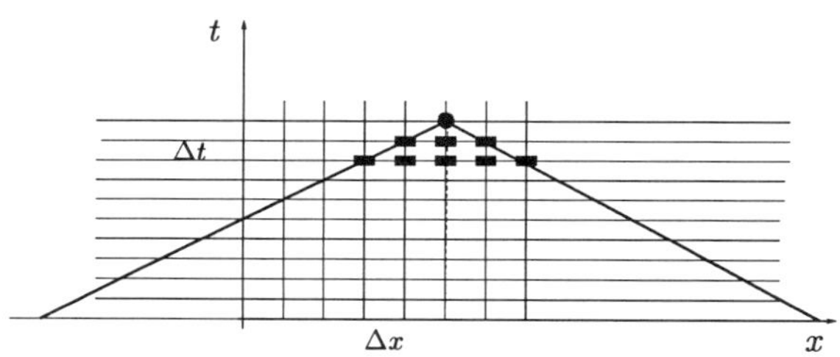

Figure 4: Leap-frog scheme (2.10).

2.2.2 The Cauchy problem. Propagation, stability, convergence

In Fig. 5, the solid line represents the numerical cone of dependence for the upwind scheme (2.9), while the dashed lined represents the characteristic line. Taking an initial value vanishing at every point but one on the grid shows that in the case where a is larger than $\Delta x/\Delta t$, the scheme cannot converge.

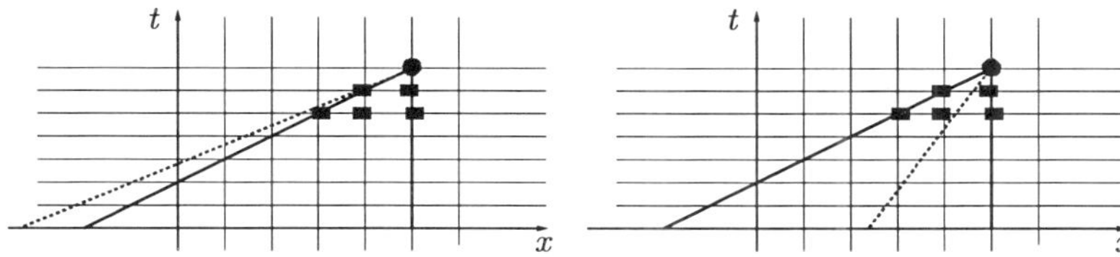

$|a| > \frac{\Delta x}{\Delta t}$ $\qquad$ $|a| < \frac{\Delta x}{\Delta t}$

Figure 5: Cone of dependence for the upwind scheme (2.9) (dashed) together with the characteristic line for the equation (solid).

So we have the following:

2.3 Theorem *If a three-points numerical scheme converges, then the following CFL condition holds :*

$$\gamma = |a| \frac{\Delta t}{\Delta x} \leq 1. \tag{2.11}$$

Here CFL is the classical shorthand for the three mathematicians Courant-Friedrichs-Lewy [35]. As for stability, we have the following definition. We shall call u_h^n the sequence $(u_j^n)_{j \in \mathbb{R}}$.

2.4 Definition The scheme is *stable* if there exists a real number α such that, for every meshsize in time Δt, and space Δx, we have, for any time step n, that

$$\|u_h^n\|_{l^2} \leq C e^{\alpha n \Delta t} \|u_h^0\|_{l^2}.$$

A simple characterization is given by a discrete Fourier transform in space: for any real number χ we consider simple waves of the type

$$u_j^n = \hat{u}^n e^{i \chi x_j}. \tag{2.12}$$

Inserting (2.12) into the scheme, we get for a one step method in time

$$\hat{u}^{n+1} = g(\gamma, \chi \Delta x) \hat{u}^n. \tag{2.13}$$

The function g is the amplification factor, and as (2.13) suggests, it depends only on two quantities, γ and $\chi \Delta x$. For the upwind scheme (2.9), for instance, we have

$$g(\gamma, \zeta) = 1 - \gamma(1 - e^{-i\zeta}).$$

For a second order scheme like the leapfrog scheme (2.10), there is an amplification matrix

$$\begin{pmatrix} \hat{u}^{n+1} \\ \hat{u}^n \end{pmatrix} = G(\gamma, \chi \Delta x) \begin{pmatrix} \hat{u}^n \\ \hat{u}^{n-1} \end{pmatrix}.$$

For instance, for the leapfrog scheme we have

$$G(\gamma, \zeta) = \begin{pmatrix} 1 & -2i\gamma \sin \zeta \\ 1 & 0 \end{pmatrix}.$$

Then the scheme is stable for step sizes Δt and Δx if and only if there exists a real number β such that for any ζ we have

$$\|G(\gamma,\zeta)\| \leq e^{\beta \Delta t},$$

where $\|\cdot\|$ is the usual euclidian norm on the space of matrices. This leads to a condition on γ. For instance, for the two schemes above, the scheme is stable under the strict CFL condition. If we go to higher orders in time and space, the stability condition becomes more restrictive. For details on these notions see [35].

Stability and consistency (i.e. order greater than one) are the key notions for a *linear scheme*, due to the Lax theorem:

$$\textit{Stability} + \textit{consistency} \iff \textit{convergence}.$$

2.2.3 The initial boundary value problem

We consider a scheme with homogeneous data and initial values, and a discrete inhomogeneous boundary condition.

We insert $u_j^n = \alpha z^n r^j$ into the homogeneous equation. Then we get the dispersion relation between z and r, which we solve as an equation in the variable r as a function of z. We now insert u_j^n into the heterogeneous boundary condition. We get an equation like $E(z,\gamma)\alpha = g$. If the preceding equation is invertible for all z such that $|z| \geq 1$, it is called *GKS stable* (for Gustafsson, Kreiss and Sundström). These notions of stability are explained in detail in the book by Strikwerda [35].

This notion of stability is stronger than the usual L^2 stability, because it implies estimates on the boundary. Some boundary conditions can be L^2 stable, but not GKS stable. The strength of the condition comes from the fact that it provides the algebraic criterium cited above.

2.2.4 Transparent boundary conditions

We suppose here that $a > 0$, and we consider the problem on the half-line $x > 0$, with a boundary condition $g(t)$ at $x = 0$. Suppose we want to compute the solution only on the *domain of interest* $]0, A[$. Since the solution propagates to the right, Theorem 2.2 gives the answer : we do not need any information on the boundary. We introduce now a discretized problem. Suppose it is the upwind scheme (2.9). Fig. 6 shows that we do not need any information on the boundary $x = A$ either.

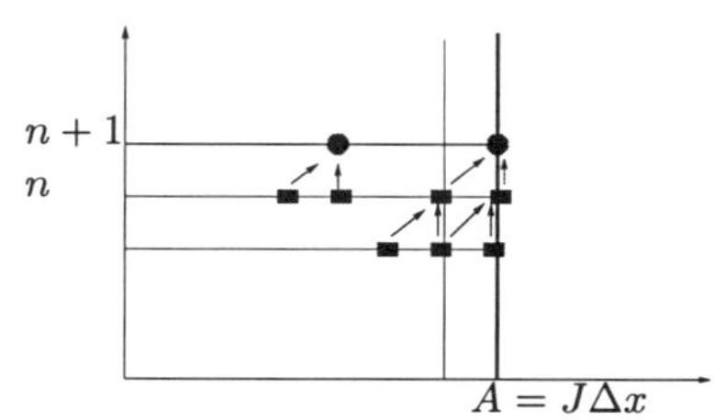

Figure 6: Upwind scheme (2.9) on a right boundary.

The situation is totally different if we deal with centered schemes, as illustrated in Fig. 7:

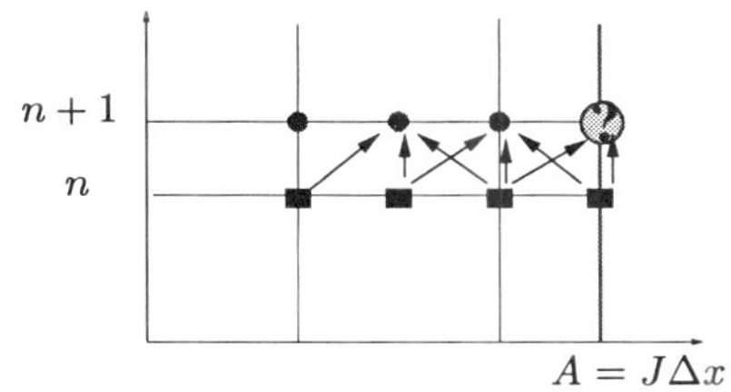

Figure 7: Leap-frog scheme (2.10) on the right boundary.

At time $n+1$, we do not have enough information to compute u_J^{n+1}. The usual procedures are either to use an upwind scheme on the boundary, to extrapolate, or to create fictitious points outside the domain [18]. The constraints are the stability of the discrete initial boundary value problem, and the consistency with the equation. Apart from the important work by B. Gustafsson in [18] on the stability, we do not know of a general analysis of this problem.

3 The one-dimensional wave equation

3.1 The homogeneous case

Consider the one-dimensional wave equation with a constant velocity c:

$$\frac{1}{c^2}\frac{\partial^2 u}{\partial t^2} - \frac{\partial^2 u}{\partial x^2} = f \quad \text{in} \quad \mathbb{R} \times (0, T). \tag{3.1}$$

3.1.1 Propagation properties

If the right-hand side vanishes, $f = 0$, the general solution is given by

$$u(x,t) = F(x+ct) + G(x-ct), \tag{3.2}$$

where F and G are arbitrary functions. The first term $(x,t) \mapsto F(x+ct)$ propagates to the left, along the characteristics with slope $-c$:

$$\left(\frac{1}{c}\frac{\partial}{\partial t} - \frac{\partial}{\partial x}\right) F(x+ct) = 0, \tag{3.3}$$

while the second term $(x,t) \mapsto G(x-ct)$ propagates to the right, along the characteristics with slope $+c$:

$$\left(\frac{1}{c}\frac{\partial}{\partial t} + \frac{\partial}{\partial x}\right) G(x-ct) = 0. \tag{3.4}$$

Let us now introduce the plane waves

$$u(x,t) = e^{i(kx-\omega t)}.$$

The dispersion relation is given by

$$\omega^2 = c^2 k^2.$$

3.1.2 The Cauchy problem

We now add initial data, given by an initial displacement and an initial velocity,

$$u(x,0) = u^{(0)}(x), \qquad \frac{\partial u}{\partial t}(x,0) = u^{(1)}(x) \tag{3.5}$$

The solution of (3.1), (3.5) can be given explicitely by the d'Alembert formula (3.6),

$$\boxed{\begin{aligned} u(x,t) = & \frac{1}{2}[u^{(0)}(x+ct) + u^{(0)}(x-ct)] \\ & + \frac{1}{2c}\int_{x-ct}^{x+ct} u^{(1)}(\zeta)\, d\zeta + \frac{c}{2}\iint_{\mathcal{C}_{x,t}} f(\zeta,\tau)\, d\zeta\, d\tau. \end{aligned}} \tag{3.6}$$

Fig. 8 represents the cone of dependence for the solution. It shows that the solution at time t and point x depends only on the data in the cone $\mathcal{C}_{x,t}$.

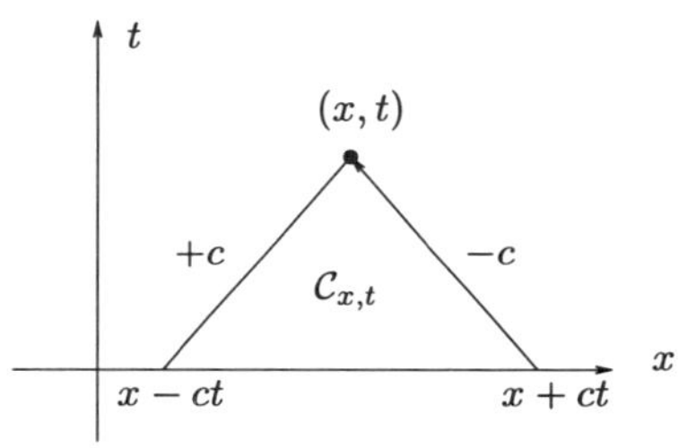

Figure 8: Cone of dependence for point (x,t).

Fig. 9 illustrates the propagation for zero right-hand side f and $u^{(1)}$, and initial values given by a gaussian function

$$u^{(0)}(x) = e^{-x^2} \tag{3.7}$$

The velocity is equal to 1. In the beginning the signal splits, and then each part goes its own way.

A direct consequence of the d'Alembert formula (3.6) is the

3.1 Theorem *The solution of the wave equation propagates with speed c: if the initial data vanish outside $[a,b]$, for any time t the solution $u(.,t)$ vanishes outside $[a-ct, b+ct]$.*

An illustration is given by the domain of influence $\mathcal{D}$ in Fig. 10.

3.1.3 The initial boundary value problem

We consider the wave equation on the half-line $(X, +\infty[$ with initial values $u^{(0)}$ and $u^{(1)}$ and vanishing right-hand side. The diagram in Fig. 11 is very illuminating. It shows the modified cone of dependence.

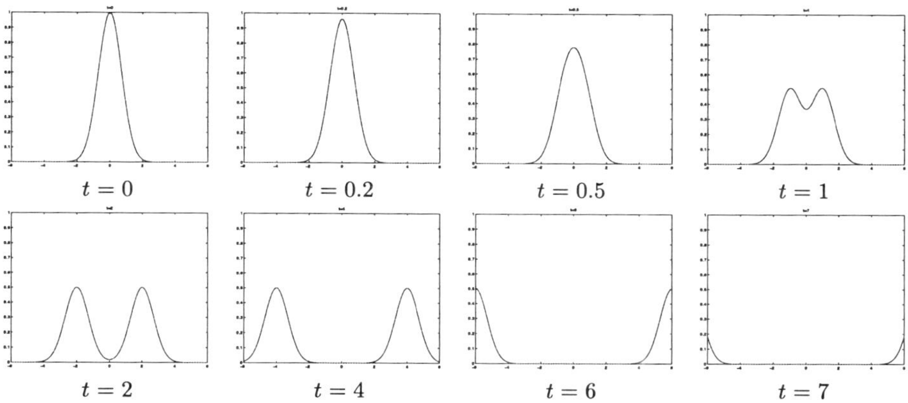

Figure 9: Evolution of the data (3.7).

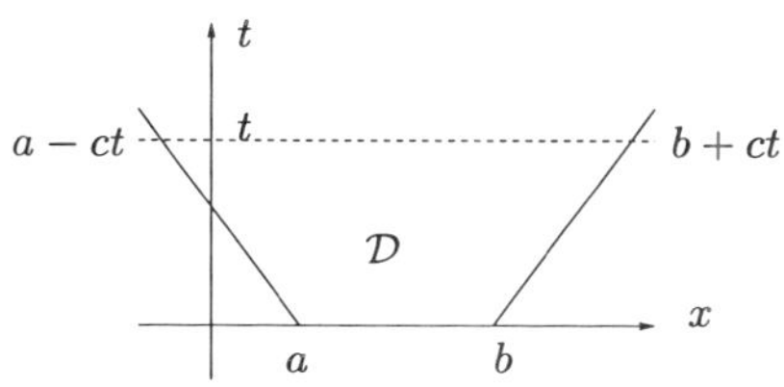

Figure 10: Domain of influence.

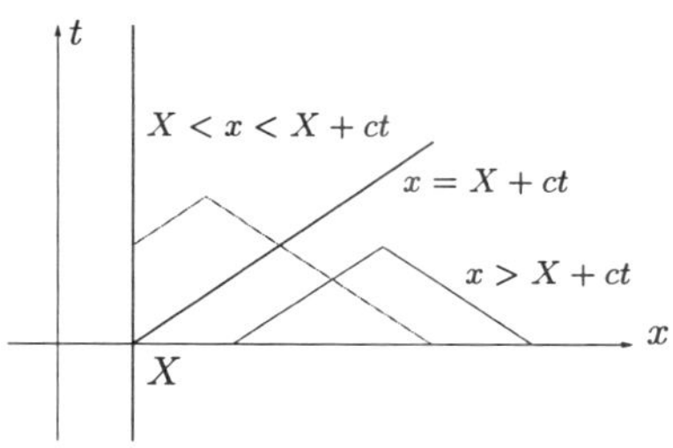

Figure 11: Domain of dependence for the problem on $(X, +\infty[$.

As in the case of the transport equation, the need for a boundary condition appears clearly. It is given for instance by an *inhomogeneous Dirichlet boundary condition* at $x = X$, which amounts to fixing the displacement of the rope at the extremity,

$$u(X, t) = g(t) \quad \text{for} \quad t \in (0, T).$$

The initial boundary value problem is given by

$$\begin{cases} \dfrac{1}{c^2}\dfrac{\partial^2 u}{\partial t^2} - \dfrac{\partial^2 u}{\partial x^2} = 0 \quad \text{in} \quad (X, +\infty) \times (0, T), \\ u(\cdot, 0) = u^{(0)}, \quad \dfrac{\partial u}{\partial t}(\cdot, 0) = u^{(1)} \quad \text{on} \quad (X, +\infty), \\ u(X, t) = g(t) \quad \text{for} \quad t \in (0, T). \end{cases} \tag{3.8}$$

3.2 Theorem *Problem* (3.8) *has a unique solution given by the formula*

$$u(x,t) = \begin{cases} \dfrac{1}{2}[u^{(0)}(x+ct) + u^{(0)}(x-ct)] + \dfrac{1}{2c}\displaystyle\int_{x-ct}^{x+ct} u^{(1)}(\zeta)\,d\zeta & \text{if } x - ct > X, \\ \dfrac{1}{2}[u^{(0)}(x+ct) - u^{(0)}(2X - (x-ct))] & \\ \quad + \dfrac{1}{2c}\displaystyle\int_{2X-(x-ct)}^{x+ct} u^{(1)}(\zeta)\,d\zeta + g\left(t + \dfrac{X-x}{c}\right) & \text{if } x - ct < X. \end{cases} \tag{3.9}$$

In Fig. 12 we illustrate the principle of images: the Dirichlet boundary condition plays the role of a mirror.

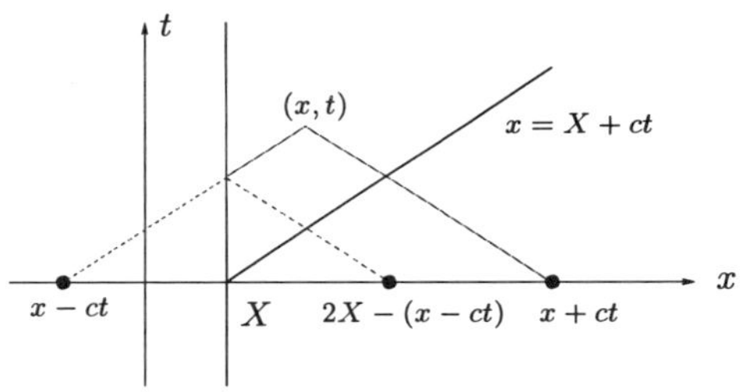

Figure 12: Principle of images on $(X, +\infty)$.

Fig. 13 represents the reflection by a homogeneous Dirichlet boundary condition on the left for the same shape of initial value $u^{(0)}(x) = e^{-3(x-2)^2}$. Of course we can write the same formulae for the problems on $(-\infty, X)$ and (X, Y).

Consider now another boundary condition, which is closely related to domain decomposition and transparent boundary conditions, and will be useful in the sequel. We consider the half-line $(X, +\infty)$, with boundary condition at $x = X$ given by

$$\left(\alpha\frac{\partial u}{\partial t} + \frac{\partial u}{\partial x}\right)(X, t) = g(t). \tag{3.10}$$

In order to make the notations lighter, we shall consider here the initial data to be zero. With the same techniques as before we obtain

$$u(x,t) = F(x+ct) + G(x-ct),$$

with, on the one hand,

$$F'(x) = G'(x) = 0 \quad \text{for all} \quad x > X, \tag{3.11}$$

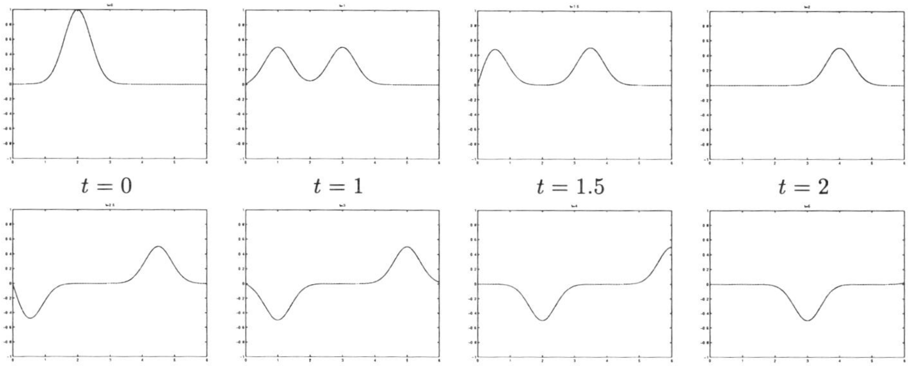

$t = 0$ $t = 1$ $t = 1.5$ $t = 2$

$t = 2.5$ $t = 3$ $t = 4$ $t = 5$

Figure 13: Complete reflection by a Dirichlet boundary condition.

and, on the other hand,

$$(\alpha c + 1)F'(X + ct) + (-\alpha c + 1)G'(X - ct) = g(t) \quad \text{for all} \quad t > 0$$

which by (3.11) gives

$$(-\alpha c + 1)G'(X - ct) = g(t) \quad \text{for all} \quad t > 0.$$

If α is different from $1/c$, this equation can be solved modulo an additive constant :

$$G(x) = \frac{1}{\alpha c - 1} \int_x^X g\left(\frac{X - \eta}{c}\right) d\eta + C \quad \text{for all} \quad x < X.$$

The solution of the initial boundary value problem with boundary condition (3.10) is defined modulo an additive constant C by

$$u(x,t) = \begin{cases} 0, & x - ct > X, \\ \dfrac{1}{\alpha c - 1} \displaystyle\int_{x-ct}^X g\left(\frac{X - \eta}{c}\right) d\eta + C, & x - ct < X. \end{cases}$$

The constant C is determined if we require the solution to be continuous

$$u(x,t) = \begin{cases} 0, & x - ct > X, \\ \dfrac{1}{\alpha c - 1} \displaystyle\int_{x-ct}^X g\left(\frac{X - \eta}{c}\right) d\eta, & x - ct < X. \end{cases}$$

In the same way, the continuous solution of the wave equation on $(-\infty, X)$ with boundary condition at $x = X$ given by

$$\left(\beta \frac{\partial u}{\partial t} + \frac{\partial u}{\partial x}\right)(X, t) = h(t)$$

is

$$u(x,t) = \begin{cases} 0, & x + ct < X, \\ \dfrac{1}{\beta c + 1} \displaystyle\int_X^{x+ct} h\left(\frac{\eta - X}{c}\right) d\eta, & x + ct > X. \end{cases}$$

On an interval (a, b) with boundary conditions

$$\begin{cases} \left(\alpha \dfrac{\partial u}{\partial t} + \dfrac{\partial u}{\partial x}\right)(a,t) = g(t), \\ \left(\beta \dfrac{\partial u}{\partial t} + \dfrac{\partial u}{\partial x}\right)(b,t) = h(t), \end{cases}$$

the continuous solution with zero initial values is

$$u(x,t) = \begin{cases} 0, & x - ct > a \text{ and } x + ct < b, \\ \frac{1}{\alpha c - 1} \int_{x-ct}^{a} g(\frac{a-\eta}{c})\, d\eta, & x - ct < a \text{ and } x + ct < b, \\ \frac{1}{\beta c + 1} \int_{b}^{x+ct} h(\frac{\eta - b}{c})\, d\eta, & x - ct > a \text{ and } x + ct > b, \\ \frac{1}{\alpha c - 1} \int_{x-ct}^{a} g(\frac{a-\eta}{c})\, d\eta + \frac{1}{\beta c + 1} \int_{b}^{x+ct} h(\frac{\eta - b}{c})\, d\eta, & x - ct < a \text{ and } x + ct > b. \end{cases}$$

3.1.4 Transparent boundary condition

Suppose the initial data are compactly supported in (a, b). Due to (3.2)–(3.4), we have

$$\left(\frac{1}{c}\frac{\partial u}{\partial t} - \frac{\partial u}{\partial x}\right)(a,t) = 0, \tag{3.12}$$

$$\left(\frac{1}{c}\frac{\partial u}{\partial t} + \frac{\partial u}{\partial x}\right)(b,t) = 0. \tag{3.13}$$

We now introduce the reflection coefficient. A plane wave propagating to the right is given by $U^I = e^{i\omega(x-ct)}$. For a given boundary condition **BC** at $x = 0$, the reflected wave is $U^R = Re^{i\omega(x+ct)}$ if the sum satisfies the boundary condition **BC**:

$$\mathbf{BC}(U^I + U^R) = 0.$$

Table 1 gives the reflection coefficient for various boundary conditions. It shows that Neumann and Dirichlet are perfectly reflecting, and the transparent boundary condition (3.13) is actually not reflecting at all. It also shows a family of absorbing boundary conditions (i.e. such that the reflection coefficient is smaller than 1) with positive parameter β.

3.1.5 Domain decomposition

For evolution problems, classical methods are based either on explicit schemes, which implies communication at every time step, or on implicit schemes, which gives an elliptic equation to solve at every time step [29, 25]. In both cases it is difficult to use a varying time step for different parts of the domain. However, for the wave equation it is often desirable to choose different time steps in different physical domains. The goals of our strategy are the following: to reduce the communications (and thus the costs), to use non-conformal discretization, and to

Table 1: Reflection coefficient for various boundary conditions.

Dirichlet boundary condition	$u = 0$	$R = -1$
Neumann boundary condition	$\dfrac{\partial u}{\partial x} = 0$	$R = 1$
Absorbing boundary condition	$\dfrac{\beta}{c}\dfrac{\partial u}{\partial t} + \dfrac{\partial u}{\partial x} = 0$	$R = \dfrac{1-\beta}{1+\beta}$
Transparent boundary condition	$\dfrac{1}{c}\dfrac{\partial u}{\partial t} + \dfrac{\partial u}{\partial x} = 0$	$R = 0$

be able to couple different codes. We will build a domain decomposition algorithm such that it works with or without overlap, is global in time, easy to implement, and the convergence is optimal. As we now show, the tool is the transparent boundary condition. For details in this section see [17, 16].

The straightforward extension of the Schwarz algorithm We present the algorithm in the case of two subdomains, $\Omega_1 = (-\infty, L)$ and $\Omega_2 = (0, +\infty)$. At step n, we solve two subproblems in $\Omega_i \times (0,T)$, with Dirichlet data on the boundary given by the previous step in the other domain. The solution in $\Omega_i \times (0,T)$ at step n is called u_i^n. The classical Schwarz algorithm extended to space-time domains is then given by:

$$\left.\begin{array}{l} \left(\dfrac{1}{c^2}\dfrac{\partial^2}{\partial t^2} - \dfrac{\partial^2}{\partial x^2}\right) u_i^n = f \text{ in } \Omega_i \times (0,T) \text{ for } i = 1,2, \\ \text{with initial data for } i = 1,2: \\ u_i^n(.,0) = u^{(0)} \text{ in } \Omega_i, \ \dfrac{\partial u_i^n}{\partial t}(.,0) = u^{(1)} \text{ in } \Omega_i, \\ \text{and transmission conditions:} \\ u_1^n(L,.) = u_2^{n-1}(L,.), \ u_2^n(0,.) = u_1^{n-1}(0,.) \text{ in } (0,T). \end{array}\right\} \tag{3.14}$$

3.3 Theorem *For the straightforward extension* (3.14) *of the Schwarz algorithm, convergence is achieved in a finite number of iterations, $n > Tc/L$.*

Proof The proof can be found in [15], but we describe it here for clarity. Consider the errors $U_i^n = u_i^n - u$. They satisfy the system (3.14) with zero data f and zero initial values. Using formula (3.9), we have

$$\begin{array}{ll} \text{for } x - ct > 0, & U_2^{n+1}(x,t) = 0, \\ \text{for } x - ct < 0, & U_2^{n+1}(x,t) = U_1^n\left(t - \dfrac{x}{c}\right), \end{array} \tag{3.15}$$

and

$$\begin{array}{ll} \text{for } x + ct < L, & U_1^{n+1}(x,t) = 0, \\ \text{for } x + ct > L, & U_1^{n+1}(x,t) = U_2^n\left(t - \dfrac{L-x}{c}\right). \end{array} \tag{3.16}$$

We can draw the diagram in Fig. 14:

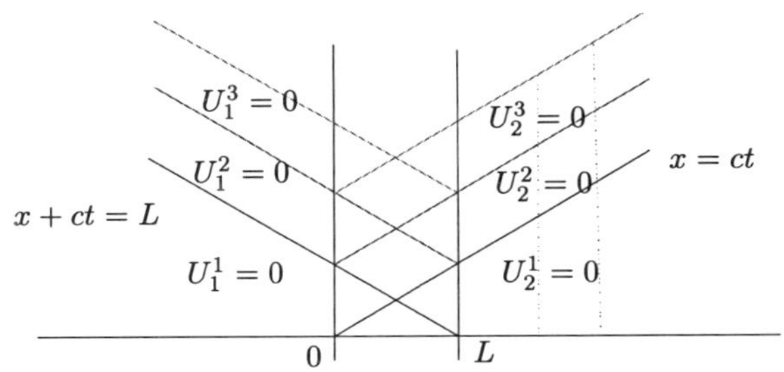

Figure 14: Evolution of the Schwarz algorithm with Dirichlet transmission conditions.

We can now read the end of the proof on Fig. 14 using (3.15) and (3.16). □

We illustrate this behaviour by an example ($c = 1$, $T = 3$). Fig. 15 shows the exact solution, Figs. 16–18 the first five iterates for $L = 0.4$.

Fig. 19 presents the convergence history in this example for two values of the overlap $L = 0.4$ and $L = 0.2$. According to Theorem 3.3, the convergence is achieved for $n > 3/0.4 = 7.5$, resp. $n > 3/0.2 = 15$.

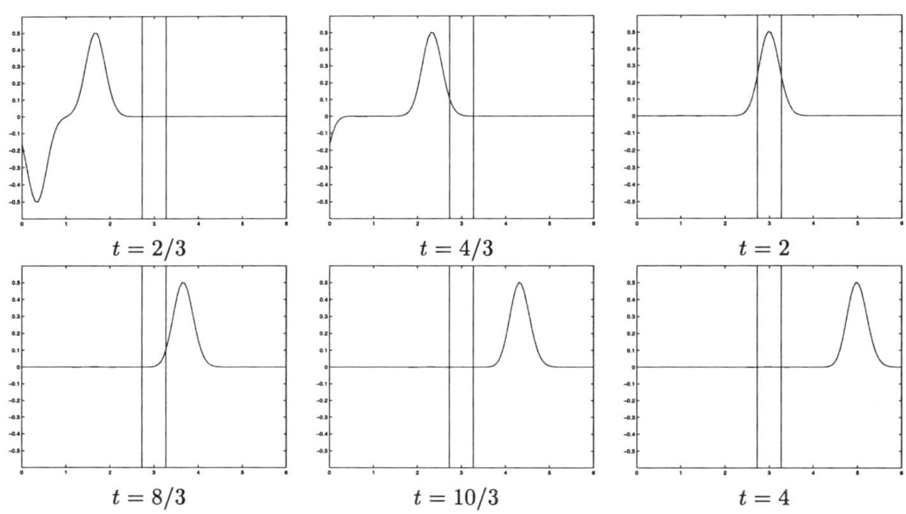

Figure 15: Exact solution.

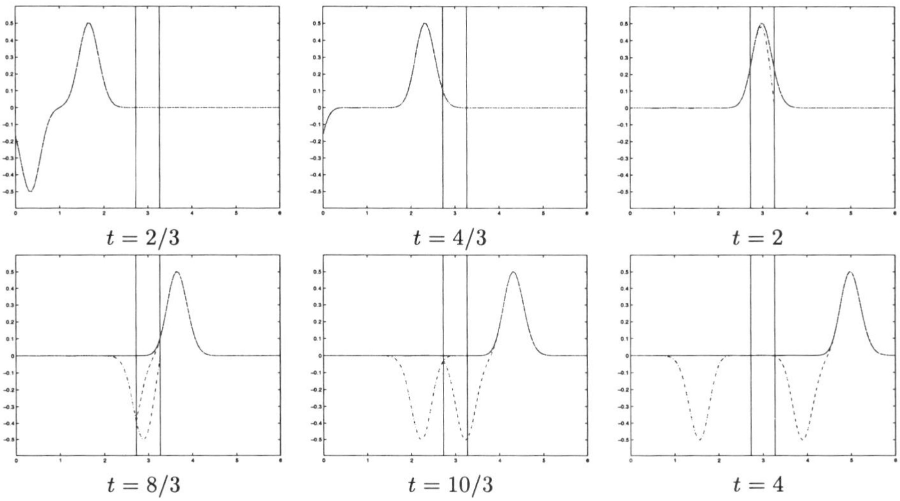

Figure 16: First iteration (dash-dot) u_1^1 and second iteration (dashed) u_2^2 for the Dirichlet transmission conditions together with the exact solution (solid).

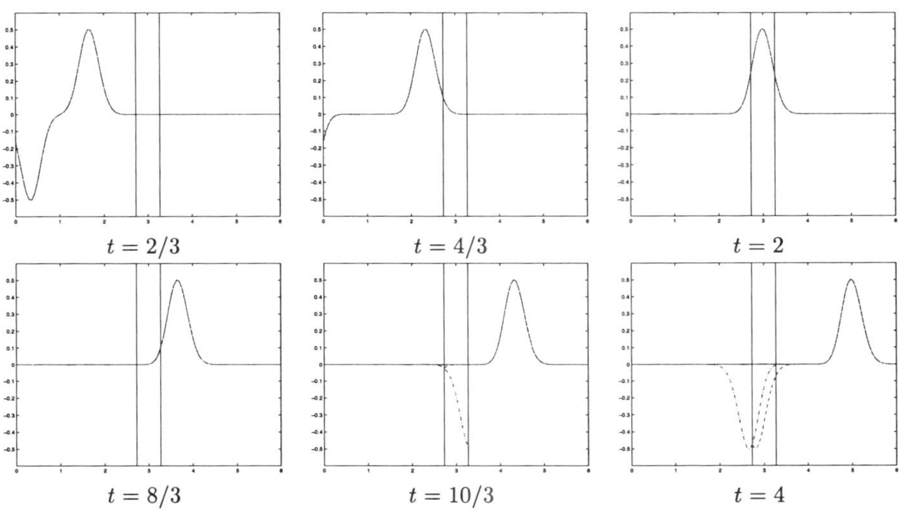

Figure 17: Third iteration (dash-dot) u_1^3 and fourth iteration (dashed) u_2^4 for the Dirichlet transmission conditions together with the exact solution (solid).

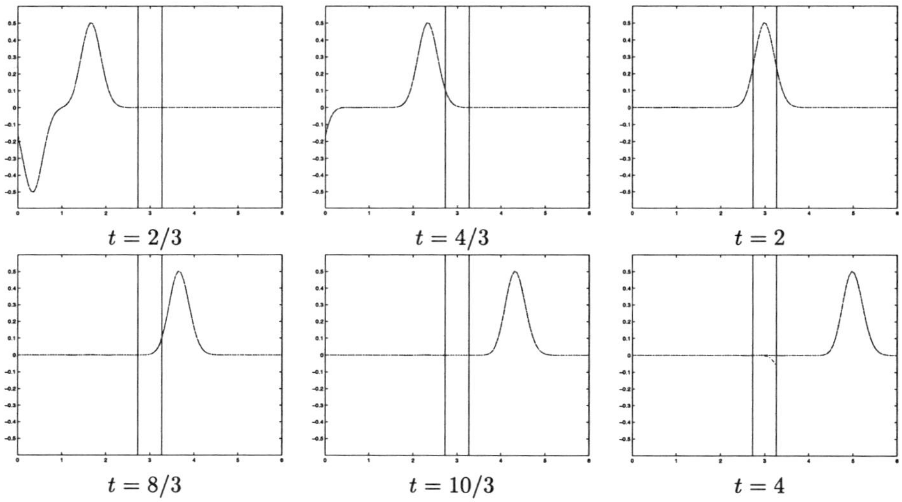

Figure 18: Fifth iteration (dash-dot) u_1^5 and sixth iteration (dashed) u_2^6 for the Dirichlet transmission conditions together with the exact solution (solid).

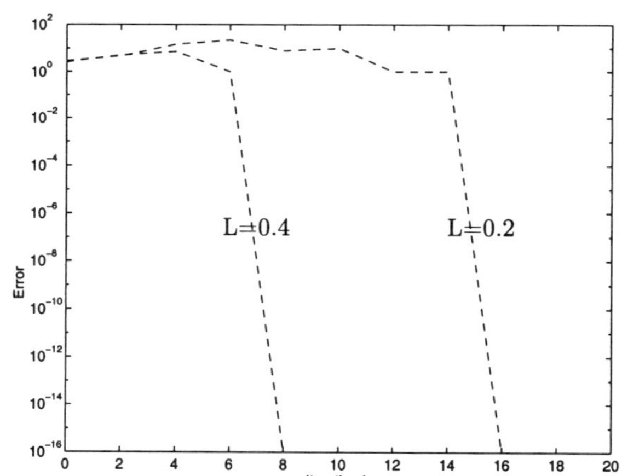

Figure 19: Convergence history for the Schwarz algorithm applied to the wave equation: error versus number of iterations.

3.1.6 The optimal Schwarz algorithm: transparent transmission condition

Note that by (3.15) and (3.16), for any transmission conditions, the error U_1^n on the left is a function of $x - ct$, while the error U_2^n on the right is a function of $x + ct$. This in turn implies the following identities for $n \geq 1$ and any positive time:

$$\begin{cases} \left(\dfrac{\partial}{\partial x} + \dfrac{1}{c}\dfrac{\partial}{\partial t}\right) U_2^n(L,t) = 0, \\ \left(\dfrac{\partial}{\partial x} - \dfrac{1}{c}\dfrac{\partial}{\partial t}\right) U_1^n(0,t) = 0. \end{cases}$$

This observation leads to the following important

3.4 Theorem *The transmission conditions defined by*

$$\left(\frac{\partial}{\partial x} + \frac{1}{c}\frac{\partial}{\partial t}\right) u_1^{n+1}(L,t) = \left(\frac{\partial}{\partial x} + \frac{1}{c}\frac{\partial}{\partial t}\right) u_2^n(L,t),$$
$$\left(\frac{\partial}{\partial x} - \frac{1}{c}\frac{\partial}{\partial t}\right) u_2^{n+1}(0,t) = \left(\frac{\partial}{\partial x} - \frac{1}{c}\frac{\partial}{\partial t}\right) u_1^n(0,t),$$

lead to well-posed initial boundary value problems even without overlap and they are optimal: convergence in the Schwarz algorithm with these transmission conditions is achieved in two iterations, i.e., u_i^2 is identical to u in Ω_i.

An example without overlap The next example shows the convergence to the accuracy of the numerical scheme without overlap. The velocity is $c = 1$, the computation is done on $(0, T)$ with $T = 2$. The initial data are $u(x, 0) = 0$, $\partial_t u(x, 0) = -100(0.5 - x)e^{-50(0.5-x)^2}$. The domain $(0, 2)$ is divided into two subdomains $(0, 1)$ and $(1, 2)$. The initial guess (u_i^0), $i = 1, 2$, is naturally chosen to be 0. The scheme is a second order finite volume scheme.

Table 2: Convergence in two iterations to the accuracy of the numerical scheme.

grid	error after 2 iterations	discretization error
50 × 50	2.6128e − 04	2.1515e − 02
100 × 100	2.7305e − 05	4.9472e − 03
200 × 200	3.2361e − 06	1.2218e − 03
400 × 400	3.9852e − 07	3.0321e − 04
800 × 800	4.9532e − 08	7.5567e − 05

3.2 The heterogeneous case

We consider now a variable velocity $c(x)$. We suppose c to be positive, and such that there exist two positive real numbers c_* and c^* satisfying

$$c_* \leq c(x) \leq c^*, \quad \text{a.e. in} \quad \mathbb{R}.$$

The wave equation is now

$$\frac{1}{c(x)^2}\frac{\partial^2 u}{\partial t^2} - \frac{\partial^2 u}{\partial x^2} = f \quad \text{in} \quad \mathbb{R} \times (0, T). \tag{3.17}$$

3.2.1 Energy estimates and well-posedness

We first give an existence result in Sobolev spaces :

3.5 Theorem *If $u^{(0)}$ is in $H^1(\mathbb{R})$, $u^{(1)}$ is in $L^2(\mathbb{R})$ and f is in $L^2(0,T;L^2(\mathbb{R}))$, then there exists a unique solution u in $H^1(0,T;L^2(\mathbb{R})) \cap L^2(0,T;H^1(\mathbb{R}))$ to* (3.5), (3.17). *Moreover u is in $C^0(0,T;L^2(\mathbb{R}))$.*

The complete proof of the theorem is beyond the scope of these lectures. It relies on the use of a finite difference scheme, through *energy estimates*:

3.6 Lemma *Let u be a smooth solution of* (3.17). *Defining the total energy at time t by*

$$\mathcal{E}(t) = \int_{\mathbb{R}} e(x,t)\,dx, \qquad e(x,t) = \frac{1}{2c^2}\left(\frac{\partial u}{\partial t}\right)^2 + \frac{1}{2}\left(\frac{\partial u}{\partial x}\right)^2,$$

we have the energy identity

$$\frac{d}{dt}\mathcal{E}(t) = \int_{\mathbb{R}} f(x,t)\frac{\partial u}{\partial t}\,dx.$$

In particular, without external forces (i.e. $f \equiv 0$), the energy is preserved in time.

In order to obtain the energy, we multiply the equation by $\partial u/\partial t$ and integrate in space on $\mathbb{R}$.

3.7 Remark The solution of the wave equation is in $L^2(0,T)$ for any finite time T but not in $L^2(0,+\infty)$.

3.2.2 Propagation properties

In a heterogeneous medium there is no explicit formula like the d'Alembert formula. However, one can still give a result for the finite speed of propagation:

3.8 Theorem *The solution of* (3.17) *with zero right-hand side propagates with at most the velocity $c^* = \sup_{x\in\mathbb{R}} c(x)$; if the initial conditions vanish outside $[a,b]$, then for any time t the solution vanishes outside $[a-c^*t, b+c^*t]$.*

Proof Let V be a positive real number. We introduce the energy on the moving half-line $(b+Vt,+\infty)$

$$\mathcal{E}_V(t) = \int_{b+Vt}^{+\infty} e(x,t)\,dx.$$

Taking the derivative in time we get

$$\frac{d}{dt}\mathcal{E}_V(t) = \int_{b+Vt}^{+\infty} \frac{\partial}{\partial t}e(x,t)\,dx - Ve(b+Vt,t).$$

We handle the first time by using the equation:

$$\int_{b+Vt}^{+\infty} \frac{\partial}{\partial t}e(x,t)\,dx = \int_{b+Vt}^{+\infty}\left[\frac{1}{c(x)^2}\frac{\partial u}{\partial t}\frac{\partial^2 u}{\partial t^2} + \frac{\partial u}{\partial x}\frac{\partial^2 u}{\partial t\partial x}\right]dx.$$

By Green's formula this transforms to

$$\int_{b+Vt}^{+\infty} \frac{\partial}{\partial t} e(x,t)\,dx = \int_{b+Vt}^{+\infty} \left(\frac{1}{c(x)^2} \frac{\partial^2 u}{\partial t^2} - \frac{\partial^2 u}{\partial x^2} \right) \cdot \frac{\partial u}{\partial t}\,dx - \frac{\partial u}{\partial t}(b+Vt,t) \cdot \frac{\partial u}{\partial x}(b+Vt,t).$$

And finally we obtain

$$\frac{d}{dt}\mathcal{E}_V(t) = \left\{ -V \left[\frac{1}{2c^2(b+Vt)} \left(\frac{\partial u}{\partial t} \right)^2 + \frac{1}{2} \left(\frac{\partial u}{\partial x} \right)^2 \right] - \frac{\partial u}{\partial t} \cdot \frac{\partial u}{\partial x} \right\} (b+Vt,t).$$

The right-hand side is a quadratic form in the two variables $\partial u/\partial t$ and $\partial u/\partial x$, whose discriminant is $1-(V/c(b+Vt))^2$. Thus, for any V such that $V \geq c^*$, we have

$$\frac{d}{dt}\mathcal{E}_V(t) \leq 0,$$

which can be rewritten as

$$\int_{b+Vt}^{+\infty} e(x,t)\,dx \leq \int_{b}^{+\infty} e(x,0)\,dx.$$

The quantity on the right vanishes, thus the quantity on the left vanishes as well. The same proof can be done on $(-\infty,a)$, which concludes the proof of the theorem. □

3.2.3 Transparent boundary condition

Here the decomposition of the solution in a part propagating to the right and a part propagating to the left becomes much more intricate. One has to introduce the theory of pseudo-differential operators, which is a very powerful but heavy tool. Instead, we shall use a PDE approach to the transparent boundary condition. We assume the data to be compactly supported in $\mathbb{R}_-$. Then the problem

$$\begin{cases} \dfrac{1}{c^2(x)} \dfrac{\partial^2 u}{\partial t^2} - \dfrac{\partial^2 u}{\partial x^2} = f \quad \text{in} \quad \mathbb{R} \times (0,T), \\ u(\cdot,0) = u^{(0)} \quad , \dfrac{\partial u}{\partial t}(\cdot,0) = u^{(1)} \quad \text{in} \quad \mathbb{R}, \end{cases}$$

is equivalent to

$$\begin{cases} \dfrac{1}{c^2(x)} \dfrac{\partial^2 u}{\partial t^2} - \dfrac{\partial^2 u}{\partial x^2} = f \quad \text{in} \quad \mathbb{R}_- \times (0,T), \\ u(.,0) = u^{(0)}, \quad \dfrac{\partial u}{\partial t}(\cdot,0) = u^{(1)} \quad \text{in} \quad \mathbb{R}_-, \\ \dfrac{\partial u}{\partial x}(0,t) - K_+(0)\dfrac{\partial u}{\partial t}(0,t) = 0 \quad \text{in} \quad (0,T), \end{cases}$$

where the transparent operator at point 0, $K_+(0)$, is defined using the exterior problem

$$\begin{cases} \dfrac{1}{c^2(x)} \dfrac{\partial^2 w}{\partial t^2} - \dfrac{\partial^2 w}{\partial x^2} = 0 \quad \text{in} \quad \mathbb{R}_+ \times (0,T), \\ w(\cdot,0) = 0, \quad \dfrac{\partial w}{\partial t}(\cdot,0) = 0 \quad \text{in} \quad \mathbb{R}_+, \\ \dfrac{\partial w}{\partial t}(0,\cdot) = g \quad \text{in} \quad (0,T). \end{cases} \tag{3.18}$$

Then the operator $K_+(0)$ is defined in a unique fashion by

$$K_+(0)g \equiv \frac{\partial w}{\partial x}(0,\cdot) \quad \text{in} \quad (0,T). \tag{3.19}$$

For general velocities, this operator cannot be written explicitly. However, when used in the context of absorbing boundary conditions, the velocity is constant in $\mathbb{R}_+$, equal to c^+, and the operator reduces to the local operator $(-1/c^+)(\partial/\partial t)$. In this case we get the transparent boundary as

$$\frac{\partial u}{\partial x}(0,t) + \frac{1}{c^+}\frac{\partial u}{\partial t}(0,t) = 0.$$

3.2.4 Domain decomposition

We cut the spatial domain $\mathbb{R}$ into I numerical domains $\Omega_i = (a_i, a_{i+1})$, $1 \le i \le I$, $a_j < a_i$ for $j < i$ and $a_1 = -\infty$, $a_{I+1} = \infty$ as described in Fig. 20.

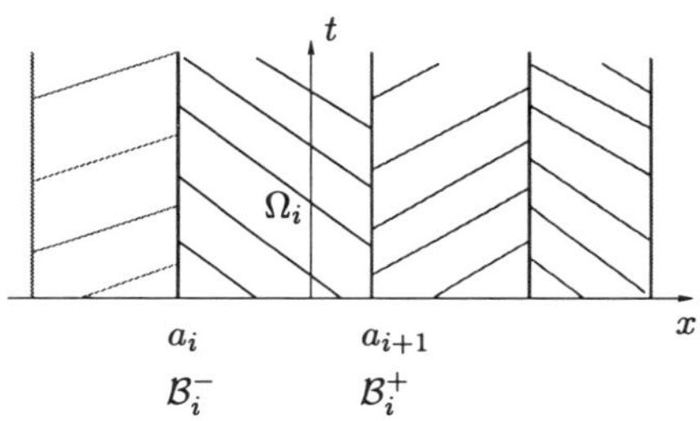

Figure 20: Decomposition into numerical domains.

We introduce the Schwarz algorithm with general transmission conditions

$$\begin{aligned}
\left(\frac{1}{c^2(x)}\partial_{tt} - \partial_{xx}\right)(u_i^{n+1}) &= f(x,t), \quad x \in \Omega_i,\ t \in (0,T),\\
\mathcal{B}_i^-(u_i^{n+1}) &= \mathcal{B}_i^-(u_{i-1}^n), \quad x = a_i,\ t \in (0,T),\\
\mathcal{B}_i^+(u_i^{n+1}) &= \mathcal{B}_i^+(u_{i+1}^n), \quad x = a_{i+1},\ t \in (0,T).
\end{aligned} \tag{3.20}$$

The optimal choice for $\mathcal{B}_i^\pm$ is the *transparent boundary operators*:

3.9 Theorem *The algorithm converges in I iterations on $(0,T)$ if the operators are given by*

$$\begin{aligned}
\mathcal{B}_i^- &:= (\partial_x - K_+(a_i)\partial_t),\\
\mathcal{B}_i^+ &:= (\partial_x - K_-(a_{i+1})\partial_t).
\end{aligned}$$

where the $K_\pm(a_i)$ are the transparent operators defined in (3.18) *and* (3.19).

In the case of a physical medium consisting of two media with constant velocities as in Fig. 21, $\mathcal{O}_1 = \mathbb{R}_-$ with velocity c_1, $\mathcal{O}_2 = \mathbb{R}_+$ with velocity c_2, the operators can be written explicitly. For example,

$$K_+(x_0)g(t) = \begin{cases} \dfrac{1}{c_1} g(t), & x_0 \in \mathcal{O}_1 \\ \dfrac{1}{c_2}\Big(g(t) + 2 \displaystyle\sum_{k=1}^{\lfloor \frac{c_2 t}{2x_0} \rfloor} r^k g(t - 2kx_0/c_2)\Big), & x_0 \in \mathcal{O}_2. \end{cases}$$

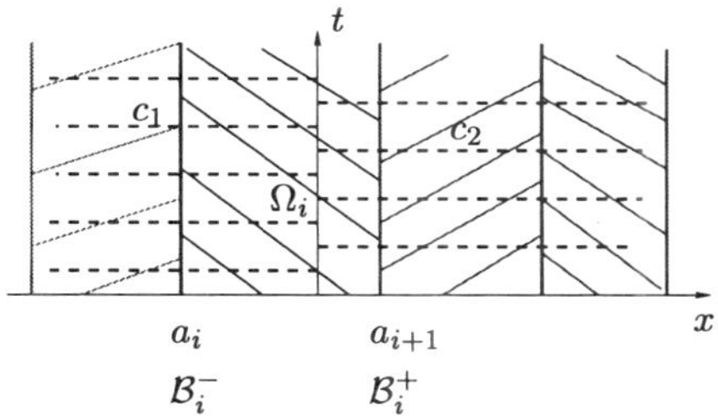

Figure 21: Physical and numerical domains.

We show now an example of convergence in three iterations for three numerical subdomains. The data are given in Table 3.

Table 3: An example with global transmission conditions.

Initial data $u(x,0) = 0$, $\dfrac{\partial u}{\partial t}(x,0) = -20(4-x)e^{-10(4-x)^2}$.
Physical domains: $\mathcal{O}_1 = \mathbb{R}_-$ with velocity $c_1 = 2$, $\mathcal{O}_2 = \mathbb{R}_+$ with velocity $c_2 = 1$.
Numerical domains: $\Omega_1 = (-\infty, 1)$, $\Omega_2 = (1,3)$, $\Omega_3 = (3, +\infty)$.
Time domain $(0,8)$.

Fig. 22 shows the convergence history.

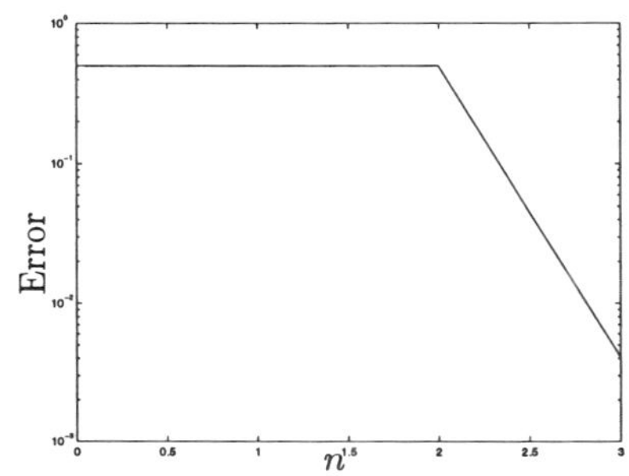

Figure 22: Error history for the optimal global transmission conditions.

Fig. 23 describes the iterates in the algorithm.

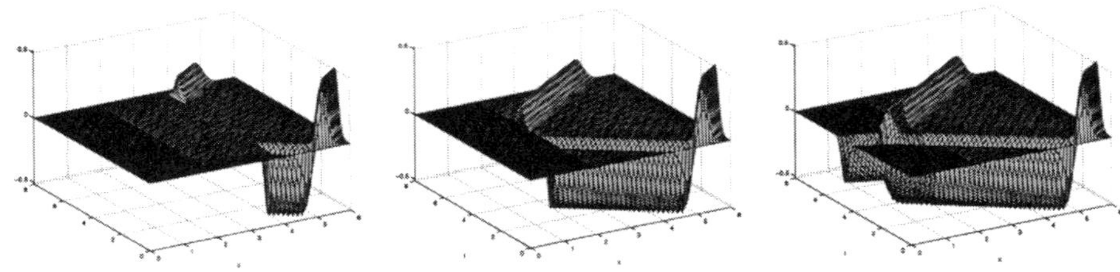

first iteration second iteration third iteration

Figure 23: Convergence in three iterations for the example in Table 3.

One can actually prove that for sufficiently small time, the optimal transmission is preserved with local operators:

3.10 Theorem *If the discontinuities lie strictly inside the domains, the algorithm converges in I iterations for*

$$T < \min_{1<i\leq I} \frac{2|a_i|}{c(a_i)}$$

with local transmission conditions

$$\mathcal{B}_i^- := \partial_x - \frac{1}{c(a_i)}\partial_t, \qquad \mathcal{B}_i^+ := \partial_x + \frac{1}{c(a_{i+1})}\partial_t. \tag{3.21}$$

In Fig. 24, we show the same example as before, except that T is chosen equal to 2. One can see that the convergence is actually achieved in two iterations.

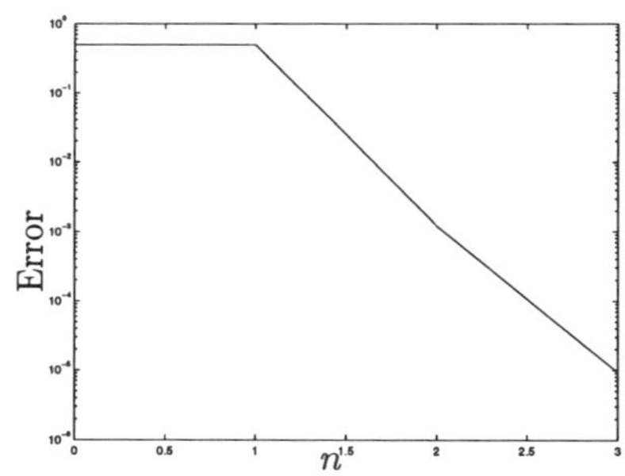

Figure 24: Error history for the local transmission conditions (3.21).

3.11 Theorem *If the velocities are constant in the numerical domains, the convergence is achieved in two iterations for*

$$T < \min_{1<i<I} \frac{|a_{i+1} - a_i|}{c((a_{i+1} + a_i)/2)}$$

with the transmission operators given by

$$\mathcal{B}_i^- := (\partial_x - \frac{1}{c^-(a_i)}\partial_t), \qquad \mathcal{B}_i^+ := (\partial_x + \frac{1}{c^+(a_{i+1})}\partial_t).$$

The proofs of Theorems 3.10 and 3.11 use the finite speed of propagation and can be found in [17].

The next example in Table 4 illustrates Theorem 3.11.

Table 4: An example with six layers and a non-uniform grid.

Physical domain $(0,6)$.
Physical subdomains with velocity $c_i \in \{1, 2/3, 1/2, 3/4, 4/5\}$.
Numerical domains aligned with discontinuities.
Time domain $(0,1)$.

Since we use a finite volumes scheme, there is a CFL condition (2.11). It varies with the velocity in each subdomain. The space step is the same everywhere, $\Delta x_i = 1/50$, the time step is chosen so that in each subdomain the CFL is close to 1.

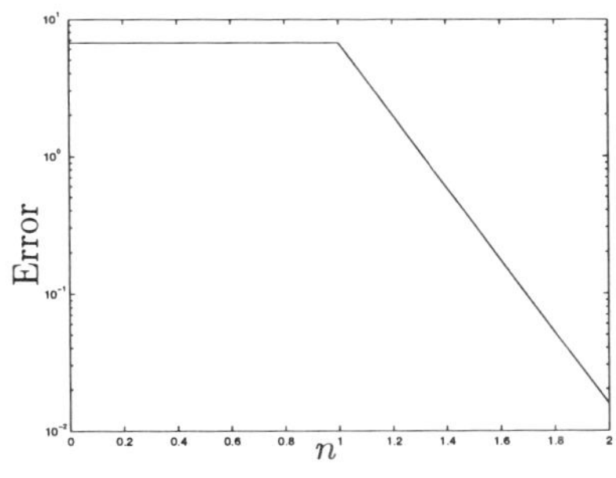

Figure 25: Convergence in two iterations with local transmission conditions on $[0,1]$.

If we summarize the last two examples, we see convergence in two iterations with local transmission conditions on $[0,1]$, and convergence in six iterations with local transmission conditions on $[0,6]$. This suggests as a conclusion to use local transmission operators as in Theorem 3.11 with subdomain boundaries aligned with physical discontinuities and time windows of length

$$\min_{1<i<I} \frac{|a_{i+1} - a_i|}{c((a_{i+1} + a_i)/2)}.$$

Variable wave speed and local transmission conditions

In the case of variable wave speed there is still convergence in I iterations with the transparent transmission condition defined from the transparent operator (3.19), but this operator is non-local. With local transmission conditions (3.21), we still have a convergence result:

3.12 Theorem *Suppose the velocity is continuous at the interfaces a_i. Then on any time interval $[0,T]$, algorithm (3.20) with local transmission conditions (3.21) is well posed and converges in the energy norm,*

$$\sum_{i=1}^{I} E_{[a_i,a_{i+1}]}(u_i^k) \to 0 \quad as \quad k \to \infty. \tag{3.22}$$

Proof We introduce the forward and backward transport operators

$$\mathcal{T}_\alpha^+ = \frac{1}{\alpha}\frac{\partial}{\partial t} + \frac{\partial}{\partial x}, \qquad \mathcal{T}_\alpha^- = \frac{1}{\alpha}\frac{\partial}{\partial t} - \frac{\partial}{\partial x},$$

where α is a positive real number. The standard energy estimate becomes in this case for positive α and β

$$\frac{d}{dt}[E_{[a,b]}(u)(t)] + \frac{\alpha}{4}[\mathcal{T}_\alpha^+ u(a,t)]^2 + \frac{\beta}{4}[\mathcal{T}_\beta^- u(b,t)]^2 = \frac{\alpha}{4}[\mathcal{T}_\alpha^- u(a,t)]^2 + \frac{\beta}{4}[\mathcal{T}_\beta^+ u(b,t)]^2.$$

By writing the energy identity in every subdomain, using the transmission conditions and adding the corresponding equations in i and n we obtain (3.22). □

We now give an example. The velocity profile, which is a typical underwater profile, was obtained from [27] and it is given as a function of depth by

$$c(x) = \begin{cases} 300 \text{ m/s} & x < 0 \quad \text{(above ground)} \\ 1500 - x/12 \text{ m/s} & 0 < x < 120 \\ 1320 + x/12 \text{ m/s} & 120 < x < 240 \\ 1505 \text{ m/s} & x > 240. \end{cases} \tag{3.23}$$

The numerical domains are $\Omega_1 = [0, 300]$ and $\Omega_2 = [300, 600]$. We start with the errors in Table 5. Fig. 26 shows the first ans second iteration.

Table 5: Convergence behavior of the algorithm for the variable sound speed profile (3.23).

Iteration	0	1	2	3
$\|\|u - u_k\|\|_\infty$	5e + 00	5e + 00	9.9e − 03	6.2e − 04

The algorithm converges again to the accuracy of the scheme in two iterations, even though the sound speed is variable in this example. This is because the variation is small in scale and thus the local approximations to the transmission conditions are sufficiently accurate to lead to the convergence in two steps. Note also that, continuing the iteration, the error is further reduced.

3.3 The discrete domain decomposition algorithm

We first decribe the scheme, and then give a convergence result.

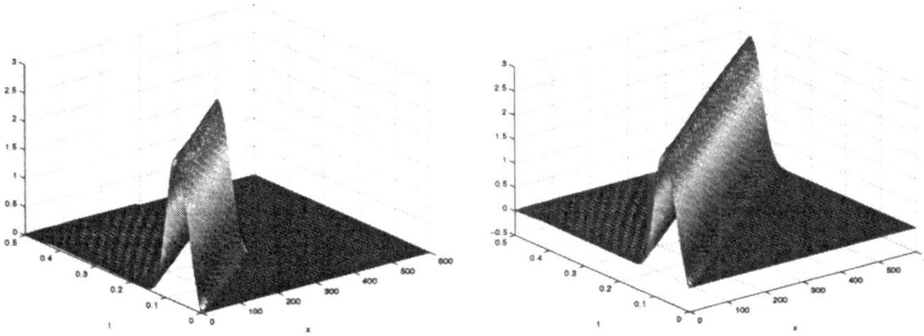

first iteration second iteration

Figure 26: Convergence in two iterations for the underwater example (3.23).

3.3.1 Description of the scheme

In order to take the transmission conditions into account, we use a vertex-centered finite volume method [14], as described in Fig. 27. We discretize the wave equation (3.17) on each subdomain $\Omega_i \times (0,T)$, $i = 1,\ldots,I$, separately, using a finite volume discretization on rectangular grids. For simplicity of the exposition we set $f = 0$. We allow non-matching grids on different subdomains, with $J_i + 2$ points in space numbered from 0 up to $J_i + 1$ and $\Delta x_i = (a_{i+1} - a_i)/(J_i + 1)$, and $N_i + 1$ grid points in time with $\Delta t_i = T/N_i$, numbered from 0 up to N_i. Note that for the exposition here we choose uniform spacing in time per subdomain, but the techniques developed are not limited to this special case. We denote the numerical approximation to $u_i^k(a_i + j\Delta x_i, n\Delta t_i)$ on Ω_i at iteration step k by $U_i^k(j,n)$.

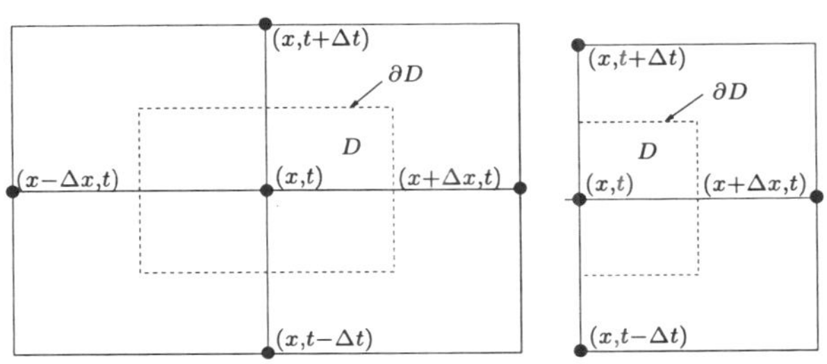

interior point boundary point

Figure 27: Control volumes for the scheme.

In the interior of the domains, this leads to the leapfrog scheme

$$\left(\frac{1}{C_i^2(j)} D_t^+ D_t^- - D_x^+ D_x^-\right)(U_i^{k+1})(j,n) = 0 \quad 1 \le j \le J_i,\ 1 \le n \le N_i,$$

with the usual notations

$$D_t^+(U)(j,n) = \frac{U(j,n+1) - U(j,n)}{\Delta t},$$
$$D_t^-(U)(j,n) = \frac{U(j,n) - U(j,n-1)}{\Delta t},$$
$$D_t^0 = \frac{D_t^+ + D_t^-}{2},$$

and the same for the variable x. $C_i(j)$ is a discrete function approximating $c(a_i + j\Delta x_i)$. The discrete transmission conditions are obtained by integrating the equation on the half-cell, and using the continuous transmission condition.

We first glue together the grids on two neighbouring domains. Let $I_n =]t_n, t_{n+1}[$ be a sequence of intervals such that $\bigcup_{n=0}^{N} \bar{I_n} = [0,T]$. The space $\mathbb{R}^{N+1}$ is equipped with the scalar product

$$(\boldsymbol{v}, \boldsymbol{w})_{N+1} = \sum_{n=0}^{N} |I_n| v_n w_n,$$

where $|I_n|$ is the length of the interval I_n. We define the map $\mathbb{F}$ from $\mathbb{R}^{N+1}$ to $L^2(0,T)$ which maps $\boldsymbol{v}$ to the function f equal to v_n in I_n for $0 \le n \le N$, and the operator $\mathbb{E}$ from $L^2(0,T)$ to $\mathbb{R}^{N+1}$ which maps f to the sequence $v_n = 1/|I_n| \int_{I_n} f(t)\,dt$.

We denote by $\mathbb{F}_i$ and $\mathbb{E}_i$ the operators corresponding to the grid in Ω_i, and we define the operator $\mathbb{P}_{i,j} : \mathbb{R}^{N_i+1} \to \mathbb{R}^{N_j+1}$ as

$$\mathbb{P}_{i,j} := \mathbb{E}_j \circ \mathbb{F}_i. \tag{3.24}$$

The discrete algorithm is now defined by

$$\begin{aligned}
&\left(\frac{1}{C_i^2(j)} D_t^+ D_t^- - D_x^+ D_x^-\right)(U_i^{k+1})(j,n) = 0 \quad 1 \le j \le J_i,\ 1 \le n \le N_i,\\
&\left(\frac{\Delta x}{C_i^2(j)} D_t^+ - \frac{\Delta t \Delta x}{2} D_x^+ D_x^-\right)(U_i^{k+1})(j,0) \frac{\Delta x}{C_i^2(j)} U_t(j) = 0 \quad 1 \le j \le J_i,\\
&B_i^-(U_i^{k+1})(0,\cdot) = \mathbb{P}_{i-1,i} \widetilde{B}_i^-(U_{i-1}^k)(J_{i-1}+1,\cdot),\\
&B_i^+(U_i^{k+1})(J_i+1,\cdot) = \mathbb{P}_{i+1,i} \widetilde{B}_i^+(U_{i+1}^k)(0,\cdot),
\end{aligned} \tag{3.25}$$

where the operators $\mathbb{P}_{i\pm1,i}$ are defined in (3.24), the discrete transmission operators $B_i^\pm$ are for $n \ge 1$ given by

$$\begin{aligned}
B_i^+(U_i)(J_i+1,n) &= \left(\frac{\Delta x_i}{2C_i^2(J_i+1)} D_t^+ D_t^- + D_x^- + \frac{1}{C_{i+1}(0)} D_t^0\right)(U_i)(J_i+1,n),\\
B_i^-(U_i)(0,n) &= \left(\frac{\Delta x_i}{2C_i^2(0)} D_t^+ D_t^- - D_x^+ + \frac{1}{C_{i-1}(J_{i-1}+1)} D_t^0\right)(U_i)(0,n),
\end{aligned} \tag{3.26}$$

and the extraction operators $\widetilde{B}_i^\pm$ are for $n \ge 1$

$$\begin{aligned}
&\widetilde{B}_i^+(U_{i+1})(0,n) = \left(-\frac{\Delta x_{i+1}}{2C_{i+1}^2(0)} D_t^+ D_t^- + D_x^+ + \frac{1}{C_{i+1}(0)} D_t^0\right)(U_{i+1})(0,n),\\
&\widetilde{B}_i^-(U_{i-1})(J_{i-1}+1,n) =\\
&\left(-\frac{\Delta x_{i-1}}{2C_{i-1}^2(J_{i-1}+1)} D_t^+ D_t^- - D_x^- + \frac{1}{C_{i-1}(J_{i-1}+1)} D_t^0\right)(U_{i-1})(J_{i-1}+1,n).
\end{aligned} \tag{3.27}$$

The transmission conditions for $n = 0$ are obtained by replacing D_t^0 by D_t^+ and $D_t^+ D_t^- /2$ by $D_t^+ /\Delta t$, and introducing the initial velocity. Note that in contrast to the continuous case, the transmission and extraction operators are different. In addition, for non-matching grids, we introduced projection operators which are non-linear. The four operators involved in (3.26), (3.27) are Lax-Wendroff type approximations of the continuous operators (they correspond to the discrete absorbing boundary conditions described in [19]).

3.3.2 Stability and convergence

All the proofs here are very long and intricate, so we refer to [17] for details. For continuous velocities on the interfaces, we have a first convergence result using discrete energy estimates.

3.13 Theorem *Assume that the velocity is continuous on the interfaces a_i. If the CFL condition is satisfied by the discretization in each subdomain, then the non-overlapping discrete Schwarz algorithm with projections* (3.25) *converges on any time interval* $[0,T]$ *in the energy norm:*

$$\sum_{i=1}^{I} E_{N_i}(U_i^k) \to 0 \quad \textit{as} \quad k \to +\infty.$$

Let us recall that in the case of piecewise continuous velocities we recommended that the numerical domain be aligned with the physical discontinuities. In the case of two velocities, when the time step is constant, we have a partial result using the discrete Laplace transform, or z transform (see [35]),

$$U_i^k(j,n) = a_i^k z^n r_i^j, \qquad z = e^{(\eta + i\omega)\Delta t}.$$

3.14 Theorem *If* $(c_1 - c_2)(\gamma_1 - \gamma_2) \geq 0$, *there exists a positive real constant $C > 0$ such that for $\eta\Delta t \ll 1$, we have*

$$||U_i^k(j,n)||_{\Omega_i,\eta,\Delta t} \leq (1 - C\eta\Delta t)^{\lfloor p/2 \rfloor} \max_l (||U_l^0(j,n)||_{\Omega_l,\eta,\Delta t}), \quad i = 1,2.$$

In particular, there is an interface wave: if $z = e^{i\omega\Delta t}$, for $\sin^2(\omega\Delta t/2) > \gamma_i$, there is a solution

$$U_i^{2k}(0,n) = De^{i\omega\Delta t(n+Vk)},$$

where V depends on γ_1, γ_2, and c_1/c_2.

4 Paraxial operators and applications in two dimensions

We mainly concentrate here on the techniques stemming from the theory of pseudodifferential operators [12, 13, 4], but we shall also call on other theories leading to useful absorbing boundary conditions.

4.1 The paraxial problem

The goal is to approximate the equation by a simpler one whose solution propagates in a particular direction. There are many applications, like for instance electromagnetic waves in atmospherical layers, acoustic waves in the ocean, techniques of migration in oil recovery, and the computation of guided waves in optical fibers. We start with the homogeneous wave equation in two dimensions

$$\frac{1}{c^2}\frac{\partial^2 u}{\partial t^2} - \frac{\partial^2 u}{\partial x_1^2} - \frac{\partial^2 u}{\partial x_2^2} = 0.$$

We look at the functions of the form

$$u = e^{i(\boldsymbol{k}\cdot\boldsymbol{x} - \omega t)} \tag{4.1}$$

which is a plane wave solution if and only if the frequency ω and the wave vector $\boldsymbol{k} = (k_1, k_2)$ are linked through the dispersion relation

$$\frac{\omega^2}{c^2} = |\boldsymbol{k}|^2 = k_1^2 + k_2^2.$$

If we introduce the angle θ defined by

$$\sin\theta = c\frac{k_1}{\omega}, \qquad \cos\theta = c\frac{k_2}{\omega},$$

we get the *slowness curve* depicted in Fig. 28.

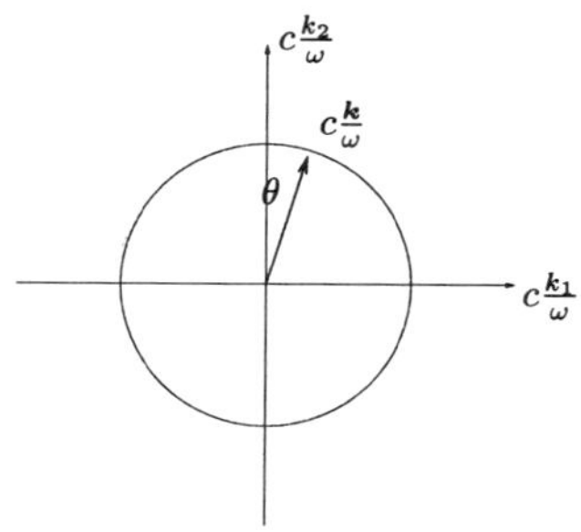

Figure 28: Slowness curve for the wave equation.

4.1.1 The paraxial approximation

Suppose we want to approximate as well as possible the waves propagating upwards, i.e. in the $x_2 > 0$ direction (in geophysics it is usually the depth direction). The dispersion relation of the waves propagating upwards is

$$c\frac{k_2}{\omega} = \sqrt{1 - \left(c\frac{k_1}{\omega}\right)^2}, \tag{4.2}$$

which corresponds to the upper half of the slowness plane. We want to approximate the square root in (4.2) by a function of ck_1/ω which can be, modulo a denominator, the dispersion relation of a partial differential operator. The only way is to use rational fractions [28]

$$(1-X)^{1/2} \approx r(X) = \frac{p_m(X)}{q_n(X)},$$

where p_m and q_n are polynomials in x of degree m and n, respectively. The classical approximations [8, 13, 4] are the Taylor or Padé expansion of low order for small θ.

$$\begin{aligned} &\textit{Taylor of order } 0: && c\frac{k_2}{\omega} \approx 1, \\ &\textit{Taylor of order } 1: && c\frac{k_2}{\omega} \approx 1 - \frac{1}{2}\left(c\frac{k_1}{\omega}\right)^2, \\ &\textit{Padé of order } 1: && c\frac{k_2}{\omega} \approx \frac{1-3/4(ck_1/\omega)^2}{1-1/4(ck_1/\omega)^2}. \end{aligned}$$

To recover the partial differential equations, we clear the denominators, and get the following table, where on the right we put the usual name of these equations.

$$\frac{1}{c}\frac{\partial u}{\partial t} + \frac{\partial u}{\partial x_2} = 0, \quad \textit{transport equation}, \tag{4.3}$$

$$\frac{1}{c}\frac{\partial^2 u}{\partial t^2} + \frac{\partial^2 u}{\partial x_2 \partial t} - \frac{c}{2}\frac{\partial^2 u}{\partial x_1^2} = 0, \quad \textit{parabolic or } 15^\circ \textit{ equation}, \tag{4.4}$$

$$\frac{1}{c}\frac{\partial^3 u}{\partial t^3} + \frac{\partial^3 u}{\partial x_2 \partial t^2} - \frac{3c}{4}\frac{\partial^3 u}{\partial x_1^2 \partial t} - \frac{c^2}{4}\frac{\partial^3 u}{\partial x_1^2 \partial x_2} = 0, \quad 45^\circ \textit{ equation}. \tag{4.5}$$

These equations were called 15° (resp. 45°) because the geophysicists considered them to be accurate for propagation angles up to 15° (resp. 45°) [8]. This strategy can be generalized to higher degrees [13, 4, 28, 20] by introducing the approximate equation

$$c\frac{k_2}{\omega} = r((ck_1/\omega)^2) = \frac{p_m((ck_1/\omega)^2)}{q_n((ck_1/\omega)^2)} = R(ik_1, i\omega). \tag{4.6}$$

Let $d = \max(2m, 2n+1)$ be the total order. By clearing the denominators we obtain a *dispersion relation*

$$\omega^{d-1}k_2 q_n((ck_1/\omega)^2) - \omega^d p_m((ck_1/\omega)^2) = 0,$$

which is of the general form $\mathcal{L}(-i\omega, ik_1, ik_2) = 0$ with $d^\circ\mathcal{L} = d$. It is the dispersion relation of

$$\mathcal{L}(\partial_t, \partial_{x_1}, \partial_{x_2})u = 0. \tag{4.7}$$

Note that the operator $\mathcal{L}$ is of first order in the x_2-direction. The absorbing boundary conditions obtained by Taylor approximations of order three or more lead to ill-posed problems [13]. Various strategies can be applied to approximate the symbol (4.2) depending on the applications, for instance:

- *Padé*: $\sqrt{1-y^2} - r(y^2) = O(y^{2m+2n+1})$ [13],
- *Tchebyshev*: minimize $\|\sqrt{1-y^2} - r(y^2)\|_\infty$ [20],
- *Least squares*: minimize $\|\sqrt{1-y^2} - r(y^2)\|_2$ [28],
- *Interpolation* at arbitrary points [20, 22].

4.1.2 The paraxial problem

The physical problem we start with is the wave equation in two dimensions with given initial conditions

$$\begin{cases} \dfrac{1}{c^2}\dfrac{\partial^2 u}{\partial t^2} - \dfrac{\partial^2 u}{\partial x_1^2} - \dfrac{\partial^2 u}{\partial x_2^2} = 0 \quad \text{in} \quad \mathbb{R}^2 \times (0,T), \\ u(0,.) = u^{(0)}, \quad \dfrac{\partial u}{\partial t}(0,.) = u^{(1)} \quad \text{in} \quad \mathbb{R}^2. \end{cases} \tag{4.8}$$

The paraxial operator (4.7) related to the paraxial approximation is of order d in time. If d is greater than 3 we need more initial values. They can be obtained by differentiating the wave equation. This can be expressed by

$$\mathcal{P}_C \begin{cases} \mathcal{L}u = 0 \quad \text{in} \quad \mathbb{R}^2 \times (0,T) \\ \dfrac{\partial^p u}{\partial t^p} u(0,.) = u^{(p)} \quad \text{in} \quad \mathbb{R}^2,\ 0 \leq p \leq d-1. \end{cases} \tag{4.9}$$

In fact, these operators are never used as high degree operators, but rather as a system of second degree operators, as we will see in Section 4.3.1.

4.2 Absorbing boundary conditions

As we saw in Section 3, the goal here is to truncate the computational domain : suppose we want to compute the solution of the wave equation only in a part Ω_{ref} of $\mathbb{R}^2$. We introduce a computational domain Ω_{comp} and boundary conditions on the boundary $\partial\Omega_{\text{comp}}$ of Ω_{comp} such that the waves can leave the computational domain with as low reflection as possible, which in turn implies that we can place the computational domain close to the reference domain Ω_{ref}. The applications are countless in engineering like geophysics, aeronautics, radars, etc.

4.2.1 Absorbing boundary conditions for a half-plane

We consider waves propagating in the half-plane $x_2 < 0$. The waves propagating upwards are given by (4.2), of the form $e^{i(\boldsymbol{k}\cdot\boldsymbol{x}-\omega t)}$, with $ck_2/\omega = +\sqrt{1-(ck_1/\omega)^2}$. Suppose a general boundary condition is given at $x_2 = 0$. We isolate the derivative in the x_2-direction,

$$\mathcal{M}(\partial_t, \partial_{x_1}, \partial_{x_2})u \equiv \frac{\partial u}{\partial x_2} + \mathcal{B}(\partial_t, \partial_{x_1})u = 0.$$

Let a wave U^{Inc} be impinging on the boundary. By the Snell or Descartes law shown in Fig. 29, the reflected wave is $U^{\text{Ref}} = Re^{i(-k_1x_1+k_2x_2-\omega t)}$, such that

$$\mathcal{M}(U^{\text{Inc}} + U^{\text{Ref}}) = 0,$$

which can be rewritten as

$$ik_2 + \mathcal{B}(-i\omega, ik_1) + R(-ik_2 + \mathcal{B}(-i\omega, ik_1)) = 0.$$

Thus the boundary condition $\mathcal{M}u = 0$ is a transparent boundary condition if its symbol is given by

$$\mathcal{B}(-i\omega, ik_1) \equiv -i\frac{\omega}{c}\sqrt{1-\left(c\frac{k_1}{\omega}\right)^2},$$

and the *transparent boundary operator is the exact paraxial operator*. As absorbing boundary condition we will impose the approximations $\mathcal{L}$ derived in Section 4.1.1.

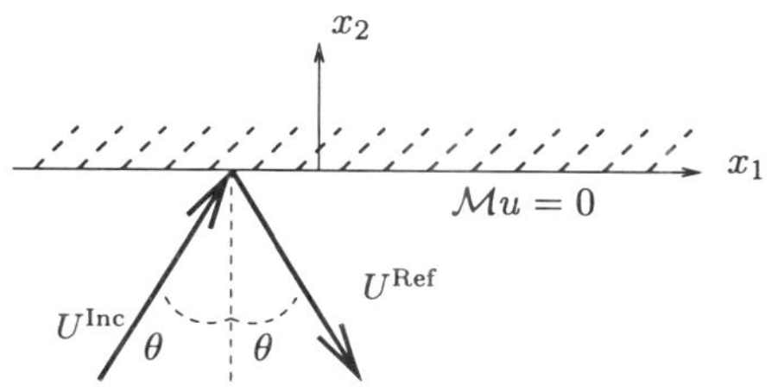

Figure 29: Reflection on the boundary $x_2 = 0$.

4.2.2 The initial boundary value problem

We suppose that the initial data are compactly supported in the half-plane $\mathbb{R}^2_- = \mathbb{R}_- \times \mathbb{R}$, and we replace the Cauchy problem (4.8) by the initial boundary value problem

$$\mathcal{P}_L \begin{cases} \dfrac{1}{c^2}\dfrac{\partial^2 u}{\partial t^2} - \dfrac{\partial^2 u}{\partial x_1^2} - \dfrac{\partial^2 u}{\partial x_2^2} = 0 \quad \text{in} \quad \mathbb{R}^2_- \times (0,T), \\ u(0,.) = u^{(0)}, \quad \dfrac{\partial u}{\partial t}(0,.) = u^{(1)} \quad \text{in} \quad \mathbb{R}^2_-, \\ \mathcal{L}u(t,0) = 0 \quad \text{in} \quad (0,T), \end{cases} \tag{4.10}$$

where $\mathcal{L}$ is any of the paraxial approximations defined in Section 4.1.1, as for instance in (4.3), (4.4) or (4.5). The quality of the approximation is measured by the reflection coefficient. It has been shown that for the absorbing boundary conditions cited above the reflection coefficient is equal to

$$R = \left(\frac{\cos\theta - 1}{\cos\theta + 1}\right)^n .$$

This is a general result for Padé approximation of type (n, n) [13].

Fig. 30 shows the historical example given by R. W. Clayton and B. Engquist in 1977 [9]. The initial value is a spherical compressional wave. On the first line the second order absorbing boundary condition (4.4) is imposed , and on the second line the perfectly reflecting Dirichlet boundary condition is imposed.

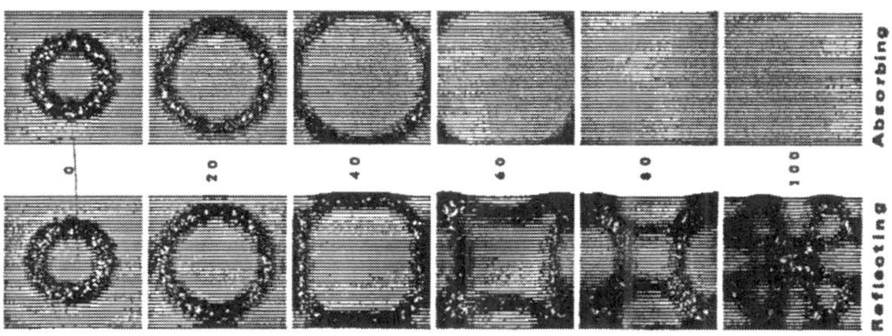

Figure 30: A historical example from [9].

4.3 Well-posedness results

We consider the paraxial problem $\mathcal{P}_C$ defined in (4.9) and the half-plane problem $\mathcal{P}_L$ defined in (4.10). $\mathcal{L}$ is the operator $\mathcal{L}(\partial t, \partial x_1, \partial x_2)$. Its symbol is $l(\omega, \boldsymbol{k}) = \mathcal{L}(i\omega, i\boldsymbol{k})$.

4.1 Remark This is the sign choice made by the analysts. It differs from the convention chosen by the physicists in (4.1).

We first give a classical result on well-posedness for Cauchy problems (for all these basic results see for instance [26]):

$$\begin{gathered} \mathcal{P}_C \textit{ well-posed} \\ \Updownarrow \\ \mathcal{L} \textit{ hyperbolic} \\ \Updownarrow \\ \forall \boldsymbol{k} \in \mathbb{R}^2, \quad l(\omega, \boldsymbol{k}) = 0 \Longrightarrow \omega \in \mathbb{R} \end{gathered}$$

There is also a characterisation of well-posed initial boundary value problems:

$$\begin{gathered} \mathcal{P}_L \textit{ well-posed} \\ \Updownarrow \\ \forall k_1 \in \mathbb{R}, \ \forall k_2 \in \mathbb{C}, \ \Im k_2 \geq 0, \\ \left.\begin{aligned} l(\omega, \boldsymbol{k}) &= 0 \\ \omega^2 = k_1^2 + k_2^2, &\quad \Re\omega \leq 0 \end{aligned}\right\} \Longrightarrow \quad (\omega, \boldsymbol{k}) = (0, \boldsymbol{0}) \end{gathered}$$

The following results have been proved in [20]:

4.2 Theorem *Problem $\mathcal{P}_C$ is well-posed if and only if the rational fraction $r(X)$ in (4.6) can be expanded as*

$$r(X) = a - \beta X - \sum_{k=1}^{n} \frac{\beta_k X}{1 - \gamma_k^2 X}, \tag{4.11}$$

where the coefficients a, β_k are such that $a > 0$, $\beta \geq 0$, $\beta_k > 0$ for $1 \leq k \leq n$.

4.3 Theorem *Problem $\mathcal{P}_L$ is well-posed if and only if we have property (4.11), and furthermore for all X in $[-1, 1]$, the value of the function $r(X)$ defined in (4.11) is positive.*

These results give a necessary condition for well-posedness:

4.4 Corollary *If problems $\mathcal{P}_L$ and $\mathcal{P}_C$ are well-posed, then $n \leq m \leq n + 1$.*

The well-posedness for the two problems are related by the following result:

4.5 Corollary *If problem $\mathcal{P}_L$ is well-posed, then problem $\mathcal{P}_C$ is well-posed.*

In the same paper the following useful criterium is also derived:

$$\begin{gathered} \mathcal{P}_L \textit{ well-posed} \\ \Updownarrow \\ \left\{\begin{array}{l} n \leq m \leq n+1 \quad \textit{and} \\ r(y) - \sqrt{1-y^2} \textit{ has } 2(m+n+1) \textit{ roots in } \mathbb{C} -]-\infty, -1] - [1, +\infty[. \end{array}\right. \end{gathered}$$

4.3.1 Writing the operator as a system

As mentioned before, the expression $\mathcal{L}(\partial_t, \partial_{x_1}, \partial_{x_2})u = 0$ is not very easy to handle, especially when it comes to discretization. Instead, following [28], we turn back to the formula

$$\left[-c\frac{k_2}{\omega} + 1 - \sum_{k=1}^{m} \frac{\beta_k c^2 k_1^2}{\omega^2 - \gamma_k c^2 k_1^2} \right] \hat{u} = 0,$$

and we introduce auxiliary unknowns φ_k by their Fourier transforms

$$\frac{c^2 k_1^2}{\omega^2 - \gamma_k c^2 k_1^2} \hat{u} = \hat{\varphi}_k,$$

$$\left(-c\frac{k_2}{\omega} + 1 \right) \hat{u} - \sum_{k=1}^{m} \beta_k \hat{\varphi}_k = 0,$$

which leads to the system of partial differential equations

$$\begin{cases} \dfrac{\partial u}{\partial t} + c\dfrac{\partial u}{\partial x_2} - \displaystyle\sum_{k=1}^{m} \beta_k \frac{\partial \varphi_k}{\partial t} = 0, \\ \dfrac{1}{c^2}\dfrac{\partial^2 \varphi_k}{\partial t^2} - \gamma_k^2 \dfrac{\partial^2 \varphi_k}{\partial x_1^2} = \dfrac{\partial^2 u}{\partial x_1^2}, \quad 1 \leq k \leq m, \end{cases}$$

with initial conditions on the functions φ_k:

$$\varphi_k(0, x) = 0.$$

Notice that the new unknowns φ_k are active only in the transverse direction, which makes this formula very easy to discretize [6].

There is an alternative theory by Higdon on absorbing boundary conditions [22]. He writes any well-posed absorbing boundary condition as a product of first-order transport operators, which makes the discretization even easier. But this involves a high order operator in the x_2-direction.

4.3.2 Bayliss and Turkel operators

We describe here another approach of absorbing boundary conditions, relying on a far-field expansion of the solution of the wave equation in three dimensions [5],

$$u(r, \theta, \varphi, t) = r^{-1} e^{ik(r-ct)} \sum_{n=0}^{\infty} r^{-n} u_n(\theta, \varphi). \tag{4.12}$$

Defining the Sommerfeld operator $\mathcal{S}$ by

$$\mathcal{S} = \left(\frac{\partial}{\partial r} + \frac{1}{c}\frac{\partial}{\partial t} \right),$$

we note, using (4.12), that we have in the three-dimensional case

$$\lim_{r\to\infty} r\left(\frac{\partial u}{\partial r} + \frac{1}{c}\frac{\partial u}{\partial t}\right) = 0,$$

and more generally in dimension d

$$\lim_{r\to\infty} r^{(d-1)/2}\left(\frac{\partial u}{\partial r} + \frac{1}{c}\frac{\partial u}{\partial t}\right) = 0.$$

A. Bayliss and E. Turkel [5] proposed an iterative improvement of the Sommerfeld condition, in the following way. For the Sommerfeld operator we have $\mathcal{S}u = O(r^{-2})$. Introducing an "augmented" version of the Sommerfeld operator

$$\mathcal{B}_1 = \mathcal{S} + \frac{1}{r},$$

we have by (4.12)

$$\mathcal{B}_1 u = O\left(\frac{1}{r^3}\right).$$

The operator $\mathcal{B}_1$ is the first order Bayliss and Turkel operator. Following the same approach, Bayliss and Turkel define a hierarchy of operators

$$\mathcal{B}_k = \left(\mathcal{S} + \frac{2k-1}{r}\right)\mathcal{B}_{k-1}.$$

A short calculation shows that for any k we have

$$\mathcal{B}_k u = \sum_{n=1}^{\infty} a_n^k u_n r^{-(n+k)},$$

with the recursion relation

$$u_n = e^{ik(r-ct)}u_{n-1}, \qquad a_n^k = (-1)^k(n-1)\cdots(n-k).$$

Thus $\mathcal{B}_k$ absorbs the first k terms of the expansion (4.12):

$$\mathcal{B}_k u = O\left(\frac{1}{r^{2k+1}}\right).$$

4.4 Domain decomposition method

This part is work in progress, see [16], and we sketch ideas only. Let $\mathbb{R}^2$ be divided into two subdomains $\Omega_1 = \mathbb{R}\times(-\infty, L)$ and $\Omega_2 = \mathbb{R}\times(0,\infty)$ with an overlap as depicted in Fig. 31. The boundaries are $\Gamma_1 = \mathbb{R}\times\{L\}$ and $\Gamma_2 = \mathbb{R}\times\{0\}$.

The natural extension of the classical Schwarz algorithm is then

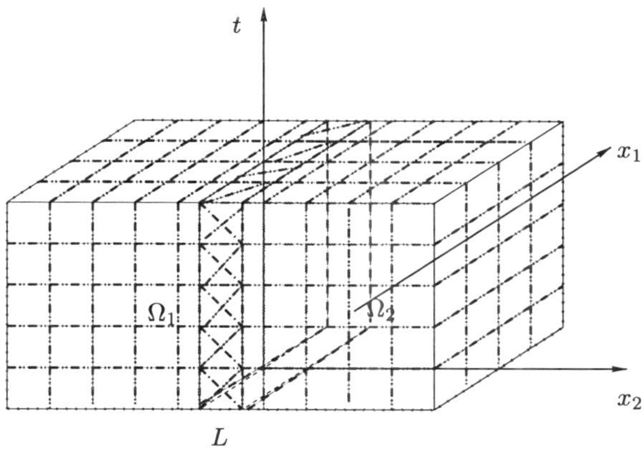

Figure 31: Domain decomposition in two half-planes.

$$\left.\begin{array}{l}\dfrac{1}{c^2}\dfrac{\partial^2 u_i^{n+1}}{\partial t^2}-\dfrac{\partial^2 u_i^{n+1}}{\partial x_1^2}-\dfrac{\partial^2 u_i^{n+1}}{\partial x_2^2}=0 \quad \text{in} \quad \Omega_i\times(0,T),\\ \text{with the initial conditions}\\ u_i^{n+1}(0,.)=u^{(0)},\quad \dfrac{\partial}{\partial t}u_i^{n+1}(0,.)=u^{(1)} \quad \text{in} \quad \Omega_i,\\ \text{and the transmission conditions}\\ u_1^{n+1}=u_2^n \quad \text{on} \quad \Gamma_1\times(0,T), \qquad u_2^{n+1}=u_1^n \quad \text{on} \quad \Gamma_2\times(0,T).\end{array}\right\} \tag{4.13}$$

The following result is a straightforward consequence of the finite speed of propagation:

4.6 Theorem *The Schwarz algorithm converges for a number of iterations $n > T\overline{c}/L$ with* $\overline{c} := \sup_{\mathbf{x}\in\Omega} c(\mathbf{x})$.

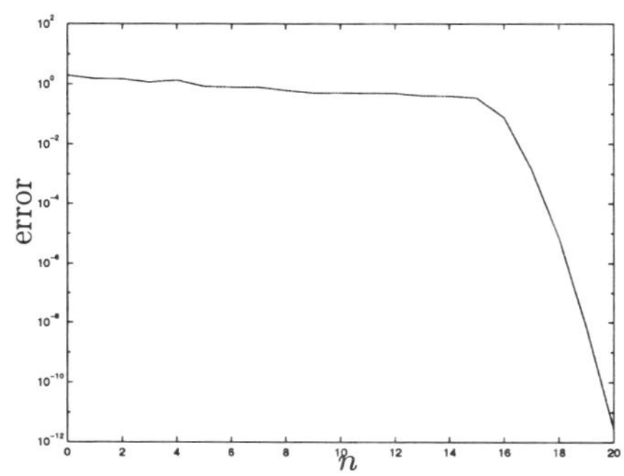

Figure 32: Convergence history for the Schwarz algorithm.

Fig. 32 shows the curve of convergence for the wave equation with speed 1 on the square $0 < x, y < 1$, for the time interval $0 < t < T = 1.2$. We have two subdomains and the overlap is $L = 0.08$. The theory predicts convergence after 15 iterations.

4.7 Remark The result in Theorem 4.6 is very general. The difficulty with many subdomains is the construction of the subdomains. In Figs. 33 and 34 we illustrate the strategy we suggest from classical Schwarz methods: first we divide the domain into squares without overlap, and then we add layers.

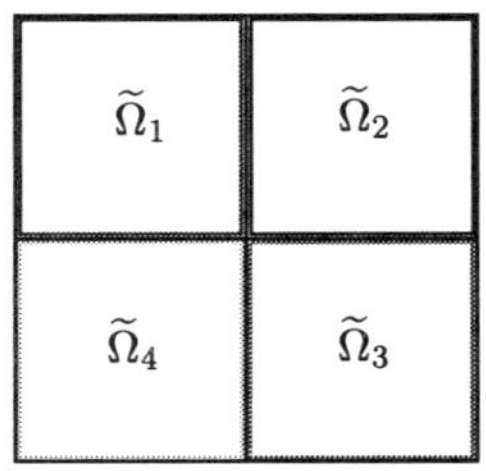

Figure 33: Non-overlapping domains.

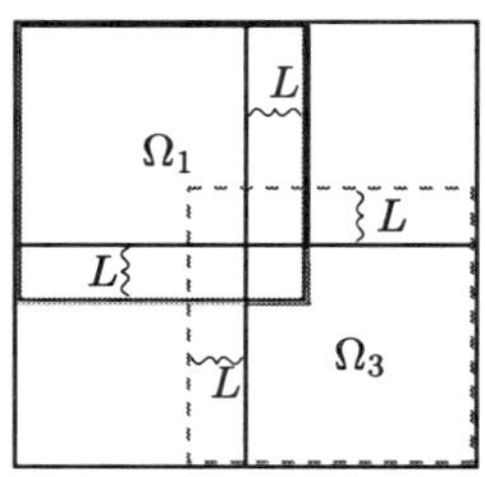

Figure 34: Overlapping domains.

4.4.1 Optimal transmission conditions

We introduce the algorithm with general transmission conditions

$$\left(\frac{\partial}{\partial x_2} + \Lambda_1(\partial_t, \partial_{x_1})\right) u_1^{n+1} = \left(\frac{\partial}{\partial x_2} + \Lambda_1(\partial_t, \partial_{x_1})\right) u_2^n, \quad x_2 = L,$$
$$\left(\frac{\partial}{\partial x_2} + \Lambda_2(\partial_t, \partial_{x_1})\right) u_2^{n+1} = \left(\frac{\partial}{\partial x_2} + \Lambda_2(\partial_t, \partial_{x_1})\right) u_1^n, \quad x_2 = 0,$$

and we use a plane wave analysis. The error (U_i^n) is given by

$$U_i^n = a_i^n e^{i(\mathbf{k}^\mathbf{i}\cdot\mathbf{x} - \omega t)},$$

with the dispersion relations

$$c\frac{k_2^2}{\omega} = \sqrt{1 - \left(c\frac{k_1^2}{\omega}\right)^2}, \qquad k_1^2 = k_1^1, \qquad k_2^2 = -k_2^1,$$

and the relation between the coefficients over a double iteration is

$$a_1^{n+1} = \frac{ik_2^1 + \lambda_1}{-ik_2^1 + \lambda_1} \cdot \frac{-ik_2^1 + \lambda_2}{ik_2^1 + \lambda_2} e^{+2ik_2^1 L} a_1^{n-1}.$$

We define the *convergence rate* of the algorithm by

$$\rho(\mathbf{k}, \omega) = \frac{-i\omega\sqrt{1 - (ck_1^1/\omega)^2} + c\lambda_1}{i\omega\sqrt{1 - (ck_1^1/\omega)^2} + c\lambda_1} \cdot \frac{i\omega\sqrt{1 - (ck_1^1/\omega)^2} + c\lambda_2}{-i\omega\sqrt{1 - (ck_1^1/\omega)^2} + c\lambda_2} e^{-2ic\omega\sqrt{1-(ck_1^1/\omega)^2}L}, \quad (4.14)$$

and the coefficients are now related by

$$a_1^{n+1} = \rho(\mathbf{k}, \omega)_1^{n-1}.$$

It is evident from (4.14) that by choosing

$$\lambda_1 = \frac{i\omega}{c}\sqrt{1 - \left(c\frac{k_1^1}{\omega}\right)^2} \quad \text{and} \quad \lambda_2 = -\frac{i\omega}{c}\sqrt{1 - \left(c\frac{k_1^1}{\omega}\right)^2},$$

the algorithm converges in two iterations. Furthermore the result is general: for these transmission conditions, with I subdomains, the convergence is achieved in I iterations. Thus: *The optimal transmission conditions correspond to the transparent operator.*

4.4.2 Approximations of the optimal transmission conditions

We approximate the symbols λ_1 and λ_2 at normal incidence, i.e. for small values of ck_1^1/ω, by $\pm i\omega/c$, and we obtain

$$\Lambda_{12} \approx \pm\frac{1}{c}\frac{\partial}{\partial t},$$

which gives as transmission conditions

$$\begin{cases} \left(\dfrac{\partial}{\partial x_2} + \dfrac{1}{c}\dfrac{\partial}{\partial t}\right) u_1^{n+1} = \left(\dfrac{\partial}{\partial x_2} + \dfrac{1}{c}\dfrac{\partial}{\partial t}\right) u_2^n, & x_2 = L, \\ \left(\dfrac{\partial}{\partial x_2} - \dfrac{1}{c}\dfrac{\partial}{\partial t}\right) u_2^{n+1} = \left(\dfrac{\partial}{\partial x_2} - \dfrac{1}{c}\dfrac{\partial}{\partial t}\right) u_1^n, & x_2 = 0. \end{cases} \quad (4.15)$$

We show in Fig. 35 the curve of convergence for this algorithm, compared to the classical Schwarz algorithm. The data are the same as in Fig. 32.

With this kind of transmission conditions, it is also possible to work without overlap. For further developments on this issue we refer to [16].

We have already seen some links between the three related topics: the paraxial approximation, the absorbing boundary conditions and domain decomposition. We summarize these in the next subsection.

4.5 Links between paraxial operators, absorbing boundary conditions and optimal domain decomposition algorithms

We start with the wave equation in the plane

$$\frac{1}{c^2}\frac{\partial^2 u}{\partial t^2} - \Delta u = 0, \quad \text{in} \quad \mathbb{R}^2 \times (0, T).$$

We introduce a rational function $R(ik_1, i\omega) = r(k_1/\omega)$ and the problems:

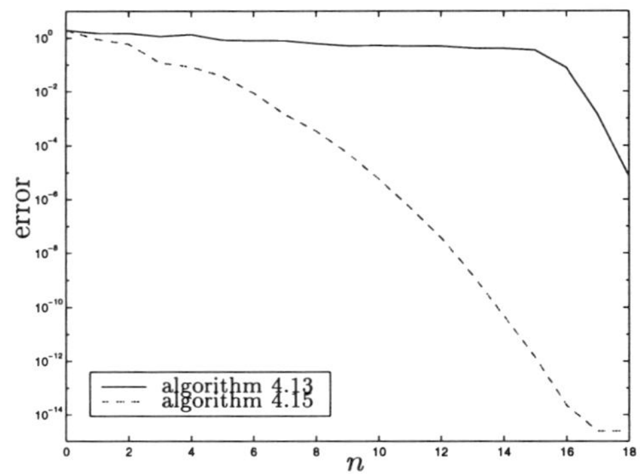

Figure 35: Convergence of the algorithm with transmission conditions (4.13) and (4.15).

(i) *The paraxial equation in the $x_2 > 0$ direction:*

$$\partial_{x_2} u + R(\partial_t, \partial_{x_1}) u = 0 \quad \text{in} \quad \mathbb{R}^2 \times (0,T)$$

(ii) *The absorbing boundary conditions:*

$$\begin{aligned} \frac{1}{c^2}\frac{\partial^2 u}{\partial t^2} - \Delta u = f \quad \text{in} \quad \mathbb{R}^2_- \times (0,T), \\ \partial_{x_2} u + R(\partial_t, \partial_{x_1}) u = 0, \quad x_2 = 0 \end{aligned}$$

(iii) *The domain decomposition algorithm without overlap:*

$$\begin{aligned} \left(\frac{1}{c^2}\frac{\partial^2}{\partial t^2} - \Delta\right) u_1^{n+1} = f \text{ in } \mathbb{R}^2_- \times (0,T), \quad \left(\frac{1}{c^2}\frac{\partial^2}{\partial t^2} - \Delta\right) u_2^{n+1} = f \text{ in } \mathbb{R}^2_+ \times (0,T), \\ (\partial_{x_2} + R(\partial_t, \partial_{x_1})) u_1^{n+1} = (\partial_{x_2} + R(\partial_t, \partial_{x_1})) u_2^n; \quad x_2 = 0, \\ (\partial_{x_2} - R(\partial_t, \partial_{x_1})) u_2^{n+1} = (\partial_{x_2} - R(\partial_t, \partial_{x_1})) u_1^n; \quad x_2 = 0. \end{aligned}$$

The error is measured in the three cases for x in $[0,1]$ as follows:

(i) *Paraxial equation: the error*

$$\text{err}(x) = \sqrt{1-x^2} - r(x)$$

(ii) *Absorbing boundary conditions: the reflection coefficient*

$$\frac{\sqrt{1-x^2} - r(x)}{\sqrt{1-x^2} + r(x)}$$

(iii) *Domain decomposition algorithm: the convergence rate*

$$\frac{\sqrt{1-x^2} - r(x)}{\sqrt{1-x^2} + r(x)}$$

5 Classical layers and perfectly matched layers for hyperbolic problems

The layer or sponge methods consist of surrounding the domain of interest by a layer. This layer should be designed in such a way that it produces as low reflection as possible and the waves are absorbed in the layer. Furthermore, to save computation, it should be made thin.

In practice, defining the layer can be very involved as it depends very much on the frequency of the incoming signal. This explains why layer methods have not been very much used. More evolved models have been proposed (coupling layers and absorbing boundary conditions) [24], but the increase in complexity lessens the interest of these methods.

Bérenger's work [7] has renewed the interest in these layer methods; the Bérenger perfectly matched layer, PML, has many attractive features: it absorbs waves of any wavelength and any frequency without spurious reflection; moreover, the corner problem is easily solved by a wise choice of the layer parameters. Finally, it is very easy to integrate into an existing code. Unfortunately, as applied to the Maxwell system, the original method leads to a system which has lost the most important properties of the Maxwell system: strong hyperbolicity and symmetry.

In the last part of this section we propose an algebraic technique leading to a new PML model which is strongly well-posed and preserves the symmetry. We start however with the classical methods.

5.1 The classical layers

The layers techniques are often inspired by well-known physical models (soundproof rooms of acoustics laboratories). As an example, we first present some models for the wave equation in two dimensions. These examples, although simple, highlight the power and the limitations of the classical approach. We summarize here some of the results by M. Israeli and S. Orszag in [24].

We consider the wave equation in two dimensions:

$$\frac{\partial^2 u}{\partial t^2} - \Delta u = 0.$$

A simple model can be obtained by adding a friction term

$$\frac{\partial^2 u}{\partial t^2} + \sigma(x)\frac{\partial u}{\partial t} - \Delta u = 0,$$

where σ is a positive function of x. In such a medium the energy is decreasing: it is a lossy medium.

More precisely, we consider the case of a propagation in the half-space $x_1 < 0$, and we build the layer of width δ on the right of the domain

$$\begin{cases} \dfrac{\partial^2 u}{\partial t^2} + \sigma(x)\dfrac{\partial u}{\partial t} - \Delta u = 0, \quad (x_1, x_2) \in (-\infty, \delta) \times \mathbb{R} \\ u(0,\cdot) = u^{(0)}, \quad \dfrac{\partial u}{\partial t}(0,\cdot) = u^{(1)} \\ u(\delta, x_2, t) = 0 \end{cases} \tag{5.1}$$

with $\sigma(x) = 0$ for $x_1 < 0$. The initial data are supported in $\mathbb{R}^2_-$.

The choice of the damping factor σ is crucial. First, the reflection at the boundary of the layer must be maintained as small as possible. Second, the decay in the layer must be sufficiently significant so that the Dirichlet condition imposed on the exterior boundary of the layer does not produce reflection. The plane wave analysis gives a more precise idea of the best choice of the damping factor.

We call u^- the solution in $\mathbb{R}^2_-$ and u^+ the solution in $]0, \delta[\times\mathbb{R}$. They are given by

$$u^- = e^{i(k_1x_1+k_2x_2-\omega t)} + Re^{i(-k_1x_1+k_2x_2-\omega t)},$$
$$u^+ = T_1e^{i(k_1^\sigma x_1+k_2x_2-\omega t)} + T_2e^{i(-k_1^\sigma x_1+k_2x_2-\omega t)}.$$

The dispersion relations are

$$k_1^2 + k_2^2 = \omega^2, \qquad (k_1^\sigma)^2 = k_1^2 + i\omega\sigma;$$

k_1 and k_1^σ are defined by

$$\frac{k_1}{\omega} > 0; \quad k_1^\sigma \in \mathbb{C}, \quad \frac{\Re k_1^\sigma}{\omega} > 0,$$

such that $e^{i(k_1x_1+k_2x_2-\omega t)}$ and $e^{i(k_1^\sigma x_1+k_2x_2-\omega t)}$ propagate to the left. For any ω the imaginary part of k_1^σ is positive, and we have the following process: the incident wave $u^I = e^{i(k_1x_1+k_2x_2-\omega t)}$ reflects at $x_1 = 0$ into $R_0e^{i(-k_1x_1+k_2x_2-\omega t)}$ and transmits in the layer into $T_0e^{i(k_1^\sigma x_1+k_2x_2-\omega t)}$. This wave is in turn reflected at the end of the layer $x_1 = \delta$ into $R_\delta T_0e^{i(-k_1^\sigma x_1+k_2x_2-\omega t)}$ and then reflected, and so on for ever. The coefficients are given by

$$R_0 = \frac{k_1 - k_1^\sigma}{k_1 + k_1^\sigma}, \qquad T_0 = 1 + R_0, \qquad R_\delta = -e^{2ik_1^\sigma\delta}.$$

For the full process, the coefficients are given by

$$R = \frac{R_0 + R_\delta}{1 + R_0R_\delta}, \qquad T_1 = \frac{1+R}{1+R_\delta}, \qquad T_2 = R_\delta T_1.$$

It is easy to see that if σ is large then $|R_0|$ is close to 1. On the other hand, if σ is small, $|R_0|$ is small and R_δ is close to 1, unless δ is large. The tuning of the coefficients is thus difficult. Other attempts have been made to improve the results: adding a term $-\mu(x)\partial^3u/\partial t\partial x_1^2$, or replacing the term $\partial u/\partial t$ by $\partial u/\partial t + \partial u/\partial x_1$[24]. In the latter case, the reflection coefficient is small for low angles of incidence which considerably improves the results and has made this strategy attractive. However, the breakthrough came later with the PML we describe now.

5.2 The PML approach of Bérenger

In the mid-90s, Bérenger [7] proposed a new layer method for Maxwell's equation, namely the *perfectly matched layer* or PML. The strength of this new method is that the layer absorbs waves regardless of their frequency and angle of incidence.

For simplicity we describe the method for the two-dimensional Maxwell equations in the transverse electric mode (TE), although all the developments in [7] have been made in the three-dimensional case. The Maxwell system then has only three unknowns, (E_1, E_2, H_3), related as follows:

$$\begin{cases} \epsilon \dfrac{\partial E_1}{\partial t} + \sigma E_1 - \dfrac{\partial H_3}{\partial x_2} = 0, \\ \epsilon \dfrac{\partial E_2}{\partial t} + \sigma E_2 + \dfrac{\partial H_3}{\partial x_1} = 0, \\ \mu \dfrac{\partial H_3}{\partial t} + \sigma^* H_3 + \dfrac{\partial E_2}{\partial x_1} - \dfrac{\partial E_1}{\partial x_2} = 0. \end{cases} \tag{5.2}$$

The coefficients ϵ and μ are the permittivity and permeability of the medium, σ is the electric conductivity and σ^* the magnetic conductivity. To establish a PML model in the region $\Omega_+ = \{x_1 > 0\}$, we split the magnetic field into two non-physical subcomponents $H_3 = H_{31} + H_{32}$, and we introduce damping factors σ_i and σ_i^*, for $i = 1, 2$, with $\sigma_1 = \sigma_2 = \sigma$ and $\sigma_1^* = \sigma_2^* = \sigma^*$ in $\Omega_- = \mathbb{R}^2 - \Omega_+$ (the domain of interest).

We then write a modified version of (5.2) in a form where only one spatial derivative appears in each equation:

$$\begin{cases} \epsilon \dfrac{\partial E_1}{\partial t} + \sigma_2 E_1 - \dfrac{\partial (H_{31} + H_{32})}{\partial x_2} = 0, \\ \epsilon \dfrac{\partial E_y}{\partial t} + \sigma_1 E_2 + \dfrac{\partial (H_{31} + H_{32})}{\partial x_1} = 0, \\ \mu \dfrac{\partial H_{31}}{\partial t} + \sigma_1^* H_{31} + \dfrac{\partial E_2}{\partial x_1} = 0, \\ \mu \dfrac{\partial H_{32}}{\partial t} + \sigma_2^* H_{32} - \dfrac{\partial E_1}{\partial x_2} = 0. \end{cases} \tag{5.3}$$

The Bérenger medium is characterized by the fictitious conductivities. We call the system $\mathbf{PML}(\sigma_1, \sigma_1^*, \sigma_2, \sigma_2^*)$. The Maxwell system appears as a particular case of the PML equations. It corresponds to $\mathbf{PML}(\sigma, \sigma^*, \sigma, \sigma^*)$:

$$\begin{aligned} \mathbf{PML}(\sigma, \sigma^*, \sigma, \sigma^*) &\longrightarrow \text{absorbing medium}, \\ \mathbf{PML}(0,0,0,0) &\longrightarrow \text{vacuum}. \end{aligned}$$

5.1 Remark The original equations are recovered by adding the equations for the split fields.

We describe now the important features of this model [7]. Suppose the conductivities satisfy the *matching conditions*

$$\frac{\sigma_i}{\epsilon} = \frac{\sigma_i^*}{\mu}, \tag{5.4}$$

Then for harmonic fields, the amplitudes decays exponentially. They are of the form

$$\Psi \exp\left(i\omega\left(t - \frac{1}{c}\boldsymbol{x} \cdot \boldsymbol{C}\right)\right) \exp\left(-\frac{\sigma_1 C_1}{\epsilon c} x_1\right) \exp\left(-\frac{\sigma_2 C_2}{\epsilon c} x_2\right)$$

where c is the speed of light, $c = 1/\sqrt{\epsilon\mu}$, and $\boldsymbol{C}$ is the unitary wave vector, $\boldsymbol{C} = (\cos\theta, \sin\theta)$.

Consider now two PML media $\mathbf{PML}(\sigma_1^1, \sigma_1^{1*}, \sigma_2^1, \sigma_2^{1*})$ and $\mathbf{PML}(\sigma_1^2, \sigma_1^{2*}, \sigma_2^2, \sigma_2^{2*})$. The reflection coefficient between these media vanishes provided

(i) *their conductivities satisfy the matching conditions* (5.4)*;*

(ii) $\Sigma^1 \cdot \tau = \Sigma^2 \cdot \tau$ *with* $\Sigma = (\sigma_1, \sigma_2)$ *and* τ *is tangent to the interface.*

These results lead to the now famous diagram 36 extracted from [7] which provides the complete strategy for PML, including the question of corners, which has never been satisfactorily solved for absorbing boundary conditions :

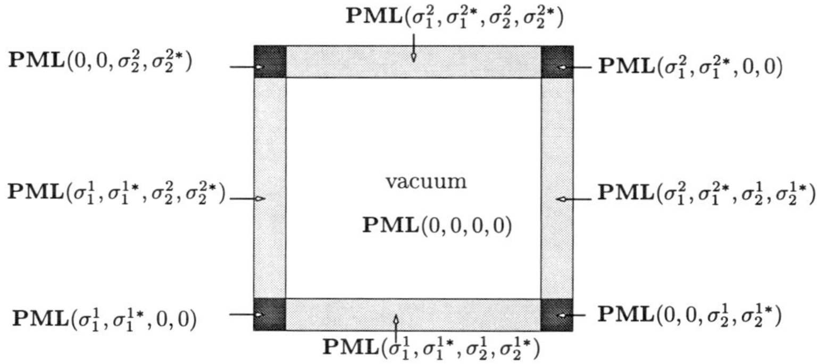

Figure 36: The **PML** technique [7].

In the harmonic case, the model **PML**$(\sigma, \sigma, 0, 0)$ becomes:

$$\left\{\begin{aligned} i\epsilon\omega\hat{E}_1 - \frac{\partial(\hat{H}_{31}+\hat{H}_{32})}{\partial x_2} &= 0,\\ i\epsilon\omega\hat{E}_2 + \sigma\hat{E}_2 + \frac{\partial(\hat{H}_{31}+\hat{H}_{32})}{\partial x_1} &= 0,\\ i\mu\omega\hat{H}_{31} + \sigma\hat{H}_{31} + \frac{\partial\hat{E}_2}{\partial x_1} &= 0,\\ i\mu\omega\hat{H}_{32} - \frac{\partial\hat{E}_1}{\partial x_2} &= 0. \end{aligned}\right. \tag{5.5}$$

We can add the last two equations and keep only one unknown $\hat{H}_3 = \hat{H}_{31} + \hat{H}_{32}$:

$$\frac{i\omega}{i\omega\mu + \sigma(x)}\frac{\partial}{\partial x_1}\left(\frac{i\omega}{i\omega\epsilon + \sigma(x)}\frac{\partial}{\partial x_1}\hat{H}_3\right) + c^2\frac{\partial^2}{\partial x_2^2}\hat{H}_3 + \omega^2\hat{H}_3 = 0.$$

In [10], Collino and Monk show that this problem is well posed except for a discrete set of exceptional frequencies (which may be empty), and that the PML formulation amounts to the continuation of the Green's function into the complex plane. As a matter of fact, making the complex change of variables

$$x_1' = x_1 - \frac{i}{\omega}\int_0^{x_1}\sigma(\xi)\,d\xi,$$

we retrieve the Helmholtz equation, so that the PML model can be viewed as a complex change of variables applied to the original system [34].

More generally, we consider a general first order system written in the form

$$\frac{\partial \varphi}{\partial t} + A_1 \frac{\partial \varphi}{\partial x_1} + A_2 \frac{\partial \varphi}{\partial x_2} = 0, \tag{5.6}$$

where φ lives in $\mathbb{R}^N$. In order to obtain the PML model we split the unknown φ into two non-physical parts $\varphi = \varphi_1 + \varphi_2$, and we write the system in a form where only one spatial derivation appears in each equation:

$$\begin{cases} \dfrac{\partial \varphi_1}{\partial t} + A_1 \dfrac{\partial \varphi}{\partial x_1} = 0, \\ \dfrac{\partial \varphi_2}{\partial t} + A_2 \dfrac{\partial \varphi}{\partial x_2} = 0, \end{cases}$$

then we introduce damping factors $\sigma_1(x_1)$ and $\sigma_2(x_2)$, and the PML system becomes :

$$\begin{cases} \dfrac{\partial \varphi_1}{\partial t} + A_1 \dfrac{\partial \varphi}{\partial x_1} + \sigma_1(x_1)\varphi_1 = 0, \\ \dfrac{\partial \varphi_2}{\partial t} + A_2 \dfrac{\partial \varphi}{\partial x_2} + \sigma_2(x_2)\varphi_2 = 0, \\ \varphi = \varphi_1 + \varphi_2. \end{cases} \tag{5.7}$$

5.2.1 Propagation properties

We start with the propagation in free space. From a plane wave solution of the original system (5.6), we can construct a plane wave solution of (5.7). To this end let

$$\boldsymbol{\Phi} = \boldsymbol{\Phi_0} e^{i(\boldsymbol{k}\cdot\boldsymbol{x} - \omega t)}$$

be a solution of (5.6). Then we obtain a solution of (5.7) as

$$\boldsymbol{\varphi}_i = \boldsymbol{\alpha}_i e^{i(\boldsymbol{k}\cdot\tilde{\boldsymbol{x}}(\boldsymbol{x}) - \omega t)}$$

with

$$\tilde{x}_j = \int_0^{x_j} \left(1 - \frac{i\sigma_j(\xi)}{\omega}\right) d\xi \tag{5.8}$$

and

$$\boldsymbol{\alpha}_i = \frac{k_i}{\omega} A_i \boldsymbol{\Phi_0}.$$

Thus if σ_i is chosen such that it vanishes in the domain of interest, and is strictly positive in the layer wich is orthogonal to x_i, the plane wave solutions of (5.7) coincide with those of the original system (5.6) in the domain of interest, and the amplitude of the wave is exponentially decaying in the layer.

Consider now a PML of thickness δ orthogonal to the x_1-direction. Imposing a homogeneous Dirichlet condition $\varphi_j = 0$ at the external boundary of the PML (at $x_1 = \delta$) and writing $\varphi_{j,\text{inc}} + R\varphi_{j,\text{ref}} = 0$ at $x_1 = \delta$, we find the amplitude of the reflected wave

$$R = -\exp\left(-2ik_1 \int_0^{\delta} \left(1 - \frac{i\sigma_1(\xi)}{\omega}\right) d\xi\right).$$

Since $k_1/\omega > 0$, the reflection coefficient is exponentially small and it can be made as small as desired by increasing the thickness of the layer; in the numerical application it is necessary to find a good compromise between the thickness of the layer and the "acceptable" error.

5.2.2 Well-posedness

The time-domain PML model is mainly designed for propagation models, i.e. hyperbolic equations. In this respect, Bérenger's approach presents some inconveniences: even if the original model is strongly well-posed, the PML model becomes weakly well-posed. This is the case for most systems, which are in fact symmetric, as for instance the Maxwell system, the Euler system, the wave equation and the elastodynamics system. The weak well-posedness can generate instabilities in the numerical schemes, and spurious reflection at the external boundary. The loss of symmetry is a disadvantage for storage reasons.

To make these notions more precise (for details see [26]), let $P(\partial/\partial x)$ be a constant-coefficient operator. We introduce the Cauchy problem:

$$\begin{cases} \varphi_t - P(\partial/\partial x)\varphi = 0 \\ \varphi(x,0) = \varphi_0(x). \end{cases} \tag{5.9}$$

We make the following definitions:

(1) Problem (5.9) is *strongly well-posed* if there is a real number α and a positive real number K such that for any smooth initial data we have

$$\|\varphi(.,t)\|_{L^2} \leq K e^{\alpha t} \|\varphi_0\|_{L^2}, \quad t \geq 0.$$

(2) Problem (5.9) is *weakly well-posed* if there is a strictly positive q, a real number α and a positive real number K such that for any smooth initial data

$$\|\varphi(.,t)\|_{L^2} \leq K e^{\alpha t} \|\varphi_0\|_{H^q}, \quad t \geq 0.$$

There is an algebraic criterium for this definition of well-posedness. Let q be a positive integer and α a real number. The following conditions are equivalent:

(i) There is $K_1 > 0$ such that

$$|e^{P(i\omega)t}| \leq K_1(1 + |\omega|^q)e^{\alpha t}, \quad \omega \in \mathbb{R}^n, \ t \geq 0;$$

(ii) There is K_2 such that

$$\|\varphi(.,t)\|_{L^2} \leq K_2 e^{\alpha t} \|\varphi_0\|_{H^q}, \quad t \geq 0$$

for all φ_0 sufficiently smooth.

Suppose now the operator P to be of first order in space, $P(\partial/\partial x) = \sum_{j=1}^n A_j \partial/\partial x_j$. We define the symbol $P(i\zeta) = i\sum_{j=1}^n \zeta_j A_j$. System (5.9)) is called:

(i) *weakly hyperbolic* if for any ζ in $\mathbb{R}^N$ the eigenvalues of $P(i\zeta)$ are purely imaginary;

(ii) *strongly hyperbolic* if for any ζ in $\mathbb{R}^N$, $|\zeta| = 1$, the eigenvalues of $P(i\zeta)$ are purely imaginary, and furthermore there is a uniform transformation to diagonal form;

(iii) *symmetric hyperbolic* if $A_j = A_j^*$ for any j;

(iv) *strictly hyperbolic* if for any ζ in $\mathbb{R}^N$, $\zeta \neq 0$, the eigenvalues of $P(i\zeta)$ are purely imaginary and distinct.

We have the following relations between well-posedness and hyperbolicity:

(i) strongly hyperbolic $\Longleftrightarrow$ strongly well-posed;

(ii) weakly hyperbolic $\Longleftrightarrow$ weakly well-posed;

(iii) symmetric hyperbolic $\Longrightarrow$ strongly hyperbolic $\Longrightarrow$ strongly well-posed;

(iv) strictly hyperbolic $\Longrightarrow$ strongly hyperbolic $\Longrightarrow$ strongly well-posed.

As an example, the Maxwell system in general and the (TE) system (5.2) in particular are symmetric hyperbolic. The eigenvalues are simple: $0, \pm i|\zeta|$. However, the Bérenger system (5.3) is not symmetric nor even strongly well-posed: 0 is a double eigenvalue, and the dimension of the corresponding eigenspace is still 1. This inconvenience is inherent in the decomposition. One can prove for this system that if (φ_1, φ_2) is a solution of the related Cauchy problem, then

$$||(\varphi_1, \varphi_2)||_{L^2} \leq Ce^{\alpha t}||(\varphi_1, \varphi_2)(0, \cdot)||_{H^1}.$$

Moreover, this estimate is optimal [33].

5.2.3 Some examples

The Bérenger technique has been used for various hyperbolic problems [33, 32, 31, 30, 3, 2]. We focus here on the two-dimensional Euler system. The compressible Euler system linearized at constant flow $(\boldsymbol{U}, R = c^2)$ can be written as a symmetric system

$$\frac{\partial \varphi}{\partial t} + \sum_{j=1}^{2} A_j \frac{\partial \varphi}{\partial x_j} = 0$$

with

$$\varphi = \begin{pmatrix} u_1 \\ u_2 \\ \rho \end{pmatrix}, \qquad A_1 = \begin{pmatrix} U_1 & 0 & c \\ 0 & U_1 & 0 \\ c & 0 & U_1 \end{pmatrix}, \qquad A_2 = \begin{pmatrix} U_2 & 0 & 0 \\ 0 & U_2 & c \\ 0 & c & U_2 \end{pmatrix}.$$

We shall consider in the sequel the case where the linearization is made about velocity 0. For the general case, a change of variables along the flow produces moving PMLs [33, 2]. The simplest strategy is to split the density into $\rho = \rho_1 + \rho_2$, and then the continuity equation becomes:

$$\frac{\partial \rho_i}{\partial t} + c\frac{\partial u_i}{\partial x_i} = 0, \quad i = 1, 2.$$

Next, we introduce the damping factor $\sigma(x)$, with $\sigma(x) > 0$ in the PML region $\{x_1 > 0\}$, and zero elsewhere. The PML model is then given by

$$\begin{cases} \dfrac{\partial \rho_1}{\partial t} + \sigma(x)\rho_1 + c\dfrac{\partial u_1}{\partial x_1} = 0 \\ \dfrac{\partial \rho_2}{\partial t} + c\dfrac{\partial u_2}{\partial x_2} = 0 \\ \dfrac{\partial u_1}{\partial t} + \sigma(x)u_1 + c\dfrac{\partial \rho}{\partial x_1} = 0 \\ \dfrac{\partial u_2}{\partial t} + c\dfrac{\partial \rho}{\partial x_2} = 0 \end{cases} \tag{5.10}$$

Again, this problem is only weakly well-posed. We introduce now a new model, based on (5.10), which will be symmetric hyperbolic and thus strongly well-posed.

5.2 Remark For the wave equation

$$\frac{\partial^2 \rho}{\partial t^2} - \Delta \rho = 0,$$

we write the problem as a first order hyperbolic system using the Euler form, and we proceed as before [32]. Another approach, used in [10], is based on the complex change of variables (5.8) and a fast Fourier transform.

5.3 Remark For the system of isotropic elastodynamics in two dimensions, one can obtain Berenger-type PMLs as well [11]. The difficulty comes for anisotropic behaviors.

5.3 A well-posed PML model

Various improvements of PML models have been proposed to circumvent the problems mentioned in the previous section; in [21] and [2], the authors (following the approach developed for electromagnetism in [1]) modify the Euler equations by introducing low-order terms only. Thus by construction the resulting model is strongly well-posed. We describe here an algebraic technique leading to a new PML model in the primitive variables, which is symmetric and preserves the propagation and absorption advantages of the Berenger model [31]. This model is similar to the one obtained in [21]. However, the approaches are different since we start from Bérenger's model.

We consider the general harmonic PML model in the frequency domain

$$\begin{cases} (i\omega + \sigma_1(x_1))\hat{\varphi}_1 + A_1\dfrac{\partial \hat{\varphi}}{\partial x_1} = 0 \\ (i\omega + \sigma_2(x_2))\hat{\varphi}_2 + A_2\dfrac{\partial \hat{\varphi}}{\partial x_2} = 0. \end{cases}$$

Adding these equations, we retrieve the physical unknowns

$$i\omega\hat{\varphi} + \left(S_1 A_1 \frac{\partial \hat{\varphi}}{\partial x_1} + S_2 A_2 \frac{\partial \hat{\varphi}}{\partial x_2}\right) = 0, \tag{5.11}$$

with $S_i = i\omega/(i\omega + \sigma_i(x_i))$.

A PML medium is thus characterized by (S_1, S_2) or (σ_1, σ_2) . Here the matching conditions of Bérenger are assumed from the beginning. One can prove that by choosing σ_1 and σ_2 as in Fig. 36, one can ensure the exponential decay in the layer, and no reflection occurs between two layers.

In order to retrieve a well-posed PML model in the time-domain, we make a change of basis in the spatial operator $\sum S_i A_i \partial/\partial x_i$ which relies on an algebraic lemma:

5.4 Lemma *There exist two invertible operators M and N such that*

$$S_1 A_1 \frac{\partial}{\partial x_1} + S_2 A_2 \frac{\partial}{\partial x_2} = M \left(A_1 \frac{\partial}{\partial x_1} + A_2 \frac{\partial}{\partial x_2} \right) N \tag{5.12}$$

which satisfy

$$A_1 \frac{\partial}{\partial x_1} N + A_2 \frac{\partial}{\partial x_2} N = 0 \quad \textit{and} \quad \lim_{\sigma_1, \sigma_2 \to 0} N = I. \tag{5.13}$$

Proof The proof is constructive. We first seek invertible operators M and N such that

$$\begin{aligned} M A_1 N - S_1 A_1 &= 0 \\ M A_2 N - S_2 A_2 &= 0. \end{aligned} \tag{5.14}$$

Using the special form of the matrices A_i, we find the general form of M and N solving (5.14) to be:

$$N = \begin{pmatrix} S_1 S_2^{-1} \lambda & 0 & n_{13} \\ 0 & \lambda & n_{23} \\ 0 & 0 & n_{33} \end{pmatrix}, \qquad M = \begin{pmatrix} \dfrac{S_1}{n_{33}} & 0 & 0 \\ 0 & \dfrac{S_2}{n_{33}} & 0 \\ -\dfrac{n_{13} S_2}{\lambda n_{33}} & -\dfrac{n_{23} S_2}{\lambda n_{33}} & \dfrac{S_2}{\lambda} \end{pmatrix},$$

where n_{13}, n_{23}, n_{33} are free parameters. We now choose N as simple as possible:

$$N = \begin{pmatrix} S_2^{-1} & 0 & 0 \\ 0 & S_1^{-1} & 0 \\ 0 & 0 & 1 \end{pmatrix}, \qquad M = \begin{pmatrix} S_1 & 0 & 0 \\ 0 & S_2 & 0 \\ 0 & 0 & S_1 S_2 \end{pmatrix},$$

and since S_i depends only on x_i, (5.13) is satisfied. □

With the above lemma, (5.11) becomes

$$i\omega\hat{\varphi} + M \left(A_1 \frac{\partial}{\partial x_1} + A_2 \frac{\partial}{\partial x_2} \right) (N\hat{\varphi}) = 0,$$

and the change of unknowns $\tilde{\varphi} = N\varphi$ gives

$$i\omega M^{-1} N^{-1} \hat{\tilde{\varphi}} + \left(A_1 \frac{\partial}{\partial x_1} + A_2 \frac{\partial}{\partial x_2} \right) \hat{\tilde{\varphi}} = 0.$$

In order to establish the time-domain model, we start by writing the operator $i\omega M^{-1}N^{-1}$ in the form:

$$i\omega M^{-1}N^{-1} = i\omega I + C + R(U_\omega)^{-1}$$

with

$$C = \begin{pmatrix} \sigma_1 - \sigma_2 & 0 & 0 \\ 0 & \sigma_2 - \sigma_1 & 0 \\ 0 & 0 & \sigma_1 + \sigma_2 \end{pmatrix}, \qquad R = \begin{pmatrix} \sigma_2(\sigma_2 - \sigma_1) & 0 & 0 \\ 0 & \sigma_1(\sigma_1 - \sigma_2) & 0 \\ 0 & 0 & \sigma_1\sigma_2 \end{pmatrix},$$

$$U_\omega = \begin{pmatrix} (i\omega + \sigma_2) & 0 & 0 \\ 0 & (i\omega + \sigma_1) & 0 \\ 0 & 0 & i\omega \end{pmatrix}.$$

Finally the time-harmonic model is reduced to

$$i\omega\widehat{\tilde{\varphi}} + C\widehat{\tilde{\varphi}} + R(U_\omega)^{-1}\widehat{\tilde{\varphi}} + \left(A_1\frac{\partial}{\partial x_1} + A_2\frac{\partial}{\partial x_2}\right)\widehat{\tilde{\varphi}} = 0. \tag{5.15}$$

5.5 Remark The new unknown $\tilde{\varphi}$ is equal to the original one, φ, in the domain of interest (i.e. when $\sigma_1 = \sigma_2 = 0$). Moreover, when transmitted to the PML the new unknown $\tilde{\varphi}$ does not "see" a contrast between the two mediums (due to $\lim_{\sigma_1,\sigma_2\to 0} N = I$).

Finally, this change of unknowns can be interpreted as a complex change of basis ($S_1A = MAN$), thus generalizing the change of variables in [34].

By an inverse Fourier transform of (5.15), we would obtain a model which is not local in time. Instead, we increase the number of unknowns (which is a classical technique when studying dispersive materials [36]) and introduce a new vector-valued variable $\tilde{\psi}$ such that $U_\omega\widehat{\tilde{\psi}} = \widehat{\tilde{\varphi}}$. Back to the time domain we obtain

$$\begin{cases} \dfrac{\partial\tilde{\varphi}}{\partial t} + C\tilde{\varphi} + R\tilde{\psi} + \left(A_1\dfrac{\partial}{\partial x_1} + A_2\dfrac{\partial}{\partial x_2}\right)\tilde{\varphi} = 0, \\ \dfrac{\partial\tilde{\psi}}{\partial t} + D\tilde{\psi} - \tilde{\varphi} = 0, \\ \tilde{\psi}(0,\cdot) = 0, \end{cases}$$

with

$$D = \begin{pmatrix} \sigma_2 & 0 & 0 \\ 0 & \sigma_1 & 0 \\ 0 & 0 & 0 \end{pmatrix}$$

or

$$\frac{\partial\Phi}{\partial t} + \left(\mathcal{C} + \mathcal{A}\left(\frac{\partial}{\partial x_1}, \frac{\partial}{\partial x_2}\right)\right)\Phi = 0 \tag{5.16}$$

where we have set

$$\Phi = \begin{pmatrix} \tilde{\varphi} \\ \tilde{\psi} \end{pmatrix}, \quad \mathcal{C} = \begin{pmatrix} C & R \\ -I & D \end{pmatrix} \quad \text{and} \quad \mathcal{A}\left(\frac{\partial}{\partial x_1}, \frac{\partial}{\partial x_2}\right) = \begin{pmatrix} A_1\dfrac{\partial}{\partial x_1} + A_2\dfrac{\partial}{\partial x_2} & 0 \\ 0 & 0 \end{pmatrix} \tag{5.17}$$

This model preserves the absorption and decay properties of the Bérenger model. Furthermore, it is hyperbolic symmetric, as shown by (5.16), (5.17). We summarize.

5.6 Theorem *The model given by* (5.16), (5.17) *is a* PML *model for the linearized Euler equations. Moreover, the Cauchy problem associated to this model is strongly well-posed.*

5.7 Remark For a layer starting at $x_1 = 0$, if we choose σ_1 to be a constant (non-zero) in the layer, the model is no longer perfectly matched. Indeed, in this case we cannot simultaneously ensure the continuity of all fields since for this we must have $\sigma_1(0) = 0$. In practice it is preferable to choose a damping factor σ_1 which does not become rapidly significant (for example parabolic). This remark is also valid for all layer methods.

5.4 Numerical results

To verify the efficiency of our PML model, we present numerical results on a benchmark problem in computational aeroacoustics [21, 23]. The initial condition is in the isentropic case an acoustic pulse centered at point x_a,

$$P_0 = \rho_0 = \exp\left(-(\ln 2)\frac{|x - x_a|^2}{9}\right), \qquad u_0 = v_0 = 0.$$

The computational domain is $[-60, 60] \times [-60, 60]$, the domain of interest is $[-50, 50] \times [-50, 50]$ surrounded by a layer of thickness $N = 10$. The damping factors $\sigma_i(x_i)$ are chosen as $\sigma_i(x_i) = \sigma_0(d_i(x_i)/D_i)^3$, where D_i is the thickness of the layer in the x_i-direction and $d_i(x_i)$ is the distance from the interface to the point in the layer. According to the results observed in [23], we take $\sigma_0 D/N \cong 8$.

The scheme is fourth order Runge-Kutta in time, and a seven point finite difference scheme in space. The contour plots of the pressure p are shown for several times in Fig. 37. We can see that the field is well absorbed by the PML layer (the wave arrives at $t = 30$ in the PML) with no noticeable reflection. The Dirichlet condition imposed on the exterior boundary of the PML does not generate spurious reflection, and the wave is totally absorbed after $t = 100$.

5.5 Conclusion

The methods developed in this section are a very promising alternative to absorbing boundary conditions for linear hyperbolic problems with constant coefficients. Progress needs to be made for more complicated models, as for instance Euler models with non-constant mean flows, or anisotropic elasticity models, and of course non-linear problems.

Acknowlegments The authors are very grateful to Martin Gander for his invitation to Montreal and his contribution to this work.

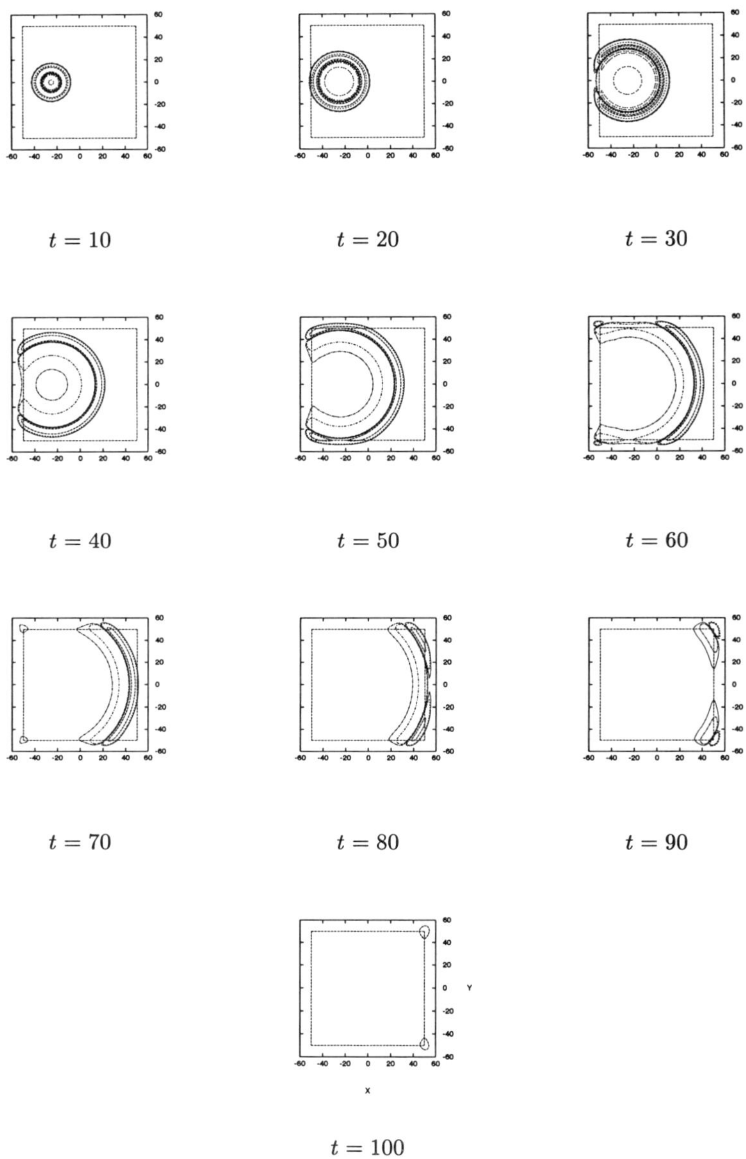

Figure 37: Pressure contours for the new PML model from [33].

References

[1] S. Abarbanel and D. Gottlieb, A mathematical analysis of the PML method, *J. Comput. Phys.* **134** (1997), 357–363.

[2] S. Abarbanel, D. Gottlieb, and S. Hesthaven, Well-posed perfectly matched layers for advective acoustics, *J. Comput. Phys.* **154** (1999), 266–283.

[3] A. Bah, Sur la modélisation de milieux fictifs absorbants de type couche de Bérenger, Ph.D. thesis, Université de Toulouse, 2001.

[4] A. Bamberger, B. Engquist, L. Halpern, and P. Joly, Higher order wave equation approximation in heterogenous media, *SIAM J. Appl. Math.* **48** (1) (1988), 129–154.

[5] A. Bayliss and E. Turkel, Radiation boundary conditions for wave-like equations, *Comm. Pure Appl. Math.* **33** (1980), 707–725.

[6] E. Bécache, F. Collino, and P. Joly, Higher-order variational finite difference schemes for solving 3-d paraxial wave equations using splitting techniques, *Wave Motion* **31** (2) (2000), 101–116.

[7] J. P. Bérenger, Three-dimensional perfectly matched layer for the absorption of electromagnetic waves, *J. Comput. Phys.* **114** (1994), 185–200.

[8] R. W. Clayton and B. Engquist, Absorbing boundary conditions for acoustic and elastic wave equations, *Bull. Seismic Soc. America* **67** (6) (1977), 1529–1540.

[9] R. W. Clayton and B. Engquist, Absorbing boundary conditions for wave equation migration, *Geophysics* **45** (5) (1980), 895–904.

[10] F. Collino and P. Monk, The perfectly matched layer in curvilinear coordinates, *SIAM J. Sci. Comput.* **19** (6) (1998), 2061–2090.

[11] F. Collino and C. Tsogka, Application of the PML absorbing layer model to the linear elastodynamic problem in anisotropic heterogenous media, Technical Rep. 3471, INRIA, August 1998.

[12] B. Engquist and A. Majda, Absorbing boundary conditions for the numerical simulation of waves, *Math. Comp.* **31** (1977), 629–651.

[13] B. Engquist and A. Majda, Radiation boundary conditions for acoustic and elastic calculations, *Comm. Pure Appl. Math.* **32** (1979), 313–357.

[14] T. Gallouet, R. Herbin, and M.-H. Vignal, Error estimates for the approximate finite volume solution of convection diffusion equations with general boundary conditions, *SIAM J. Numer. Anal.* **37** (6) (2000), 1935–1972.

[15] M. J. Gander, Analysis of parallel algorithms for time dependent partial differential equations, Ph.D. thesis, Stanford University, 1997.

[16] M. J. Gander, L. Halpern, and F. Nataf, Domain decomposition methods for wave propagation, in: *Proc. Fifth Internat. Conf. on Mathematical and Numerical Aspects of Wave Propagation Phenomena*, INRIA-SIAM, 2000.

[17] M. J. Gander, L. Halpern, and F. Nataf, Optimal Schwarz waveform relaxation for the one dimensional wave equation, Internal Report 469, École Polytechnique, Palaiseau, Sept. 2001.

[18] B. Gustafsson, Numerical boundary conditions, *Lectures in Appl. Math.* **22** (1985), 279–308.

[19] L. Halpern, Absorbing boundary conditions for the discretization schemes of the one-dimensional wave equation, *Math. Comp.* **38** (158) (1982), 415–429.

[20] L. Halpern and L. Trefethen, Well-posedness of one-way wave equations and absorbing boundary conditions, *Math. Comp.* **47** (176) (1986), 421–435.

[21] J. S. Hesthaven, On the analysi s and construction of perfectly matched layers for the linearized Euler equations, *J. Comput. Phys.* **142** (1) (1998), 129–147.

[22] R. Higdon, Initial-boundary value problems for linear hyperbolic systems, *SIAM Rev.* **28** (2) (1986), 177–217.

[23] F. Hu, On absorbing boundary conditions for linearized Euler equations by a perfectly matched layer, *J. Comput. Phys.* **129** (1996), 201–219.

[24] M. Israeli and S. Orszag, Approximation of radiation boundary conditions, *J. Comput. Phys.* **41** (1981), 115–135.

[25] C. Japhet, F. Nataf, and F. Rogier, The optimized order 2 method. Application to convection-diffusion problems, *Future Generation Computer Systems FUTURE* **18** (2001).

[26] H.-O. Kreiss and J. Lorenz, *Initial-Boundary Value Problems and the Navier-Stokes Equations*, Academic Press, New York, 1989.

[27] D. Lee, G. Botseas, and J. S. Papadakis, Finite difference solution to the parabolic wave equation, *J. Acoustic Soc. America* **70** (3) (1981), 798–799.

[28] E. Lindman, Free space boundary conditions for the time dependent wave equation, *J. Comput. Phys.* **18** (1975), 66–78.

[29] P. L. Lions, On the Schwarz alternating method. III: a variant for nonoverlapping subdomains, in: *Proc. Third International Symposium on Domain Decomposition Methods for Partial Differential Equations* (T. F. Chan, R. Glowinski, J. Périaux, and O. Widlund, eds.), SIAM, Philadelphia, PA, 1990.

[30] P. Mazet, S. Paintandre, and A. Rahmouni, Interprétation dispersive du milieu PML de Bérenger, *C. R. Acad. Sci. Paris Sér. I Math.* **327** (1998), 59–64.

[31] A. Rahmouni, Un modèle PML bien posé pour les équations d'Euler linéarisées, *C. R. Acad. Sci. Paris Sér. I Math.* **331** (2000), 159–164.

[32] A. Rahmouni, A well-posed PML model for elastodynamic, in: *Mathematical and Numerical Aspects of Wave Propagation*, SIAM, Philadelphia, PA, 2000, 916–920.

[33] A. Rahmouni, Well-posed PML models for various hyperbolic systems, Ph.D. thesis, Université Paris XIII, 2000.

[34] C. M. Rappaport, Interpreting and improving the PML absorbing boundary condition using anisotropic lossy mapping of space, *IEEE Microwave Guided Wave Letter* (1995), no. 5, 90–92.

[35] J. C. Strikwerda, *Finite Difference Schemes and Partial Differential Equations*, Chapman & Hall, 1989.

[36] A. Taflove (ed.), *Advances in Computational Electrodynamics: The Finite-Difference Time-Domain Method*, Artech House, Boston, MA, 1998.

Deterministic and random dynamical systems: theory and numerics

A. R. HUMPHRIES

School of Mathematical Sciences
University of Sussex
Brighton BN1 9QH
England

A. M. STUART

Mathematics Institute
Warwick University
Coventry CV4 7AL
England

Abstract

The theory of (random) dynamical systems is a framework for the analysis of large time behaviour of time-evolving systems (driven by noise). These notes contain an elementary introduction to the theory of both dynamical and random dynamical systems. The subject matter is made accessible by means of very simple examples and highlights relationships between the deterministic and the random theories.

Throughout the notes both continuous and discrete time are considered. Continuous time problems are primarily ordinary and stochastic differential equations, whilst approximating numerical schemes are used to illustrate discrete time. The central role of Lyapunov functions is highlighted, both in inferring the existence of attractors in the deterministic case, and in the construction of invariant measures and random attractors in the random case. The importance of approximations which inherit Lyapunov functions is emphasised.

1 Deterministic dynamical systems

1.1 Differential equations

Let $f : \mathbb{R}^d \to \mathbb{R}^d$ be a continuous function. Consider the following ordinary differential equation (ODE)

$$\frac{dx}{dt} = f(x(t)), \quad x(0) = x_0. \tag{1.1.1}$$

Here the problem is to find $x \in C(\mathbb{R}^+, \mathbb{R}^d)$ which satisfies (1.1.1). We will often write $\dot{x}$ for dx/dt. We use the notation $\mathbb{R}^+ = [0, \infty)$ and we will later need $\mathbb{Z}^+ = \{0, 1, 2, 3, \dots\}$.

A. Bourlioux and M.J. Gander (eds.), Modern Methods in Scientific Computing and Applications, 211–254.

Equation (1.1.1) can also be written as an integral equation:

$$x(t) = x_0 + \int_0^t f(x(s))\,ds, \quad x(0) = x_0. \tag{1.1.2}$$

This formulation will be useful when we introduce stochastic differential equations.

1.1.1 Example (i) A scalar linear problem:

$$\dot{x} = \lambda x, \quad x(0) = x_0 \in \mathbb{R}. \tag{1.1.3}$$

This has exact solution $x(t) = e^{\lambda t}x_0$. Note that $x(t) \to 0$ as $t \to \infty$ if $\lambda < 0$. Since we know the exact solution this is a useful test problem for numerical algorithms.

(ii) A scalar nonlinear equation:

$$\dot{x} = f(x) = x - x^3, \quad x(0) = x_0. \tag{1.1.4}$$

This has the exact solution

$$x(t) = \frac{x_0}{[e^{-2t} + x_0^2(1 - e^{-2t})]^{1/2}}. \tag{1.1.5}$$

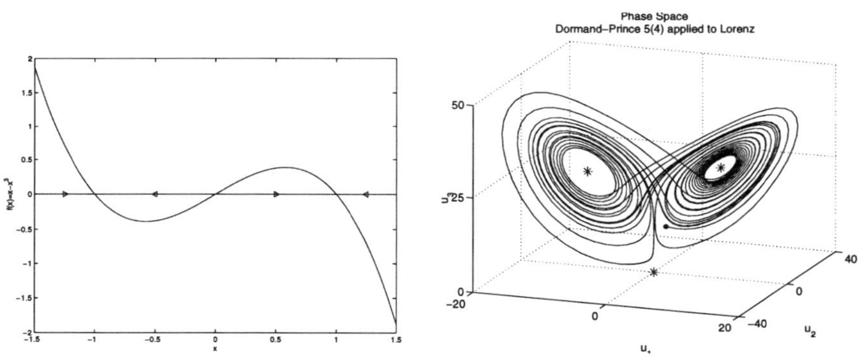

Figure 1: (i) Graph of $f(x) = x - x^3$ against x with the dynamics of (1.1.4) indicated on the x-axis. (ii) A numerically computed trajectory of the Lorenz equations (1.1.6).

From the graph of $f(x)$ (see Fig. 1), or the exact solution, it is easy to infer that for any $x_0 > 0$, $x(t) \to 1$ as $t \to \infty$, similarly $x(t) \to -1$ as $t \to \infty$ if $x_0 < 0$, and if $x_0 = 0$ then $x(t) = 0$ for all $t \in \mathbb{R}$.

(iii) The Lorenz equations in $\mathbb{R}^3$:

$$\begin{aligned} \dot{x} &= \sigma(y - x), \quad x(0) = x_0 \\ \dot{y} &= rx - y - xz, \quad y(0) = y_0 \\ \dot{z} &= xy - bz, \quad z(0) = z_0. \end{aligned} \tag{1.1.6}$$

These equations have no general closed form solution.

In Fig. 1 a numerical approximation to a trajectory over a very long time period is plotted. Since the classical error bound for the numerical approximation grows exponentially in time it is important to question the relationship of this figure to the true dynamics.

Lyapunov functions

It is often of importance to understand how functions of $x(t)$ change with time. By using the *generator* $\mathcal{L}$:

$$\mathcal{L}V = \sum_{i=1}^{d} f_i \frac{\partial V}{\partial x_i} = \langle f(x), \nabla V(x) \rangle, \tag{1.1.7}$$

we may achieve this. If $x(t)$ solves (1.1.1) and $V \in C^2(\mathbb{R}^d, \mathbb{R})$ then

$$\frac{d}{dt}\Big\{V(x(t))\Big\} = \mathcal{L}V(x(t)). \tag{1.1.8}$$

Note that $\mathcal{L}V(x(t)) = \langle \dot{x}, \nabla V(x) \rangle$, and often for deterministic dynamical systems $\frac{d}{dt}V(x(t))$ is calculated directly, without recourse to $\mathcal{L}$. We introduce $\mathcal{L}$ here in the deterministic case because of the central role of its generalisation in the stochastic case.

Also important is the adjoint operator $\mathcal{L}^*$, defined by $\langle f, \mathcal{L}g \rangle = \langle \mathcal{L}^* f, g \rangle$ with $\langle \cdot, \cdot \rangle$ the standard inner product on $\mathbb{R}^d$, so that in this case

$$\mathcal{L}^* V = -\sum_{i=1}^{d} \frac{\partial}{\partial x_i}\Big(f_i V\Big).$$

This leads us to the Liouville equation for propagation of probability densities:

1.1.2 Lemma *If x_0 is a random variable with density $\rho_0(x)$ then $x(t)$ is a random variable which, under appropriate differentiability conditions on f and ρ_0, has density $\rho(x,t)$ which is a classical solution of the equation*

$$\begin{aligned} \frac{\partial \rho}{\partial t} &= \mathcal{L}^* \rho, \quad (x,t) \in \mathbb{R}^d \times (0,T], \\ \rho(x,0) &= \rho_0(x), \quad x \in \mathbb{R}^d. \end{aligned}$$

Numerical methods

We consider numerical methods for the ODE (1.1.1) which, given $\Delta t > 0$, define approximations x_n to $x(t_n)$ at the equally spaced time intervals $t_n = n\Delta t \in [0,T]$.

Initially we introduce two numerical methods for the ODE (1.1.1). The first is the *forward Euler* method, sometimes called the *explicit Euler* method, and is defined as

$$x_{n+1} = x_n + \Delta t f(x_n). \tag{1.1.9}$$

There are many derivations of this method. From (1.1.2) we have

$$x(t_{n+1}) = x(t_n) + \int_{t_n}^{t_{n+1}} f(x(s))\, ds,$$

so letting $x_n = x(t_n)$ and approximating the integral

$$\int_{t_n}^{t_{n+1}} f(x(s))\, ds \approx \int_{t_n}^{t_{n+1}} f(x(t_n))\, ds = \Delta t f(x_n)$$

we obtain the forward Euler method.

The second method considered is the *backward Euler* method defined as

$$x_{n+1} = x_n + \Delta t f(x_{n+1}). \tag{1.1.10}$$

This method is sometimes called the *implicit Euler* method. Since x_{n+1} appears twice in (1.1.10) in general we have to solve a nonlinear equation to find x_{n+1} at each step. This can make the method much more computationally expensive to implement than the forward Euler method; nevertheless there are situations where the extra work is justified and we will illustrate some of these in these notes.

The backward Euler method can be derived in a similar fashion to the forward Euler method. Many other simple numerical methods such as the *trapezoidal rule* or *implicit midpoint* rule can also be derived by applying the corresponding numerical integration formula to the integral in (1.1.2). These methods are all examples of Runge-Kutta methods.

Also of interest are so-called *split step* or *partitioned methods.* Suppose $f(x) = f_1(x) + f_2(x)$, then we can define the *split-step Euler method* by

$$\left.\begin{aligned} x_* &= x_n + \Delta t f_1(x_n), \\ x_{n+1} &= x_* + \Delta t f_2(x_{n+1}). \end{aligned}\right\} \tag{1.1.11}$$

Such splitting methods are important in the solution of Hamiltonian systems and stochastic problems, and we will see examples of both.

1.1.3 Example (i) Applying the forward Euler method to the linear test problem (1.1.3) we obtain

$$x_{n+1} = x_n + \Delta t f(x_n) = x_n + \Delta t \lambda x_n = (1 + \Delta t \lambda) x_n.$$

Since this is true for arbitrary n we obtain the general solution

$$x_n = (1 + \Delta t \lambda)^n x_0.$$

(ii) Applying the forward Euler method to (1.1.4) we obtain

$$x_{n+1} = x_n + \Delta t x_n (1 - x_n^2).$$

In this case, as in general for nonlinear problems, there is no simple formula linking x_n directly to x_0.

(iii) Applying the backward Euler method to equation (1.1.4) we obtain

$$x_{n+1} = x_n + \Delta t x_{n+1}(1 - x_{n+1}^2).$$

Thus we need to solve a cubic equation to find x_{n+1}, given x_n.

(iv) Consider the separable Hamiltonian system

$$\dot{p} = \frac{\partial V}{\partial q}, \qquad \dot{q} = -\frac{\partial T}{\partial p},$$

where $p, q \in \mathbb{R}^d$ are generalised position and momenta, and $H(p,q) = T(p) + V(q)$ is preserved. We can apply the forward or backward Euler method directly to these problems. Alternatively write

$$\dot{x} = \begin{pmatrix} \dot{p} \\ \dot{q} \end{pmatrix} = \begin{pmatrix} \frac{\partial H}{\partial q} \\ -\frac{\partial H}{\partial p} \end{pmatrix} = \begin{pmatrix} \frac{\partial H}{\partial q} \\ 0 \end{pmatrix} + \begin{pmatrix} 0 \\ -\frac{\partial H}{\partial p} \end{pmatrix} := f_1(x) + f_2(x).$$

Now apply the split-step Euler method (1.1.11) which, after a little calculation, leads to

$$\left.\begin{aligned} p_{n+1} &= p_n + \Delta t \tfrac{\partial V}{\partial q}(q_n) \\ q_{n+1} &= q_n - \Delta t \tfrac{\partial T}{\partial p}(p_{n+1}). \end{aligned}\right\} \tag{1.1.12}$$

In this context the method is known as the *symplectic Euler method.* It essentially preserves the Hamiltonian structure of the dynamical system and has much better approximation properties for this type of problem than forward or backward Euler. Moreover, note that although the general formulation (1.1.11) is implicit, the symplectic Euler method (1.1.12) is explicit for this separable Hamiltonian problem [21].

In general when applying the backward Euler method, or any other implicit method, there are nonlinear equations to solve at each step. We will briefly consider how these equations can be solved in practice by considering the backward Euler method in one dimension:

$$x_{n+1} = x_n + \Delta t f(x_{n+1}), \quad x_n \in \mathbb{R}. \tag{1.1.13}$$

We could try to solve this by the simple iteration

$$x_{n+1}^{k+1} = x_n + \Delta t f(x_{n+1}^k), \quad x_{n+1}^0 = x_n,$$

and let $x_{n+1} = \lim_{k\to\infty} x_{n+1}^k$ (if the limit exists). Of course, in practice we cannot iterate to infinity so we stop iterating if $|x_{n+1}^k - x_{n+1}^{k-1}| < \tau$ for some small tolerance $\tau > 0$. However, this iteration has the drawback that $\lim_{k\to\infty} x_{n+1}^k$ need not exist if Δt is too large. Indeed ensuring that this iteration is convergent often imposes quite severe restrictions on the step-size Δt.

Thus it is often preferable to use the more sophisticated Newton iteration. For $g \in C(\mathbb{R}, \mathbb{R})$, Newton's method for solving $g(x) = 0$ is given by iterating for $k = 0, 1, 2, \ldots$,

$$x^{k+1} = x^k - \frac{g(x^k)}{g'(x_k)}.$$

So to solve (1.1.13) we let

$$g(x) = x - x_n - \Delta t f(x),$$

so that finding x_{n+1} is equivalent to solving $g(x) = 0$ which by Newton's method we can do by iterating

$$x_{n+1}^{k+1} = x_{n+1}^k - \frac{x_{n+1}^k - x_n - \Delta t f(x_{n+1}^k)}{1 - \Delta t f'(x_{n+1}^k)}.$$

Thus when applying an implicit method we have to perform an inner iteration in k at each step of the outer iteration in n.

One way to avoid this is to set $x_{n+1}^0 = x_n$ and take $k = 1$, to obtain the *linearised Euler method*

$$x_{n+1} = x_n + \Delta t \frac{f(x_n)}{1 - \Delta t f'(x_n)}.$$

These ideas generalise to arbitrary dimension.

Bibliographical remark Throughout the following we minimise the size of the bibliography by often referring to [24]. Although it is often not the original source of the material we reference, it has a comprehensive bibliography and pointers to the relevant literature at the end of every chapter.

1.2 Dynamical systems

There are several definitions of "dynamical system" in use. In the context of the ODE (1.1.1), many definitions require both a unique solution for all $t \in \mathbb{R}$ and all initial conditions $x_0 \in \mathbb{R}^d$, and also that for any fixed $t \in \mathbb{R}$ the solution is continuous with respect to the initial data x_0. However, many interesting problems such as (1.1.4) and the Lorenz equations, which possess bounded solutions in the limit as $t \to \infty$, have solutions which become unbounded in finite negative time, and so by such definitions would not be dynamical systems.

1.2.1 Example Consider (1.1.4) with $|x_0| > 1$. Then using the exact solution (1.1.5) we see that $|x(t)| \to \infty$ as $t \to \frac{1}{2}\ln(1 - \frac{1}{x_0^2}) < 0$.

Thus this definition is too restrictive, and we will work with a weaker concept. The differential equation (1.1.1) is said to define a *semi-dynamical system* if it has a unique solution for all $t \geqslant 0$ and all initial conditions $x_0 \in \mathbb{R}^d$. We will work with semi-dynamical systems throughout, but for brevity will omit the "semi" and refer to them as dynamical systems. Note that we make no reference to continuity with respect to initial data, though for (1.1.1) with continuous f uniqueness of solutions implies continuity with respect to initial data [7].

Dynamical systems defined by ODEs are characterised by having a continuous time variable. We can also have dynamical systems with a discrete time variable. The continuous map $g : \mathbb{R}^d \to \mathbb{R}^d$ defines a discrete dynamical system via the iteration

$$x_{n+1} = g(x_n). \tag{1.2.1}$$

Since numerical methods are iterated in this way they are good candidates to be considered as discrete dynamical systems.

We will be interested in both dynamical systems defined by ODEs and in discrete dynamical systems. Our discrete dynamical systems will arise from numerical solutions of ODEs. We will analyse the behaviour of numerical methods by using dynamical systems techniques to compare the dynamical systems defined by an ODE and its numerical solution.

Evolution operator

To be able to analyse both continuous and discrete time dynamical systems together, we now set up general notation for a dynamical system evolving in a state space $\mathbb{X}$, which will usually be $\mathbb{R}^d$, and time space $\mathbb{T}$, where usually $\mathbb{T} = \mathbb{R}^+$ for the continuous time case, and $\mathbb{T} = \mathbb{Z}^+$ in the discrete time case; the choices $\mathbb{T} = \mathbb{R}$ and $\mathbb{T} = \mathbb{Z}$ will also occur occasionally. Our primary motivation here is the study of (1.1.1) and its discrete time counterparts (1.1.9) and (1.1.10).

1.2.2 Definition A *dynamical system* on a space $\mathbb{X}$ with time $\mathbb{T}$ is a mapping

$$\varphi : \mathbb{T} \times \mathbb{X} \to \mathbb{X}, \quad (t, x_0) \mapsto \varphi(t, x_0), \tag{1.2.2}$$

where the mapping φ satisfies the *semi-group property* that $\varphi(0, x) = x$ for all $x \in \mathbb{X}$ and

$$\varphi(t + s, x_0) = \varphi(t, \varphi(s, x_0)) \quad \forall s, t \in \mathbb{T},\ x \in \mathbb{X}.$$

It will sometimes be helpful to write

$$\varphi(t,x) = \varphi^t x,$$

and we write φ for φ^1. In this notation the semi-group property reads $\varphi^0 = \mathrm{id}_{\mathbb{X}}$ and

$$\varphi^{t+s} = \varphi^t \circ \varphi^s.$$

For a set $B \subset \mathbb{X}$ we write $\varphi(t,B)$ for

$$\varphi(t,B) = \bigcup_{x\in B} \varphi(t,x).$$

The mapping φ is known as the *evolution operator*. For the ODE (1.1.1) we have $\varphi(t,x_0) = \varphi^t x_0 = x(t)$, the solution at time t with initial condition $x(0) = x_0$. For the map (1.2.1) we have $\varphi(n,x_0) = \varphi^n x_0 = x_n$, the solution after n iterates of the map with initial condition x_0.

Examples

1.2.3 Example For (1.1.3) we have $\mathbb{X} = \mathbb{R}$ and $\mathbb{T} = \mathbb{R}$ with

$$\varphi(t,x_0) = \varphi^t x_0 = e^{\lambda t} x_0 = x(t).$$

1.2.4 Example From Example 1.1.3(i), applying the forward Euler method to (1.1.3), we have $\mathbb{X} = \mathbb{R}$ and $\mathbb{T} = \mathbb{Z}$ with

$$\varphi(n,x_0) = \varphi^n x_0 = (1+\Delta t\lambda)^n x_0 = x_n.$$

1.2.5 Example For (1.1.4) we have $\mathbb{X} = \mathbb{R}$ and $\mathbb{T} = \mathbb{R}^+$ with

$$\varphi(t,x_0) = \varphi^t x_0 = \frac{x_0}{[e^{-2t} + x_0^2(1-e^{-2t})]^{1/2}} = x(t).$$

1.2.6 Example For the forward Euler method applied to (1.1.4) we have $\mathbb{X} = \mathbb{R}$ and $\mathbb{T} = \mathbb{Z}^+$ with

$$\varphi(1,x_n) = \varphi^1 x_n = \varphi x_n = x_n + \Delta t f(x_n) = x_n + \Delta t(x_n - x_n^3) = x_{n+1}.$$

1.2.7 Example For the backward Euler method applied to (1.1.4) we have $\mathbb{X} = \mathbb{R}$ and $\mathbb{T} = \mathbb{Z}^+$ with

$$\varphi(1,x_n) = \varphi^1 x_n = \varphi x_n = x_n + \Delta t f(\varphi x_n) = x_n + \Delta t(\varphi x_n - (\varphi x_n)^3) = x_{n+1}.$$

It is not immediately apparent whether φ is well-defined in this case. We discuss this issue below.

Existence and uniqueness

In defining dynamical systems via the mapping (1.2.2) we are implicitly assuming existence and uniqueness of solutions. In practice, existence and uniqueness of solutions will need to be verified in different scenarios.

1.2.8 Example Consider the ODE

$$\dot{x} = x^2, \quad x(0) = 1,$$

which has exact solution

$$x(t) = \frac{1}{1/x_0 - t}.$$

Thus $x(t) \to \infty$ as $t \to 1/x_0$ for $x_0 > 0$, so this differential equation does not define a dynamical system.

1.2.9 Definition A function $f : \mathbb{X} \to \mathbb{X}$ is *Lipschitz* on $B \subset \mathbb{X}$ with Lipschitz constant L_B if

$$\|f(x) - f(y)\| \leqslant L_B \|x - y\|, \quad \forall x, y \in B.$$

If f is Lipschitz on $\mathbb{X}$ then f is *globally Lipschitz.* If f is Lipschitz on every bounded subset of $\mathbb{X}$, then f is *locally Lipschitz.*

Clearly globally Lipschitz implies locally Lipschitz. Note also that if f is differentiable on $\mathbb{X}$, then f must be locally Lipschitz. The following lemmas will be useful.

1.2.10 Lemma (Continuous Gronwall Lemma) *Let $z(t)$ satisfy*

$$\dot{z} \leqslant az + b, \quad z(0) = z_0,$$

for constants $a \neq 0$ and $b \in \mathbb{R}$. Then

$$z(t) \leqslant e^{at} z_0 + \frac{b}{a}(e^{at} - 1).$$

Proof Integrate the differential inequality, see for example [24]. □

1.2.11 Lemma (Discrete Gronwall Lemma) *Let $C \in (0,1)$ and $D_n \in \mathbb{R}$ and suppose $Z_n \in \mathbb{R}$ satisfies*

$$Z_{n+1} \leqslant C Z_n + D_n, \quad \forall n \geqslant 0.$$

Then

$$Z_n \leqslant Z_0 C^n + \sum_{j=1}^{n} C^{n-j} D_{j-1}.$$

If $D_n \equiv D$ is constant then

$$Z_n \leqslant \frac{D}{1-C}(1 - C^n) + Z_0 C^n. \tag{1.2.3}$$

Proof By induction, see [24] for example. □

1.2.12 Theorem *If f is globally Lipschitz on $\mathbb{R}^d$ then the ODE* (1.1.1) *has a unique solution for all $t \in \mathbb{R}$ and any $x_0 \in \mathbb{R}^d$. Thus φ^t is well defined for all $t \in \mathbb{R}$, and* (1.1.1) *defines a dynamical system on $\mathbb{X} = \mathbb{R}^d$, with $\mathbb{T} = \mathbb{R}$ or $\mathbb{R}^+$. Moreover,*

$$e^{-Lt}\|x_0 - y_0\| \leqslant \|\varphi^t x_0 - \varphi^t y_0\| \leqslant e^{Lt}\|x_0 - y_0\|. \tag{1.2.4}$$

Proof Uses Picard iteration, see for example [24]. The final bound follows from

$$\frac{1}{2}\frac{d}{dt}\|x(t) - y(t)\|^2 = \langle f(x) - f(y), x - y\rangle,$$

and

$$-L\|x - y\|^2 \leqslant \langle f(x) - f(y), x - y\rangle \leqslant L\|x - y\|^2,$$

using the Continuous Gronwall Lemma 1.2.10. □

Note that (1.2.4) bounds the rate of both convergence and divergence of trajectories. In particular it means that trajectories cannot cross or merge as, if $x_0 \neq y_0$, then $x(t) \neq y(t)$ for all $t \in \mathbb{R}$.

Unfortunately this theorem is of limited application as few f encountered in practice are globally Lipschitz, though many are locally Lipschitz. For example, both (1.1.4) and the Lorenz equations (1.1.6) have differentiable and hence locally Lipschitz f, but f is not globally Lipschitz.

However, if f is locally Lipschitz it can be shown that solutions are unique whilst they exist, and a solution $x(t)$ can only fail to exist if it blows up in finite time, i.e. $\|x(t)\| \to \infty$ as $t \to T < \infty$. Often it is possible to prevent such blow up. The generator (1.1.7) will be useful for this purpose.

1.2.13 Assumption (Generator Assumption) There is a function $V \in C(\mathbb{X}, \mathbb{R})$ with $\lim_{\|x\|\to\infty} V(x) = \infty$, and real numbers $a, b, d \in (0, \infty)$, $c \in \mathbb{R}$ such that $V(0) < d/a$ and

$$\mathcal{L}\{V(x)\} \leqslant b\|x\|^c[d - aV(x)], \tag{1.2.5}$$

where $\mathcal{L}$ is the generator for (1.1.1) given by (1.1.7).

1.2.14 Theorem *If f is locally Lipschitz and satisfies Assumption* 1.2.13 *then the ODE* (1.1.1) *has a unique solution for all $t \geqslant 0$, and any $x_0 \in \mathbb{R}^d$. Thus φ^t is well defined for all $t \in \mathbb{R}^+$, and* (1.1.1) *defines a dynamical system on $\mathbb{X} = \mathbb{R}^d$ with $\mathbb{T} = \mathbb{R}^+$.*

Proof

$$\frac{d}{dt}\{V(x(t)\} = \mathcal{L}V(x(t)) \leqslant b\|x(t)\|^c[d - aV(x(t))].$$

Hence $\dot{V} \leqslant 0$ when $V(x(t)) \geqslant d/a$. It follows that

$$\limsup_{t\to\infty} V(x(t)) \leqslant \max\left\{V(x(0)), \frac{d}{a}\right\} := M.$$

But since $\lim_{\|x\|\to\infty} V(x) = \infty$ there exists $R \geqslant 0$ such that $V(x) > M$ for all x such that $\|x\| > R$. Thus $\limsup_{t\to\infty} \|x(t)\| \leqslant R$, and solutions cannot blow up. The rest of the proof mimics the globally Lipschitz case. □

We will also consider f under the following assumption.

1.2.15 Assumption (One-Sided Lipschitz Condition) There exists $c > 0$ such that

$$\langle f(u) - f(v), u - v\rangle \leqslant c\|u - v\|^2 \tag{1.2.6}$$

for all $u, v \in \mathbb{R}^d$.

1.2.16 Example Let $f(x) = x - x^3$ as in (1.1.4). Then f is differentiable and hence locally Lipschitz, but is not globally Lipschitz. Now

$$\begin{aligned}\langle f(u) - f(v), u - v\rangle &= (u - v)(u - u^3 - v + v^3)\\ &= (u-v)^2\left(1 - \frac{1}{2}u^2 - \frac{1}{2}(u+v)^2 - \frac{1}{2}v^2\right)\\ &\leqslant (u-v)^2,\end{aligned}$$

so this function does satisfy a one-sided Lipschitz condition.

1.2.17 Theorem *If f is locally Lipschitz and satisfies Assumption* 1.2.15, *then* (1.1.1) *defines a dynamical system on $\mathbb{X} = \mathbb{R}^d$ with $\mathbb{T} = \mathbb{R}^+$, and φ^t satisfies*

$$\|\varphi^t x_0 - \varphi^t y_0\| \leqslant e^{ct}\|x_0 - y_0\|.$$

Proof Note that the final bound follows from

$$\frac{1}{2}\frac{d}{dt}\|x(t) - y(t)\|^2 = \langle f(x) - f(y), x - y\rangle \leqslant c\|x - y\|^2,$$

using the Continuous Gronwall Lemma 1.2.10. □

The situation for maps defined by explicit numerical methods is quite different. Given $f : \mathbb{R}^d \to \mathbb{R}^d$ the forward Euler method has $\varphi x = x + \Delta t f(x)$ for any $x \in \mathbb{R}^d$ and hence the map can be iterated infinitely many times, so defines a dynamical system.

However, for implicit numerical methods such as the backward Euler method we need to verify existence and uniqueness of solutions.

1.2.18 Theorem *Suppose f is globally Lipschitz with Lipschitz constant L and $L\Delta t < 1$, then the backward Euler method applied to* (1.1.1) *defines a dynamical system on $\mathbb{X} = \mathbb{R}^d$ with $\mathbb{T} = \mathbb{Z}$ or $\mathbb{Z}^+$.*

Proof Let $g(x) = x_n + \Delta t f(x)$. Then $g(x_{n+1}) = x_{n+1}$, and so we are required to show that for any $x_n \in \mathbb{R}^d$ the function $g(x)$ has a unique fixed point. Now

$$\|g(x) - g(y)\| = \Delta t\|f(x) - f(y)\| \leqslant L\Delta t\|x - y\|,$$

and since $L\Delta t < 1$, g is a contraction mapping of $\mathbb{R}^d$ into itself, and so has a unique fixed point. Backward in time existence and uniqueness follow because x_n is given explicitly from x_{n+1}. □

As already noted, not many f are globally Lipschitz. Hence the following is important.

1.2.19 Lemma *Suppose f satisfies a one-sided Lipschitz condition* (1.2.6) *and $c\Delta t < 1$, then the equation*

$$a - \Delta t f(a) = b, \tag{1.2.7}$$

has a unique solution $a \in \mathbb{R}^d$ for any $b \in \mathbb{R}^d$.

Proof Let $g(x) = x - b - \Delta t f(x)$. Then

$$\langle g(x) - g(y), x - y\rangle = \langle x - y - \Delta t(f(x) - f(y)), x - y\rangle \geqslant (1 - c\Delta t)\|x - y\|^2.$$

Now since $1 - c\Delta t > 0$ we can apply the uniform monotonicity theorem to determine the existence of a unique solution. See [24] for details. □

1.2.20 Corollary *Suppose f satisfies a one-sided Lipschitz condition* (1.2.6) *and $c\Delta t < 1$, then the backward Euler method applied to* (1.1.1) *defines a dynamical system with $\mathbb{X} = \mathbb{R}^d$, $\mathbb{T} = \mathbb{Z}^+$.*

1.2.21 Example Corollary 1.2.20 combined with Example 1.2.16 implies that the evolution operator φ defined in Example 1.2.7 for the backward Euler method applied to (1.1.4) is well-defined on $\mathbb{R}$ provided $\Delta t < 1$.

In some situations it is only possible to prove existence, and not uniqueness, of a numerical solution. See Lemma 1.4.13 for example.

1.3 Invariant sets

Consider two dynamical systems defined by mappings φ, $\varphi_{\Delta t}$ (where φ is the evolution operator for an ODE of the form (1.1.1) and $\varphi_{\Delta t}$ is the evolution operator for its numerical approximation). To compare these dynamical systems we compare their dynamic properties. (This contrasts with the traditional numerical analysis approach of comparing errors between trajectories: $\|\varphi^n x_0 - \varphi^n_{\Delta t} x_0\|$.) The simplest features of dynamical systems to compare are invariant sets, and the simplest invariant sets are fixed points.

1.3.1 Definition A point $\bar{u} \in \mathbb{X}$ is a *fixed point* of $\varphi : \mathbb{T} \times \mathbb{X} \to \mathbb{X}$ if $\varphi^t \bar{u} = \varphi(t, \bar{u}) = \bar{u}$ for all $t \in \mathbb{T}$.

1.3.2 Example (i) Consider (1.1.3), with $\lambda \neq 0$. Then it follows from Example 1.2.3 that $\bar{u} = 0$ is the unique fixed point.

(ii) Consider the forward Euler method applied (1.1.3), with $\lambda \neq 0$. Again $\bar{u} = 0$ is the unique fixed point.

It is clear from (1.1.1) that $\bar{u} \in \mathbb{R}^d$ is a fixed point of the differential equation if and only if $f(\bar{u}) = 0$. Also from (1.1.9) and (1.1.10) we see that $\bar{u}$ is a fixed point for the forward Euler or backward Euler methods if and only if $f(\bar{u}) = 0$. So in the case of numerical approximation by forward or backward Euler the fixed points of φ and $\varphi_{\Delta t}$ agree. This is not true for all numerical methods: many methods admit so-called *spurious fixed points* $\bar{u}_{\Delta t}$ with $\varphi^n_{\Delta t}\bar{u}_{\Delta t} = \bar{u}_{\Delta t}$ for all $n \in \mathbb{Z}$ but $f(\bar{u}_{\Delta t}) \neq 0$.

We now consider more general invariant sets.

1.3.3 Definition A bounded set $B \subset \mathbb{X}$ is *positively invariant* if $\varphi^t B \subset B$ for all $t \in \mathbb{T}$, such that $t \geqslant 0$ and *invariant* if also $B \subset \varphi^t B$ for all $t \in \mathbb{T}$ such that $t \geqslant 0$.

Note that a set B is thus invariant if $\varphi^t B = B$ for all $t \in \mathbb{T}$. If B is an invariant set then for any $u_0 \in B$ there exists $u : \mathbb{T} \to \mathbb{X}$ such that for any $0 < t \in \mathbb{T}$ and $0 > s \in \mathbb{T}$, $u(t) = \varphi^t u_0$, $\varphi^{-s} u(s) = u_0$ and $\varphi^t u(s) = u(s+t)$. The set $\{u(t) : t \in \mathbb{T}\}$ is called a *complete orbit* of u_0. Indeed the complete orbit of any point $u_0 \in \mathbb{X}$ is itself an invariant set, and a set $B \in \mathbb{X}$ is invariant if and only if every point $x \in B$ is on a complete orbit which is contained in B; see for example [24].

1.3.4 Example Consider (1.1.4). The fixed points are $\{-1\}$, $\{0\}$, $\{1\}$. Additionally $(-1, 0)$ and $(0, 1)$ are both invariant sets, as is the union of any combination of these five invariant sets.

1.3.5 Example Consider

$$\left.\begin{array}{ll} \dot{r} = r(1 - r^2), & r \geqslant 0, \\ \dot{\theta} = 1, & \theta \in [0, 2\pi]. \end{array}\right\} \tag{1.3.1}$$

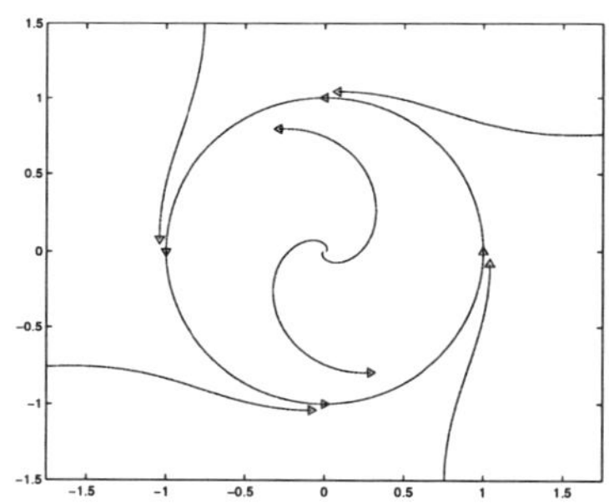

Figure 2: Trajectories of (1.3.1).

Here the bounded invariant sets include $\{(r, \theta) : r = 1\}$, $\{(r, \theta) : 0 < r < 1\}$, $\{(r, \theta) : r < 1\}$, and the complete orbit through any point on the unit disc. There are thus uncountably many bounded invariant sets in this case.

The set $\{(r, \theta) : r = 1\}$ in the previous example is a *periodic orbit.* Other invariant sets include invariant tori and strange attractors. Often invariant sets which are "stable" are of more interest than unstable invariant sets and this leads us to the concept of attractors.

1.4 Attractors

An attractor is defined as an invariant set $\mathcal{A}$ such that $\varphi^t B$ approaches $\mathcal{A}$ as $t \to \infty$ for all bounded $B \subset \mathbb{X}$.

1.4.1 Definition For sets $A, B \subseteq \mathbb{X}$ the Hausdorff semi-distance $\operatorname{dist}(A, B)$ is defined by

$$\operatorname{dist}(A, B) = \sup_{a \in A} \inf_{b \in B} \|a - b\|.$$

Note that Definition 1.4.1 only gives a semi-distance: $\operatorname{dist}(A, B)$ will be small if every point $a \in A$ is close to some point $b \in B$; but there may be points $b \in B$ which are not close to A. In particular, if $A \subset B$ then $\operatorname{dist}(A, B) = 0$ but $\operatorname{dist}(B, A)$ may be large.

1.4.2 Example Let $A = \{0\}$, and $B = [0, 1]$. Then $\operatorname{dist}(A, B) = 0$, but $\operatorname{dist}(B, A) = 1$.

1.4.3 Definition The set $\mathcal{A} \subset \mathbb{X}$ is a *global attractor* if,

- $\mathcal{A}$ is a compact set;
- $\varphi^t \mathcal{A} = \mathcal{A}$ for all $t \geqslant 0$ (invariance);
- for all bounded B
$$\lim_{t \to \infty} \operatorname{dist}(\varphi^t B, \mathcal{A}) = 0.$$

1.4.4 Definition K is an *absorbing set* if, for all bounded $B \subset \mathbb{X}$, there exists $t^* = t^*(B) \geqslant 0$:

$$\varphi^t B \subseteq K \quad \forall t \geqslant t^*.$$

The following theorem (see [6, 28, 24]) is valid in both discrete and continuous time, with $t, T \in \mathbb{T}$.

1.4.5 Theorem *If a dynamical system has a compact absorbing set K, then*

$$\mathcal{A} = \bigcap_{t \geqslant 0} \varphi^t K$$

is a global attractor.

Note that in this case

$$\mathcal{A} = \bigcap_{t \geqslant 0} \varphi^t K = \bigcap_{t \geqslant 0} \overline{\bigcup_{T \geqslant t} \varphi^T K} := \Lambda(K),$$

where $\Lambda(K)$ is the *ω-limit set* of K. Thus the global attractor is the ω-limit set of the absorbing set. Note that the form of the ω-limit set $\Lambda(K)$ simplifies, because the absorbing set maps the dynamics into itself.

Finding invariant sets directly is usually very hard, especially in the chaotic case. However, Theorem 1.4.5 allows us to infer the existence of an attractor whenever we have an absorbing set.

1.4.6 Example For Example 1.2.3 with $\lambda < 0$ we have $\varphi^t B = e^{\lambda t} B$, and it is easy to verify that any interval $B = [-a, a]$ is an absorbing set. Hence

$$\mathcal{A} = \Lambda(B) = \bigcap_{t \geqslant 0} \varphi^t B = \bigcap_{t \geqslant 0} e^{\lambda t}[-a, a] = \{0\},$$

and thus the global attractor is the (unique) fixed point.

1.4.7 Example For Example 1.2.4 with $\lambda < 0$ and $|1+\Delta t\lambda| < 1$ we have $\varphi^n B = (1+\Delta t\lambda)^n B$, and again any interval $B = [-a, a]$ is an absorbing set. Hence

$$\mathcal{A} = \Lambda(B) = \bigcap_{n \geqslant 0} \varphi^n B = \bigcap_{n \geqslant 0} (1 + \Delta t\lambda)^n [-a, a] = \{0\},$$

and thus the global attractor is the (unique) fixed point. Notice that the condition for the numerical solution to have the same attractor as in the previous example is that $\Delta t < -2/\lambda$.

1.4.8 Example For Example 1.2.5 with $B = [a, b]$ we have

$$\varphi^t B = \left[\frac{a}{[e^{-2t} + a^2(1 - e^{-2t})]^{1/2}}, \frac{b}{[e^{-2t} + b^2(1 - e^{-2t})]^{1/2}}\right],$$

and provided $a < -1$ and $b > 1$, $B = [a, b]$ is an absorbing set. Hence

$$\begin{aligned}\mathcal{A} = \Lambda(B) = \bigcap_{t \geqslant 0} \varphi^t B &= \bigcap_{t \geqslant 0} \left[\frac{a}{[e^{-2t} + a^2(1 - e^{-2t})]^{1/2}}, \frac{b}{[e^{-2t} + b^2(1 - e^{-2t})]^{1/2}}\right] \\ &= [-1, 1],\end{aligned}$$

and thus the global attractor is the interval $[-1, 1]$.

1.4.9 Example For Example 1.3.5 the global attractor is the disc $\{r \leqslant 1\}$.

Now we study the ODE (1.1.1), together with the backward Euler approximation of it. Recall the Generator Assumption 1.2.13.

1.4.10 Theorem *Let f be locally Lipschitz and satisfy Assumption* 1.2.13. *Then the ODE* (1.1.1) *defines a dynamical system on $\mathbb{R}^d$ and has a global attractor.*

Proof By Theorem 1.2.14 the ODE (1.1.1) defines a dynamical system on $\mathbb{R}^d$, and it remains to prove that it has an absorbing set to establish that it has a global attractor.

Since $V(0) < d/a$, there exists $R > 0$ such that $V(x) < d/a$ for all x such that $\|x\| < R$. Now let $Y(t) = V(x(t))$ and suppose $Y(t) \geqslant d/a$. Then

$$\dot{Y}(t) \leqslant b\|x\|^c[d - aY(t)] \leqslant bR^c[d - aY(t)].$$

This implies that $B_\varepsilon = \{x \in \mathbb{R}^d : V(x) \leqslant d/a + \varepsilon\}$ is forward invariant for any $\varepsilon > 0$. To establish the existence of an absorbing set, and hence a global attractor, it remains to show that trajectories enter such a set in finite time.

Let B be a bounded set and let $x_0 \in B$ with $V(x_0) > d/a$. Then, whilst $x(t) \notin B_\varepsilon$,

$$\dot{Y} + abR^cY(t) \leqslant bdR^c,$$

and integrating from $[0, t]$

$$Y(t) \leqslant \frac{d}{a} + e^{-abR^ct}\left[Y(0) - \frac{d}{a}\right].$$

A simple calculation shows that if

$$t > \frac{1}{abR^c} \ln\left(\frac{1}{\varepsilon}\left[Y(0) - \frac{d}{a}\right]\right),$$

then $Y(t) < d/a + \varepsilon$, hence

$$\left\{x : V(x) \leqslant \frac{d}{a} + \varepsilon\right\}$$

is an absorbing set for any $\varepsilon > 0$. Finally note that the attractor $\mathcal{A} \subset \{x : V(x) \leqslant d/a\}$ (since if not we get a contradiction by choosing ε sufficiently small.) □

Note that $Y(t)$ enters B_ε exponentially fast. The random analogue of this fact will give exponential convergence to invariant measures. See Section 2.3.

We will also employ the following assumption which implies Assumption 1.2.13.

1.4.11 Assumption (Dissipativity Condition) There exists α, $\beta > 0$ and $p \geqslant 1$ such that

$$\langle x, f(x)\rangle \leqslant \alpha - \beta\|x\|^{2p}, \quad \forall x \in \mathbb{R}^d.$$

1.4.12 Theorem *Let f be locally Lipschitz and satisfy Assumption* 1.4.11. *Then the ODE* (1.1.1) *defines a dynamical system on $\mathbb{R}^d$ and has a global attractor contained in $\{x : \|x\|^{2p} \leqslant \alpha/\beta\}$.*

Proof Let $V(x) = \|x\|^{2p} = (\sum_{i=1}^d x_i^2)^p$. Then $\partial V/\partial x_i = 2p\|x\|^{2p-2}x_i$, and

$$\mathcal{L}V(x) = \langle f(x), \nabla V(x)\rangle = 2p\|x\|^{2p-2}\langle f(x), x\rangle \leqslant 2p\|x\|^{2p-2}[\alpha - \beta V(x)].$$

Thus Assumption 1.2.13 holds and the result follows by Theorem 1.4.10. □

We now move on to consider the backward Euler method (1.1.10). One complication we will encounter here is that, whilst a solution is guaranteed to exist for all time-steps positive, it is no longer unique. To be precise we need to extend the concept of dynamical system to allow set-valued sequences, which we refer to as *generalised dynamical systems* [24], generalizing Definition 1.2.2 in a fairly straightforward way. The following lemma underpins this existence theory.

1.4.13 Lemma *Let Assumption* 1.4.11 *hold, and let $b \in \mathbb{R}^d$ be given. Then the equation*

$$a - \Delta t f(a) = b,$$

has at least one solution $a \in \mathbb{R}^d$.

Proof Let $g(x) = x - b - \Delta t f(x)$. Then

$$\begin{aligned}\langle g(x), x\rangle &= \|x\|^2 - \langle b, x\rangle - \Delta t\langle f(x), x\rangle,\\ &\geqslant \Delta t\beta\|x\|^{2p} + \|x\|^2 - \|b\|\|x\| - \alpha\Delta t.\end{aligned}$$

Thus for $R > 0$ sufficiently large $\langle g(x), x\rangle > 0$ for all x such that $\|x\| = R$ and the result follows on applying the Brouwer fixed-point theorem. See [24] for details. □

The definitions of global attractors and related concepts for generalised dynamical systems are analogous to those for dynamical systems and similar properties hold [24].

1.4.14 Theorem *Let Assumption* 1.4.11 *hold, then the backward Euler method* (1.1.10) *defines a generalised dynamical system on* $\mathbb{R}^d$ *and has a global attractor* $\mathcal{A}_{\Delta t}$.

Proof The existence of a generalised dynamical system follows from Lemma 1.4.13. To deduce the existence of an attractor, consider

$$x_{n+1} - \Delta t f(x_{n+1}) = x_n$$

hence

$$\begin{aligned}\|x_{n+1}\|^2 &= \|x_n\|^2 + 2\Delta t\langle x_{n+1}, f(x_{n+1})\rangle - \Delta t^2\|f(x_{n+1})\|^2 \\ &\leqslant \|x_n\|^2 + 2\Delta t(\alpha - \beta\|x_{n+1}\|^{2p}).\end{aligned}$$

Now given any $\varepsilon > 0$, assume there exists $x_0 \in \mathbb{R}^d$ such that $\|x_n\|^{2p} > \alpha/\beta + \varepsilon$ for all $n \geqslant 0$. Then

$$\|x_{n+1}\|^2 - \|x_n\|^2 \leqslant -2\Delta t\beta\varepsilon,$$

hence

$$\|x_n\|^2 \leqslant \|x_0\|^2 - 2n\Delta t\beta\varepsilon,$$

which supplies a contraction for n sufficiently large. Thus

$$\{x : \|x\|^{2p} \leqslant \alpha/\beta + \varepsilon\},$$

is an absorbing set for any $\varepsilon > 0$ and there is a global attractor $\mathcal{A}_{\Delta t} \subset \{x : \|x\|^{2p} \leqslant \alpha/\beta\}$. □

An important point to note in the previous theorem is that the backward Euler method has preserved the Lyapunov function of the continuous problem. More can be said about the attractor from Theorem 1.4.14:

1.4.15 Theorem *Let Assumptions* 1.4.11 *hold and let* $\mathcal{A}$ *be the global attractor of the dynamical system defined by* (1.1.1) *and* $\mathcal{A}_{\Delta t}$ *the global attractor for the backward Euler method applied to* (1.1.1) *then*

$$\lim_{\Delta t \to 0} \operatorname{dist}(\mathcal{A}_{\Delta t}, \mathcal{A}) = 0.$$

Proof See for example [24]. □

Recall that in general the backward Euler method only defines a generalised dynamical system for this problem. However, for Δt sufficiently small we can show that it defines a dynamical system on a neighbourhood of the attractor.

1.4.16 Theorem *Let Assumption* 1.4.11 *hold and let* L *be the Lipschitz constant for* f *on* $K = \{x : \|x\|^{2p} \leqslant \alpha/\beta\}$. *If* $L\Delta t < 1$ *then the backward Euler method applied to* (1.1.1) *defines a dynamical system on* K, *with* $\mathbb{T} = \mathbb{Z}^+$, *and for* $x_0,\, y_0 \in K$

$$(1 - L\Delta t)^n\|x_0 - y_0\| \leqslant \|x_n - y_n\| \leqslant (1 + L\Delta t)^n\|x_0 - y_0\|. \tag{1.4.1}$$

The inequalities (1.4.1) are a discrete analogue of (1.2.4) and bound the rate of convergence/divergence of trajectories, and prevent trajectories from merging or crossing in K.

Note that other conditions, for example the one-sided Lipschitz condition, can be used to infer that the backward Euler method defines a dynamical system on $\mathbb{R}^d$.

1.4.17 Example Consider the backward Euler method applied to (1.1.4). We showed in Example 1.2.21 that for $\Delta t < 1$, the map φ is well-defined, so that this defines a dynamical system on $\mathbb{R}$. Note that

$$\langle f(x), x\rangle = x^2 - x^4 = \frac{1}{2} - \frac{1}{2}(1-x^2)^2 - \frac{1}{2}x^4 \leqslant \frac{1}{2}(1-x^4)$$

and so Assumption 1.4.11 is satisfied, and hence Theorem 1.4.14 implies the existence of an attractor $\mathcal{A}_{\Delta t}$. Now Theorem 1.4.15 implies that

$$\lim_{\Delta t\to 0} \operatorname{dist}(\mathcal{A}_{\Delta t}, \mathcal{A}) = 0.$$

as $\Delta t \to 0$. In fact, a little algebra shows that $\mathcal{A}_{\Delta t} = \mathcal{A} = [-1, 1]$ for any $\Delta t < 1$.

The behaviour of the forward Euler method applied to the same ODE is very different as the following example shows.

1.4.18 Example Consider the forward Euler method applied to (1.1.4). Then

$$x_{n+1} = x_n + \Delta t(x_n - x_n^3) = [1 + \Delta t - \Delta t x_n^2]x_n.$$

Thus if $x_n^2 > 1 + 2/\Delta t$, then $[1 + \Delta t - \Delta t x_n^2] < -1$ and $|x_{n+1}| > |x_n|$. It follows that given any $\Delta t > 0$ for all x_0 with

$$x_0^2 > 1 + \frac{2}{\Delta t},$$

then $\lim_{n\to\infty} \|x_n\| = \infty$. Thus in this case there is no absorbing set and hence no global attractor.

In contrast if $x_0^2 \in (1, 1 + 2/\Delta t)$ then $|1 + \Delta t - \Delta t x_n^2| < 1$, then $\limsup_{n\to\infty} \|x_n\| \leqslant 1$, so that these trajectories do converge to the attractor $\mathcal{A} = [-1, 1]$ of (1.1.4). A general theory of local convergence for attractors has been developed, but we will not pursue it here; see [24] for details. We will see that when we add noise, all trajectories for the forward Euler method can escape to infinity no matter how small Δt is.

The previous example is a specific case of the forward Euler method applied to an equation satisfying Assumption 1.4.11. The behaviour seen, namely that some trajectories escape to infinity, happens for any f satisfying the same assumption.

1.5 Case study 1: The Langevin equation

Let $F : \mathbb{R}^d \to \mathbb{R}$, and consider the following ODE for $q, p \in \mathbb{R}^d$ the position and momenta of a particle of unit mass, namely

$$\dot{q} = p, \tag{1.5.1}$$
$$\dot{p} = -\gamma p - \nabla F(q). \tag{1.5.2}$$

The Langevin SDE, which we will consider later, is a generalisation of this ODE, which includes noise.

For (1.5.1), (1.5.2) setting $\gamma = 0$ gives a conservative Hamiltonian system, and the choice $\gamma > 0$ gives a damped Hamiltonian. To see this, note that, if we set

$$H(q,p) = \frac{1}{2}\|p\|^2 + F(q),$$

then

$$\dot{H} = \mathcal{L}H(q,p) = -\gamma\|p\|^2. \tag{1.5.3}$$

A generalisation of the Lyapunov function H will be useful in the following, and for this we require that $F(q) \geqslant 0$ for all $q \in \mathbb{R}^d$, and that there exists $\beta \in (0,1)$ and $\alpha \geqslant 0$ such that

$$\frac{1}{2}\langle \nabla F(q), q\rangle \geqslant \beta F(q) + \gamma^2 \frac{\beta(2-\beta)}{8(1-\beta)}\|q\|^2 - \alpha. \tag{1.5.4}$$

For example, if

$$F(q) = \frac{1}{4}(1 - \|q\|^2)^2, \tag{1.5.5}$$

then $F(q) \geqslant 0$ and, moreover, since

$$\nabla F(q) = -(1 - \|q\|^2)q,$$

we have

$$\langle \nabla F(q), q\rangle = (\|q\|^2 - 1)\|q\|^2, \tag{1.5.6}$$

and so the right and left hand sides of (1.5.4) are both quartics in $\|q\|$ and for any $\beta \in (0,1)$ and sufficiently large $\alpha > 0$ the inequality (1.5.4) holds.

Although H is non-increasing in time for $\gamma > 0$, its rate of change is zero whenever $p = 0$. In order to find a function whose rate of decrease is bounded from below outside a bounded set in phase space we introduce

$$V(q,p) = \frac{1}{4}\|p\|^2 + \frac{1}{4}\|p + \gamma q\|^2 + F(q). \tag{1.5.7}$$

Then

$$\begin{aligned}
\dot{V} &= \mathcal{L}V, \\
&= \frac{1}{2}\langle p, \dot{p}\rangle + \frac{1}{2}\langle \dot{p} + \gamma\dot{q}, p + \gamma q\rangle + \langle \nabla F(q), \dot{q}\rangle \\
&= -\frac{1}{2}\langle p, \gamma p + \nabla F(q)\rangle - \frac{1}{2}\langle \nabla F(q), p + \gamma q\rangle + \langle \nabla F(q), p\rangle \\
&= -\frac{\gamma}{2}\|p\|^2 - \frac{\gamma}{2}\langle \nabla F(q), q\rangle,
\end{aligned}$$

and using (1.5.4) gives

$$\dot{V} \leqslant -\gamma\left[\frac{1}{2}\|p\|^2 + \beta F(q) + \gamma^2\frac{\beta(2-\beta)}{8(1-\beta)}\|q\|^2\right] + \gamma\alpha.$$

But, for $\beta \in (0,1)$, straightforward calculation shows

$$\frac{1}{2}\|p\|^2 + \gamma^2\frac{\beta(2-\beta)}{8(1-\beta)}\|q\|^2 \geqslant \frac{\beta}{4}\|p\|^2 + \frac{\beta}{4}\|p + \gamma q\|^2.$$

Hence

$$\dot{V} \leqslant \gamma(\alpha - \beta V). \tag{1.5.8}$$

Thus applying the continuous Gronwall Lemma 1.2.10 we see that we approach any open neighbourhood of $\{x : V(x) \leqslant \alpha/\beta\}$ exponentially fast, and

$$\limsup_{t\to\infty} V(q(t), p(t)) \leqslant \frac{\alpha}{\beta}.$$

Thus by (1.5.7),

$$\limsup_{t\to\infty} \|p(t)\|^2 \leqslant \frac{4\alpha}{\beta}, \quad \text{and} \quad \limsup_{t\to\infty} \|p(t) + \gamma q(t)\|^2 \leqslant \frac{4\alpha}{\beta}.$$

Using $\|\gamma q\| \leqslant \|p + \gamma q\| + \|p\|$, we deduce further that

$$\limsup_{t\to\infty} \|q(t)\|^2 \leqslant \frac{16\alpha}{\gamma^2\beta}.$$

But we can say more. Because $H(q(t), p(t))$ is bounded below and monotonically decreasing, $\lim_{t\to\infty} H$ exists and

$$0 = \lim_{t\to\infty} \dot{H} = \lim_{t\to\infty} p(t) = \lim_{t\to\infty} \dot{q}(t),$$

using (1.5.3) and (1.5.1). It follows further that $\lim_{t\to\infty} \dot{p}(t) = 0$ and thus all solutions $(p(t), q(t))$ tend to the set of fixed points in the limit as $t \to \infty$ and, moreover, by (1.5.1)–(1.5.2) these fixed points satisfy

$$p = 0, \qquad \nabla F(q) = 0.$$

This problem is an example of a generalised gradient system, in the sense defined in [6]. The Lyapunov structure (1.5.8) is preserved by the backward Euler method—see [16]. In the next chapter we will study the effect of noise on this system.

1.6 Case study 2: particles in a velocity field

Consider the following model for particles moving in a two-dimensional velocity field:

$$\tau \ddot{x} = v(x) - \dot{x}. \tag{1.6.1}$$

Here $x \in \mathbb{T}^2$ (in this section $\mathbb{T}^2$ denotes the two-dimensional unit torus) denotes the particle position and $\dot{x} \in \mathbb{R}^2$ its velocity; we assume $\tau > 0$. The force on the particle is proportional to the difference between the fluid velocity at the particle site, $v(x)$, and the particle velocity, $\dot{x}$. The velocity field is given by $v = \nabla^{\perp}\psi$, where ψ is a stream-function, $\nabla^{\perp}$ denotes skew-gradient and hence $\nabla \cdot v = 0$. Thus x denotes the position of a particle moving according to Stokes' law in a two-dimensional incompressible velocity field v. Later on we will consider time-dependent random velocity fields $v(x, t)$.

Letting $y = \dot{x} \in \mathbb{R}^2$, we can write this as a system $\dot{u} = f(u)$ with $u = (x, y) \in \mathbb{X} = \mathbb{T}^2 \times \mathbb{R}^2$:

$$\begin{aligned} \dot{x} &= y, \\ \dot{y} &= \frac{1}{\tau}(v(x) - y). \end{aligned}$$

We assume that v is Lipschitz on $\mathbb{T}^2$. Then f is globally Lipschitz on $\mathbb{T}^2 \times \mathbb{R}^2$, and (1.6.1) defines a dynamical system.

We show that this dynamical system possesses a global attractor. The choice of $\mathbb{X} = \mathbb{T}^2 \times \mathbb{R}^2$ implies that $\|x(t)\| \leqslant 1$ for all $t \in \mathbb{R}$, and so it only remains to show that $\|y(t)\|$ is ultimately bounded independently of initial data to get an absorbing set. Now

$$\begin{aligned}\tau \frac{1}{2}\frac{d}{dt}\|y\|^2 &= \tau\langle y, \dot{y}\rangle \\ &= \langle v(x), y\rangle - \|y\|^2 \\ &\leqslant \frac{1}{2}[v(x)]^2 + \frac{1}{2}\|y\|^2 - \|y\|^2 \\ &\leqslant \frac{1}{2}c_v^2 - \frac{1}{2}\|y\|^2.\end{aligned}$$

Here $c_v = \max_{x\in\mathbb{T}^2}\|v(x)\| < \infty$, since a continuous function on a compact set achieves its supremum. Thus it follows that $\limsup_{t\to\infty}\|y(t)\| \leqslant c_v$ and so the system possesses an absorbing set and hence an attractor. On the attractor the particle velocity does not exceed the maximum fluid velocity in $\mathbb{T}^2$.

A prototypical example of a velocity field is given by the Taylor-Green explicit solution of the forced Navier-Stokes equations namely, for $x = (x_1, x_2)$,

$$\psi(x) = \sin(2\pi x_1)\sin(2\pi x_2).$$

Fig. 3 shows the direction field and streamlines induced by this velocity field; notice the four vortices. In the case $\tau = 0$, where particles follow fluid stream-lines, all solutions are periodic (except for a Lebesgue measure zero set of initial data lying on *heteroclinic orbits* connecting fixed points at $t = \pm\infty$) and there is no global attractor. For $\tau > 0$, however, there is a global attractor. Empirically we observe this to be made up of a finite number of periodic orbits, a finite number of fixed points, and orbits connecting them. This is illustrated in Fig. 4 which shows the distribution of 5000 particles, after sufficient time has elapsed for transient behaviour to disappear; what is seen is hence a visualisation of the global attractor in position space, excluding a number of heteroclinic orbits connecting pairs of periodic orbits and/or fixed points at $t = \pm\infty$.

The calculations are performed using the split-step Euler method

$$\begin{aligned}x_{n+1} &= x_n + \Delta t y_n, \\ y_{n+1} &= y_n + \frac{\Delta t}{\tau}\left(v(x_{n+1}) - y_{n+1}\right).\end{aligned}$$

This gives

$$\left(1 + \frac{\Delta t}{\tau}\right) y_{n+1} - \frac{\Delta t}{\tau} v(x_{n+1}) = y_n,$$

and so

$$\left(1 + \frac{\Delta t}{\tau}\right)^2 \|y_{n+1}\|^2 - 2\frac{\Delta t}{\tau}\left(1 + \frac{\Delta t}{\tau}\right)\langle y_{n+1}, v(x_{n+1})\rangle + \frac{\Delta t^2}{\tau^2}\|v(x_{n+1})\|^2 = \|y_n\|^2,$$

implying

$$\left(1 + \frac{\Delta t}{\tau}\right)^2 \|y_{n+1}\|^2 \leqslant \|y_n\|^2 + \frac{\Delta t}{\tau}\left(1 + \frac{\Delta t}{\tau}\right)[\|y_{n+1}\|^2 + \|v(x_{n+1})\|^2] - \frac{\Delta t^2}{\tau^2}\|v(x_{n+1})\|^2,$$

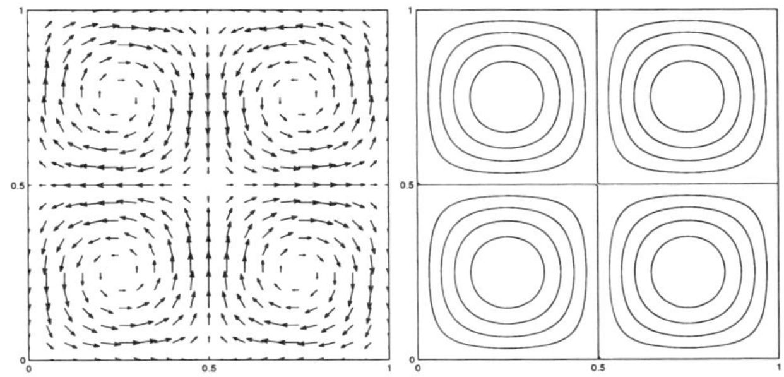

Figure 3: Taylor-Green velocity field: (i) direction field, (ii) streamlines.

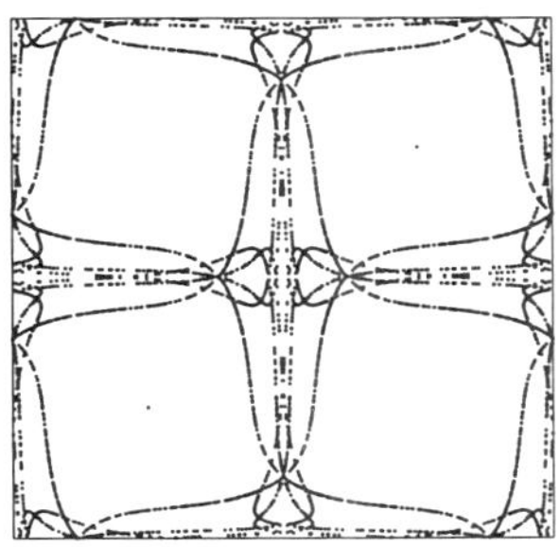

Figure 4: Global attractor for (1.6.1).

so that

$$\left(1+\frac{\Delta t}{\tau}\right)\|y_{n+1}\|^2 \leqslant \|y_n\|^2 + \frac{\Delta t}{\tau}\,c_v^2.$$

Thus, applying the discrete Gronwall Lemma 1.2.3, we obtain

$$\|y_n\|^2 \leqslant c_v^2 + \frac{1}{(1+\Delta t/\tau)^n}\,[\|y_0\|^2 - c_v^2],$$

and the existence of an absorbing set and global attractor follows. As for the exact solution, on the attractor the particle velocity does not exceed the maximum fluid velocity in $\mathbb{T}^2$.

2 Random dynamical systems

Here we take many of the ideas developed for dynamical systems and generalise them to situations where noise is present. This leads to the subject of random dynamical systems, recently given firm foundations in the book [1]. The subject can be rather technical upon first encounter. We aim to give an accessible introduction to some of it, in particular to those parts of relevance in numerical analysis. By being accessible we will miss out on many of the subtleties of the subject and the reader is encouraged to study [1], and the references therein, for a thorough treatment.

2.1 Stochastic differential equations

Let $W(t)$ denote m-dimensional Brownian motion (see the discussion at the end of the section), $f : \mathbb{R}^d \to \mathbb{R}^d$ a smooth function and $\Sigma \in \mathbb{R}^{d\times m}$ a fixed matrix. Consider the following integral equation for $x \in C([0,T], \mathbb{R}^d)$:

$$x(t) = x(0) + \int_0^t f(x(s))\, ds + \Sigma W(t), \quad x(0) = X. \tag{2.1.1}$$

Underlying Brownian motion is *Wiener measure*, under which $W(t)$ is, with probability 1, in the space $C([0,T], \mathbb{R}^d)$ for any $T > 0$. However, also with probability 1, Brownian motion is nowhere differentiable. Thus it is not possible to differentiate (2.1.1) and find a non-autonomous differential equation in the usual sense. Nonetheless the equation is often written formally as the *stochastic differential equation* (SDE)

$$dx = f(x)dt + \Sigma dW, \quad x(0) = X. \tag{2.1.2}$$

If Σ depends upon t, explicitly and/or through $x(t)$, then it is necessary to introduce a notion of *stochastic integral* to interpret the term $\Sigma(t, x(t))dW(t)$ since, with probability 1, $W(t)$ is not of bounded variation. However, for the purpose of these notes we will consider the case of *additive noise* with Σ constant. The precise interpretation of (2.1.2) is then simply the integral equation (2.1.1).

Since $W(t)$ is, with probability 1, in $C([0,T], \mathbb{R}^d)$, it follows, under suitable conditions on f[1], that the solution $x(t)$ is itself a random function which is, with probability 1, in $C([0,T], \mathbb{R}^d)$. The objective of solving (2.1.2) is to find the properties of this random function, given the properties of Brownian motion. We denote Wiener measure by $\mathbb{P}$ and expectation under it by $\mathbb{E}$. The book [19] is a suitable introduction to the theory of SDEs; the books [8, 15] provide further useful development of the subject in the dynamical context of interest here.

Itô formula

In the following we assume that Σ is constant and either

(1) f is globally Lipschitz; or

(2) f is locally Lipschitz and satisfies a structural assumption inducing boundedness of solutions.

This is exactly what we did for ODEs. It is often of importance to understand how functions of $x(t)$ change with time. By using the *generator*:

$$\mathcal{L}V = \sum_{i=1}^{d} f_i \frac{\partial V}{\partial x_i} + \frac{1}{2} \sum_{i,j=1}^{d} [\Sigma\Sigma^T]_{ij} \frac{\partial^2 V}{\partial x_i \partial x_j}, \tag{2.1.3}$$

we may achieve this through the Itô formula [19]:

[1]The simplest being that f is globally Lipschitz; see [15] for generalisations.

2.1.1 Lemma *Let $x(t)$ solve (2.1.2) and let $V \in C^2(\mathbb{R}^d, \mathbb{R})$. Then the process $Y(t) = V(x(t))$ satisfies*

$$d\{V(x(t))\} = \mathcal{L}V(x(t))dt + \langle \nabla V(x(t)), \Sigma dW \rangle, \quad Y(0) = g(X).$$

To interpret this result we would need to introduce Itô stochastic integrals in order to make sense of the last term. However, for our purposes the following consequence is all we will need:

2.1.2 Lemma *Let $x(t)$ solve (2.1.2) and let $V \in C_0^2(\mathbb{R}^d, \mathbb{R})$ (C^2 functions with compact support). Then for $0 \leqslant s < t < \infty$ we have that $Y(t) = V(x(t))$ satisfies*

$$\mathbb{E}Y(t) = \mathbb{E}Y(s) + \mathbb{E}\int_s^t \mathcal{L}V(x(\tau))\, d\tau.$$

By means of *Dynkin's formula* [19] this result can be extended to $V \in C^2$ which are not compactly supported, provided sufficient boundedness of the solutions $x(t)$ is established. We do not provide details here, but freely apply Lemma 2.1.2 to non-compactly supported V when the necessary boundedness conditions on $x(t)$ hold.

A second (indirect) consequence of Lemma 2.1.2 is the *Fokker-Planck* equation for propagation of densities. For this we need the adjoint $\mathcal{L}^*$:

$$\mathcal{L}^*V = -\sum_{i=1}^d \frac{\partial}{\partial x_i}(f_i V) + \frac{1}{2}\sum_{i,j=1}^d \frac{\partial^2}{\partial x_i \partial x_j}\{[\Sigma\Sigma^T]_{ij}V\}.$$

2.1.3 Lemma *Assume that the law of $x(t)$, $\mathbb{P}\{x(t) \in A\}$, has a density $\rho(x,t) \in C^{2,1}(\mathbb{R}^d \times (0,T], \mathbb{R}^+)$. Then ρ satisfies the Fokker-Planck equation*

$$\begin{aligned} \frac{\partial \rho}{\partial t} &= \mathcal{L}^*\rho, \quad (x,t) \in \mathbb{R}^d \times (0,T], \\ \rho(x,0) &= \delta(x - x_0), \quad x \in \mathbb{R}^d. \end{aligned}$$

Numerical methods

We introduce three numerical methods for the SDE (2.1.2). The first, called the *Euler-Maruyama* method [11], is simply an SDE analogue of the *forward Euler* method (1.1.9) for deterministic systems. Given $t_n = n\Delta t$, and defining

$$\Delta W_n = W(t_{n+1}) - W(t_n),$$

the approximation x_n to $x(t_n)$ satisfies:

$$x_{n+1} = x_n + \Delta t f(x_n) + \Sigma \Delta W_n. \tag{2.1.4}$$

This may be derived, as in the deterministic case, by approximating the integral in (1.1.2). The second method considered we call the *stochastic backward Euler* method and is defined by

$$x_{n+1} = x_n + \Delta t f(x_{n+1}) + \Sigma \Delta W_n. \tag{2.1.5}$$

The final method, called the *split-step stochastic backward Euler* method, is defined by

$$\begin{aligned} x_* &= x_n + \Delta t f(x_*), \\ x_{n+1} &= x_* + \Sigma \Delta W_n. \end{aligned} \tag{2.1.6}$$

The idea of this method is similar to that underlying (1.1.11) with noise playing the role of f_2. For all three methods we choose $x_0 = X$ from (2.1.2).

Note that the Euler-Maruyama method is explicit, whereas the remaining two methods considered are implicit, requiring solution of a nonlinear equation for x_{n+1}, given x_n and ΔW_n. Here ΔW_n is an m-dimensional Gaussian random variable with mean 0 and covariance matrix $\Delta t I$; we denote this $\mathcal{N}(0, \Delta t I)$.

Brownian motion

Brownian motion is a continuous time analogue of the random walk. It is a process $\{W(t)\}_{t\geqslant 0}$, with $W(0) = 0$, and satisfying three basic properties:

(1) *time-homogeneity:* $W(t)$ and $W(t+s) - W(s)$ have the same distribution $\forall s, t \geqslant 0$;

(2) *independent increments:* $\{W(t_i) - W(s_i)\}_{i\geqslant 1}$ are independent random variables whenever the intervals $(t_i, s_i]$ are disjoint;

(3) *Gaussian increments:* $W(t) \sim \mathcal{N}(0, \sigma^2 t)$.

Standard Brownian motion denotes the choice $\sigma = 1$. Standard Brownian motion in $\mathbb{R}^m$ is a function $W : \mathbb{R}^+ \to \mathbb{R}^m$ with each component $W_i(t)$ being an independent standard Brownian motion.

It is possible to construct an underlying probability space in which such random functions live, and then to study the regularity of such functions under the resulting *Wiener measure.* See [9, 10] for details. Such analysis shows that, with probability 1, Brownian paths are α-Hölder continuous for any $\alpha \in [0, 1/2)$. This (lack of) regularity makes rates of convergence of approximations a more subtle issue than for ODEs.

Bibliographical remark The numerical analysis of SDEs is a subject in its infancy, though several books on the subject are now available, for example [2, 11]. Numerical analysis of SDEs in the context of random dynamical systems has not yet been developed in a systematic fashion and this fact will be manifest in our presentation. We will concentrate on two important dynamical concepts: (i) ergodicity and invariant measures and (ii) random attractors. Foundational work on the numerical analysis of SDEs according to their ability to reproduce ergodic properties is due to Talay [25, 26, 27]; see also [22, 16]. The study of random attractors, and the effect of discretisation, is being developed by Kloeden and co-workers (see [3, 12] for example), with recent extension to numerical approximation of SDEs by Robinson [20].

2.2 Random dynamical systems

We now generalise the concept of dynamical systems to allow for non-autonomous problems where the evolution in the space $\mathbb{X}$ depends upon time. Such time-dependence can be given a variety of differing structures. Here we study *random dynamical systems* where the time-dependence is introduced through a *stationary* driving noise; by this we mean that the finite dimensional distributions of the random process are unaffected by translation in time. Our primary motivation is the study of the SDE (2.1.2) and its time-discrete counterparts, the forward and backward Euler methods (2.1.4), (2.1.5), (2.1.6). We start by giving the structure of the underlying noise model, and then show how to incorporate this into the definition of a random dynamical system. Our presentation is a curtailed version of [1, Chapter 1], which the reader should consult for greater depth and precision.

Noise model

A family $(\theta^t)_{t\in\mathbb{T}}$ of mappings of a measurable space $(\Omega, \mathcal{F})$ into itself is called a *measurable dynamical system* with time $\mathbb{T}$ if it satisfies the following conditions:

(1) $(\omega, t) \in (\Omega, \mathbb{T}) \mapsto \theta^t\omega \in \Omega$ is measurable;

(2) $\theta^0 = \mathrm{id}_\Omega$;

(3) $\theta^{s+t} = \theta^s \circ \theta^t \ \forall s, t \in \mathbb{T}$.

We will only consider the choices $\mathbb{T} = \mathbb{R}, \mathbb{Z}, \mathbb{R}^+, \mathbb{Z}^+$ for the time variable.

We now introduce a probability measure on the space $(\Omega, \mathcal{F})$ and consider the probability space $(\Omega, \mathcal{F}, \mathbb{P})$. We assume that $\theta^t\mathbb{P} = \mathbb{P} \quad \forall t \in \mathbb{T}$: that is, $\mathbb{P}\{(\theta^t)^{-1}(A)\} = \mathbb{P}\{A\}$ for all $(t, A) \in (\mathbb{T}, \mathcal{F})$. The measure $\mathbb{P}$ is then said to be *invariant* with respect to θ. We have a *measure preserving* or *metric* dynamical system, which we denote by $(\Omega, \mathcal{F}, \mathbb{P}, \theta^t)$. This is the underlying noise model used in the theory of random dynamical systems. The noise is stationary, because of the invariance of $\mathbb{P}$ under θ.

Evolution driven by noise

In the following, $\mathcal{B}(\mathbb{T})$ denotes the Borel sets in $\mathbb{T}$.

2.2.1 Definition A *measurable random dynamical system* on a measurable space $(\mathbb{X}, V)$ over a metric dynamical system $(\Omega, \mathcal{F}, \mathbb{P}, \theta^t)$ with time $\mathbb{T}$ is a mapping

$$\varphi : \mathbb{T} \times \mathbb{X} \times \Omega \to \mathbb{X}, \ (t, x, \omega) \mapsto \varphi(t, x, \omega),$$

where the mapping φ is $\mathcal{B}(\mathbb{T}) \otimes V \otimes \mathcal{F}$ measurable and satisfies the *co-cycle property* that $\varphi(0, x, \omega) = x$ for all $(x, \omega) \in (\mathbb{X}, \Omega)$ and, for all $t, s \in \mathbb{T}$, $x \in \mathbb{X}$, $\omega \in \Omega$,

$$\varphi(t + s, x, \omega) = \varphi(t, \varphi(s, x, \omega), \theta^s\omega).$$

It will sometimes be helpful to write

$$\varphi(t, x, \omega) = \varphi^t(\omega)x$$

as we will be interested in mapping sets in $\mathbb{X}$. In this notation the co-cycle property reads $\varphi^0(\omega) = \mathrm{id}_\mathbb{X}$ and

$$\varphi^{t+s}(\omega) = \varphi^t(\theta^s\omega) \circ \varphi^s(\omega).$$

φ is a generalisation of the evolution operator, now acting on noise as well as $\mathbb{X} \times \mathbb{T}$. We may think of such a process as a *skew-product* over the noise-process. A skew-product is a dynamical system such that

$$
\begin{aligned}
(Y, Z) &\to (Y, Z), \\
y &\mapsto f(y), \\
z &\mapsto g(y, z).
\end{aligned}
$$

The mapping f over Y is called the *base transformation*, and the mapping g over Z the *fibre transformation.*

In the case of random dynamical systems, the noise-process is the base, driving the evolution of real interest taking place in the fibre. We have $(Y, Z) = (\Omega, \mathbb{X})$ and define the measurable dynamical system Θ^t on $(\Omega \times \mathbb{X}, \mathcal{F} \otimes V)$ by

$$\Theta^t(x, \omega) = (\varphi^t(\omega)x, \theta^t\omega).$$

What distinguishes the random dynamical systems set-up from an arbitrary skew-product is the structure induced by $\mathbb{P}$. Without the stationarity assumption on the noise the framework would be too large to admit useful mathematical development; at the same time many problems arising in applications fit into the framework developed here. Together these two reasons suggest that developing the subject of random dynamical systems is a worthwhile enterprise.

Examples

We illustrate some of the concepts above by means of a number of simple examples. We will frequently invoke the *Markov property* [10]. Roughly, a random sequence or function $\{u_t\}_{t\in\mathbb{T}}$ is Markovian if the statistics of $u_t, t > s$ given u_s are independent of knowledge of u_τ for $\tau < s$. We will also frequently use the fact that a Gaussian measure is specified by its mean and covariance [10].

2.2.2 Example Consider a linearly damped map driven by white noise:

$$x_{n+1} = \lambda x_n + \eta_n,$$

where $|\lambda| < 1$ and the η_n form an independent identically distributed (i.i.d.) sequence of random variables with $\eta_0 \sim \mathcal{N}(0, \sigma^2)$.

In this case $\mathbb{T} = \mathbb{Z}^+$, $\mathbb{X} = \mathbb{R}$, $\Omega = \{\omega \in \mathbb{R}^{\mathbb{Z}^+} : \omega = (\eta_0, \eta_1, \dots)\}$ and θ^t is the shift on such sequences defined by $\theta^1\omega = (\eta_1, \eta_2, \dots)$. Then $\mathbb{P}$ is the measure induced on such i.i.d. sequences by assuming $\eta_0 \sim \mathcal{N}(0, \sigma^2)$. The fact that the dynamical system for the noise is measure preserving follows since the η_j are i.i.d. The process $\{x_n\}$ is known as a discrete Ornstein-Uhlenbeck (OU) process.

2.2.3 Example We generalise the previous example to a linearly damped map driven by a coloured noise, namely the discrete-time OU process $\{\eta_n\}$:

$$
\begin{aligned}
x_{n+1} &= \lambda x_n + \eta_n, \\
\eta_{n+1} &= \gamma \eta_n + \xi_n,
\end{aligned}
$$

where $|\lambda|, |\gamma| < 1$ and ξ_n forms an i.i.d. sequence with $\xi_0 \sim \mathcal{N}(0,1)$. Again $\mathbb{T} = \mathbb{Z}^+$, $\mathbb{X} = \mathbb{R}$, $\Omega = \{\omega \in \mathbb{R}^{\mathbb{Z}^+} : \omega = (\eta_0, \eta_1, \dots)\}$ and θ^t is the shift map on sequences from the previous example. To ensure stationarity of the noise ω we choose η_0 independently of the i.i.d. sequence $\{\xi_j\}$ and set

$$\eta_0 \sim \mathcal{N}(0, \sigma^2),$$

where $\sigma^2 = (1-\gamma^2)^{-1}$. Then

$$\mathbb{E}\eta_1^2 = \gamma^2 \mathbb{E}\eta_0^2 + 1 = \mathbb{E}\eta_0^2.$$

By the Markov property the sequence ω is then stationary. To see the measure $\mathbb{P}$ explicitly we employ the formula

$$\eta_n = \gamma^n \eta_0 + \sum_{j=0}^{n-1} \gamma^j \xi_{n-1-j}.$$

The sequence $\{\eta_j\}$ is then seen to be Gaussian (since it is a linear combination of Gaussians) with mean zero and covariance

$$\mathbb{E}\eta_l \eta_m = \sigma^2 \gamma^{|m-l|}.$$

This completely specifies the invariant measure $\mathbb{P}$. Here we are using the fact that the mean and covariance define a Gaussian measure.

The preceding examples can be extended to $\mathbb{T} = \mathbb{Z}$ with bi-infinite noise sequences $\omega \in \mathbb{R}^{\mathbb{Z}}$. This extension will be necessary and relevant when studying random attractors.

2.2.4 Example Consider the SDE (2.1.2), together with the assumption that the drift f is a locally Lipschitz vector field satisfying

$$\exists \alpha > 0, \beta > 0 : \ \langle f(x), x \rangle \leqslant \alpha - \beta \|x\|^2. \tag{2.2.1}$$

Here $\mathbb{T} = \mathbb{R}^+$ and $\mathbb{X} = \mathbb{R}^d$. The set Ω comprises Brownian paths with invariant measure induced by the property of independent increments.[2]

Under our assumptions on f and Σ the solution of (2.1.2) exists for all $t > 0$ (essentially because the Lyapunov condition Assumption 2.3.6 holds for $V(x) = \|x\|^2$, see [8] or Theorem 36 in Chapter 2 of [15] for a precise statement of this result).

The preceding example can be extended to $\mathbb{T} = \mathbb{R}$ by considering Brownian motions with $t \in \mathbb{R}$. However, to do this requires choice of random initial data for $x(0)$ which is dependent on $\{W(t), t \leqslant 0\}$ [1]. The construction generalises the fact that, for deterministic dynamical systems, complete bounded orbits exist for all points on the global attractor [6, 24].

2.2.5 Example Consider the Euler-Maruyama method (2.1.4) for the approximation of (2.1.2). This can be formulated as a random dynamical system with $\mathbb{T} = \mathbb{R}^+$, $\mathbb{X} = \mathbb{R}^d$ and $\Omega = \{\omega \in \{\mathbb{R}^m\}^{\mathbb{Z}^+} : \omega = (\Delta W_0, \Delta W_1, \dots)\}$. Recall that $\{\Delta W_j\}$ forms an i.i.d. sequence so that, with θ being the shift defined in Example 2.2.2, we have the underlying measure preserving dynamical system representing the noise.

[2]More precisely, we work with equivalence classes of paths of the form $\{a + W(t), a \in \mathbb{R}^m\}$, where $W(t)$ is standard Brownian motion. See [1, Appendix A2] for details.

2.2.6 Example To pose the implicit methods (2.1.5), (2.1.6) as random dynamical systems it is necessary to prove existence and uniqueness for the implicit equations. This cannot be done in general, but can be achieved for certain natural structural assumptions, such as the one-sided Lipschitz condition Assumption 1.2.15. See [16] for details. Once this is done the formal set-up is similar to that of the previous example.

2.3 Invariant measures

The simplest way to introduce the concept of invariant measure is through the skew-product picture. Define $\pi_\Omega : \mathbb{X} \times \Omega \to \Omega$ by $\pi_\Omega(x, \omega) = \omega$. Notice that the structure of the skew-product means that

$$\pi_\Omega \circ \Theta^\bullet = \theta^\bullet \circ \pi_\Omega.$$

If μ is invariant for Θ, so that $\Theta^t \mu \equiv \mu$, then we see that

$$\pi_\Omega \circ \mu = \pi_\Omega \circ \Theta^t \circ \mu = \theta^t \circ (\pi_\Omega \circ \mu),$$

so that $\pi_\Omega \mu$ is $\theta(\cdot)$ invariant. Recalling that $\mathbb{P}$ is invariant for θ, these considerations motivate the following definition:

2.3.1 Definition Given a measurable random dynamical system φ, a probability measure μ on $(\mathbb{X} \times \Omega, V \times \mathcal{F})$ is an *invariant measure* for φ if

- $\Theta^t \mu = \mu \quad \forall t \in \mathbb{T}$;
- $\pi_\Omega \mu = \mathbb{P}$.

The measure μ will not be a product measure in general. However, the *disintegration*

$$\mu(A) = \int_A \mu_\omega(dx) \mathbb{P}(d\omega)$$

is often useful. If $\mu_\omega(dx)$ is independent of ω then the measure is product.

In all the examples we study here, the noise and the variable of interest which it drives will, together, form a Markov chain (discrete time) or Markov process (continuous time). We denote this Markov chain/process by (x_t, η_t) with $\eta_t \in \mathbb{Y}$ being the noise and $x_t \in \mathbb{X}$ being the noise driven process. Notice that ω is the *complete path* $\{\eta_t\}_{t \in \mathbb{T}} \in \Omega$. The preceding definition of invariant measure applies in the space $\mathbb{X} \times \Omega$. However, by virtue of the Markovian nature of our problems, it will be natural to seek invariant measures in $\mathbb{X} \times \mathbb{Y}$—to find a measure ν so that, if $(x_0, \eta_0) \sim \nu$ then $(x_t, \eta_t) \sim \nu$ for all $t \in \mathbb{T}$. Having found such a measure, additional statistics relating x_0 to $\{\eta_t\}_{t \in \mathbb{T}}$ are required to obtain the invariant measure μ in $\mathbb{X} \times \Omega$. For the first examples we will carry out this step of going from $\mathbb{X} \times \mathbb{Y}$ to $\mathbb{X} \times \Omega$. For the general theory of SDEs we will omit it. Indeed for such problems the measure ν is itself product measure in $\mathbb{X} \times \mathbb{Y}$, with the measure in $\mathbb{Y}$ being Gaussian measure on Brownian increments. So it is only necessary to look for an invariant measure in $\mathbb{X}$, and this is the viewpoint we take when looking at the general theory of ergodicity for SDEs.[3]

[3]Here we are implicitly assuming that $\mathbb{T} = \mathbb{R}^+$. In the case $\mathbb{T} = \mathbb{R}$ the invariant measure at $t = 0$ is correlated to the noise path for $t < 0$ [4].

2.3.2 Example Consider Example 2.2.2. We first find the invariant measure in $\mathbb{X}\times\mathbb{Y}$ where, here $\mathbb{X} = \mathbb{Y} = \mathbb{R}$. In fact the invariant measure in $\mathbb{X} \times \mathbb{Y}$ is product since the $\{\eta_n\}$ are i.i.d. To compute the invariant measure in the $\mathbb{X}$ coordinate, observe that $\{x_n\}$ is a Markov chain with Gaussian invariant measure $\mathcal{N}(0, \sigma^2/(1-\lambda^2))$. To see this we seek a Gaussian invariant measure in $\mathbb{X}$ with mean 0. Note that η_0 is independent of x_0 so that

$$\mathbb{E}|x_1|^2 = \lambda^2\mathbb{E}|x_0|^2 + \mathbb{E}|\eta_0|^2.$$

Equating $\mathbb{E}|x_1|^2$ and $\mathbb{E}|x_0|^2$ gives the variance of the invariant measure in $\mathbb{X}$. The invariant measure in $\mathbb{Y}$ is independent Gaussian $\mathcal{N}(0, \sigma^2)$.

We now lift this Gaussian invariant measure from $\mathbb{X} \times \mathbb{Y}$ to $\mathbb{X} \times \Omega$, seeking a Gaussian measure in

$$\{x_0, \{\eta_0, \eta_1, \eta_2, \dots\}\} \in \mathbb{X} \times \Omega.$$

We find that $\mathbb{E}x_0 = \mathbb{E}\eta_j = 0$, and

$$\begin{aligned} \mathbb{E}x_0^2 &= \frac{\sigma^2}{1-\lambda^2}, \qquad & \mathbb{E}\eta_j^2 &= \sigma^2 \\ \mathbb{E}x_0\eta_i &= 0 \quad \forall i, \qquad & \mathbb{E}\eta_i\eta_j &= 0 \quad \forall i \neq j. \end{aligned}$$

We have verified the variance of x_0, the η_j are i.i.d. with the stated variance, and are therefore uncorrelated, whilst the condition on $\mathbb{E}x_0\eta_i$ follows provided x_0 is chosen independently of $\omega = (\eta_0, \eta_1, \dots)$. Because of this independence it follows that the disintegration yields product measure: $\mu_\omega(dx)$ is independent of ω.

2.3.3 Example Now we generalise this approach to finding the invariant measure to Example 2.2.3. Again we start in $\mathbb{X} \times \mathbb{Y} = \mathbb{R} \times \mathbb{R}$. The pairs $\{(x_n, \eta_n)\}$ form a Markov chain and its invariant measure is Gaussian with mean 0 and covariance matrix calculated as follows. We seek σ, μ and α:

$$\eta_0 \sim \mathcal{N}(0, \sigma^2), \qquad x_0 \sim \mathcal{N}(0, \mu^2), \qquad \mathbb{E}x_0\eta_0 = \alpha.$$

Note that

$$\begin{aligned} \mathbb{E}\eta_1^2 &= \gamma^2\mathbb{E}\eta_0^2 + 1 \\ \mathbb{E}x_1\eta_1 &= \lambda\gamma\mathbb{E}x_0\eta_0 + \gamma\mathbb{E}\eta_0^2 \\ \mathbb{E}x_1^2 &= \lambda^2\mathbb{E}x_0^2 + 2\lambda\mathbb{E}x_0\eta_0 + \mathbb{E}\eta_0^2. \end{aligned}$$

Invoking invariance (equating $\mathbb{E}\eta_1^2$ with $\mathbb{E}\eta_0^2$ and so forth) gives

$$\sigma^2 = \frac{1}{(1-\gamma^2)}, \qquad \alpha = \frac{\gamma\sigma^2}{(1-\gamma\lambda)}, \qquad \mu^2 = \frac{2\lambda\alpha + \sigma^2}{(1-\lambda^2)}.$$

Eliminating α from the expression for μ gives

$$\mu^2 = \frac{(1+\gamma\lambda)\sigma^2}{(1-\gamma\lambda)(1-\lambda^2)}.$$

Thus we have an invariant measure on $\mathbb{X} \times \mathbb{Y}$. We now lift this to $\mathbb{X} \times \Omega$.

In the previous example x_0 was not correlated with $\omega = (\eta_0, \eta_1, \dots)$. In this example there is correlation and we must calculate this in order to completely specify the Gaussian invariant measure on $\mathbb{X} \times \Omega$ for the skew-product. If we let $a_j = \mathbb{E}x_0\eta_j$, then straightforward calculation shows that

$$\begin{aligned}\mathbb{E}x_1\eta_{j+1} &= \mathbb{E}[\lambda x_0 + \eta_0]\eta_{j+1} \\ &= \lambda \mathbb{E}x_0\eta_{j+1} + \mathbb{E}\eta_0\eta_{j+1}.\end{aligned}$$

Invoking invariance we see that

$$a_j = \lambda a_{j+1} + \sigma^2\gamma^{j+1}.$$

In order to obtain a probability measure we need a bounded solution (as $j \to \infty$) and, since $|\lambda| < 1$, the only bounded solution of this recursion is

$$a_j = \mathbb{E}x_0\eta_j = \frac{\sigma^2\gamma^{j+1}}{(1-\lambda\gamma)}.$$

In summary we have now found a Gaussian measure for

$$\{x_0, \{\eta_0, \eta_1, \eta_2, \dots\}\} \in \mathbb{X} \times \Omega$$

which is specified by $\mathbb{E}x_0 = \mathbb{E}\eta_j = 0$ and

$$\begin{aligned}\mathbb{E}x_0^2 &= \mu^2, \qquad & \mathbb{E}\eta_j^2 &= \sigma^2 \\ \mathbb{E}x_0\eta_i &= a_j \quad \forall i, \qquad & \mathbb{E}\eta_i\eta_j &= \sigma^2\gamma^{|i-j|} \quad \forall i \neq j.\end{aligned}$$

For this problem the disintegration is not that of product measure, due to the correlation of x_0 with $\omega = (\eta_0, \eta_1, \dots)$, manifest in non-zero a_j.

In both of these examples we have calculated the invariant measure by finding the invariant measure of a Markov chain on $\mathbb{X} \times \mathbb{Y}$, and then finding additional correlations between the solution in the fibre and the noise in the base space; this then gives the invariant measure on $\mathbb{X}\times\Omega$. This approach can also be used for studying SDEs, and their time discretisations, using a Markov process in the case of the SDE. Because the calculation of correlations between the fibre and base solutions is straightforward, but tedious, we will omit it in the developments which follow Assumption 2.3.6. The essence of the calculation is to find the invariant measure for the underlying Markov process itself in $\mathbb{X}\times\mathbb{Y}$, and we concentrate on this in our examples of SDEs.

For SDEs we note that Lemma 2.1.3 holds the key to understanding the existence of invariant measures. We are interested in finding conditions under which there is a unique solution $\bar{\rho}(x)$ of

$$\mathcal{L}^*\bar{\rho} = 0, \qquad \int_{\mathbb{R}^d} \bar{\rho}(x)\, dx = 1 \tag{2.3.1}$$

and that ρ converges to $\bar{\rho}$ in the weak-$*$ topology (identifying ρ with the measure it induces). This means that for a suitable class of functions g,

$$\int_{\mathbb{R}^d} \rho(x,t)g(x)\, dx \to \int_{\mathbb{R}^d} \bar{\rho}(x)g(x)\, dx$$

as $t \to \infty$ or, in more abstract notation,

$$\mathbb{E}g(x(t)) \to \pi(g),$$

where π is the measure with density $\bar{\rho}$.

2.3.4 Example We generalise Example 2.3.2 to continuous time by studying the Ornstein-Uhlenbeck (OU) process

$$dx = -xdt + \sigma dW, \quad x(0) = x_0. \tag{2.3.2}$$

Thus $\mathbb{X} = \mathbb{R}$, $\mathbb{T} = \mathbb{R}^+$. A straightforward calculation, using the representation of the solution as an Itô stochastic integral, shows that

$$x(t) \sim \mathcal{N}(e^{-t}y_0, \sigma^2[1 - e^{-2t}]/2),$$

indicating convergence to the Gaussian invariant measure $\mathcal{N}(0, \sigma^2/2)$ as $t \to \infty$ in $\mathbb{X}$. The invariant measure in $\mathbb{Y}$ (where the driving Brownian motion lies) is independent of that in $\mathbb{X}$ and we do not discuss it explicitly.

It is readily verified that the density associated with the Gaussian measure for $x(t)$ is a steady state solution of the Fokker-Planck equation of Lemma 2.1.3 which reduces to

$$\frac{\partial \rho}{\partial t} + \frac{\partial}{\partial x}(-x\rho) = \frac{\sigma^2}{2}\frac{\partial^2 \rho}{\partial x^2}, \quad \rho(x, 0) = \delta(x).$$

2.3.5 Example Now we generalise the Example 2.3.3 to continuous time. Consider the equations

$$\begin{aligned} dx &= [-ax + \eta]dt, \quad x(0) = x_0 \\ d\eta &= -\eta dt + \sigma dW, \quad \eta(0) = \eta_0. \end{aligned}$$

We assume $a > 0$. Here $\mathbb{X} = \mathbb{Y} = \mathbb{R}$, $\mathbb{T} = \mathbb{R}^+$. We are viewing η as the driving noise, not W.

The Fokker-Planck equation takes the form

$$\frac{\partial \rho}{\partial t} + \frac{\partial}{\partial x}\{[-ax + \eta]\rho\} + \frac{\partial}{\partial \eta}\{-\eta\rho\} = \frac{\sigma^2}{2}\frac{\partial^2 \rho}{\partial \eta^2}.$$

The previous example shows that η has Gaussian distribution and converges to a Gaussian invariant measure. Since

$$x(t) = e^{-at}x(0) + \int_0^t e^{-a(t-s)}\eta(s)\, ds$$

and η is Gaussian we deduce that x too is Gaussian. Note, however, that it is correlated to η, unlike the previous example.

These considerations suggest that we seek a steady solution of the Fokker-Planck equation in the form

$$\bar{\rho}(x, \eta) \propto \exp\{-\alpha x^2 + \beta x\eta - \gamma\eta^2\}.$$

(The constant of proportionality should be chosen so that $\bar{\rho}$ integrates to 1 on $\mathbb{R}^2$.) Substitution shows that

$$\alpha = \frac{a(a+1)^2}{\sigma^2}, \qquad \beta = \frac{2a(a+1)}{\sigma^2}, \qquad \gamma = \frac{(a+1)}{\sigma^2}$$

and the Gaussian invariant measure for $(x, \eta) \in \mathbb{X} \times \mathbb{Y}$ is found; note that x and η are correlated as $\beta \neq 0$.

To find the invariant measure for the skew-product $\{x_0, \{\eta(t)\}_{t\in\mathbb{R}^+}\} \in \mathbb{X}\times\Omega = \mathbb{R}\times\mathbb{R}^{\mathbb{R}^+}$, it is necessary to calculate how x_0 correlates with the entire path $\{\eta(t)\}_{t\in\mathbb{R}^+}$. A straightforward, but tedious, calculation shows that

$$\mathbb{E}x(t)\eta(t+\xi) = \frac{\sigma^2}{2(1+a)}\, e^{-\xi} \quad \forall t, \xi \in \mathbb{R}^+.$$

We now describe an abstract theory which is useful for finding ergodic invariant measures for SDEs where explicit solutions of the Fokker-Planck equation are not known. We look only for invariance in $\mathbb{X}$, viewing the Brownian motion evolving in $\mathbb{Y}$, with complete path in Ω, as the noise.

The following two assumptions encode two basic properties needed for ergodicity: *tightness* to ensure that probability does not leak out to infinity, and *reachability* showing that neighbourhoods of every point in phase space can be reached by appropriate choice of noise. There are two versions of the Lyapunov condition, one for continuous time $\mathbb{T} = \mathbb{R}$ or $\mathbb{R}^+$, and one for discrete time $\mathbb{T} = \mathbb{Z}$ or $\mathbb{Z}^+$.

2.3.6 Assumption (Lyapunov Condition—Continuous Time) There is a function $V : \mathbb{X} \to [1, \infty)$, with $\lim_{\|x\|\to\infty} V(x) = \infty$, and real numbers $a \in (0, \infty)$, $d \in (0, \infty)$ such that

$$\mathcal{L}\{V(x)\} \leqslant -a\{V(x)\} + d, \tag{2.3.3}$$

where $\mathcal{L}$ is the generator for (2.1.2) given by (2.1.3).

Let $\mathcal{F}_n$ denote the sub σ-algebra of all events up to time n (see [15]).

2.3.7 Assumption (Lyapunov Condition—Discrete Time) There is a function $V : \mathbb{X} \to [1, \infty)$, with $\lim_{\|x\|\to\infty} V(x) = \infty$, and real numbers $\alpha \in (0, 1)$, and $\beta \in [0, \infty)$ such that

$$\mathbb{E}[V(x_{n+1})|\mathcal{F}_n] \leqslant \alpha V(x_n) + \beta.$$

2.3.8 Assumption The Markov chain or process $\{x_t\}$, $t \in \mathbb{T}$, with transition kernel $P_t(x, A)$ satisfies, for some fixed compact set $C \in \mathcal{B}(\mathbb{X})$, the following:

(1) (*reachability*) for some $y^* \in \mathrm{int}(C)$ and for any $\delta > 0$, there is a $t_1 = t_1(\delta) \in \mathbb{T}$ such that

$$P_{t_1}\Big(x, B(y^*, \delta)\Big) > 0 \quad \forall x \in C;$$

(2) (*smoothness*) for $t \in \mathbb{T}$ the transition kernel possesses a density $p_t(x, y)$, precisely

$$P_t(x, A) = \int_A p_t(x, y)\, dy \quad \forall x \in C,\ A \in \mathcal{B}(\mathbb{X}) \cap \mathcal{B}(C),$$

and $p_t(x, y)$ is jointly continuous in $(x, y) \in C \times C$.

In what follows, we will use the shorthand notation $\|f\| \leqslant V$ to mean $\|f(x)\| \leqslant V(x)$ for all $x \in \mathbb{X}$, and define

$$\mathcal{G} = \{\text{measurable } g : \mathbb{X} \to \mathbb{R} \text{ with } \|g\| \leqslant V\}.$$

The following ergodic theorem, which follows from a straightforward development of ideas in [8, 18], will give invariant measures for a variety of SDEs and their discretisations.

2.3.9 Theorem *Let $x(t)$ denote the solution of the SDE* (2.1.2) (*resp. a discrete time Markov chain*) *with transition kernel $P_t(x, A)$. Assume that there is a $T > 0$ for which the following holds: the Markov process* (*resp. chain*) *satisfies Assumptions* 2.3.6 (*resp.* 2.3.7) *and* 2.3.8 *with C given by*

$$C = \left\{x : V(x) \leqslant \frac{2\beta}{\gamma - \alpha}\right\}$$

for some $\gamma \in (\alpha^{1/2}, 1)$. Then there exists a unique invariant measure π. Furthermore, there is $\mu(\gamma) \in (0,1)$ and $\kappa(\gamma) \in (0, \infty)$ such that for all measurable $g \in \mathcal{G}$

$$|\mathbb{E}^{x_0} g(x(t)) - \pi(g)| \leqslant \kappa e^{-\mu t} V(x_0).$$

Proof See [16]. □

2.3.10 Example Consider Example 2.2.4 which generates a random dynamical system on $\mathbb{X} = \mathbb{R}^d$. If $V(x) = \frac{1}{2}\|x\|^2 + 1$ then (2.2.1) shows that

$$\mathcal{L}V(x) = -2\beta V(x) + \alpha + \beta + \frac{1}{2}\sigma^2,$$

where $\sigma^2 = \|\Sigma\|_F^2$ and $\|\cdot\|_F$ denotes the Frobenius norm. Thus Assumption 2.3.6 holds. Assumption 2.3.8 may also be established easily if $m \geqslant d$, and we assume this here. By Theorem 2.3.9 the SDE then has a unique exponentially attracting invariant measure on $\mathbb{R}^d$ [16]; this can be lifted to an invariant measure for the skew-product on $\mathbb{R}^d \times \Omega$ but we omit the details.

The harder case $m < d$ is also studied in [16]. It is harder because the Fokker-Planck equation of Lemma 2.1.3 is no longer uniformly parabolic—there is diffusion only in m directions for a PDE in d space dimensions. This makes both smoothness and reachability impossible to establish in general: it is necessary to study the interaction of the deterministic flow (manifest in the hyperbolic part of the Fokker-Planck equation) with the noise (manifest in the directions in which parabolic behaviour is found) to find when they hold. See [16] for examples where this can be done, and for references to the literature.

2.3.11 Example Consider the SDE

$$dx = -x^3 dt + dW, \quad x(0) = X.$$

This equation satisfies (2.2.1) and is hence geometrically ergodic. In particular, for all polynomials V,

$$\mathbb{E}^{x_0} V(x(t)) \to \pi(g),$$

exponentially fast in time. In fact π is explicitly known and has density proportional to $\exp(-x^4/2)$, as can be verified from the Fokker-Planck equation. However, the Euler-Maruyama scheme (2.1.4) is not ergodic for any $\Delta t > 0$, however small. In particular, for any $\Delta t > 0$ there is x_0 such that $\mathbb{E}|x_n|^2 \to \infty$ as $n \to \infty$. Furthermore, $\mathbb{P}\{\limsup_{n\to\infty} |x_n| = \infty\} > 0$ for any $\Delta t > 0$ and any $x_0 \in \mathbb{R}^d$. (See [16] and references to related results contained therein).

This is the random analogue of our observations in Example 1.4.18. The situation here is much more severe, however, since for *any* $\Delta t > 0$, $x_0 \in \mathbb{R}^d$ there is positive probability of blow-up. For the deterministic case blow-up is avoided for fixed x_0 by reduction of Δt.

2.3.12 Example To overcome the difficulties of the previous example, consider the split-step backward Euler method (2.1.6) applied to (2.1.2) under (2.2.1) which implies Assumption 1.4.11 with $p = 1$. Since

$$x_* - \Delta t f(x_*) = x_n$$

it follows that

$$\begin{aligned}\|x_*\|^2 &= \|x_n\|^2 + 2\Delta t\langle x_*, f(x_*)\rangle - \Delta t^2\|f(x_*)\|^2 \\ &\leqslant \|x_n\|^2 + 2\Delta t(\alpha - \beta\|x_*\|^2).\end{aligned}$$

Rearranging we have that

$$\|x_*\|^2 \leqslant \frac{2\Delta t\alpha}{1+2\Delta t\beta} + \frac{1}{1+2\Delta t\beta}\|x_n\|^2,$$

and hence that

$$\|x_{n+1}\|^2 \leqslant \frac{2\Delta t\alpha}{1+2\Delta t\beta} + \frac{1}{1+2\Delta t\beta}\|x_n\|^2 + 2\langle x_*, \Sigma\Delta W_n\rangle + \|\Sigma\Delta W_n\|^2.$$

Now the ΔW_n are i.i.d. random variables with

$$\mathbb{E}\|\Sigma\Delta W_n\|^2 = \sigma^2\Delta t.$$

Thus

$$\mathbb{E}(\|x_{n+1}\|^2|\mathcal{F}_n) \leqslant \frac{1}{1+2\Delta t\beta}\|x_n\|^2 + \frac{2\Delta t\alpha}{1+2\Delta t\beta} + \sigma^2\Delta t.$$

This follows since x_* is independent of ΔW_n so that

$$\mathbb{E}(\langle x_*, \Sigma\Delta W_n\rangle|\mathcal{F}_n) = 0.$$

We have established that Assumption 2.3.7 (the Lyapunov condition) holds for this numerical method. If $m = d$ then, as for the SDE itself, reachability and smoothness are straightforward. Hence by Theorem 2.3.9, the numerical method is ergodic. Problems with $m < d$ can be studied similarly. For details see [16].

2.4 Random attractors

In the previous sections we generalised deterministic invariant sets to consider invariant measures for random dynamical systems. Attractors can also be usefully generalised to random dynamical systems. To do this it is important to observe that $\varphi(t, \theta^{-t}\omega)x_0$ denotes the solution at time 0 starting from x_0 at time $-t$. A random attractor is defined as a set $\mathcal{A}$ such that $\varphi(t, \theta^{-t}\omega)B$ approaches $\mathcal{A}$ as $t \to \infty$ for all bounded B. It has the interpretation as the set in which all solutions will lie at time $t = 0$ when the initial data at time $t = -\infty$ is in a bounded set. This kind of convergence, involving initial data at time $-\infty$, is known as *pull-back* convergence. Because of this, it is necessary in this section to consider $\mathbb{T} = \mathbb{R}$ or $\mathbb{Z}$ (and not $\mathbb{R}^+$ or $\mathbb{Z}^+$). In the deterministic case, where $\varphi = \varphi(t)$ only, the definition of random attractor coincides with the deterministic definition (Definition 1.4.3) of an attractor.

A *random closed (compact) set* is a mapping $A : \Omega \to V$, where $A(\omega)$ is closed (compact) for all $\omega \in \Omega$ and the mapping $\omega \mapsto \text{dist}(x, A(\omega))$ is measurable for each $x \in \mathbb{X}$. The mapping $U : \Omega \to V$ is a *random open* set if $U^c : \Omega \to V$ is a random closed set.

2.4.1 Definition A random set K is a *random absorbing set* if, for all bounded $B \subset X$ and with probability one, there exists $t^* = t^*(B, \omega) > 0$ such that

$$\varphi(t, \theta^{-t})B \subset K(\omega) \quad \forall t \geqslant t^*.$$

2.4.2 Definition The set $\mathcal{A}(\omega) \subset X$ is a *random global attractor* if, with probability one,

- $\mathcal{A}(\omega)$ is a random compact set;
- $\varphi(t, \omega)\mathcal{A}(\omega) = \mathcal{A}(\theta^t\omega)$ $\forall t \geqslant 0$ (invariance);
- for bounded non-random B

$$\lim_{t\to\infty} \text{dist}(\varphi(t, \theta^{-t}\omega)B, \mathcal{A}(\omega)) = 0,$$

  where dist is defined in Definition 1.4.1.

The following theorem from [1] is valid in both discrete and continuous time, with $t, T \in \mathbb{T}$.

2.4.3 Theorem [4] *If a random dynamical system has a random absorbing set, then*

$$\mathcal{A}(\omega) := \overline{\bigcup_{B\subset X} \Lambda(B, \omega)}$$

is a random global attractor, where the union is over all bounded $B \subset X$, and

$$\Lambda(B, \omega) = \bigcap_{T\geqslant 0} \overline{\bigcup_{t\geqslant T} \varphi(t, \theta^{-t}\omega)B}$$

is the ω-limit set *of* B.

The notion of attractor based on data at time $-\infty$ is not always what is needed in application. Hence the following is of interest:

2.4.4 Theorem [4] *If $\mathcal{A}$ is a random global attractor, then*

$$\operatorname{dist}(\varphi(t,\omega)B, \mathcal{A}(\theta^t\omega)) \to 0$$

in probability.

2.4.5 Example For both the Examples 2.2.2, 2.2.3 we have

$$x_{l+m} = \lambda^m x_l + \sum_{j=0}^{m-1} \lambda^j \eta_{l+m-1-j}.$$

For both these examples the random attractor is a (random) point. To see the pull-back convergence, set $l+m=0$ and $x_l = y$ to obtain

$$x_0 = \lambda^m y + \sum_{j=0}^{m-1} \lambda^j \eta_{-1-j}.$$

Letting $m \to \infty$ gives a natural guess for the random attractor, namely

$$\mathcal{A}(\omega) = \sum_{j=0}^{\infty} \lambda^j \eta_{-1-j},$$

and this infinite sum can be shown to converge, with probability one, for both examples. Furthermore, convergence of x_0 to $\mathcal{A}(\omega)$ with probability one can also be shown. Thus the random attractor is a point for each $\omega \in \Omega$.

To illustrate the forward convergence we set $l=0$, $x_0 = y$ to obtain

$$x_m = \lambda^m y + \sum_{j=0}^{m-1} \lambda^j \eta_{m-1-j}.$$

Now

$$\mathcal{A}(\theta^m\omega) = \sum_{j=0}^{\infty} \lambda^j \eta_{m-1-j}.$$

Hence

$$\begin{aligned}
|x_m - \mathcal{A}(\theta^m\omega)| &= |\lambda^m y + \sum_{j=m}^{\infty} \lambda^j \eta_{m-1-j}| \\
&= |\lambda^m| \cdot |y + \sum_{j=m}^{\infty} \lambda^{j-m} \eta_{m-1-j}| \\
&= |\lambda|^m |y + \mathcal{A}(\omega)| \\
&\to 0
\end{aligned}$$

as $m \to \infty$ with probability 1. Thus, in this particular case, the forward convergence to the random attractor is almost sure; this is stronger than the convergence in probability guaranteed by Theorem 2.4.4 and is a consequence of the simple (point) structure of the random attractor here.

Now we study the SDE (2.1.2), together with the two implicit backward Euler approximations of it. We employ Dissipativity Assumption 1.4.11. It is straightforward to see that, for continuous f, this implies that there exist $\alpha_0, \beta_0, \gamma_0, \eta_0 > 0$ and $p \geqslant 1$ such that

$$\begin{aligned} \langle x, f(x)\rangle &\leqslant \alpha_0 - \beta_0\|x\|^{2p} \qquad \forall x \in \mathbb{R}^d, \\ \|f(x)\| &\leqslant \eta_0 + \gamma_0\|x\|^{2p-1} \quad \forall x \in \mathbb{R}^d. \end{aligned}$$

Note that the first of these expressions implies that there are $\alpha_1, \beta_1 > 0$ such that

$$\langle x, f(x)\rangle \leqslant \alpha_1 - \beta_1\|x\|^2 - \frac{1}{2}\beta_0\|x\|^{2p} \quad \forall x \in \mathbb{R}^d. \tag{2.4.1}$$

A central lemma for both the continuous and discrete time cases is:

2.4.6 Lemma *Under Assumption* 1.4.11 *there exists* $C > 0$ *such that*

$$I := \langle f(v+z) + \beta_1(v+z), v\rangle \leqslant C[1 + \|z\|^2 + \|z\|^{2p}].$$

Proof From Assumptions 1.4.11 (and the implied equation (2.4.1))

$$\begin{aligned} I &= \langle f(v+z) + \beta_1(v+z), v+z\rangle - \langle f(v+z) + \beta_1(v+z), z\rangle \\ &\leqslant \alpha_1 - \frac{1}{2}\beta_0\|v+z\|^{2p} + \|f(v+z)\|\,\|z\| + \beta_1\|v+z\|\,\|z\| \\ &\leqslant \alpha_1 - \frac{1}{2}\beta_0\|v+z\|^{2p} + \eta_0\|z\| + \gamma_0\|v+z\|^{2p-1}\|z\| + \frac{\beta_1\delta^2}{2}\|v+z\|^2 + \frac{\beta_1}{2\delta^2}\|z\|^2. \end{aligned}$$

By the Young inequality, for all $\varepsilon > 0$,

$$\|v+z\|^{2p-1}\|z\| \leqslant \varepsilon\|v+z\|^{2p} + \frac{1}{2p\varepsilon^{2p-1}}\|z\|^{2p}$$

and so, by choosing ε and then δ sufficiently small, the result follows. □

2.4.7 Example We now outline how to prove the existence of a random global attractor for the SDE; the method will be mimicked for the backward Euler methods. First let $z(t), t \in \mathbb{R}$ be the stationary solution of the continuous time OU process

$$dz + \beta_1 z dt = \Sigma dW \tag{2.4.2}$$

whose solution we studied in Example 2.3.4 in the case $d = 1$.

An important observation in what follows is that z has growth which is at most logarithmic in $t \to \pm\infty$ [15]: with probability one there are constants $A, B > 0$ such that

$$\|z(t)\|^2 \leqslant A + B\log[1 + |t|]. \tag{2.4.3}$$

This generalises the law of the iterated logarithm from Brownian motion to the OU process. It is proved using the exponential Martingale inequality (see [15, Chapter 1, Theorem 7.4]). If we define $v := x - z$ then

$$\frac{dv}{dt} + \beta_1 v = f(v+z) + \beta_1(v+z). \tag{2.4.4}$$

We will show that v, and hence x, has a random absorbing set, leading to the existence of the random global attractor. The motivation for introducing v is that it satisfies a standard ODE with random right hand side, rather than an SDE. It is hence easier to directly obtain uniform in t bounds for v than for x.

2.4.8 Theorem *Let the Dissipativity Assumption* 1.4.11 *hold. Then x solving* (2.1.2) *generates a random dynamical system on $\mathbb{R}^d$ with $\mathbb{T} = \mathbb{R}$ and has a random global attractor.*

Proof The existence and uniqueness of solutions follows by application a Theorem 3.6 in Chapter 2 of [15], after noting that (2.4.1) implies that

$$\langle x, f(x)\rangle \leqslant \alpha_1 - \beta_1\|x\|^2 \quad \forall x \in \mathbb{R}^d. \tag{2.4.5}$$

By Lemma 2.4.6 we have, from (2.4.4),

$$\frac{1}{2}\frac{d}{dt}\|v\|^2 + \beta_1\|v\|^2 \leqslant C[1 + \|z\|^2 + \|z\|^{2p}].$$

Thus

$$\begin{aligned}
\|v(0)\|^2 - e^{-2\beta_1 t}\|v(-t)\|^2 &\leqslant \int_{-t}^{0} Ce^{2\beta_1 s}[1 + \|z(s)\|^2 + \|z(s)\|^{2p}]\,ds \\
&\leqslant \int_{-\infty}^{0} Ce^{2\beta_1 s}[1 + \|z(s)\|^2 + \|z(s)\|^{2p}]\,ds \\
&= R(\omega).
\end{aligned}$$

Now

$$\begin{aligned}
\|x(0)\|^2 &\leqslant 2\|v(0)\|^2 + 2\|z(0)\|^2 \\
&\leqslant 4e^{-2\beta_1 t}[\|x(-t)\|^2 + \|z(-t)\|^2] + 2R(\omega) + 2\|z(0)\|^2.
\end{aligned}$$

Note that, by (2.4.3), almost surely

$$R(\omega) < \infty.$$

Thus if $\|x(-t)\|^2 \leqslant \rho$ then, almost surely, there exists $t^*(\rho,\omega) > 0$ such that $\|x(0)\|^2 \leqslant r(\omega) < \infty$, where

$$r(\omega) := 2R(\omega) + 2\|z(0)\|^2 + \varepsilon$$

for any $\varepsilon > 0$. By Theorem 2.4.3 the proof is complete. □

2.4.9 Example We now move on to consider the approximation schemes (2.1.5) and (2.1.6). In addition to the Dissipativity Assumption 1.4.11, we also assume the one-sided Lipschitz condition (1.2.6). This enables us to prove existence and uniqueness of the random dynamical system generated by these two methods, for all Δt sufficiently small.

2.4.10 Theorem *Let Assumptions* 1.4.11 *and* 1.2.15 *hold. Then both the backward Euler method* (2.1.5) *and the split-step backward Euler method* (2.1.6) *define random dynamical systems on $\mathbb{R}^m$ with $\mathbb{T} = \mathbb{Z}$ and have a random global attractor.*

Proof The existence of a random dynamical system follows from Lemma 1.2.19 for both methods. For the backward Euler method we have

$$(1 + \Delta t\beta_1)x_{n+1} = x_n + \Delta t[f(x_{n+1}) + \beta_1 x_{n+1}] + \Sigma\Delta W_n.$$

Define Z_n to be the discrete OU process

$$(1 + \Delta t\beta_1)Z_{n+1} = Z_n + \Sigma\Delta W_n$$

and $V_n := x_n - Z_n$. Then

$$(1+\Delta t\beta_1)V_{n+1} = V_n + \Delta t[f(x_{n+1}) + \beta_1 x_{n+1}],$$

and hence, using Lemma 2.4.6,

$$\begin{aligned}(1+\Delta t\beta_1)\|V_{n+1}\|^2 &\leqslant \langle V_n, V_{n+1}\rangle + C[1+\|Z_{n+1}\|^2 + \|Z_{n+1}\|^{2p}] \\ &\leqslant \frac{1}{2}\|V_n\|^2 + \frac{1}{2}\|V_{n+1}\|^2 + C[1+\|Z_{n+1}\|^2 + \|Z_{n+1}\|^{2p}].\end{aligned}$$

Re-arranging gives

$$(1+2\Delta t\beta_1)\|V_{n+1}\|^2 \leqslant \|V_n\|^2 + 2C[1+\|Z_{n+1}\|^2 + \|Z_{n+1}\|^{2p}].$$

Using the discrete Gronwall Lemma 1.2.11 gives the desired absorbing set, and hence attractor, after using a discrete time analogue of (2.4.3); this can be drived by use of the exponential Martingale inequality (see [15, Theorem 7.4, Chapter 1]).

For the split-step backward Euler method (2.1.6),

$$\begin{aligned}(1+\Delta t\beta_1)x_* &= x_n + \Delta t[f(x_*) + \beta_1 x_*] \\ x_{n+1} &= x_* + \Sigma\Delta W_n.\end{aligned}$$

Define Z_n by

$$\begin{aligned}(1+\Delta t\beta_1)Z_* &= Z_n \\ Z_{n+1} &= Z_* + \Sigma\Delta W_n.\end{aligned}$$

If $V_n := x_n - Z_n$ and $V_* := x_* - Z_*$ then

$$(1+\Delta t\beta_1)V_* = V_n + \Delta t[f(V_*+Z_*) + \beta_1(V_*+Z_*)].$$

But $V_* = V_{n+1}$ and so the analysis now proceeds as for the backward Euler method, replacing Z_{n+1} by $Z_* = (1+\Delta t\beta_1)^{-1}Z_n$. □

2.5 Case study 1: the Langevin equation

The Langevin equation plays a central role in statistical physics. It describes how a mechanical system, with Hamiltonian $\frac{1}{2}\|p\|^2 + F(q)$, behaves when it is placed in contact with a heat bath: a larger Hamiltonian system. The overall coupled system is assumed Hamiltonian and then a variety of arguments, most using the large relative size of the heat bath, lead to an equation for p, q which includes the effect of energy exchange with the heat bath. This energy exchange is in the form of damping, from (p,q) to the bath, and noise which adds energy from the bath to (p,q), in mean square. The paper [5] gives a derivation of the equation from a simple mechanical system. Other derivations are less precise [29], invoking empirical assumptions about the effect of coupling, but have led to a model which is useful in many contexts.

Let $W(t)$ be standard d-dimensional Brownian motion, $F : \mathbb{R}^d \to \mathbb{R}$, $\sigma \in \mathbb{R}^{d\times d}$ and $\rho_i \in \mathbb{R}^d$ be the ith column of σ; we assume that the ρ_i are linearly independent so that σ is invertible. The Langevin SDE for $q, p \in \mathbb{R}^d$, the position and momenta of a particle of unit mass, is then

$$dq = p\,dt, \tag{2.5.1}$$
$$dp = -\gamma p\,dt - \nabla F(q)\,dt + \sigma\,dW. \tag{2.5.2}$$

Here $\gamma > 0$ to ensure a damped-driven Hamiltonian. In the case $d = 1$ and $\sigma = \sqrt{2}$, for example, there is a known invariant measure with density

$$\rho(p, q) \propto \exp\left\{-\gamma\left[\frac{p^2}{2} + F(q)\right]\right\}.$$

We assume that (1.5.4) holds with (1.5.5) being a prototypical example. This problem may be viewed as an analogue of Example 2.2.2 in that the noise process dW/dt is white—uncorrelated in time.

The Langevin equation (2.5.1), (2.5.2) is ergodic: the assumptions of Theorem 2.3.9 can be established, using the Lyapunov function (1.5.7) and the random analogue of (1.5.8) to establish Assumption 2.3.6. Because noise is only present in the momentum equation (2.5.2) it is not immediate that the smoothness and reachability (Assumption 2.3.8) follow, and certain vector field commutators need to be checked. See [16] for details.

A simple calculation, indicating why ergodicity might hold, is as follows. By Lemma 2.1.2

$$\mathbb{E}H(q(t), p(t) = \mathbb{E}H(q(0), p(0)) + \mathbb{E}\int_0^t \left[\frac{\sigma^2}{2} - \gamma\|p(s)\|^2\right] ds,$$

and so

$$\limsup_{t\to\infty} \frac{1}{t}\int_0^t \mathbb{E}\,\|p(s)\|^2\,ds \leqslant \frac{\sigma^2}{2\gamma}.$$

This shows that the *Césaro average* of $\mathbb{E}\,\|p(s)\|^2$ is uniformly bounded in time. In fact

$$\mathbb{E}\,\|p(t)\|^2 \to \frac{\sigma^2}{2\gamma}$$

by ergodicity [16]. The split-step backward Euler method is also ergodic because it preserves Lyapunov structure; see [16].

2.6 Case study 2: particles in a random velocity field

Random velocity fields are frequently used in turbulence modelling to give analytically tractable caricatures of complex flows. In particular they can be fitted to predictions about energy spectra. The idea was pioneered by Kraichnan [13] and a useful overview of the subject is given in Section 6 of [14]. We describe a model for the motion of particles in a random two-dimensional incompressible velocity field.

Let $x \in \mathbb{T}^2$ (in this section $\mathbb{T}^2$ denotes the two-dimensional unit torus) denote the particle position and $\dot{x} \in \mathbb{R}^2$ its velocity. We assume that

$$\tau\ddot{x} = v(x, t) - \dot{x}, \tag{2.6.1}$$

where $v = \nabla^{\perp}\psi$. Here ψ is the stream-function, $\nabla^{\perp}$ denotes skew-gradient and hence $\nabla\cdot v = 0$. Thus x denotes the position of a particle moving according to Stokes' law in a two-dimensional incompressible velocity field v. We assume that

$$\psi = \sum_{k\in\mathcal{K}} y_k \cos(k\cdot x) + z_k \sin(k\cdot x),$$

where

$$\mathcal{K} = \{2\pi(k_1, k_2), k_i \in \{1, \ldots, M\}\}$$

and

$$\begin{aligned} dy_k &= -\alpha_k y_k\, dt + \sqrt{\lambda_k}\, dB_k^y, \\ dz_k &= -\alpha_k z_k\, dt + \sqrt{\lambda_k}\, dB_k^z. \end{aligned}$$

Here the families $\{B_k^y\}_{k\in\mathcal{K}}$, $\{B_k^z\}_{k\in\mathcal{K}}$ are mutually independent families of i.i.d. standard Brownian motions. We assume that

$$\alpha := \min_{k\in cK}\{\alpha_k\} > 0.$$

The stream-function ψ may be viewed as the solution of a stochastic PDE of Ornstein-Uhlenbeck type, leading to the equation

$$d\psi + A\psi\, dt = dW,$$

where $A = -\Delta$ subject to periodic boundary conditions on $\mathbb{T}^2$, and W is a Wiener process in $L^2(\mathbb{T}^2)$ with covariance operator Q. With this notation the α_k are the eigenvalues of A whilst the λ_k are the eigenvalues of Q. We have taken M finite but infinite M can be handled.

We view ψ as the noise process in $C(\mathbb{R}, C^2(\mathbb{T}^2))$, driving the random dynamical system for particle motions in $\mathbb{T}^2 \times \mathbb{R}^2$. In this sense the problem is an analogue of Example 2.2.3 since the driving noise is coloured—correlated in time. See [23] for details of conditions on Q leading to sufficient regularity of ψ to obtain existence and uniqueness for (2.6.1).

Fig. 5 shows a typical velocity field at a fixed instant of time. (This should be compared with Fig. 3 shown earlier.) Fig. 6 shows the distribution of particles at time $t = 1$, starting from an initial configuration in which the particles are at rest and uniformly distributed on a grid.

This problem has a random attractor (see [23]), and the absorbing set calculation is very similar to that presented in the deterministic case, where $v(x,t) = v(x)$. The problem is also ergodic; again Theorem 2.3.9 can be used to establish this (see [17]).

Acknowledgments The authors are grateful to James Robinson, Hersir Sigurgeirsson, John Terry, Roger Tribe, Paul Tupper and Jon Warren for help in preparation of these notes. Hersir Sigurgeirsson created all the graphics related to Case Study 2. We are grateful to Anne Bourlioux, Martin Gander and Gert Sabidussi for giving us the opportunity, and necessary support, to enable us to give lectures based on these notes.

ARH has received support from the Engineering and Physical Sciences Research Council of the UK under grant GR/M06925, the Leverhulme Trust of the UK and the Fields Institute of Canada. AMS is supported by the Engineering and Physical Sciences Research Council of the UK under grant GR/N00340.

Figure 5: Random velocity field: (i) direction field (ii) streamlines.

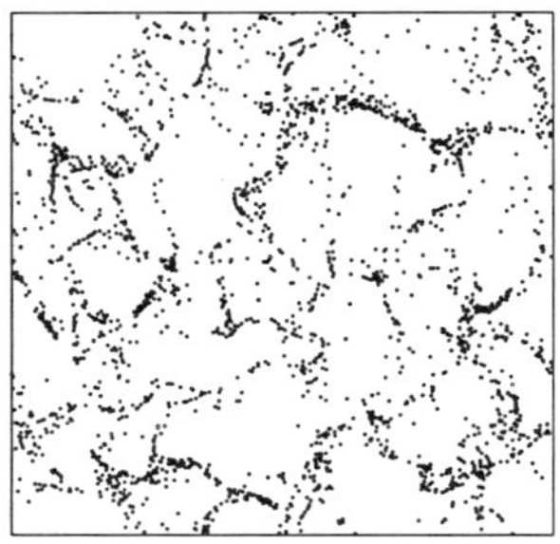

Figure 6: Particle distribution for (2.6.1).

References

[1] L. Arnold, *Random Dynamical Systems*, Springer-Verlag, New York, 1998.

[2] S. S. Artemiev and T. A. Averina, *Numerical Analysis of Systems of Ordinary and Stochastic Differential Equation*, VSP Press, Utrecht, 1997.

[3] D. Cheban, P. E. Kloeden and B. Schmalfuß, Pullback attractors in dissipative nonautonomous differential equations under discretization, *J. Dynam. Differential Equations* **13** (2001), 185–213.

[4] H. Crauel and F. Flandoli, Attractors for random dynamical systems, *Probab. Theory Related Fields* **100** (1994), 365–393.

[5] G. W. Ford and M. Kac, On the quantum Langevin equation, *J. Statist. Phys.* **46** (1987), 803–810.

[6] J. K. Hale, *Asymptotic Behaviour of Dissipative Systems*, Math. Surveys Monogr. **25**, Amer. Math. Soc., Providence, RI, 1988.

[7] P. Hartman, *Ordinary Differential Equations*, Academic Press, New York, 1978.

[8] R. Z. Has'minskii, *Stochastic Stability of Differential Equations*, Sijthoff and Noordhoff, 1980.

[9] I. Karatzas and S. E. Shreve, *Brownian Motion and Stochastic Calculus*, Springer-Verlag, New York, 1991.

[10] S. Karlin and H. M. Taylor, *A First Course in Stochastic Processes*, Academic Press, New York, 1975.

[11] P. E. Kloeden and E. Platen, *Numerical Solution of Stochastic Differential Equations*, Appl. Math. **23**, Springer-Verlag, New York, 1992.

[12] P. Kloeden and B. Schmalfuß, Nonautonomous systems, cocycle attractors and variable time-step discretization, *Numer. Algorithms* **14** (1997), 141–152.

[13] R. H. Kraichnan, Diffusion by a random velocity field, *Phys. Fluids* **13** (1970), 22–31.

[14] A. Majda and P. R. Kramer, Simplified models for turbulent diffusion: theory, numerical modelling and physical phenomena, *Phys. Rep.* **314** (1999), 237–574.

[15] X. Mao, *Stochastic Differential Equations and Applications*, Horwood, Chichester, 1997.

[16] J. Mattingly, D. J. Higham, and A. M. Stuart, Ergodicity for SDEs and approximations: locally Lipschitz vector fields and degenerate noise, preprint, 2001.

[17] J. Mattingly and A. M. Stuart, Geometric ergodicity of some hypo-elliptic diffusions for particle motions, to appear in *Markov Process. Related Fields* (2002).

[18] S. Meyn and R. L. Tweedie, *Markov Chains and Stochastic Stability*, Springer-Verlag, New York, 1992.

[19] B. Øksendal, *Stochastic Differential Equations*, Springer-Verlag, Berlin, 1998.

[20] J. C. Robinson, Stability of random attractors under perturbation and approximation, submitted to *J. Differential Equations* (2001).

[21] J. M. Sanz-Serna and M. P. Calvo, *Numerical Hamiltonian Problems*, Chapman & Hall, 1994.

[22] T. Shardlow and A. M. Stuart, A perturbation theory for ergodic Markov chains with application to numerical approximation, *SIAM J. Numer. Anal.* **37** (2000), 1120–1137.

[23] H. Sigurgeirsson and A. M. Stuart, Inertial particles in a random velocity field, submitted to *Stochastics Dynamics* (2002).

[24] A. M. Stuart and A. R. Humphries, *Dynamical Systems and Numerical Analysis*, Cambridge University Press, Cambridge, 1996.

[25] D. Talay, Second-order discretization schemes for stochastic differential systems for the computation of the invariant law, *Stochastics Stochastics Rep.* **29** (1990), 13–36.

[26] D. Talay, Approximation of upper Lyapunov exponents of bilinear stochastic differential systems, *SIAM J. Numer. Anal.* **28** (1991), 1141–1164.

[27] D. Talay, Approximation of the invariant probability measure of stochastic Hamiltonian dissipative systems with non globally Lipschitz coefficients, in: *Progress in Stochastic Structural Dynamics* (R. Bouc and C. Soize, eds.), Publ. Laboratoire Math. Appl. CNRS **152**, 1999.

[28] R. Temam, *Infinite Dimensional Dynamical Systems in Mechanics and Physics*, Appl. Math. Sci. **68**, Springer-Verlag, New York, 1989.

[29] G. E. Uhlenbeck, G. W. Ford, and E. W. Montroll, *Lectures in Statistical Mechanics*, Lectures in Appl. Math. **1**, Amer. Math. Soc., Providence, RI, 1963.

Optimal investment problems and volatility homogenization approximations

Mattias JONSSON

Department of Mathematics
University of Michigan
Ann Arbor, MI 48109–1109
USA

Ronnie SIRCAR

Operations Research & Financial Engineering Department
Princeton University
Princeton, NJ 08544
USA

Abstract

We describe some stochastic control problems in financial engineering arising from the need to find investment strategies to optimize some goal. Typically, these problems are characterized by nonlinear Hamilton-Jacobi-Bellman partial differential equations, and often they can be reduced to linear PDEs with the Legendre transform of convex duality. One situation where this cannot be achieved is in a market with stochastic volatility. In this case, we discuss an approximation using asymptotic analysis in the limit of fast mean-reversion of the process driving volatility. Simulations illustrate that marginal improvement can be achieved with this approach even when volatility is not fluctuating that rapidly.

1 Introduction

The purpose of this article to introduce to students and researchers schooled in differential equations and scientific computing some problems involving optimal investment strategies that arise in financial engineering. These can often be characterized by nonlinear second-order Hamilton-Jacobi-Bellman (HJB) equations in three or more space dimensions, so that it is difficult to solve them numerically within the time-frame that the solution is required. We discuss an approximation method based on asymptotic analysis that is effective for models in which asset prices have randomly varying volatility. This method supposes that volatility is fluctuating rapidly with respect to the timescale of the optimization problem. The upshot of the singular perturbation analysis is to approximate the changing volatility by two averages when forecasting future expected performance. One homogenized volatility is the root-mean-square average that is common in financial calculations; the second is a harmonic average that arises from the particular structure of the HJB equations in this context. We show from

A. Bourlioux and M.J. Gander (eds.), Modern Methods in Scientific Computing and Applications, 255–281.

simulation that even when the volatility is not fluctuating so rapidly, as the analysis assumes, the homogenization approximations serve to improve average performance, albeit marginally, over the competing constant volatility theory.

1.1 Background

In the financial services industry, stochastic modeling of prices has long been recognized as crucial to everyday management of risk, particularly that associated with derivative instruments. By derivatives, we refer to contracts such as options whose payoffs depend in some way on the behaviour of the price of some underlying asset (stock, commodity, exchange rate, for example) over some period of time. The canonical example, a call option on, say, a stock, pays nothing if the stock price X_T on a specified *expiration date* T in the future is less than the specified *strike price* K, and the difference $X_T - K$ otherwise. It is completely characterized by its payoff function

$$h(X_T) = (X_T - K)^+,$$

where $x^+ = \max\{x, 0\}$, the positive-part function.

Such securities with potentially unbounded payouts (for example if X_T becomes very large in the case of a call) can bring huge and rapid returns on relatively small investments if the stock moves the right way for the holder. But they are often associated with huge losses (for example for the writer of the call option). Large trading organizations have been using (to varying degrees) mathematical models for hedging their derivatives risk, at least since the Black-Scholes-Merton (BSM) methodology introduced in 1973 [2, 15].

In this context, let us remark that in some mathematical market models, it is possible to replicate the payoff of any derivative by continuous trading in the underlying asset. Such markets, which include the BSM model, are called *complete* and are much easier to analyze than the incomplete ones. In this article we will be largely interested in stochastic volatility, a feature that renders the market incomplete.

A recurrent problem in managing risks associated with derivatives positions is to maximize over suitably defined trading strategies

$$\mathbb{E}\{U(X_T, V_T)\},$$

the expectation of a *utility function* U of the future value of a controlled random *wealth* (or portfolio value) process $(V_t)_{0 \leq t \leq T}$ where the utility also depends on the (possibly vector-valued) state variable (X_t) at the future time T. In the situations of interest here, (X_t) is an asset price, possibly driven by other stochastic variables like volatility, U is a concave function of V_T, and the state-dependence of the utility function arises from risk associated with a derivative security $h(X_T)$.

The control variable, denoted (π_t), represents the amount the investor has in the stock. The rest $V_t - \pi_t$ is in the bank earning interest at rate r. The wealth process (V_t) therefore evolves according to

$$dV_t = \frac{\pi_t}{X_t}\, dX_t + r(V_t - \pi_t)\, dt, \tag{1.1}$$

which reads that the change in the portfolio value is the number of stocks held times the change in the stock price, plus interest earned on the rest in the bank. We assume the initial wealth v is given. Recent work on existence and uniqueness of optimal solutions under various models and constraints is summarized in [13].

We first give some practical examples that can be described in this framework, and then describe one approach to this problem using dynamic programming.

1.2 Examples

In a complete market, there exists an amount $v^\star$ of initial capital and a strategy $(\pi_t^\star)$ such that

$$V_T = h(X_T) \quad \text{with probability 1.}$$

In this case, $(\pi_t^\star)$ is called the replicating strategy for the derivative. Alternatively, it is the *hedging* strategy for a short position in the derivative since it is often used to cancel *exactly* (in theory) the risk of a short derivative position. The quantity $v^\star$ is called the *Black-Scholes price* of the derivative.

The problems we describe here are concerned with the gap

$$h(X_T) - V_T$$

which may not be zero with probability 1 either because the initial capital $v < v^\star$ or because of market incompleteness.

1.2.1 Partial hedging

Suppose the investor has written (sold) a derivative, but wants to insure or hedge against his future (uncertain) liability with a trading strategy in the stock. The problem is to find a strategy $(\pi_t)_{0 \le t \le T}$ such that, starting with initial capital v, the terminal value of the *hedging portfolio* V_T comes as close as possible to $h(X_T)$. We want the performance of the strategy to be penalized for falling short, with the actual size of the shortfall taken into account, but when the strategy overshoots the target, the size of the overshoot has no bearing on the measure of *risk*. This problem has been studied under various assumptions recently by Föllmer-Leukert [7] and Cvitanić and Karatzas [3], where existence and uniqueness results are established. In Section 7, we discuss *computation* of optimal strategies when volatility is changing randomly.

If we are given enough initial capital v, we can make sure that the terminal wealth V_T exceeds $h(X_T)$ with probability 1. The smallest such v is called the *superhedging price* of the derivative $h(X_T)$. Clearly, the partial hedging problem is only interesting if v is strictly smaller than the superhedging price.

For the explicit computations and examples here, the penalty function we use is

$$\mathbb{E}\left\{\frac{1}{p}((h(X_T) - V_T)^+)^p\right\},$$

the *one-sided* pth moment. The parameter $p > 1$ allows control of one's risk aversion. We will present simulations corresponding to the case $p = 1.1$ in Section 7. The limiting case $p = 1$ has a particular economic significance since the penalty function then corresponds to a coherent measure of risk (see [3]), but also leads to certain implementation difficulties. This is why we pick $p = 1.1$.

We also insist that $V_t \geq 0$ for all $0 \leq t \leq T$ almost surely for the strategy to be admissible. That is, the hedging portfolio is bounded below by zero, a portfolio constraint. The problem can be reformulated as a (state-dependent) utility maximization with

$$U(x,v) = \frac{1}{p}[h(x)^p - ((h(x)-v)^+)^p].$$

For fixed x, this is concave in v on $(0,\infty)$, strictly concave on $(0,h(x))$ and satisfies $U(x,0) = 0$, $U(x,v) = (1/p)h(x)^p$ for $v \geq h(x)$, see Fig. 3.

We call this problem *partial hedging* of the derivative risk. It is the main example that we pick to illustrate the asymptotic analysis in Section 6; it is also treated in [12].

1.2.2 Mean-square pricing

The problem is to find the initial wealth level such that the minimum variance

$$\inf_{(\pi_t)} \mathbb{E}\{(h(X_T) - V_T)^2\}$$

is minimized as a function of v. This is then defined to be the option price. In our framework, this corresponds to the utility function

$$U(x,v) = h(x)^2 - (h(x)-v)^2.$$

Such a quadratic state-dependent utility is considered in [5, 1], for example. In complete markets the expectation above is zero for the optimal v because the option can be replicated exactly with initial capital $v^\star$, the Black-Scholes price. The technique of mean-square pricing is most interesting in incomplete markets, such as markets with stochastic volatility.

1.2.3 Options based portfolio insurance

Fund managers are interested in selling trading opportunities in which the clients can take advantage of a rising market but are insured against losses in the sense that the value of the fund is bounded below by a number that is either constant or depends on a future asset price. More precisely, the client is guaranteed that the portfolio value V_T at time $T > 0$ is at least $h(X_T)$, i.e. the value of an option with payoff $h(X_T) \geq 0$. This is studied in complete markets in [6], for example, and corresponds to the utility function

$$U(x,v) = \begin{cases} \tilde{U}(v) & \text{if } v \geq h(x) \\ -\infty & \text{otherwise} \end{cases},$$

where $\tilde{U}$ is an increasing, concave (state-independent) utility function.

1.2.4 Utility-indifference pricing

In this mechanism for pricing, described in [4] for example, the (selling) price of a derivative with payoff $h(X_T)$ is defined to be the extra initial compensation the seller would have to receive so that he/she is *indifferent with respect to maximum expected utility* between having the liability of the short derivative position or just trading in the stock.

Let

$$H(v) = \sup_{\pi} \mathbb{E}\{\tilde{U}(V_T - h(X_T)) \mid \text{initial capital is } v\},$$

where $\tilde{U}$ is an increasing concave utility function and we show only the v dependence of H. Here the seller can trade in the stock, but at time T has liability $-h(X_T)$. Thus the relevant state-dependent utility function is

$$U(x, v) = \tilde{U}(v - h(x)),$$

where $\tilde{U}$ is an increasing, concave (state-independent) utility function. Let

$$M(v) = \sup_{\pi} \mathbb{E}\{\tilde{U}(V_T) \mid \text{initial capital is } v\},$$

the solution to the same problem without the derivative position. Then the (seller's) price P of the derivative security is defined by

$$H(v + P) = M(v).$$

In a complete market, this mechanism recovers the Black-Scholes price $v^\star$ which does not depend on the wealth level v of the seller. The same desirable wealth-independent property can be attained in an incomplete market by assuming an exponential utility $\tilde{U}(v) = -e^{-\gamma v}$, where γ is a risk-aversion coefficient.

2 The Merton problem

To begin our exposition of the methods for dealing with utility maximization problems, we begin with the simplest and original continuous-time stochastic control problem in finance, the Merton optimal asset allocation problem. The original reference is [14] and this is reprinted in the book [16]. Here we concentrate on maximizing expected utility at a final time horizon T, mainly because we are later interested in problems associated with derivative contracts that have a natural terminal time; other versions of the asset allocation problem consider optimizing utility of consumption over all times, possibly with an infinite horizon.

An excellent mathematical introduction to this type of stochastic control problem, as well as stochastic modelling and connections to differential equations, is the book by Øksendal [17].

2.1 The BSM model

The Black-Scholes-Merton (BSM) model takes the price of the underlying asset (for example a stock) to have a deterministic growth component (measured by an average rate of return μ) and uncertainty or risk generated (for mathematical convenience) by a Brownian motion and quantified by a volatility parameter σ. The price X_t at time t satisfies the stochastic differential equation

$$\frac{dX_t}{X_t} = \mu\, dt + \sigma\, dW_t, \tag{2.1}$$

where (W_t) is a standard Brownian motion. This model is simple and extremely tractable for a number of important problems (particularly the problem of *pricing* and hedging derivative securities), and as such, it has had enormous impact over the past thirty years. It is the primary example of a complete, continuous-time market.

An investor starts with capital v at time $t = 0$ and has the choice to invest in the stock or put his/her money in the bank earning interest at the (assumed constant) interest rate r. He continuously balances his portfolio, adjusting the weights between these two choices so as to maximize his expected utility of wealth at time T. That is, he has a utility function U that is increasing (he prefers more wealth to less) and concave (representing risk-aversion) through which he values one investment strategy over another.

If π_t is the dollar amount of stock that the investor holds at time t, we define his wealth process (V_t) by (1.1). The investor looks for a strategy $\pi_t^\star$ that maximizes

$$\mathbb{E}\{U(V_T)\}$$

starting with initial wealth v and under the constraint of nonbankruptcy: $V_t \geq 0$ for all $0 \leq t \leq T$ (almost surely).

Without loss of generality, we shall henceforth take the interest rate r to be zero. This can be justified simply by measuring all capital values *in units of the money market account*, i.e. replacing (X_t), (V_t) and (π_t) by $(X_t e^{-rt})$, $(V_t e^{-rt})$ and $(\pi_t e^{-rt})$, respectively. We recover the same equations as long as we relabel $\mu - r$ as μ. With this convention, the equations for (X_t, V_t) in terms of the control π_t are

$$\frac{dX_t}{X_t} = \mu\, dt + \sigma\, dW_t, \tag{2.2}$$

$$dV_t = \mu\pi_t\, dt + \sigma\pi_t\, dW_t, \tag{2.3}$$

where μ is now the excess growth over the risk free rate. Notice that the process (V_t) is autonomous, i.e. it does not directly involve (X_t).

2.2 Dynamic programming and the Bellman equation

One approach to this problem is to observe that if π_t were chosen to be a (nice) function of V_t, the controlled process (V_t) defined by (2.3) would be a Markov process. Therefore, *if* the optimal strategy were of this so-called *feedback form*, we could take advantage of the structure and study the evolution of the optimal expected utility as a function of the starting wealth and starting time. To this end, we define the *value function*

$$H(t,v) = \sup_{(\pi_s)_{s\in[t,T)}} \mathbb{E}\{U(V_T) \mid V_t = v\},$$

where by hypothesis, H is a function of only the starting time t and wealth v (and not, for instance, details of the path). We shall also use the shorthand

$$\mathbb{E}_{t,v}\{\cdot\} = \mathbb{E}\{\cdot \mid V_t = v\}.$$

Notice that (X_t) has vanished from the problem because everything can be stated in terms of the process (V_t). This is a direct consequence of the geometric Brownian motion model (2.2) assumed here. If μ or σ were general functions of X_t, the reduction in dimension would not be possible. In the optimization problems involving derivative contracts, X_T appears in the utility function and, again, we cannot eliminate the x variable.

The following is a loose derivation of the Bellman equation for the value function H. The idea is to divide the control process (π_s) over the the time interval $[t,T)$ into the (constant) control π_t over $[t,t+dt)$ and the rest (π_s) over $s \in [t+dt,T)$. We then optimize over these two parts separately. Formally, conditioning on the wealth at time $t+dt$ and using the iterated expectations formula, this looks like

$$\begin{aligned}
H(t,v) &= \sup_{(\pi_s)_{s\in[t,T)}} \mathbb{E}_{t,v}\{\mathbb{E}\{U(V_T) \mid V_{t+dt}\}\} \\
&= \sup_{\pi_t} \mathbb{E}_{t,v}\left\{ \sup_{(\pi_s)_{s>t}} \mathbb{E}\{U(V_T) \mid V_{t+dt}\}\right\} \\
&= \sup_{\pi_t} \mathbb{E}_{t,v}\{H(t+dt, V_{t+dt})\} \\
&= \sup_{\pi_t} \mathbb{E}_{t,v}\{H(t,v) + H_t(t,v)\,dt + \mathcal{L}_v H(t,v)\,dt + \sigma\pi_t H_v(t,v)\,dW_t\},
\end{aligned}$$

where $\mathcal{L}_v$ denotes the infinitesimal generator of (V_t), namely

$$\mathcal{L}_v = \frac{1}{2}\sigma^2\pi_t^2\frac{\partial^2}{\partial v^2} + \mu\pi_t\frac{\partial}{\partial v},$$

and we have used Itô's formula in the last step. Note that $\mathcal{L}_v$ depends on the control π, but we do not denote the dependence in this notation. The last term is zero since $\mathbb{E}\{dW_t\} = 0$. We therefore obtain

$$H_t + \sup_{\pi_t} \mathcal{L}_v H = 0. \tag{2.4}$$

Notice that the computation exploits the Markovian structure to reduce the optimization over the whole time period to successive optimizations over infinitesimal time intervals. This description of the optimal strategy for the long run, that is to do as well as one can over the short run, is called the Bellman principle.

The Bellman partial differential equation (2.4) applies in the domain $t < T$ and $v > 0$ with the terminal condition

$$H(T,v) = U(v)$$

and the boundary condition $H(t,0) = 0$ enforcing the bankruptcy constraint.

If we can find a smooth solution $H(t,v)$ to which Itô's formula can be applied, it follows from a verification theorem that H gives the maximum expected utility and the optimal strategy is given by a Markov control $\pi_t = \pi(V_t)$. For details, see [8].

In our case, the Bellman equation is

$$H_t + \sup_{\pi}\left(\frac{1}{2}\sigma^2\pi^2 H_{vv} + \mu\pi H_v\right) = 0.$$

The internal optimization is simply to find the extreme point of a quadratic. We shall assume (and it can be shown rigorously) that the value function inherits the concavity of the utility function. Moreover, under reasonable conditions, it is *strictly* concave for $t < T$ even if U is not. This follows from the diffusive part of the equation. Therefore $H_{vv} < 0$. The optimization over π is not constrained because we have not assumed any constraints on the trading strategies themselves, as long as the wealth stays positive, and therefore the quadratic is maximized by

$$\pi^\star = -\frac{\mu H_v}{\sigma^2 H_{vv}}. \tag{2.5}$$

Substituting with this π, we can rewrite the Bellman equation as

$$H_t - \frac{\mu^2}{2\sigma^2}\frac{H_v^2}{H_{vv}} = 0. \tag{2.6}$$

2.3 Solution by Legendre transform

We further take advantage of the assumed convexity of the value function to define the Legendre transform

$$\hat{H}(t,z) = \sup_{v>0}\{H(t,v) - zv\},$$

where $z > 0$ denotes the dual variable to v. The value of v where this optimum is attained is denoted by $g(t,z)$, so that

$$g(t,z) = \inf\{v > 0 \mid H(t,v) \geq zv + \hat{H}(t,z)\}.$$

The two functions $g(t,z)$ and $\hat{H}(t,z)$ are closely related and we shall refer to either one of them as the (convex) dual of H. In this article, we will work mainly with the function g, as it is easier to compute numerically and suffices for the purposes of computing optimal trading strategies. The function $\hat{H}$ is related to g by $g = -\hat{H}_z$.

At the terminal time, we denote

$$\hat{U}(x,z) = \sup\{U(v) - zv \mid 0 < v < \infty\}$$
$$G(x,z) = \inf\{v > 0 \mid U(v) \geq zv + \hat{U}(z)\}.$$

This is illustrated in Fig. 1.

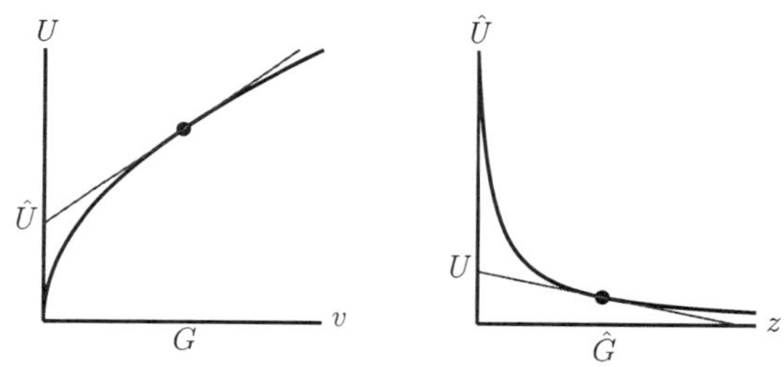

Figure 1: The Legendre transform. The left curve is the graph of $U(v)$ the utility function and the tangent line has slope z for $v = G(z)$ and has vertical intercept $\hat{U}(z)$. The right curve is the graph of $\hat{U}(z)$ and the tangent line has slope $-v$ at the point $z = \hat{G}(v)$ and has vertical intercept $U(v)$.

Assuming that H is strictly concave and smooth as a function of v, we have that

$$H_v(t, g(t,z)) = z,$$

or $g = H_v^{-1}$, whereby, in economics parlance, g is sometimes referred to as the inverse of marginal utility.

By differentiating this with respect to t and z we recover the following rules relating the derivatives of the value function to the dual function g:

$$H_{tv} = -\frac{g_t}{g_z} \qquad H_{vv} = \frac{1}{g_z} \qquad H_{vvv} = -\frac{g_{zz}}{g_z^3},$$

where the left is evaluated at $(t, g(t,z))$. Differentiating equation (2.6) with respect to v and substituting gives an autonomous equation for g:

$$g_t + \frac{\mu^2}{2\sigma^2} z^2 g_{zz} + \frac{\mu^2}{\sigma^2} z g_z = 0,$$
$$g(T,z) = G(z),$$

in $t < T$ and $z > 0$. The key observation is that this is now a *linear* PDE.

2.4 Optimal strategy

We are not usually interested in the value function, but rather in the optimal investment strategy. From (2.5), we can compute the optimal stock holding as a feedback formula in terms of derivatives of the value function. In terms of the dual function g, it is given by

$$\pi_t^\star = -\frac{\mu}{\sigma^2} z g_z(t,z).$$

Thus for a given Merton problem, we solve the linear PDE for g and recover the investment amount $\pi^\star$ from its first derivative. All that remains is to use the value of the dual variable z that corresponds to the current time and wealth level (t, v). This is obtained from the relation

$$g(t,z) = v.$$

2.5 Example: power utility

The success of the Merton approach is due primarily to an appealing explicit formula for certain simple utility functions. For an isoelastic (or power) utility function of the form

$$U(v) = \frac{v^\gamma}{\gamma}, \qquad 0 < \gamma < 1,$$

the Legendre duals G and $\hat{U}$ are given by

$$G(z) = z^{1/(\gamma-1)} \quad \text{and} \quad \hat{U}(z) = \frac{1-\gamma}{\gamma} z^{\gamma/(\gamma-1)}.$$

These are illustrated in Fig. 2. In this case, the linear PDE for g admits a separable solution of the form

$$g(t,z) = z^{1/(\gamma-1)} u(t),$$

for some function $u(t)$ we can compute. It follows that for a given (t, v),

$$g(t,z) = v$$

Figure 2: Terminal conditions for the Merton problem with power utility.

and the optimal strategy is given by

$$\pi_t^\star = -\frac{\mu}{\sigma^2} z g_z = -\frac{\mu}{\sigma^2}\frac{1}{\gamma - 1} g = \frac{\mu}{\sigma^2(1-\gamma)} v.$$

That is, the optimal strategy is to hold the fraction

$$M = \frac{\mu}{\sigma^2(1-\gamma)}$$

of current wealth in the risky asset (the stock) and to put the rest in the bank. As the stock price rises, this strategy says to sell some stock so that the fraction of the portfolio comprised of the risky asset remains the same.

The fraction M is known as the Merton ratio. More importantly, as we see next, this *fixed-mix* result generalizes to multiple securities as long as they are also assumed to be geometric Brownian motions.

2.6 The multi-dimensional Merton problem

We have a market with $n \geq 1$ stocks where the ith stock price $(X_t^{(i)})$ (in units of the money market account) is modeled by

$$\frac{dX_t^{(i)}}{X_t^{(i)}} = \mu_i \, dt + \sum_{j=1}^{n} \sigma_{ij} \, dW_t^{(j)},$$

where the μ_i are the growth rates less the risk free rate and $(W_t^{(i)})$ are independent Brownian motions. The price processes are correlated through the diffusion terms. We denote by (X_t) the vector price process, μ is the vector of return rates, and σ is the $n \times n$ volatility matrix.

An investor chooses amounts $\pi_t = (\pi_t^{(1)}, \ldots, \pi_t^{(n)})$ to invest in each stock, and puts the rest in the bank. Again (π_t) denotes the vector control process. His wealth process is

$$dV_t = \pi_t^T \mu \, dt + \pi_t^T \sigma \, dW_t,$$

where ^{T} denotes transpose. Recall that in the one-dimensional Merton problem, the stock price process disappeared from the problem. This happens in the multidimensional case, too, and the Bellman equation for the value function $H(t, v)$ becomes

$$H_t + \sup_{\pi} \left(\frac{1}{2} \pi^T \sigma \sigma^T \pi H_{vv} + \pi^T \mu H_v \right) = 0.$$

This can be transformed to a linear PDE using the Legendre transform. In the case of power utility, $U(v) = v^\gamma/\gamma$, the optimal strategy is given by

$$\pi_t^\star = \frac{1}{1-\gamma}(\sigma\sigma^T)^{-1}\mu v.$$

Thus, the optimal strategy is again to hold fixed fractions of current wealth in the risky assets (the stocks) and to put the rest in the bank. The fractions are given by the vector of Merton ratios $1/(1-\gamma)(\sigma\sigma^T)^{-1}\mu$.

3 State-dependent utility maximization: constant volatility

In this section, we describe briefly the generalization of the utility maximization problem to the case when the process modeling the price of the stock cannot be eliminated. We assume still that it is a geometric Brownian motion, but that the utility depends on its final value, as in the examples outlined in Section 1.2.

The value function is

$$H(t,x,v) = \sup_\pi \mathbb{E}\{U(X_T, V_T) \mid X_t = x, V_t = v\}$$

and it is conjectured to satisfy the Bellman equation

$$\begin{aligned} 0 &= H_t + \mu x H_x + \frac{1}{2}\sigma^2 x^2 H_{xx} + \sup_\pi \left(\pi\mu H_v + \pi\sigma^2 x H_{xv} + \frac{1}{2}\pi^2\sigma^2 H_{vv}\right) \\ &= H_t + \mu x H_x + \frac{1}{2}\sigma^2 x^2 H_{xx} - \frac{(\mu H_v + \sigma^2 H_{xv})^2}{2\sigma^2 H_{vv}}. \end{aligned} \tag{3.1}$$

3.1 Convex duals

Proceeding as in the Merton problem, we introduce the duals

$$\begin{aligned} \hat{H}(t,x,z) &= \sup\{H(t,x,v) - zv \mid 0 < v < \infty\} \\ g(t,x,z) &= \inf\{v > 0 \mid H(t,x,v) \geq zv + \hat{H}(t,x,z)\}. \end{aligned}$$

Notice that the we take the convex duals with respect to the variable v only. What is of interest is how the extra state variable x affects the transformation of the Bellman PDE. The transformation rules for the derivatives are given by

$$H_v = z \qquad \begin{aligned} H_t &= \hat{H}_t \\ H_x &= \hat{H}_x \end{aligned} \qquad \begin{aligned} H_{vv} &= -\frac{1}{\hat{H}_{zz}} \\ H_{xv} &= -\frac{\hat{H}_{xz}}{\hat{H}_{zz}} \end{aligned} \qquad H_{xx} = \hat{H}_{xx} - \frac{\hat{H}_{xz}^2}{\hat{H}_{zz}}.$$

This implies that

$$\hat{H}_{xx} = H_{xx} - \frac{H_{xv}^2}{H_{vv}}. \tag{3.2}$$

If we rewrite (3.1) as

$$H_t + \mu x H_x + \frac{1}{2}\sigma^2 x^2 \left(H_{xx} - \frac{H_{xv}^2}{H_{vv}} \right) - \mu x \frac{H_v H_{xv}}{H_{vv}} - \frac{\mu^2}{2\sigma^2}\frac{H_v^2}{H_{vv}} = 0,$$

we see that the first nonlinear term is transformed into a linear term because of (3.2), and the third is as in the Merton problem described in Section 2.3 where it became linear in the dual variables. Moreover the same happens to the second nonlinear term and we get

$$\hat{H}_t + \frac{1}{2}\sigma^2 x^2 \hat{H}_{xx} + \frac{1}{2}\frac{\mu^2}{\sigma^2} z^2 \hat{H}_{zz} + \mu x H_x - \mu x z \hat{H}_{xz} = 0. \tag{3.3}$$

Note that the first nonlinear term would not have been transformed to a linear one if the (V_t) process was not perfectly correlated with the (X_t) process. This is the situation in the stochastic volatility models we discuss later because volatility is not a traded asset.

We can recover the PDE for the other dual function $g(t, x, z)$ using

$$g = -\hat{H}_z$$

and differentiating the PDE for $\hat{H}$ with respect to z to give

$$g_t + \frac{1}{2}\sigma^2 x^2 g_{xx} + \frac{\mu^2}{\sigma^2} z g_z + \frac{1}{2}\frac{\mu^2}{\sigma^2} z^2 g_{zz} - \mu x z g_{xz} = 0. \tag{3.4}$$

3.2 Optimal strategy

Having solved the *linear* PDE for either g or $\hat{H}$, we recover the optimal strategy from

$$\begin{aligned}\pi^* &= -\frac{(\mu H_v + \sigma^2 x H_{xv})}{\sigma^2 H_{vv}} \\ &= \frac{\mu}{\sigma^2} z \hat{H}_{zz} - x \hat{H}_{xz} \\ &= x g_x - \frac{\mu}{\sigma^2} z g_z.\end{aligned}$$

The value of z to be used when the stock and wealth are (x, v) at time t are found from

$$g(t, x, z) = v.$$

3.3 Example: partial hedging

In this problem, we are partially hedging a derivative with payoff $h(X_T) \geq 0$ at time T, as explained in Section 1.2.1. We assume that v is strictly less than the superhedging price of the derivative. In the present case of constant volatility, the market is complete and the superhedging price is the Black-Scholes price $v^\star$. It is the (uniquely determined) initial capital for a replicating strategy for the derivative.

Recall that the state-dependent utility function is

$$U(x, v) = \frac{1}{p}(h^p - ((h - v)^+)^p),$$

where $h = h(x)$ and $p > 1$. This has convex duals

$$G(x,z) = \left(h - z^{1/(p-1)}\right)^+$$
$$\hat{U}(x,z) = \left(\frac{1}{p}h^p + \frac{p-1}{p}z^{p/(p-1)} - zh\right)\mathcal{H}\left(h - z^{1/(p-1)}\right),$$

where $\mathcal{H}$ is the Heaviside (step) function. See Fig. 3.

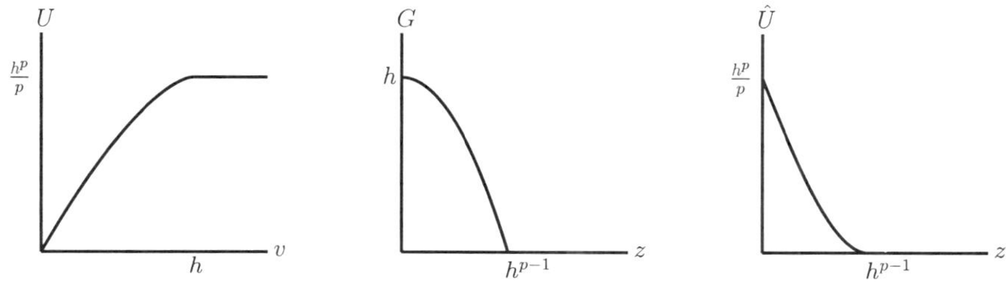

Figure 3: Terminal conditions for partial hedging.

The linear PDEs (3.3) and (3.4) have a probabilistic interpretation as the expectation of a function of a (two-dimensional) Markov process, but this can be reduced to a function of just the terminal stock price (because of the degeneracy of (3.4)). It can be shown that

$$g(t,x,z) = \mathbb{E}^\star_{t,x}\{(h(X_T) - \tilde{c}X_T^{-\kappa})^+\}, \tag{3.5}$$

where $\kappa = \mu/\sigma^2(p-1)$, and $\mathbb{E}^\star$ denotes expectation with respect to the (so-called risk-neutral) probability measure $\mathbb{P}^\star$ under which

$$dX_t = \sigma X_t dW_t^\star,$$

with $(W_t^\star)$ a $\mathbb{P}^\star$-Brownian motion, and

$$\tilde{c} = \tilde{c}(t,x,z) = zx^{-\mu/\sigma^2}\exp\left(\frac{1}{2}\left(\frac{\mu^2}{\sigma^2} - \mu\right)(T-t)\right).$$

This representation is useful in finance where linear diffusion PDEs are associated with pricing equations for derivative securities. We do not stress this interpretation here, but merely comment that the optimal strategy is to trade the stock in such a way so as to replicate the *target wealth*, which here is a European derivative contract with the modified payoff function

$$(h(X_T) - \tilde{c}X_T^{-\kappa})^+.$$

This is illustrated in Fig. 4. The number z is determined by

$$g(t,x,z) = \text{current wealth}.$$

For computational purposes, the only requirement is to solve the PDE for g or $\hat{H}$. It turns out that g is more amenable to explicit computation.

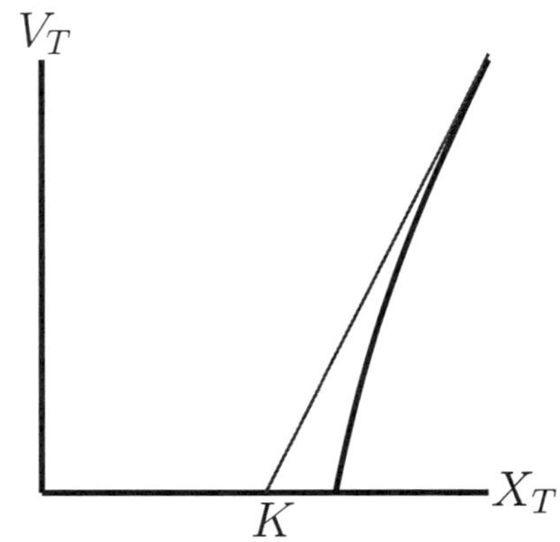

Figure 4: Target wealth for partial hedging of a European call option. The thin line is the payoff function of the original call. When partial hedging, they payoff is replaced by the thick line.

3.4 Explicit computation for a call option

Sometimes we can get closed formulas for the function $g(t,x,z)$ and its derivatives, and hence the optimal strategy in the utility maximization. This greatly increases computational speed and is the case, for instance, when we are partially hedging a call option $h(x) = (x-K)^+$. Indeed, let us define $\tilde{d}_2 = \tilde{d}_2(t,x,z)$ as the (unique) solution to

$$xe^{-\frac{1}{2}\sigma^2\tau - \sigma\sqrt{\tau}\tilde{d}_2} - K - z^{\frac{1}{p-1}} e^{\frac{\mu^2\tau}{2(p-1)\sigma^2} + \frac{\mu\sqrt{\tau}\tilde{d}_2}{(p-1)\sigma}} = 0,$$

Given K and values of t, x, z, this can be solved numerically, since the left hand side is a strictly decreasing function of $\tilde{d}_2$. We then get from (3.5):

$$\begin{aligned} g(t,x,z) &= \int_{-\infty}^{\tilde{d}_2} \left(xe^{-\frac{1}{2}\sigma^2\tau - \sigma\sqrt{\tau}\xi} - K - z^{\frac{1}{p-1}} e^{\frac{\mu^2\tau}{2(p-1)\sigma^2} + \frac{\mu\sqrt{\tau}\xi}{(p-1)\sigma}} \right) \frac{1}{\sqrt{2\pi}} e^{-\frac{1}{2}\xi^2}\, d\xi \\ &= xN(\tilde{d}_1) - KN(\tilde{d}_2) - z^{\frac{1}{p-1}} e^{\frac{p\mu^2\tau}{2(p-1)^2\sigma^2}} N(\tilde{d}_3), \end{aligned} \tag{3.6}$$

where

$$\tilde{d}_1 = \tilde{d}_2 + \sigma\sqrt{\tau} \qquad \tilde{d}_3 = \tilde{d}_2 - \frac{\mu\sqrt{\tau}}{(p-1)\sigma}.$$

This is similar to the celebrated Black-Scholes formula for the price of a call option, and in fact reduces to this as $z \to 0$.

In special cases such as $p = 2$ one can also get a closed formula for $\hat{H}(x,z)$, but we only need g in order to find the optimal strategy:

$$\begin{aligned} \frac{\pi^*(t,x,z)}{x} &= g_x - \frac{\mu z}{\sigma^2 x} g_z \\ &= N(\tilde{d}_1) + \frac{\mu z^{1/(p-1)}}{(p-1)\sigma^2 x} \exp\left(\frac{p\mu^2\tau}{2(p-1)^2\sigma^2}\right) N(\tilde{d}_3). \end{aligned}$$

This strategy was derived in [7] and we shall refer to it as the Föllmer-Leukert strategy. It is implemented in the simulations of Section 7.

4 Stochastic volatility models

The BSM model supposes that asset price volatility is constant. This is contradicted by empirical evidence of fluctuating historical volatility. More crucially, volatility as a measure of risk is seen by traders as the most important variable (after the price of the underlying asset itself) in driving probabilities of profit or loss.

Stochastic volatility models which replace σ in (2.1) by a random process (σ_t) arise not just because of empirical evidence of historical volatility's "random characteristics", but also from considerations of market hysteria, uncertainty in estimation, or they could be used to simulate non-Gaussian (heavy-tailed) returns distributions. They describe a much more complex market than the Black-Scholes model, and this is reflected in the increase in difficulty of the derivative hedging problems we shall describe. They were introduced in the academic literature in the late 1980's [11] and are popular in the industry today.

A model for stock prices in which volatility (σ_t) is a random process starts with the stochastic differential equation

$$\frac{dX_t}{X_t} = \mu\, dt + \sigma_t\, dW_t, \tag{4.1}$$

the analogue of (2.1).

A key aim in the modeling is to say as little about volatility as possible so that we are not tied to a specific model. Since volatility is not observed directly, there is a paucity of consistent econometric information about its behaviour. One feature that most empirical studies point out, and which squares with common experience, is that volatility is *mean-reverting*: it is not wandering into far-flung excursions, but seems to be pulled upwards when it is low and downwards when it is high. It is convenient to model volatility as a function of a simple mean-reverting (ergodic) Markov diffusion process (Y_t), for example an Ornstein-Uhlenbeck (OU) process. We discuss the class of models

$$\sigma_t = f(Y_t); \qquad dY_t = \alpha(m - Y_t)\, dt + \beta\, d\hat{Z}_t, \tag{4.2}$$

where f is a positive bounded function through which the generality of possible volatility models is obtained. In fact the asymptotic results for derivative pricing in [10] are insensitive to all but a few general features of f, and the way the method there is calibrated means that this function never has to be chosen.

In (4.2), $(\hat{Z}_t)$ is a Brownian motion modeling the fine-scale volatility fluctuations that is correlated with the other Brownian motion (W_t). Itô diffusions provide a simple way to model the much observed "leverage effect" that volatility and stock price shocks are negatively correlated: when volatility goes up, stock prices tend to fall. The instantaneous correlation coefficient ρ, where

$$\mathbb{E}\{dW_t\, d\hat{Z}_t\} = \rho\, dt$$

measures this asymmetry in the probability distribution of future stock prices: $\rho < 0$ generates a fatter left tail.

4.1 Fast mean-reverting stochastic volatility

The effects of *fast mean-reversion* in volatility were studied in [10]. Mean-reversion is mathematically described by ergodicity and refers to the characteristic *time* it takes for an ergodic process to get back to the mean-level of its long-run distribution. The separation of scales that we shall exploit is that while stock prices change on a tick-by-tick time-scale (minutes), volatility changes more slowly, on the scale of a few days, but still fluctuates rapidly over the time-scale (months) of a derivative contract. This phenomenon of bursty or clustering volatility is characterized as *fast mean-reversion* in the models we look at. That is, the volatility process is *fast* mean-reverting with respect to the long time-scale (months) of reference. (It is slow mean-reverting with respect to the tick time-scale, by which it is sometimes described.)

The important parameter in (4.2) is α, the rate of mean-reversion of the volatility-driving process (Y_t). Fast mean-reversion describes the limit α tending to infinity with $\beta^2/2\alpha$, the variance of the long-run distribution of the OU process, fixed. A detailed study of high-frequency $S\&P500$ data, where the large rate of volatility mean-reversion was established by a variety of methods, appears in [9]. As an illustration, Fig. 5 shows simulated volatility paths

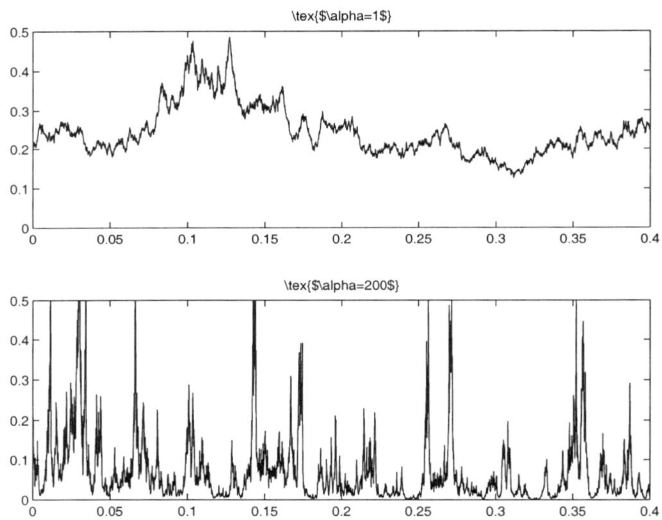

Figure 5: Simulated volatility for small ($\alpha = 1$, top) and large ($\alpha = 200$, bottom) rates of mean-reversion for the OU model, with the choice $f(y) = e^y$. Note how high volatility appears in short bursts in the latter case.

for one of the models above in which $\alpha = 1$ in the top graph and $\alpha = 200$ in the bottom graph. Of course volatility is not directly observable as a time-series, which complicates estimation issues and makes it desirable to have a theory that is robust to volatility modeling.

In Section 6, we exploit this separation of scales to construct an approximation for the state-dependent utility optimization problem under stochastic volatility, which is discussed in the next section.

5 Utility maximization under stochastic volatility

In this section, we look at the state-dependent utility maximization problem and how it is modified for a market with stochastic volatility. Under the class of mean-reverting stochastic volatility models (4.1)–(4.2), we study as before the Bellman equation of dynamic programming: the value function

$$H(t,x,y,v) = \sup_{\pi} \mathbb{E}\{U(X_T,V_T) \mid X_t = x, V_t = v, Y_t = y\}$$

is conjectured to satisfy:

$$H_t + \mathcal{L}_{x,y}H - \frac{(\mu H_v + f(y)^2 x H_{xv} + \rho\beta f(y) H_{yv})^2}{2f(y)^2 H_{vv}} = 0, \tag{5.1}$$

where

$$\mathcal{L}_{x,y}H = \mu x H_x + \alpha(m-y)H_y + \frac{1}{2}f(y)^2x^2H_{xx} + \rho\beta f(y) x H_{xy} + \frac{1}{2}\beta^2 H_{yy},$$

the infinitesimal generator of the process (X_t, Y_t). The domain is $x > 0$, $-\infty < y < \infty$, $v > 0$ and $t < T$ and the terminal condition is

$$H(T,x,y,v) = U(x,v).$$

If a smooth solution (to which Itô's formula can be applied) can be found, a verification theorem shows that the optimal strategy is given in feedback form by

$$\pi_t^\star = -\frac{\mu H_v + f(y)^2 x H_{xv} + \rho\beta f(y) H_{yv}}{f(y)^2 H_{vv}}. \tag{5.2}$$

We proceed to study the dual optimization problem for the Legendre transform (with respect to v) of the value function. Defining

$$\begin{aligned}
\hat{H}(t,x,y,z) &= \sup\{H(t,x,y,v) - zv \mid 0 < v < \infty\} \\
g(t,x,y,z) &= \inf\{v > 0 \mid H(t,x,y,v) \geq zv + \hat{H}(t,x,y,z)\},
\end{aligned}$$

the dual functions g and $\hat{H}$ satisfy a simpler-looking equation. In the case of g it reads

$$\begin{aligned}
g_t + \mathcal{L}_{x,y}g + \frac{1}{2}\frac{\mu^2}{f(y)^2}z^2 g_{zz} - \mu x z g_{xz} - \frac{\rho\beta\mu}{f(y)} z g_{yz} + \frac{\mu^2}{f(y)^2} z g_z \\
- \mu x g_x - \frac{\rho\beta\mu}{f(y)} g_y - \frac{1}{2}\beta^2(1-\rho^2)\left[\frac{2g_y g_{yz}}{g_z} - \frac{g_y^2}{g_z^2} g_{zz}\right] = 0,
\end{aligned} \tag{5.3}$$

which is still nonlinear, but only because of the last bracketed term. In the case of a complete market, meaning nonrandom ($\beta = 0$) or perfectly correlated ($|\rho| = 1$) volatility, notice that the nonlinear term disappears and the work done by the Legendre transform is to reveal that the optimization problem is simply a linear pricing problem in disguise. However, this complete reduction is not possible with stochastic volatility. Nonetheless, the transform has done some work in isolating the nonlinearity due to the fact that volatility is not traded.

6 Asymptotics for utility maximization

In this section, we study the effect of uncertain volatility on the optimal strategies for state-dependent utility maximization. We take advantage of *fast mean-reversion* and use a singular perturbation analysis to find a relatively simple trading strategy that approximates the optimal one. The analysis in Section 5 still has not yielded a way to compute the optimal strategy short of solving one of the nonlinear PDEs (5.1) or (5.3) which have three spatial dimensions. One of the benefits of the approach described here is easing of this dimensional burden. Another one is robustness—as we will see we do not need to know all the parameters in the model for the purposes of the approximate strategy.

In the zero-order approximation derived here, two kinds of average (or homogenized) volatilities emerge: $\bar{\sigma} := \langle f^2 \rangle^{1/2}$ and $\sigma_\star := \langle f^{-2} \rangle^{-1/2}$, where $\langle \cdot \rangle$ denotes a particular averaging procedure described below.

6.1 Singular perturbation analysis

We introduce the scaling

$$\alpha = 1/\varepsilon$$
$$\beta = \sqrt{2}\,\nu/\sqrt{\varepsilon},$$

where $0 < \varepsilon \ll 1$ and ν is fixed, to model fast mean-reversion (clustering) in market volatility. Recall that α measures the characteristic speed of mean-reversion of (Y_t) and ν^2 is the variance of the long-run distribution, measuring the typical size of the fluctuations of the volatility-driving process.

Then $g = g^\varepsilon$ satisfies the PDE (5.3), which we re-write with the new notation as:

$$\left(\frac{1}{\varepsilon}\mathcal{L}_0 + \frac{1}{\sqrt{\varepsilon}}\mathcal{L}_1 + \mathcal{L}_2\right) g^\varepsilon + \frac{\nu^2}{\varepsilon}(1-\rho^2)\mathrm{NL}^\varepsilon = 0, \tag{6.1}$$

where we define

$$\mathcal{L}_0 = \nu^2 \frac{\partial^2}{\partial y^2} + (m-y)\frac{\partial}{\partial y}, \tag{6.2}$$

$$\mathcal{L}_1 = \sqrt{2}\,\rho\nu \left(f(y) x \frac{\partial^2}{\partial x \partial y} - \frac{\mu}{f(y)} z \frac{\partial^2}{\partial y \partial z} - \frac{\mu}{f(y)}\frac{\partial}{\partial y}\right), \tag{6.3}$$

$$\mathcal{L}_2 = \frac{\partial}{\partial t} + \frac{1}{2} f(y)^2 x^2 \frac{\partial^2}{\partial x^2} + \frac{\mu^2}{f(y)^2}\left(\frac{1}{2} z^2 \frac{\partial^2}{\partial z^2} + z\frac{\partial}{\partial z}\right) - \mu x z \frac{\partial^2}{\partial x \partial z}, \tag{6.4}$$

and the nonlinear part is

$$\mathrm{NL}^\varepsilon = -\frac{\partial}{\partial z}\left(\frac{(g_y^\varepsilon)^2}{g_z^\varepsilon}\right) = -\left[\frac{2 g_y^\varepsilon g_{yz}^\varepsilon}{g_z^\varepsilon} - \left(\frac{g_y^\varepsilon}{g_z^\varepsilon}\right)^2 g_{zz}^\varepsilon\right].$$

Notice that $\mathcal{L}_0$ is the usual (scaled) OU generator and $\mathcal{L}_1$ takes derivatives in y and kills functions that do not depend on y.

The approach is now to think of the actual market (and our optimization problem) as embedded in a family of similar problems parametrized by (small) values of ε. For $\varepsilon = 0$, volatility is mean-reverting "infinitely fast" and can be replaced by some average as far as expectations are concerned. However two different averages are needed for different facets of the optimal strategy. This principal approximation may be sufficient for many purposes. It can be improved by perturbation or expansion around $\varepsilon = 0$, but the cost is greater reliance on model specification. We refer to [12] for details.

6.1.1 Expansion

We look for an expansion

$$g^{\varepsilon}(t,x,y,z) = g^{(0)}(t,x,y,z) + \sqrt{\varepsilon} g^{(1)}(t,x,y,z) + \varepsilon g^{(2)}(t,x,y,z) + \cdots$$

for small ε. There is a similar expansion for the other dual function $\hat{H}$ and the two are related by $\hat{H}_z = -g$.

6.1.2 Term of order $1/\varepsilon$

Inserting the expansion for g and comparing terms of order $1/\varepsilon$ gives

$$\nu^2 g_{yy}^{(0)} + (m-y) g_y^{(0)} + \nu^2(1-\rho^2)\frac{\partial}{\partial z}\left(\frac{(g_y^{(0)})^2}{g_z^{(0)}}\right) = 0.$$

An obvious solution to this is any smooth function $g^{(0)}$ that does not depend on y. It is an interesting and crucial fact that the solution $g^{(0)}$ that we are looking for is of this kind, i.e. does not depend on y. To see this, it is easier to work with the PDE for the zero-order term $\hat{H}^{(0)}$ of $\hat{H}$ (recall that $g = -\hat{H}_z$ so $g^{(0)} = -\hat{H}_z^{(0)}$). This reads

$$\nu^2 \hat{H}_{yy}^{(0)} + (m-y)\hat{H}_y^{(0)} - \nu^2(1-\rho^2)\frac{(\hat{H}_{yz}^{(0)})^2}{\hat{H}_{zz}^{(0)}} = 0. \tag{6.5}$$

Denoting $u = \hat{H}_y$ and using the concavity property $\hat{H}_{zz} > 0$ we may write (6.5) as an ordinary differential inequality:

$$\nu^2 u_y + (m-y)u \geq 0. \tag{6.6}$$

Let us define

$$\Phi(y) = \frac{1}{\sqrt{2\pi\nu^2}} e^{-(y-m)^2/2\nu^2}, \tag{6.7}$$

the density of the $\mathcal{N}(m,\nu^2)$ distribution. Using this, we can re-write (6.6) as

$$\frac{1}{\Phi(y)}\frac{\partial}{\partial y}(\Phi(y)u(y,z)) \geq 0.$$

By integrating we get

$$\begin{aligned} u(y,z) &\geq u(m,z)e^{(y-m)^2/2\nu^2} \quad \text{for} \quad y \geq m \\ u(y,z) &\leq u(m,z)e^{(y-m)^2/2\nu^2} \quad \text{for} \quad y \leq m. \end{aligned}$$

We conclude that $u(m, z) = 0$, because otherwise $u(m, y)$ would grow too fast as $y \to \pm\infty$. This implies that any solution to (6.5) is independent of y. This is because if we specify the value q of any solution at the point $y = m$,

$$\hat{H}^{(0)}(m, z) = q,$$

then the constant function

$$\tilde{H}(y, z) \equiv q \quad \text{for all} \quad y$$

is also a solution to (6.5) with $\tilde{H}_y(m, z) = 0$. From uniqueness of solutions it follows that $\hat{H}^{(0)} \equiv \tilde{H}$ and so $\hat{H}^{(0)}$ and $g^{(0)} = -\hat{H}_z^{(0)}$ do not depend on y, so

$$g^{(0)} = g^{(0)}(t, x, z).$$

6.1.3 Term of order $1/\sqrt{\varepsilon}$

At the order $1/\sqrt{\varepsilon}$,

$$\mathcal{L}_1 g^{(0)} + \mathcal{L}_0 g^{(1)} = 0,$$

which implies $g^{(1)}$ also does not depend on y because $\mathcal{L}_1 g^{(0)} = 0$ and $\mathcal{L}_0$ has null space spanned by constants. This is a general property of generators of "nice" ergodic processes like the OU.

Since both $g^{(0)}$ and $g^{(1)}$ do not depend on y, the nonlinear term is effectively

$$\mathrm{NL}^{\varepsilon} = O(\varepsilon^2)$$

and only contributes to the asymptotics when we compare order ε and higher. This fact is crucial to the further analysis. We will go as far as order one (ε^0) here, so we are dealing essentially with linear asymptotics (except for the very first equation).

6.1.4 Zero-order term

At order 1, we have

$$\mathcal{L}_0 g^{(2)} + \mathcal{L}_1 g^{(1)} + \mathcal{L}_2 g^{(0)} = 0.$$

The middle term is zero because $g^{(1)}$ does not depend on y. We have a Poisson equation (in y) for $g^{(2)}$. The solvability condition is that $\mathcal{L}_2 g^{(0)}$ must be centered with respect to the invariant distribution of the OU process (Y_t) (equivalently, orthogonal to the null space of the adjoint of $\mathcal{L}_0$, the Fredholm alternative). Therefore

$$\langle \mathcal{L}_2 g^{(0)} \rangle = \langle \mathcal{L}_2 \rangle g^{(0)} = 0, \tag{6.8}$$

where $\langle \cdot \rangle$ denotes the averaging

$$\langle \Psi \rangle = \int \Psi \Phi = \frac{1}{\sqrt{2\pi\nu^2}} \int_{-\infty}^{\infty} \Psi(y) e^{-(m-y)^2/2\nu^2}\, dy,$$

that is, the average with respect to the $\mathcal{N}(m, \nu^2)$ distribution, the invariant or long-run distribution of the OU process (Y_t).

The averaged operator is

$$\langle \mathcal{L}_2 \rangle = \frac{\partial}{\partial t} + \frac{1}{2}\bar{\sigma}^2 x^2 \frac{\partial^2}{\partial x^2} + \frac{\mu^2}{\sigma_\star^2}\left(\frac{1}{2}z^2\frac{\partial^2}{\partial z^2} + z\frac{\partial}{\partial z}\right) - \mu x z \frac{\partial^2}{\partial x \partial z},$$

where we define

$$\bar{\sigma}^2 = \langle f^2 \rangle \tag{6.9}$$

$$\frac{1}{\sigma_\star^2} = \left\langle \frac{1}{f^2} \right\rangle. \tag{6.10}$$

The terminal condition is

$$g^{(0)}(T, x, z) = G(x, z).$$

The problem for $g^{(0)}(t,x,z)$ is similar to the constant volatility problem (3.4), with two important differences:

(1) The zero-order approximation $g^{(0)}$ depends not just on the usual long-run average historical volatility $\bar{\sigma}$, but also on the *harmonically-averaged volatility* $\sigma_\star$ defined by (6.10). Thus the asymptotic approximation of the optimal strategy will depend on estimating this unusual volatility too. By Jensen's inequality, $\sigma_\star \leq \bar{\sigma}$ and equality holds if and only if volatility is constant a.s.

(2) The "homogenized" operator $\langle \mathcal{L}_2 \rangle$ is nondegenerate even though $\mathcal{L}_2$ is degenerate. As a result, $g^{(0)}(t,x,z)$ is the expectation of a functional of a *two-dimensional* Brownian motion, unlike the expectation in (3.5). In other words, the zero-order asymptotic approximation is not simply the complete market problem with constant averaged volatility.

The consequences of this are discussed in Section 6.2.

6.1.5 Zero-order strategy

The optimal zero-order strategy is given by

$$\pi^\star = \left(x\frac{\partial}{\partial x} - \frac{\mu}{f(y)^2} z \frac{\partial}{\partial z}\right) g^{(0)}. \tag{6.11}$$

Notice that this *does* depend on tracking volatility $f(y)$ even though the corrected minimum expected loss does not (to zero-order). However, it does not depend on the difficult to estimate parameters of a volatility model, namely α, β, ρ, m, nor on a specific choice of model f.

6.1.6 Interpretation and estimation of $\sigma_\star$

One possible way to estimate $\sigma_\star$ is to use the Taylor expansion

$$\begin{aligned}\frac{1}{\sigma_t^2} &= \frac{1}{\bar{\sigma}^2 + (\sigma_t^2 - \bar{\sigma}^2)} \\ &\approx \frac{1}{\bar{\sigma}^2}\left[1 - \left(\frac{\sigma_t^2 - \bar{\sigma}^2}{\bar{\sigma}^2}\right) + \left(\frac{\sigma_t^2 - \bar{\sigma}^2}{\bar{\sigma}^2}\right)^2\right],\end{aligned}$$

so that

$$\langle \frac{1}{\sigma_t^2} \rangle \approx \frac{\langle \sigma_t^4 \rangle}{\bar{\sigma}^6}.$$

The long-run volatility $\bar{\sigma}$ and the fourth-moment $\langle \sigma_t^4 \rangle$ can be estimated stably from the second and fourth-moments of high-frequency historical returns. There is no need to specify a volatility model $f(Y_t)$.

This rough estimator also shows that

$$\frac{\bar{\sigma}^2}{\sigma_\star^2} \approx \frac{\langle \sigma_t^4 \rangle}{\bar{\sigma}^4}, \tag{6.12}$$

and so $\bar{\sigma}/\sigma_\star$ is a measure of excess kurtosis.

As an indication of the type of volatility observed in the market, we plot in Fig. 6 the time-series of the VIX index produced by the CBOE, which is a measure of the daily volatility of US equity indices. It is produced from option implied volatilities and is only used here to gauge the order of volatility fluctuations. Taking this as a realization of volatility, we

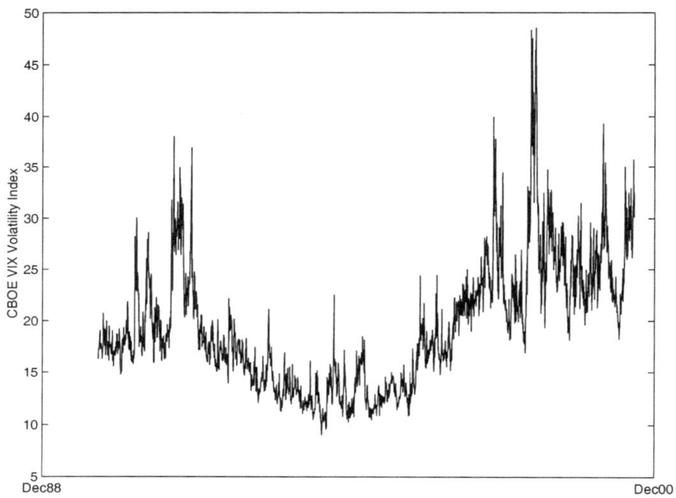

Figure 6: CBOE VIX volatility index Dec 1988–Dec 2000 (in percentage units).

estimate the two average volatilities:

$$\bar{\sigma} \approx 0.220 \qquad \sigma_\star \approx 0.177,$$

which gap describes wide fluctuations in this measure of volatility, as is seen from the picture.

6.2 Explicit computations

In order to implement the zero order strategy we need to know the function $g^{(0)}(t, x, z)$ and its derivatives with respect to x and z. While these three quantities could be computed numerically from a discretization of the linear PDE (6.8), it is quicker to use the Feynman-Kac representation (see e.g. [17]) of the solution. We refer to [12] for further details on the following argument (in the case of partial hedging). It is possible to write

$$g^{(0)}(t,x,z) = \mathbb{E}\{g(t,x,za;\bar{\sigma})\}, \tag{6.13}$$

where $g(t,x,z;\bar{\sigma})$ is the solution to the constant volatility ($=\bar{\sigma}$) problem (3.6) and the expectation is over η drawn from a standard $\mathcal{N}(0,1)$ distribution. Here

$$a = a(\eta) = e^{\frac{\mu}{\sigma_\star}\sqrt{1-\hat{\rho}^2}\sqrt{\tau}\eta + \frac{1}{2}\frac{\mu^2}{\sigma_\star^2}(1-\hat{\rho}^2)\tau},$$

where the parameter $\hat{\rho}$ is given by

$$\hat{\rho} := \frac{\sigma_\star}{\bar{\sigma}}, \tag{6.14}$$

and satisfies $0 < \hat{\rho} \leq 1$ by Jensen's inequality. This parameter can be thought of as a correlation coefficient and is a measure of how much volatility is fluctuating. In particular, $\hat{\rho} = 1$ if and only if volatility is constant a.s. In the case $f(y) = e^y$ in (4.2), known as the expOU model, for example, the approximation (6.12) turns out to be exact and $\hat{\rho} = e^{-2\nu^2}$, where $\nu^2 = \beta^2/2\alpha$ is a measure of the *size* of volatility fluctuations.

We can also use (6.13) to compute derivatives of $g^{(0)}$. For instance, we get

$$\begin{aligned} g_x^{(0)}(t,x,z) &= \mathbb{E}\{g_x(t,x,za;\bar{\sigma})\} \\ g_z^{(0)}(t,x,z) &= \mathbb{E}\{a g_z(t,x,za;\bar{\sigma})\}, \end{aligned}$$

which further yields a formula for the optimal zero-order strategy using (6.11). The main point of the formulas above is that if we have an efficient way of computing the constant-volatility solution (such as in the case of partial hedging of a call option), then we can fairly easily compute the zero-order approximation through (6.13). Specifically, we may compute $g^{(0)}(t,x,z)$ (and its derivatives) on a grid in z as weighted averages of the constant volatility solutions $g(t,x,z;\bar{\sigma})$ on a corresponding grid.

7 Simulations

In this section we present numerical results that illustrate the approximation described above. We focus on partial hedging of a call option. In particular, we demonstrate the performance of the zero-order strategy suggested by the asymptotics and the Föllmer-Leukert strategy derived in [7] for a constant volatility market (see Section 3.4). However, in the latter at every rebalancing of the portfolio, we update the level of the volatility as would be done in market practice. We assume throughout the level of volatility is perfectly observable. A reasonable approximation of current volatility can usually be obtained each day (for example from near-to-expiration implied volatilities or by averaging intraday returns), but it is harder to assemble a reasonable time-series including intraday volatility levels on which to perform detailed model calibration.

7.1 Simulation details

For the simulation, we discretized the explicit model

$$\begin{aligned}\frac{dX_t}{X_t} &= \mu\, dt + e^{Y_t}\, dW_t\\ dY_t &= \alpha(m - Y_t)\, dt + \nu\sqrt{2\alpha}\left(\rho\, dW_t + \sqrt{1-\rho^2}\, dB_t\right)\\ dV_t &= \pi_t \frac{dX_t}{X_t},\end{aligned}$$

where $r = 0.03, \mu = 0.15 - r$, $K = 100$, $\rho = -0.2$, and (W_t) and (B_t) are independent Brownian motions. In the (Y_t) process, we fixed $\bar{\sigma} = 0.1$ and $\nu = 0.25$ and ran hedging tests for various α. Changing α does not change $\bar{\sigma}$ or $\sigma_\star$ because the invariant density of the OU process does not change if we hold $\bar{\sigma}$ and ν fixed. The characteristic speed of the process varies with α, but not the characteristic size of volatility fluctuations. In this case, with f the exponential function,

$$\sigma_\star = \bar{\sigma} e^{-2\nu^2} = 0.0882.$$

We used initial stock price $X_0 = 100$ (at the money) and initial capital equal to 25% of the Black-Scholes price with constant volatility $\bar{\sigma}$.

To compute the two strategies along the path we did as follows.

- For the Föllmer-Leukert strategy, given t, X_t and V_t we solved numerically the equation $g(t, X_t, z; \bar{\sigma}) = V_t$ for z, where g satisfies (3.6). The hedging ratio $\Delta = \pi/X$ is given by

$$\Delta_t = g_x(t, X_t, z; \bar{\sigma}) - \frac{\mu}{f(Y_t)^2} \frac{z}{X_t} g_z(t, X_t, z; \bar{\sigma})$$

  and the value of the portfolio was updated using $dV_t = \Delta_t\, dX_t$. Notice that the level of volatility is updated in this hedge as would be done in practice.

- For the zero-order strategy we did as follows. Given t, X_t and V_t we solved the equation $g^{(0)}(t, X_t, z) = V_t$ for z, with $g^{(0)}$ from (6.13). The hedging ratio was then chosen as

$$\Delta_t = g_x^{(0)}(t, X_t, z) - \frac{\mu}{(\exp(Y_t)^2)} \frac{z}{X_t} g_z^{(0)}(t, X_t, z)$$

  and the value of the portfolio updated using $dV_t = \Delta_t\, dX_t$.

Throughout, we used $p = 1.1$; see Section 1.2.1 for a comment on this choice. The time horizon of the problem is $T = 0.5$ years, so we are hedging a six-month option. We used a time-discretization of 200 even intervals over this period, corresponding roughly to rehedging twice a day. This might be more frequent than in practice, but we wanted to compare the two strategies rather than deal with significant discretization issues.

7.2 Large α simulation

We first compared the strategies when the volatility *is* fast mean-reverting as assumed by the asymptotics. We set $\alpha = 200$ in annualized units which corresponds to a typical mean-reversion time of about one day. We computed estimates for the expected losses

$$L = \mathbb{E}\left\{\left(\frac{1}{p}(X_T - K - V_T)^+\right)^p\right\}$$

for the two strategies.

With 60000 paths, we got $L_1 = 2.979$ (Föllmer-Leukert) and $L_2 = 2.903$ (zero-order), so the zero-order strategy outperformed the Föllmer-Leukert strategy by approximately 2.6%. This is illustrated in Fig. 7. One natural question is whether the improvement in the expected

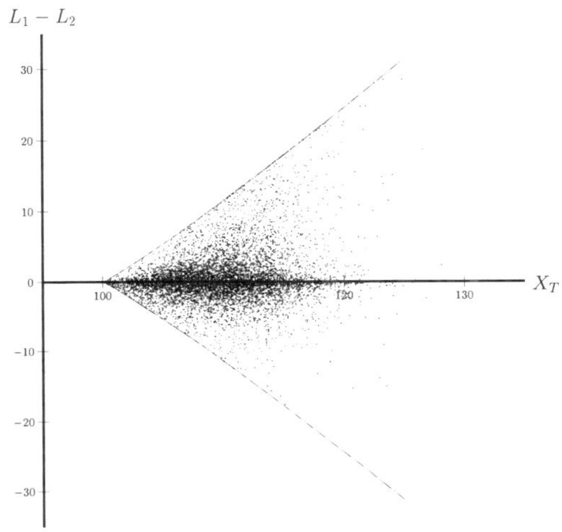

Figure 7: Difference in losses between Föllmer-Leukert strategy and zero-order strategy as a function of the terminal stock price for 60000 paths with $\alpha = 200$.

loss comes at significant cost to the variance. From these simulations, the answer is no (for the Föllmer-Leukert strategy, the standard deviation of the loss was 4.46, and it was 4.21 for the zero-order strategy.)

7.3 Small α simulation

In some markets, the rate of mean-reversion may not be so rapid. Nonetheless, the zero-order strategy may still have some effect as a crude approximation to the full stochastic volatility strategy. We compared the strategies when the volatility is mean-reverting with $\alpha = 1$ in annualized units which corresponds to a typical mean-reversion time of about one year.

With 60000 paths, we got $L_1 = 3.156$ (Föllmer-Leukert) and $L_2 = 3.111$ (zero-order), so the zero-order strategy outperformed the Föllmer-Leukert strategy by approximately 1.4%. This is illustrated in Fig. 8.

The standard deviations were 3.88 for the Föllmer-Leukert strategy and again it was lower (3.68) for the zero-order strategy.

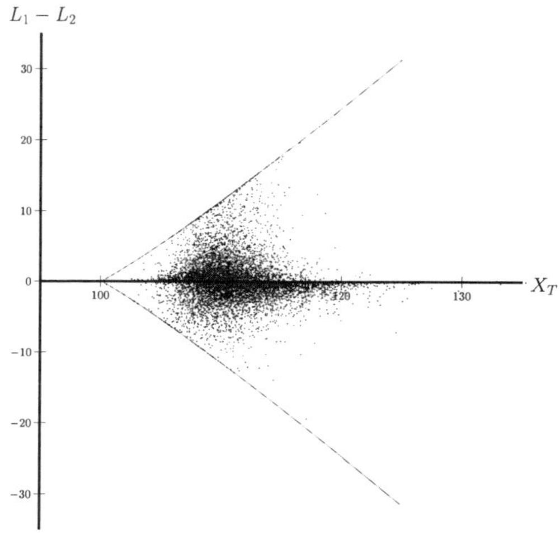

Figure 8: Difference in losses between Föllmer-Leukert strategy and zero-order strategy as a function of the terminal stock price for 60000 paths with $\alpha = 1$.

8 Conclusions

In this article we have reviewed some stochastic optimization problems that are the subject of much current research in financial mathematics. These present significant computational challenges, particularly within realistic market models such as those with stochastic volatility that we have focused on here.

Our approach is to use a homogenization approximation exploiting rapid mean-reversion in the stochastic process driving volatility. This leads to a strategy that, as demonstrated by simulation, performs better than the strategy based on assuming volatility is constant (the usual root-mean-square average $\bar{\sigma}$), but with minimal implementation overhead. The key ingredient is the harmonically averaged volatility $\sigma_\star$ which encapsulates the effect of mean-reverting stochastic volatility for these problems. Obviously, the degree of suboptimality of this approximation will be extremely model dependent, but since volatility models and parameters are not specified with much confidence in practice, this robust method is preferable to solving a high-dimensional Bellman equation whose parameters are anyway uncertain.

Additionally, as is often found with homogenization approximations, the zero-order strategy is an improvement *even outside the regime of fast mean-reversion.* This is demonstrated in the small α simulations above.

References

[1] D. Bertsimas, L. Kogan, and A. Lo, Pricing and hedging derivative securities in incomplete markets: an epsilon-arbitrage approach, technical report, MIT Sloan School,

1999.

[2] F. Black and M. Scholes, The pricing of options and corporate liabilities, *J. Political Econom.* **81** (1973), 637–659.

[3] J. Cvitanić and I. Karatzas, On dynamic measures of risk, *Finance and Stochastics* **3** (4) (1999).

[4] M. Davis, V. Panas, and T. Zariphopoulou, European pricing with transaction costs, *SIAM J. Control and Optim.* **31** (1993), 470–93.

[5] D. Duffie and H. Richardson, Mean-variance hedging in continuous time, *Ann. Appl. Probab.* **1** (1991), 1–15.

[6] N. ElKaroui, M. Jeanblanc-Picque, and V. Lacoste, Optimal portfolio management with American capital guarantee, technical report, CMAP, École Polytechnique, November 2000, preliminary version.

[7] H. Föllmer and P. Leukert, Efficient hedging: Cost versus shortfall risk, *Finance and Stochastics* **4** (2) (2000).

[8] W. H. Fleming and H. M. Soner, *Controlled Markov Processes and Viscosity Solutions*, Appl. Math. **25**, Springer-Verlag, Berlin, 1993.

[9] J.-P. Fouque, G. Papanicolaou, K. R. Sircar, and K. Solna, Mean-reversion of S&P 500 volatility, submitted, 2000.

[10] J.-P. Fouque, G. Papanicolaou, and K. R. Sircar, *Derivatives in Financial Markets with Stochastic Volatility*, Cambridge University Press, 2000.

[11] J. Hull and A. White, The pricing of options on assets with stochastic volatilities, *J. Finance* **42** (2) (1987), 281–300.

[12] M. Jonsson and K. R. Sircar, Partial hedging in a stochastic volatility environment, to appear in *Math. Finance* 2002.

[13] I. Karatzas and S. Shreve, *Methods of Mathematical Finance*, Springer-Verlag, Berlin, 1998.

[14] R. C. Merton, Lifetime portfolio selection under uncertainty: the continous-time case, *Rev. Econom. Statist.* **51** (1969), 247–257.

[15] R. C. Merton, Theory of rational option pricing, *Bell J. Econom.* **4** (1) (1973), 141–183.

[16] R. C. Merton, *Continuous-Time Finance*, Blackwell, Oxford, 1992.

[17] B. Øksendal, *Stochastic Differential Equations*, 5th ed., Springer-Verlag, Berlin, 1998.

Image processing with partial differential equations

Karol MIKULA

Department of Mathematics and Descriptive Geometry
Slovak University of Technology
813 68 Bratislava
Slovakia

Abstract

In many applications computers analyse images or image sequences which are often contaminated by noise, and their quality can be poor (e.g. in medical imaging). We discuss how nonlinear partial differential equations (PDEs) can be used to automatically produce an image of much higher quality, enhance its sharpness, filter out the noise, extract shapes, etc. The models are based on the well-known Perona-Malik image selective smoothing equation and on geometrical equations of mean curvature flow type. Since the images are given on discrete grids, PDEs are discretized by variational techniques, namely by the semi-implicit finite element, finite volume and complementary volume methods in order to get fast and stable solutions. Convergence of the schemes to variational solutions of these strongly nonlinear problems and the extension of the methods to adaptive scheme strategies improving computational efficiency are discussed. Computational results with artificial and real 2D, 3D images and image sequences are presented.

1 Introduction

The aim of this paper is to present mathematical models, numerical methods and computational results in the processing of two-dimensional (2D), three-dimensional (3D) images and image sequences. The models which we use, are based on a partial differential equations (PDEs) approach. Namely, PDEs of nonlinear (degenerate) diffusion type are applied to initially given images. From the mathematical point of view, the input processed image can be modelled by a real function $u^0(x)$, $u^0 : \Omega \to \mathbb{R}$, where $\Omega \subset \mathbb{R}^d$ represents a spatial domain. Typically Ω is rectangular and $d = 2$ or 3. In the case of an input image sequence $u^0(x, \theta)$, $u^0 : \Omega \times [0, T_A] \to \mathbb{R}$, it depends on the additional parameter θ representing a point in a real time interval of acquisition $[0, T_A]$.

Image processing operations involving PDEs are nonlinear image filtration, edge detection, deblurring and image enhancement, restoration, image inpainting, shape extraction and analysis, image segmentation, motion analysis, motion based filtering etc. [1, 59, 61, 8, 46, 55]. Typical 2D examples are given by a large variety of medical images, satellite or camera system images, old archive documents, texts pre-processed for automatic reading, old corrupted photographs or any other digital images of poor quality. 3D examples arise in bioengineering, medicine or in material quality control, where 3D volumetric aquisition methods are widely used nowadays. The processing of image sequences can be found in the restoration of movies, video sequence analysis, visual recording of growth (of human organs, leaves of plants, etc.) or

A. Bourlioux and M.J. Gander (eds.), Modern Methods in Scientific Computing and Applications, 283–321.

in improvement of the quality of medical image sequences. For example, the ultrasound acquisition of a beating heart in 3D echocardiography gives one interesting application. This article is a review of models and methods which can be used in computational image processing. The details can be found in the original papers [26, 5, 42, 22, 56, 41, 27, 6, 33, 34, 23, 24, 45].

2 Nonlinear diffusion models in image processing

The first step to use PDEs for image processing was done in the beginning of the 80s [32, 65]. By the simple observation that the Gauss function

$$G_\sigma(x) = \frac{1}{(4\pi\sigma)^{d/2}} e^{-|x|^2/4\sigma}$$

is a fundamental solution of the linear heat (diffusion) equation, it has been possible to replace the classical image processing operation—convolution of an image with G_σ with a given variance $v = \sqrt{2\sigma}$ (Gaussian smoothing)—by solving the linear heat equation for a corresponding time $t = \sigma$ with initial condition given by the processed image. It is well known that Gaussian smoothing (linear diffusion) blurs edges in the images and moves their positions. Although such a phenomenon can cause no problems in some examples of data analysis, in image processing, where the visual impression is important and a precise localization of edges is also necessary (e.g. to compute volumes of segmented objects), the linear image smoothing is generally not the best choice. A way has been found to overcome these shortcomings, namely to switch to nonlinear diffusion models.

Due to the evolutionary character of the process which controls the processing using diffusion equations, application of any PDE to an initially given image is understood as its embedding in the so-called *scale space*. In the case of nonlinear PDEs one speaks about *nonlinear scale space*. The axioms and fundamental properties of such embeddings have been given and studied in [1], where the notion of *image multiscale analysis* has been introduced. The image multiscale analysis associates with a given image $u^0(x)$ a family $u(t,x)$ of smoothed-simplified images depending on an abstract parameter $t \in [0,T]$, the *scale*. As has been proved in [1], if such a family fulfills certain basic assumptions—pyramidal structure, regularity and local comparison principle—then $u(t,x)$, $u : [0,T] \times \Omega \to \mathbb{R}$, can be represented as the unique viscosity solution (in the sense of [12]) of a general second order (degenerate) parabolic partial differential equation. This theoretical result has also an important practical counterpart. The equations of (degenerate) parabolic type have a smoothing property, so they are a natural tool for filtering (image simplification) by removing spurious structures, e.g. noise. Moreover, the simplification should be "image oriented", e.g. it should respect edges and not blur them. Or, it should recognize motion of a structure in an image sequence, and consequently the smoothing (diffusion) should respect the motion coherence in consecutive frames. Such requirements, or even more sophisticated ones related to the geometrical characteristics of the image, bring strong nonlinearity into the parabolic PDEs, and make this field interesting not only because of the applications but also from a mathematical and numerical point of view.

2.1 Anisotropic diffusion of Perona-Malik type

Since the end of the 80s, the nonlinear diffusion equations have been used for processing of 2D and 3D images. After the pioneering work of Perona and Malik [51] who modified the linear heat equation to nonlinear diffusion preserving edge positions, there has been a great deal of interest in the application and analysis of such equations. At present, the following nonlinear PDE [9] is widely used

$$u_t - \nabla.(g(|\nabla G_\sigma * u|)\nabla u) = 0, \tag{2.1}$$

where $u(t,x)$ is an unknown function defined in $Q_T \equiv [0,T]\times\Omega$. The equation is accompanied by zero Neumann boundary conditions and the initial condition

$$\frac{\partial u}{\partial \nu} = 0 \quad \text{on} \quad I \times \partial\Omega, \tag{2.2}$$

$$u(0,x) = u^0(x) \quad \text{in} \quad \Omega, \tag{2.3}$$

where ν is the unit normal vector to the boundary of Ω. We assume that $\Omega \subset \mathbb{R}^d$ is a bounded rectangular domain, $I = [0,T]$ is a scaling interval,

$$g : \mathbb{R}_0^+ \to \mathbb{R}^+ \text{ is a nonincreasing function}, \quad g(\sqrt{s}) \text{ is smooth}, \tag{2.4}$$

$$g(0) = 1, \quad \text{and we admit } g(s) \to 0 \text{ for } s \to \infty,$$

$$G_\sigma \in C^\infty(\mathbb{R}^d) \text{ is a smoothing kernel (e.g. the Gauss function)}, \tag{2.5}$$

$$\int_{\mathbb{R}^d} G_\sigma(x)\,dx = 1, \qquad \int_{\mathbb{R}^d} |\nabla G_\sigma|\,dx \le C_\sigma,$$

$$G_\sigma(x) \to \delta_x \text{ for } \sigma \to 0, \quad \delta_x \text{ is the Dirac measure at the point } x,$$

$$u^0 \in L_\infty(\Omega), \tag{2.6}$$

and

$$\nabla G_\sigma * u = \int_{\mathbb{R}^d} \nabla G_\sigma(x-\xi)\tilde{u}(\xi)\,d\xi, \tag{2.7}$$

where $\tilde{u}$ is an extension of u to $\mathbb{R}^d$. One can consider the extension of u by 0 outside Ω or the reflective periodic extension of the image [9].

The equation (2.1) represents a modification of the original Perona-Malik model [51, 29, 47]

$$u_t - \nabla.(g(|\nabla u|)\nabla u) = 0, \tag{2.8}$$

called also *anisotropic diffusion* in the computer vision community. Perona and Malik introduced (2.8) in the context of edge enhancement. The equation selectively diffuses the image in the regions where the signal has small variance in intensity in contrast with those regions where the signal changes its tendency. Such a diffusion process is governed by the shape of the diffusion coefficient given by the function g in (2.4) and by its dependence on ∇u, which is understood as an edge indicator [51]. Since $g \to 0$ for large gradients, the diffusion is strongly slowed down on edges, while outside them it provides averaging of pixel intensities as in the linear case. From a mathematical point of view, for practical choices of g (e.g. $g(s) = 1/(1+s^2)$, $g(s) = e^{-s^2}$), the original Perona-Malik equation can behave locally like the backward heat equation. It is, in general, an ill-posed problem which suffers from non-uniqueness and whose solvability is a difficult problem [29]. One way to overcome this mathematical disadvantage has been proposed by Catté, Lions, Morel and Coll in [9]. They introduced the convolution with the Gaussian kernel G_σ into the decision process for the value of the diffusion coefficient. Since convolution with the Gaussian is equivalent to linear diffusion, their model combines ideas of linear and nonlinear scale space equations. Such a slight modification made it possible to prove the existence and uniqueness of solutions for the modified equation, and to keep the practical advantages of the original formulation. Moreover, usage of the *Gaussian gradient* $\nabla G_\sigma * u$ combines the theoretical and implementation aspects of the model. The convolution (with prescribed σ) gives a unique way to compute gradients of a piecewise constant image. It also bounds (depending on σ) the gradient of the solution as input of the function g in the continuous model—which corresponds to the situation in numerical implementations where gradients evaluated on a discrete grid are finite. Also, the local edge enhancement is more understandable in the presence of noise.

Figure 1: Smoothing of the noisy image keeping the edges using anisotropic diffusion. Shown are the 0th, 10th, 20th, 30th, 40th and 50th discrete steps of a semi-implicit finite volume algorithm [41].

We present applications of the regularized Perona-Malik equation (2.1) in the following three examples. First, the artificial image (Fig. 1, 256 × 256 pixels) with additive noise is processed by our finite volume scheme [41]; then nonlinear multiscale analysis of the medical image (Fig. 2, 463 × 397 pixels) computed by the co-volume discretization [24] is given; and finally, there is an application of the 3D adaptive finite element method [5, 6] to 3D echocardiographic image of one moment of the cardiac cycle with the left ventricle in open

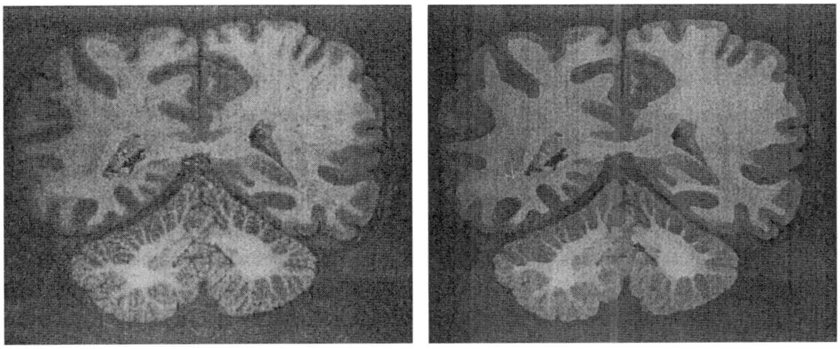

Figure 2: Initial image (left); result after 20 steps of regularized Perona-Malik filtering (right) using semi-implicit complementary volume discretization [24].

phase. In Fig. 3, one can see a visualization of the isosurface representing the boundary between blood and muscle forming an edge in 3D image intensity.

2.2 Curvature-driven equations in image processing

In rather general situations, the edges are related to level lines (level surfaces in 3D) of the image intensity function. For example, as one can see in Fig. 4, the edge representing the blood-cardiac muscle interface corresponds also to an isosurface of the greylevel image intensity function. In the three subfigures, the corresponding isolines are visible in 2D cutting slices. The isoline gives a curve in 2D plane representing the boundary of the left ventricle in the slice. The curve is non-smooth because of the errors in the acquisition. One would like to smooth it in order to remove the large acquisition errors. One way how to proceed is to move the curve (surface in 3D) in the direction of its inner normal with the velocity proportional to its (mean) curvature. The motion of convex and concave pieces is opposite due to the sign of the curvature, and the large fingers shrink much faster than the smoother parts, due to the curvature dependence of the flow. The motion by (mean) curvature is governed by a heat equation (Gaussian smoothing), but applied in the intrinsic curve (surface) geometry. In Fig. 5 we present the smoothing effect of such *geometrical diffusion*. On the left, the cuttings of unfiltered iso-surfaces are plotted, on the right the filtered ones (all after binarization with the same treshold). We can see an immediate extinction of small structures (noise) due to their high curvature, and a smoothing of the larger structures in the image.

The motion by curvature is used in numerous of applications related to the so-called free-boundary problems and interphase dynamics [59]. There exists a mathematical theory for such problems [16, 11]; also, numerical methods are available and in further development. From the computational point of view there are two main approaches for solving curve or surface evolution governed by curvature. In the so-called *Lagrangean approach* [14, 15, 43, 44, 45, 13], the curve, respectively surface, itself is discretized. Then a system of algebraic equations is derived for the new position of discrete points [14, 15, 43, 44, 31], or the equations are given for their intrinsic characteristics like curvature and tangential angle [40, 39, 45]. This system is solved to get the new curve (surface) position in the next discrete time step.

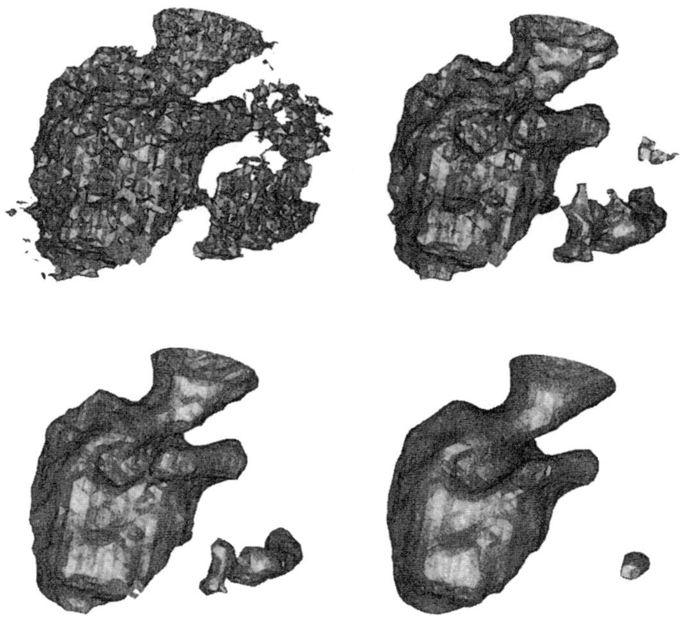

Figure 3: Smoothing of the human left ventricle by anisotropic diffusion. We visualize corresponding level surfaces in the 0th, 2nd, 4th and 8th discrete steps of the semi-implicit adaptive finite element algorithm [6].

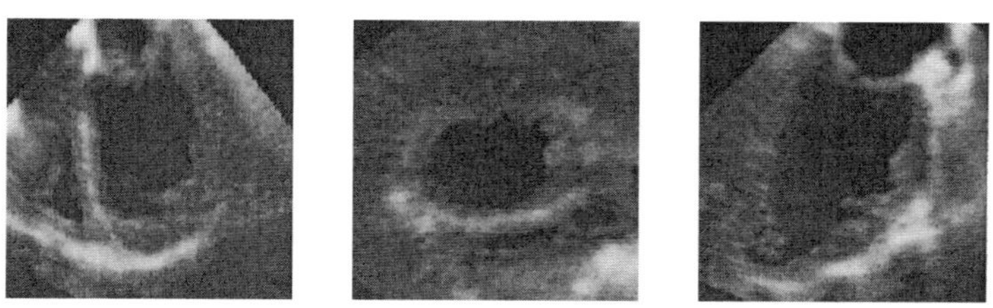

Figure 4: 2D orthogonal slices crossing in the centre of the 3D echocardiographic cube with the image of the human left ventricle.

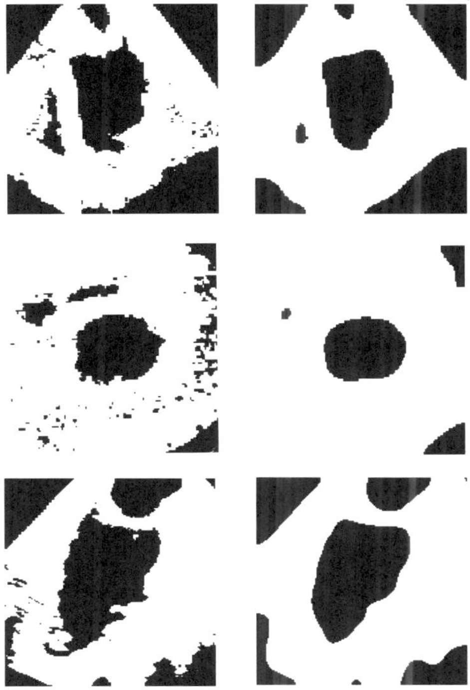

Figure 5: 2D cuts of the result of 3D processing by geometrical diffusion of mean curvature flow type.

In such a way numerically evolving curves are used e.g. in image segmentation [45]. The Lagrangean approach is an efficient and computationally fast method but, because of the parametric representation of the curves (surfaces), it can hardly handle the evolution through singularities, splitting and merging of the curves or surfaces during the evolution. In spite of that, the so-called *Eulerian approach* implicitly handles the curvature-driven motion by passing the problem to a higher dimensional space and solving there the evolution equation for a graph whose evolving level sets correspond to the evolving curve or surface. The *phase-field equations* (see e.g. [48, 7]) and *level set methods* [50, 58, 59] are approaches of that type. Especially, the *level set equation* of Osher and Sethian,

$$u_t = |\nabla u| \nabla . \left(\frac{\nabla u}{|\nabla u|} \right), \tag{2.9}$$

for curvature-dependent motion is well suited for image processing applications, since all geometrical information about image level lines or level surfaces is handled in once. As can be seen in Fig. 5 it provides denoising and silhouette smoothing at the same time. The level set equation not only moves each level set with the velocity proportional to its normal mean curvature field, but it also fulfills the so-called *morphological principle*: if u is a solution then, for any nondecreasing function φ, $\varphi(u)$ is a solution as well. This *contrast invariant* property has large significance in the axiomatic theory of image processing [1]. It also means that level sets move independently of each other; they diffuse only intrinsically, there is no diffusion across them in the normal direction. Thus it is a natural tool for directional smoothing of the image along its level lines. This idea was used in [3], where the equation

$$u_t = g(|\nabla G_\sigma * u|)|\nabla u|\nabla.\left(\frac{\nabla u}{|\nabla u|}\right) \tag{2.10}$$

has been suggested for computational image and shape analysis. It is accompanied by the same boundary and initial conditions (2.2)–(2.3) as in the case (2.1). Equation (2.10) can be used for image silhouettes smoothing ($g \equiv 1$, see e.g. [1, 2, 42, 22]), or it can be used for edge-preserving smoothing in a similar way as equation (2.1). The Perona-Malik function $g(s)$ depending on $|\nabla G_\sigma * u|$ is used to strongly slow down the motion of silhouettes which are at the same time edges. The regions between them are smoothed by the mean curvature flow.

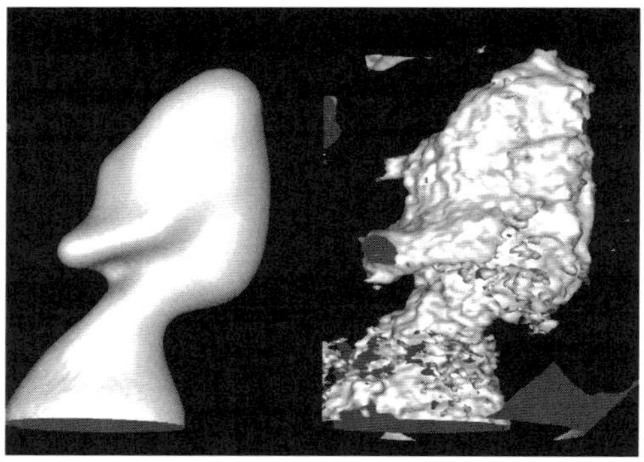

Figure 6: Ventricular shape extraction using the level set equation [42].

In Fig. 6 we present smoothing of the left ventricular iso-surface by the level set equation. In Fig. 7 we smooth an initial 321×373 pixel image (ancient coat-of-arms of the Slovak town Kremnica shown on the left), scanned from a book with neither paper nor colors of good quality. Also shown are the results after 5 and 10 discrete scale steps of the co-volume scheme for equation (2.10) with $g(s) = 1/(1+s^2)$. We also present Fig. 8, where two chromosomes are extracted from an initial noisy 3D image of a human cell by image selective smoothing (2.10) with the same g.

To end of this section, let us mention a general useful use of viewing gradient-dependent nonlinear diffusion equations. The nonlinear diffusion term $\nabla.(g(|\nabla u|)\nabla u)$ can be rewritten in 2D as

$$\nabla.(g(|\nabla u|)\nabla u) = g(|\nabla u|)u_{\xi\xi} + H'(|\nabla u|)u_{\eta\eta},$$

where $H(s) = sg(s)$ and ξ, η are tangential and orthogonal vectors to the level line, respectively. From this form one can clearly see how diffusion works along and across the image

Figure 7: Initial image (left); result of smoothing after 5 (middle) and 10 (right) scale steps using equation (2.10) and co-volume discretization [23].

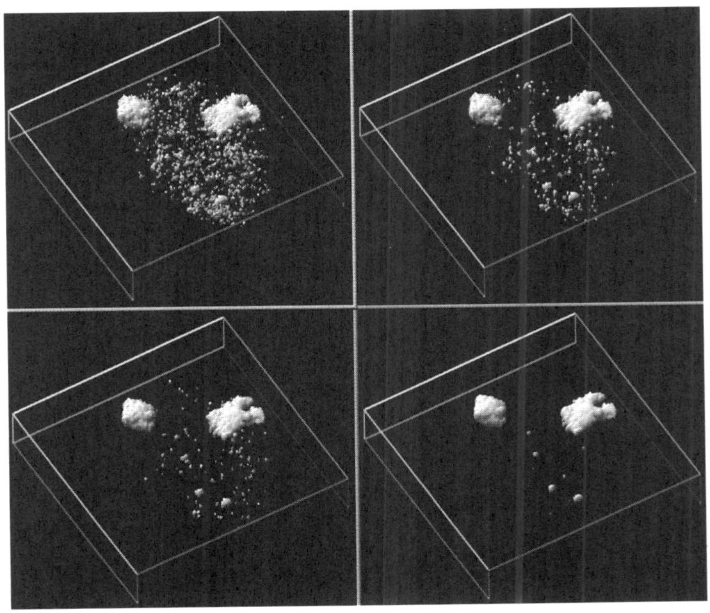

Figure 8: Extraction of two chromosomes in a human cell using geometrical diffusion (2.10) [22].

silhouettes with different choices of g. There is always positive, but possibly strongly slowed-down (depending on the shape of g) diffusion *along* level lines. *Across* level lines there can be forward diffusion (when $H'(s)$ is positive), zero diffusion (e.g. in the Rudin-Osher-Fatemi model [53] dealing with total variation denoising, and also in the mean curvature flow equation in the level set formulation), or backward diffusion (in the original Perona-Malik model [51, 47, 29]).

2.3 Some generalized models

The following generalization of the Perona-Malik equation has been introduced in [27]:

$$\partial_t b(x,u) - \nabla.(g(|\nabla G_\sigma * \beta(x,u)|)\nabla\beta(x,u)) = f(u^0 - u). \tag{2.11}$$

The functions b and β represent new nonlinearities which make the image multiscale analysis locally dependent on values of the intensity function u and on the position in the image x. Such a generalization is useful in any situation where properties of the image or requirements to the image processing operation can be expressed in dependence on x and u. For example, if a different speed of the diffusion process is desirable in different parts of the image or for different ranges of the intensity function, then equation (2.11) can be used. In the points, where the derivative β'_u is small (b'_u is large), the diffusion process is slowed down, while where β'_u is large (b'_u is small) the diffusion process is speeded up. Degenerate cases from the point of view of the theory of parabolic PDEs, when either β'_u or b'_u is equal 0 or ∞, can also be included. The degenerate cases can be interpreted as total stopping of diffusion, or as diffusion with the infinite speed in some image regions. Applying the regularized anisotropic diffusion (2.1) improves some set of edges. On the other hand, it destroys details which are under the edge threshold (given by g) or undistinguished from the noise at some scale. If such details are contained in certain ranges of greylevels, then they can be conserved by a special choice of the function β or b. As a demonstration we present Fig. 9. In that image, the colors of Flora's face are damaged only. We present the reconstruction of the original (left image) by anisotropic diffusion accompanied with the *slow diffusion effect* (image on the right). Using the proper choice of β (b is linear), which is constant for darker (lower) greylevels and linear for the upper range of u, the face is selectively smoothed and the details around it are conserved. For theory and numerical algorithm we refer to [27]. The right-hand side of (2.11) with nondecreasing function f can be used to force the solution to be close to original u^0 [49]. The Lipschitz continuous f causes no important difference in numerical analysis compared to the zero right-hand side.

Weickert (see e.g. [61, 63]) introduced a generalization of the Perona-Malik equation of the form

$$u_t - \nabla.(D\nabla u) = 0, \tag{2.12}$$

where D is a matrix depending on the eigenvalues and eigenvectors of the so-called *structure tensor* $\nabla u(\nabla u)^T$. The dependence is such that diffusion strongly prefers direction of line structures in the image. That idea has also been used by Preusser and Rumpf in multiscale flow field visualization in computational fluid dynamics [52].

There exist generalizations of the basic equations from the previous two subsections to the processing of color images. An RGB image can be viewed as a composition of three greyscale

Figure 9: Processing of a color image using slowed anisotropic diffusion [27].

images representing the levels of intensity for red, green and blue colors. Then it is natural to consider a Perona-Malik-type system of equations adapted to the RGB image. The idea is not to apply the Perona-Malik anisotropic diffusion equation to each channel independently (which would be the simplest possibility), but to synchronize the diffusion in each channel by computing a common diffusion coefficient depending on the information coming from all three colors. In [64, 63, 34], dealing with color images and vector valued diffusion, the following system of nonlinear partial differential equations has been considered:

$$\partial_t u_i - \nabla.(d \nabla u_i) = 0, \quad i = 1, 2, 3, \tag{2.13}$$

$$d = g\left(\sum_{j=1}^{3} |\nabla G_\sigma * u_j|\right). \tag{2.14}$$

The equations are accompanied by initial and zero Neumann boundary conditions for each color. In the case (2.13)–(2.14), the edges of a highly destroyed channel can be recovered by information coming from the remaining channels [34]. A geometrical model based on minimal surfaces and curvature-driven motion for processing of color images is given in [30].

2.4 Space-time filtering algorithm

A given 3D space-time image sequence (e.g. in 3D echocardiography) $u^0(x,\theta)$ is a special 4D image where a motion coherence of subsequent frames is present. Usually, the aim is to extract relevant motion information from the sequence, filter out the noise, and enhance moving structures. To that end, it seems reasonable to use additional information (in comparison with still image processing) given by the motion correspondence in the image sequence.

We assume that certain objects acquired at different times, and thus being in different frames of the sequence, are formed by points that preserve their intensity along the motion trajectory. Such objects are called *Lambertian structures.* Moreover, we assume that motion is smooth in time, and thus the motion trajectories are close to straight lines locally. Designing the model we consider the following quantity [1, 2] proposed by Guichard [19]:

$$\begin{aligned} clt_u(t,x,\theta) = \min_{w_1,w_2} \frac{1}{(\Delta\theta)^2}(\Delta\theta\ &|\langle \nabla u(t,x,\theta), w_1 - w_2\rangle| \\ &+ |u(t,x-w_1,\theta-\Delta\theta) - u(t,x,\theta)| \\ &+ |u(t,x+w_2,\theta+\Delta\theta) - u(t,x,\theta)|) \end{aligned} \tag{2.15}$$

where w_1, w_2 are arbitrary vectors in $\mathbb{R}^N$ and $\Delta\theta$ is a time increment. The scalar function clt_u will introduce a measure of coherence in time for the moving structures. It consists of the sum of three positive parts and we want to find the minimum in all possible directions w_1, w_2. The last two terms in the sum on the right-hand side of (2.15) are related to the differences in the intensities of end-points of the candidate Lambertian velocity vectors w_1, w_2. To find the directions of such vectors, we look at the points whose intensity value is closest to the intensity $u(t,x,\theta)$ in the previous frame (term $|u(t,x-w_1,\theta-\Delta\theta) - u(t,x,\theta)|$) and in the next frame (term $|u(t,x+w_2,\theta+\Delta\theta) - u(t,x,\theta)|$). Those differences are scaled by the factor $1/(\Delta\theta)^2$. Note that if we find corresponding Lambertian points, then both terms vanish. The first term in the sum, namely $|\langle \nabla u(t,x,\theta), w_1 - w_2\rangle|/(\Delta\theta)$, corresponds to the so-called *apparent acceleration*, i.e. to the difference between the candidate Lambertian velocity vectors w_1 and w_2 in the direction of ∇u. For details and some more background from the optical flow point of view we refer to [1, 2, 19]. The quantity clt_u is thus related to the curvature of the space-time level curve passing through the space-time point (x,θ) in the scale t (curvature of the Lambertian trajectory). The value of clt_u vanishes for the Lambertian points that are in uniform motion. This is consistent with the purpose not to alter such trajectories. On the other hand, for the noisy points there is no motion coherence and thus clt_u will be large there.

Concerning the space coherence, we assume that distinguished structures are located in the regions with a certain mean value of the image intensity, and that the object boundary forms an edge in the image. In order to construct a spatial diffusion process we thus require specific behavior on the edges as in the Perona-Malik anisotropic diffusion equation.

To combine time coherence of moving objects with their spatial localization we consider the following equation for the processing of image sequences [56]:

$$u_t = clt_u \nabla.(g(|\nabla G_\sigma * u|)\nabla u). \tag{2.16}$$

As an application, we are dealing with a phantom-like image sequence consisting of expanding, slightly deforming and moving ellipses with the inner structure in the form of quatrefoils. We add impulsive (salt & pepper), Gaussian noise and blurring to the frames of

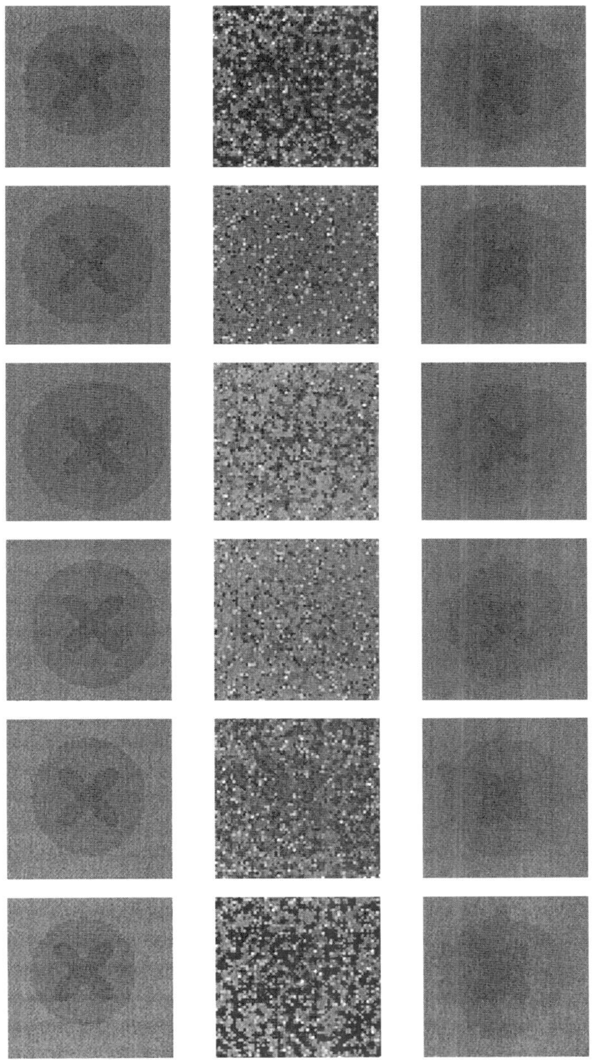

Figure 10: 2D phantom-original (left), noisy (middle) and processed (right) images.

the image sequence. The original six-frame sequence and its destroyed version are shown in the first two columns of Fig. 10. The reconstruction of any noisy frame of this sequence by a standard (still image) filtering algorithm is very difficult task (by our experience impossible). The right-hand column of Fig. 10 represents the results of (2.16) applied to the noisy sequence after 10 discrete scale steps of the numerical implementation from [56].

Next we have applied the method to 3D echocardiographic sequence. In Figs. 11–12 the iso-surfaces corresponding to the blood-muscle interface have been visualized. Figs. 11–12 consist of three sub-figures. For each row, on the left we show the echo-volume visualized using the original noisy data, in the middle the result after three discrete scale steps, and on the right after nine discrete scale steps of the model (2.16).

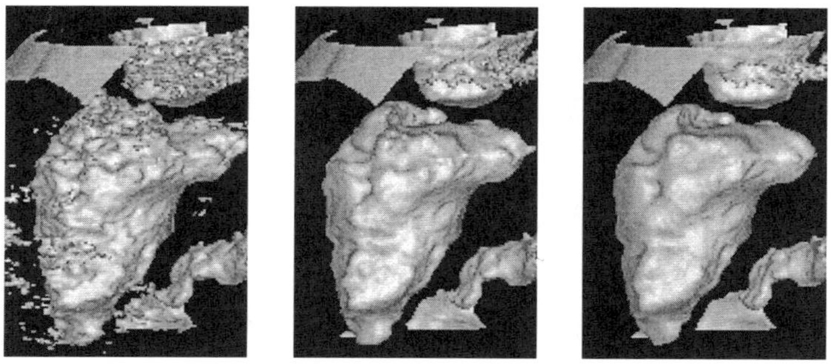

Figure 11: The multiscale analysis of the 1st frame of a 3D-echocardiographic image sequence by the equation (2.16) [56].

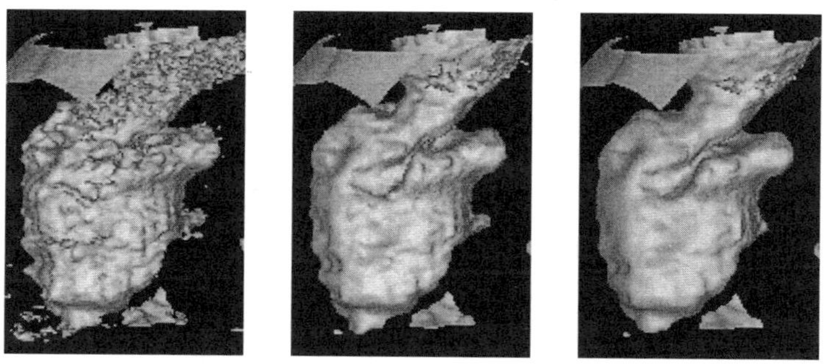

Figure 12: The multiscale analysis of the 7th frame of a 3D-echocardiographic image sequence by the equation (2.16) [56].

3 Variational computational methods

The aim of this chapter is to present numerical schemes for solving the nonlinear diffusion equations presented in the previous sections. Since images are given on a discrete grid (pixel/voxel structure), we discretize the PDEs to get a numerical scheme implemented on the computer. One can use a wide range of methods devoted to the numerical solution of PDEs. Semi-implicit schemes [9, 26, 5, 6, 62, 63, 41, 33, 34, 52, 42, 22, 23, 24, 13], where the nonlinear terms of the equation are treated from the previous discrete scale step, and linear terms are considered at the current scale level, have favorable stability and efficiency properties. For space discretization either finite element method [26, 5, 6, 52, 13], finite/complementary volume method [41, 33, 34, 23] or finite difference methods [9, 62, 63] can be used. In this paper we discuss discretization in space by variational techniques, i.e. using finite element and finite/complementary volume methods. They are based on integral (weak, variational) formulations of the initial-boundary value problems for PDEs. Variational methods have a strong physical background since they are based on principles like minimization of energy (finite element method) or conservation laws (finite and complemetary volume methods). They allow a clear and physically meaningful derivation of difference equations which are local and easy to implement. Convergence of the schemes to solutions of the PDEs can be proved.

For our presentation we have chosen two representative geometry-driven diffusion models: the regularized Perona-Malik anisotropic diffusion equation (2.1), and the nonlinear degenerate diffusion equation of mean curvature flow type (2.10) due to Alvarez, Lions and Morel. They are accompanied by the boundary and initial conditions (2.2)–(2.3), and for both equations we assume the hypotheses (2.4)–(2.6). Both models represent a similar view to image selective smoothing, in the sense that edge positions are preserved in the image multiscale analysis.

Remark In the next paragraphs, we will use the standard notations for the functional spaces $L_p(\Omega)$, $p \geq 1$, and $C^k(\overline{\Omega})$. By V we denote the Sobolev space $W^{1,2}(\Omega)$ of $L_2(\Omega)$-functions with square integrable weak derivates. The absolute value will be denoted by $|\cdot|$ and the norm by $\|\cdot\|$ with a subscript given by the corresponding functional space. By C we denote a general (large) constant.

3.1 Semi-discretizations in scale

We start by the semi-discretizations in scale of the problems given by (2.1) and (2.10), respectively. Choosing $N \in \mathbb{N}$ we obtain the length of the uniform discrete scale step $k = T/N$. We replace the scale derivative in (2.1), (2.10) by backward difference. The nonlinear terms of the equations are treated from the previous scale step while the linear terms are considered at the current scale level—this means semi-implicitness of the method. Let us start with approximation in scale of equation (2.1) (see [9, 26, 41]).

Semi-discrete linear scheme for solving equation (2.1) *Let $N \in \mathbb{N}$, $k = T/N$ and $\sigma > 0$ be fixed numbers, and let u^0 be given by* (2.3). *For every $n = 1, \ldots, N$, we look for a function u^n which is a solution of the equation*

$$\frac{u^n - u^{n-1}}{k} - \nabla.(g(|\nabla G_\sigma * u^{n-1}|)\nabla u^n) = 0. \tag{3.1}$$

It is not difficult to see that there exist unique variational solutions u^n of (3.1) at every discrete scale step for which the stability estimates

$$\|u^n\|_{L_2(\Omega)} \le \|u^0\|_{L_2(\Omega)}, \qquad \|u^n\|_{L_\infty(\Omega)} \le \|u^0\|_{L_\infty(\Omega)}, \qquad n = 1, \dots, N,$$
$$\sum_{n=1}^{N} \|\nabla u^n\|^2_{L_2(\Omega)} k \le C, \qquad \sum_{n=1}^{N} \|u^n - u^{n-1}\|^2_{L_2(\Omega)} \le C,$$

hold [26, 6]. Moreover, if we construct the so-called Rothe's step function $\bar{u}^{(N)}(t) = u^n$, $(n-1)k < t \le nk$, one can prove its convergence in $L_2(Q_T)$ to the unique weak solution of (2.1) ([26]; see also [25] for the comprehensive theory and applications of the so-called Rothe's method in the numerical analysis of PDEs).

In spite of equation (2.1), the equation (2.10) is not in divergence form. For second order partial derivatives, as is usual in variational methods, we would like to use integration by parts or the divergence theorem to get an integral formulation (see the next sections). Thus, we first move the term in front of the divergence to the time derivative, and then we write semi-implicit discretization of (2.10) in scale [23].

Semi-discrete linear scheme for solving equation (2.10) *Let $N \in \mathbb{N}$, $k = T/N$ and $\sigma > 0$ be fixed numbers, and let u^0 be given by (2.3). For every $n = 1, \dots, N$, we look for a function u^n which is a solution of the equation*

$$\frac{1}{g(|\nabla G_\sigma * u^{n-1}|)|\nabla u^{n-1}|} \frac{u^n - u^{n-1}}{k} - \nabla . \left(\frac{\nabla u^n}{|\nabla u^{n-1}|} \right) = 0. \tag{3.2}$$

Since in the general situation there can be zero in the denominators of (3.2), we will regularize this equation in the sense of Evans and Spruck [16]. We approximate $|\nabla u^{n-1}| \approx \sqrt{\varepsilon + |\nabla u^{n-1}|^2}$ (with ε a small real number) in the implementation, and study the behavior of this regularization for ε tending to zero [23]. Let us denote

$$g^{n-1} := g(|\nabla G_\sigma * u^{n-1}|). \tag{3.3}$$

To see stability of (3.2) (formally, without regularization at this point), multiply it by $u^n - u^{n-1}$, then integrate over Ω. Using

$$a(a-b) = \frac{1}{2}a^2 - \frac{1}{2}b^2 + \frac{1}{2}(a-b)^2, \tag{3.4}$$

we get

$$\int_\Omega \frac{(u^n - u^{n-1})^2}{k g^{n-1} |\nabla u^{n-1}|} dx + \frac{1}{2} \int_\Omega \frac{|\nabla u^n|^2 - |\nabla u^{n-1}|^2 + |\nabla u^n - \nabla u^{n-1}|^2}{|\nabla u^{n-1}|} dx = 0.$$

Since

$$|\nabla u^n - \nabla u^{n-1}|^2 = (|\nabla u^n| - |\nabla u^{n-1}|)^2 + \left(\frac{\nabla u^n}{|\nabla u^n|} - \frac{\nabla u^{n-1}}{|\nabla u^{n-1}|} \right)^2 |\nabla u^n|\, |\nabla u^{n-1}|,$$

we get

$$\int_\Omega \frac{(u^n - u^{n-1})^2}{kg^{n-1}|\nabla u^{n-1}|}\,dx + \frac{1}{2}\int_\Omega \frac{|\nabla u^n|^2 - |\nabla u^{n-1}|^2 - (|\nabla u^n| - |\nabla u^{n-1}|)^2}{|\nabla u^{n-1}|}\,dx$$
$$+ \int_\Omega \frac{(|\nabla u^n| - |\nabla u^{n-1}|)^2}{|\nabla u^{n-1}|}\,dx + \frac{1}{2}\int_\Omega \left(\frac{\nabla u^n}{|\nabla u^n|} - \frac{\nabla u^{n-1}}{|\nabla u^{n-1}|}\right)^2 |\nabla u^n|\,dx = 0.$$

Due to the positivity of the other terms we get for the second term

$$\int_\Omega \frac{|\nabla u^n|\,|\nabla u^{n-1}| - |\nabla u^{n-1}|^2}{|\nabla u^{n-1}|}\,dx \le 0,$$

which gives

$$\|\nabla u^n\|_{L_1(\Omega)} \le \|\nabla u^{n-1}\|_{L_1(\Omega)}, \tag{3.5}$$

and thus

$$\|\nabla u^n\|_{L_1(\Omega)} \le \|\nabla u^0\|_{L_1(\Omega)}, \quad 1 \le n \le N, \tag{3.6}$$

which is a $W^{1,1}$ stability estimate for the linear semi-implicit scheme (3.2), i.e. an estimate of the decay of total variation of the semi-discrete solutions. This estimate is a basic property of the flow by mean curvature and of a solution of the level set equation as well [60], and can also be interpreted as a curve shortening property [14, 40] of level sets.

3.2 Space discretizations

A discrete image is given on structure of pixels/voxels with rectangular shape in general (but it is not necessary for the methods presented in this paper). We will use this image structure to create a computational grid for the spatial discretization methods. Concerning the relation of the computational grid to the pixel structure, there will be a difference between the finite element and the complementary volume methods on the one hand, and the finite volume method on the other. The difference is related to the type of approximation of the solution of the partial differential equations assumed in those methods. While in the finite volume method the approximation of the solution is assumed to be piecewise constant [17], in the complementary volume and finite element methods it is assumed to be continuous piecewise linear. Thus, in the finite volume method, the computational grid will directly be taken to be the pixel structure of the image. The initially given and subsequently computed values of discrete intensity are considered as approximations of average of continuous intensity function on pixels. On the other hand, in the finite element and complementary volume methods, the initially given values of discrete intensity, and also the computed values, are considered as approximations of the continuous intensity function in the centers of pixels. The centers of pixels then correspond to the nodes of the finite element or the complementary volume triangulation. We can get such a triangulation by connecting the centers of pixels by a new rectangular mesh and then dividing every rectangle into two triangles (or six tetrahedra in 3D). It also means that in these two methods the computational domain Ω is given as the union of all triangles constructed in this way (Ω thus corresponds to the image domain minus the outer half of every boundary pixel)—see Fig. 13. Let us note that the splitting of every pixel into two triangles, as depicted in Fig. 13, is not the only possibility. The orientation of triangles can change locally, e.g. following an edge direction, or it can be given by a refinement procedure in a bisection algorithm [5]. We will assume that the constructed triangulation has no interior angle larger than $\pi/2$. Then, for the complementary volume method, we will construct a dual mesh. This dual mesh will again, in a sense, copy the pixel structure of the image. Let us note that in the finite element method we will use just the triangulation, in the complementary volume method we use both the triangulation and the dual mesh, and in the finite volume method we use just the pixel structure of the image corresponding to the dual mesh. Let us note also the possibility to use the center of pixels (and the corresponding rectangular mesh) to construct a bilinear finite element solution of the problem [52].

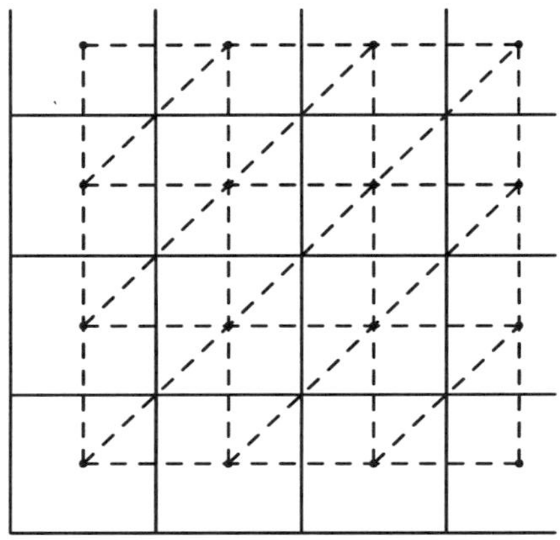

Figure 13: The image pixels (solid lines) corresponding to the finite volume mesh and to the dual mesh for the complementary volume method. Triangulation (dashed lines) for the finite element and complementary volume methods, with nodes (round points) corresponding to the centers of pixels.

We now define quantities which will be used in descriptions of fully discrete schemes (see also [60, 17, 10]). Let us assume that we are given a triangulation $\mathcal{T}$ (e.g. given by a previous construction). The dual mesh will consist of cells p (also called complementary or finite volumes) associated with the pth node of the triangulation $\mathcal{T}$, $p = 1, \ldots, M$. The dual mesh will be denoted by τ. The co-volume p is bounded by the lines (planes in 3D) that bisect and are perpendicular to the edges emanating from the node. By $m(p)$ we denote measure in $\mathbb{R}^d$ of p. We will denote the edge of $\mathcal{T}$ connecting the pth node to the qth by σ_{pq}, and its length by d_{pq}. We denote by E_{pq} the set of simplices having σ_{pq} as an edge, i.e., $E_{pq} = \{T \in \mathcal{T} | \sigma_{pq} \subset T\}$. Let e_{pq} denote the co-edge that is the perpendicular bisector of σ_{pq}, and let x_{pq} be a point of intersection of e_{pq} and σ_{pq}. By $m(e_{pq})$ we denote the measure of e_{pq} in $\mathbb{R}^{d-1}$. For each $T \in E_{pq}$, let c_{pq}^T be the length of the portion of e_{pq} that is in T, i.e., $c_{pq}^T = m(e_{pq} \cap T)$. Let $S(p)$ be the set of simplices that have the pth node as a vertex, and for each node of $\mathcal{T}$ let $N(p)$ denote the set of nodes q, $q \neq p$, which are connected to the pth node by an edge and for which $m(e_{pq}) \neq 0$. In the situation depicted in Fig. 13, $N(p)$ consists of four neighbouring nodes q in the direction of the coordinate axes, for every inner node p of the triangulation. We denote by $\mathcal{E}$ the set of pairs of adjacent finite volumes,

$$\mathcal{E} = \{(p,q) \in \tau^2,\ p \neq q,\ m(e_{pq}) \neq 0\}.$$

Given a triangulation $\mathcal{T}$, we define the set $V_h \subset V$ of piecewise linear finite elements, i.e.,

$$V_h = V_h(\mathcal{T}) := \{v \in C^0(\overline{\Omega})|\ v|_T \in \mathcal{P}_1 \text{ for all } T \in \mathcal{T}\}.$$

For any $v_h \in V_h$ we will use the notation $v_p := v_h(x_p)$, where x_p is the coordinate vector of the pth node of the triangulation. Let $u_h^0 = I_h(u^0) \in V_h(\mathcal{T})$ be the nodal interpolant of u^0. This will be the initial function for the finite element and complementary volume methods. In the finite volume method we denote by $\overline{u}_p$ the representative value for the cell (the bar indicates that we mean the average value in the cell and not a nodal value). The initial condition for the finite volume scheme is then assumed to be

$$\overline{u}_p^0 = \frac{1}{m(p)} \int_p u^0(x)\, dx, \quad p \in \tau. \tag{3.7}$$

Before going to spatial discretizations, let us make the following remark on the realization of the convolution included in the evaluation of the Perona-Malik function g in either (3.1) or (3.2). We use two strategies. The first is the following: Using the Gauss function G_σ as the smoothing kernel, one can replace the term $G_\sigma * u^{n-1}$ by solving the linear heat equation for time σ with the initial condition given by u^{n-1}. This linear equation can be solved numerically at the same grid by just one implicit step with length σ. Thus, as a realization of the convolution, we look for a function u^c which is a solution of the heat equation discretized in time by the backward Euler method with step σ:

$$\frac{u^c - u^{n-1}}{\sigma} = \Delta u^c, \tag{3.8}$$

where Δ denotes the Laplace operator. This strategy is very suitable for the finite element and complementary volume methods, since in this case, the numerical solution u^c of (3.8) is piecewise linear on triangles, its gradients are constant, and thus we can simply evaluate the Perona-Malik function on every triangle.

In the finite volume method we can use another approach. Since we will consider piecewise constant spatial approximations $\overline{u}_p^n$ in every discrete scale step n, we can replace the convolution (integral) by a sum over pixels. For the gradient of the convolution term at any point x we get

$$\nabla G_\sigma * u^{n-1}(x) = \sum_r \overline{u}_r^{n-1} \int_r \nabla G_\sigma(x-s)\, ds, \tag{3.9}$$

where we used the convolution derivative property

$$\frac{\partial}{\partial x_i}(G_\sigma * u^{n-1}(x)) = \frac{\partial G_\sigma}{\partial x_i} * u^{n-1}(x)$$

and

$$\frac{\partial G_\sigma}{\partial x_i} * u^{n-1}(x) = \int_{\mathbb{R}^d} \frac{\partial G_\sigma}{\partial x_i}(x-s) u^{n-1}(s)\, ds = \sum_r \overline{u}_r^{n-1} \int_r \frac{\partial G_\sigma}{\partial z_i}(x-s)\, ds.$$

The sum in (3.9) is evaluated over control volumes r which surround the point x. If we choose a compactly supported smoothing kernel with support in a ball $B_{\sigma^*}(0)$ with radius σ^*, e.g. the function

$$G_\sigma(x) = \frac{1}{Z} e^{|x|^2/(|x|^2-\sigma^2)},$$

where $\sigma^* = \sigma$ and the constant Z is chosen so that G_σ has unit mass, then the sum is restricted only to the control volumes contained in $B_{\sigma^*}(x)$, the ball centered at x. If the point x is close to the boundary of the image domain, we use an extension of a discrete solution. The coefficients $\int_r \nabla G_\sigma(x-s)\,ds$ in (3.9) can be computed in advance using a computer algebra system, e.g. Mathematica [66]. The same situation arises when we use the Gauss function and consider as the ball $B_{\sigma^*}(0)$ a "numerical support" of the Gauss function (i.e., we consider a domain in which the values of the Gauss function are above some treshold). Then, again, just a finite sum is evaluated in (3.9).

3.3 Finite element method in image processing

To describe the ideas of finite element space discretization, let us consider the equation (2.1) [26, 5]. Let k, σ be given numbers. Before the discretization we use approach (3.8) for the realization of the convolution, and then we write (3.1) as a couple of integral identities (weak formulations):

$$\int_\Omega u^n v\,dx + k\int_\Omega g(|\nabla u^c|)\nabla u^n \nabla v\,dx = \int_\Omega u^{n-1} v\,dx, \tag{3.10}$$

$$\int_\Omega u^c v\,dx + \sigma\int_\Omega \nabla u^c \nabla v\,dx = \int_\Omega u^{n-1} v\,dx \tag{3.11}$$

that hold for all $v \in V$. Then, at each scale level n, we look for a continuous piecewise linear function $u_h^n \in V_h(\mathcal{T})$ satisfying

$$\int_\Omega u_h^n v_h\,dx + k\int_\Omega g(|\nabla u_h^c|)\nabla u_h^n \nabla v_h\,dx = \int_\Omega u_h^{n-1} v_h\,dx \tag{3.12}$$

for all $v_h \in V_h(\mathcal{T})$, with $u_h^c \in V_h(\mathcal{T})$ being the solution of

$$\int_\Omega u_h^c v_h\,dx + \sigma\int_\Omega \nabla u_h^c \nabla v_h\,dx = \int_\Omega u_h^{n-1} v_h\,dx, \quad \forall v_h \in V_h(\mathcal{T}). \tag{3.13}$$

Considering the standard Lagrangian base functions $\varphi_q \in V_h(\mathcal{T})$, $q = 1, \dots, M$, given by $\varphi_q(x_p) = \delta_{qp}$ (Kronecker delta) for all nodes of $\mathcal{T}$, the functions u_h^n, u_h^c are given by

$$u_h^n = \sum_{p=1}^{M} u_p^n \varphi_p, \qquad u_h^c = \sum_{p=1}^{M} u_p^c \varphi_p. \tag{3.14}$$

Then using (3.14)) in (3.12)–(3.13) and taking $v_h = \varphi_q$, $q = 1, \dots, M$, as test functions, we get two Ritz-Galerkin systems of linear equations for the nodal values u_p^n, u_p^c, $p = 1, \dots, M$, of the piecewise linear functions u_h^n, u_h^c, respectively:

$$\sum_{p=1}^{M}\left(\int_\Omega \varphi_p\varphi_q\,dx + k\int_\Omega g(|\nabla u_h^c|)\nabla\varphi_p\nabla\varphi_q\,dx\right)u_p^n = \int_\Omega u_h^{n-1}\varphi_q\,dx, \quad q=1,\dots,M, \tag{3.15}$$

$$\sum_{p=1}^{M}\left(\int_\Omega \varphi_p\varphi_q\,dx + \sigma\int_\Omega \nabla\varphi_p\nabla\varphi_q\,dx\right)u_p^c = \int_\Omega u_h^{n-1}\varphi_q\,dx, \quad q=1,\dots,M. \tag{3.16}$$

Thus, in every discrete scale step we need to solve two linear systems with the matrices $\mathbf{M}+k\mathbf{A}(g(|\nabla u_h^c|))$, $\mathbf{M}+\sigma\mathbf{A}(1)$, respectively, where $\mathbf{M}_{q,p} = \int_\Omega \varphi_p\varphi_q\,dx$ is the so-called mass matrix and $\mathbf{A}(w)_{q,p} = \int_\Omega w\nabla\varphi_p\nabla\varphi_q\,dx$ is the stiffness matrix, which are symmetric and positive definite. The discrete solutions can be found efficiently by the preconditioned conjugate gradient method. It is also customary to use the so-called lumped (diagonalized) mass matrix $\mathbf{M}_{q,p} = \sum_{k=1}^{M}\varphi_p(x_k)\varphi_q(x_k)\int_\Omega \varphi_k\,dx$ in the systems (3.15)–(3.16). In order to improve the efficiency of the finite element method, a choice of different (coarsened) triangulations $\mathcal{T}_n$ in subsequent scale steps $n = 1,\dots,N$, is possible (see [5, 6, 52] and the next subsection).

The same ideas as above can be used for the finite element discretization of equation (2.10). The only difference will be in the facts that the mass matrix $\mathbf{M}(w)$ will depend on $w = 1/(g(|\nabla u_h^c|)|\nabla u_h^{n-1}|)$, the stiffness matrix $\mathbf{A}(w)$ will depend on $w = 1/|\nabla u_h^{n-1}|$, and that Evans-Spruck regularization is used in the denominators.

3.4 Adaptivity in the finite element method

We can improve efficiency of the finite element method by the use of adaptively chosen grids at each scale step. Usually, for time-dependent problems, a modification consisting of refinement and coarsening steps is necessary to adjust the grid at a given time step [4]. However, for the problem (2.1) it is sufficient to coarsen the initial grid successively. There is no spatial *movement* of edges, hence no refinements of the grids are needed. This access may reduce the computational effort considerably, since with the increasing scale the solution tends to be more flat in large sub-regions of the image. The coarsening of the computational grids rapidly reduces the number of unknowns in the linear systems to be solved at the discrete scale steps of the method.

Concerning the adaptive algorithm, first we generate a triangulation $\mathcal{T}_0$ corresponding to a pixel/voxel structure of the image by (globally) refining a coarse grid $\mathcal{T}^0$, the so-called *macro triangulation*. The refinement procedure generates a sequence $\mathcal{T}^0, \mathcal{T}^1, \mathcal{T}^2, \dots$ of finer and finer meshes until the desired structure is reached. Then $\mathcal{T}_0 := \mathcal{T}^{k_0}$, where k_0 is the last refinement step. For the refinement we have chosen the so called *bisection method*, which allows easily for coarsening [4].

Refinement of the grid by bisection [4] Before starting the refinement process, one edge of every triangle of the triangulation at kth refinement level is marked (Fig. 14). This edge is called the *refinement edge* (a good choice is the longest one). To divide a single triangle, it is cut through the midpoint of the refinement edge and the vertex opposite to the refinement edge. The new refinement edges are chosen opposite to the new vertex (Fig. 15).

Let us start with $\mathcal{T}^0$. Then for every bisection level k let Σ^+ be the set of those triangles which have to be divided ($\Sigma^+ = \mathcal{T}^k$ in case of uniform refinement). Then *one bisection step* (see also Fig. 16) at level k is given by:

while $\Sigma^+ \neq \emptyset$ **do**

- `bisect all` $T \in \Sigma^+$ `as described above, obtain an intermediate triangulation` $\hat{\mathcal{T}}^k$ `(possibly non-conforming)`
- `let now` Σ^+ `be the set of those tetrahedra with a non-conforming node.`

endwhile
$\mathcal{T}^{k+1} := \hat{\mathcal{T}}^k$

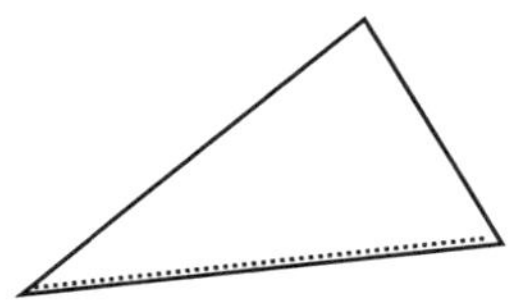

Figure 14: Triangle with refinement edge

Figure 15: Bisection of a single triangle

To provide a *local coarsening*, we make the following definitions:

(i) A simplex $T \in \mathcal{T}$ has level l if T was obtained after l refinement steps.

(ii) A simplex T is said to have *locally finest level* if the levels of all neighbours are less than or equal to the level of T.

(iii) Let $T \in \mathcal{T}$, and let T' be the father of T. A vertex P of T which was inserted while bisecting T' is called the *coarsening node* of T.

(iv) Let K be an edge of the triangulation $\mathcal{T}$ and K' the "father"-edge of K with midpoint Q. Set $M := \{T \in \mathcal{T} | T \cap K' \neq \emptyset\}$. If Q is the coarsening node for all $T \in M$, then M is called a *resolvable patch* (Fig. 17).

If M is a resolvable patch, then all $T \in M$ can be coarsened without interfering with $T' \in \mathcal{T}$ outside M. Therefore resolvable patches are the configurations which we allow to be coarsened. This guarantees that the coarsening process stays local.

Coarsening of the grid [4] Let $\mathcal{T}_n$ be a triangulation obtained by refinement and coarsening steps. Let $\Sigma^- \subset \mathcal{T}_n$ be the set of triangles to be derefined. Then *one coarsening step* consists of:

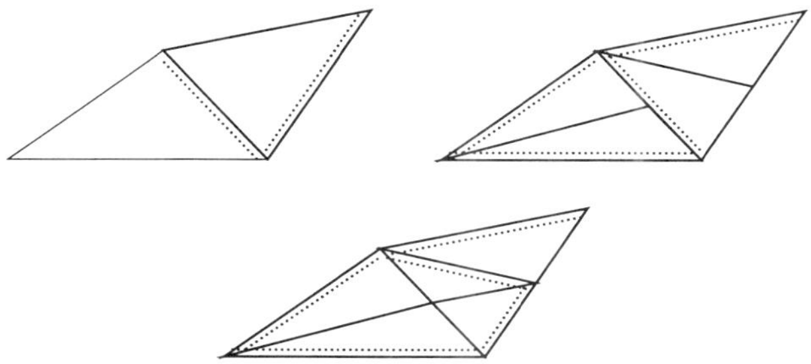

Figure 16: $\mathcal{T}^0$, $\hat{\mathcal{T}}^0$ and $\mathcal{T}^1$

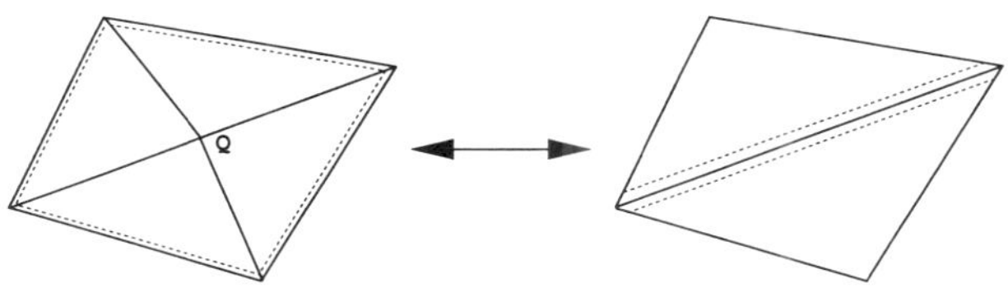

Figure 17: Resolvable patch M with coarsening node Q and coarsened patch

for each $T \in \Sigma^-$ **do**
    if T `belongs to a resolvable patch` M **then**
        if $T' \in \Sigma^-$ `for all` $T' \in M$ **then**
            `derefine` M
        endif
    endif
enddo

Coarsening criterion and our adaptive method [5] As the local behavior of $\nabla G_\sigma * u$ determines the evolution process and is an indicator for edges, the *coarsening criterion* is based on this value. Let $\varepsilon > 0$ be a given tolerance. For the nth scale step and u_h^n, the corresponding numerical solution on the grid $\mathcal{T}_n$, we allow all triangles $T \in \mathcal{T}_n$ to be coarsened, if

$$h_T |\nabla u_h^n| \leq \varepsilon \quad \text{on} \quad T,$$

where h_T is the diameter of triangle T. Thus we have the following *adaptive scheme*:

```
Let $\mathcal{T}_0, u_h^0$ be given.
for $n = 1, 2, \dots N$ do
    define $\Sigma^- := \{T \in \mathcal{T}_{n-1}; h_T|\nabla u_h^{n-1}| \le \epsilon$ on $T\}$
    derefine $\mathcal{T}_{n-1}$ according to $\Sigma^-$ to obtain $\mathcal{T}_n$
    set up the matrix $\mathbf{M} + \sigma\mathbf{A}(1)$
    compute $u_h^c \in V_h(\mathcal{T}_n)$
    set up the matrix $\mathbf{M} + k\mathbf{A}(g(|\nabla u_h^c|))$
    compute $u_h^n \in V_h(\mathcal{T}_n)$
enddo
```

The same adaptive algorithm can be used in the 3D case, just changing triangles to tetrahedra. For 3D implementation and computational results we refer to [6]. There exist finite element software packages (see e.g. [57]) based on the bisection refinement and coarsening method, thus the adaptivity described above can be implemented also in such an environement.

3.5 Complementary volume spatial discretization

In this subsection we will discretize equation (2.10) or, more precisely, the semi-discrete approximation (3.2) by means of the complementary volume method [23]. In order to derive the complementary volume spatial discretization [60], we integrate (3.2) over a co-volume p

$$\int_p \frac{u^n - u^{n-1}}{g^{n-1}|\nabla u^{n-1}|k}\, dx = \int_p \nabla.\left(\frac{\nabla u^n}{|\nabla u^{n-1}|}\right) dx. \tag{3.17}$$

Using the divergence theorem on the right-hand side we get

$$\int_p \nabla\left(\frac{\nabla u^n}{|\nabla u^{n-1}|}\right) dx = \int_{\partial p} \frac{1}{|\nabla u^{n-1}|}\frac{\partial u^n}{\partial \nu}\, ds = \sum_{q\in N(p)} \int_{e_{pq}} \frac{1}{|\nabla u^{n-1}|}\frac{\partial u^n}{\partial \nu}\, ds. \tag{3.18}$$

If $u_h^n \in V_h(\mathcal{T})$ is a continuous piecewise linear function on a triangulation $\mathcal{T}$ and if its nodal values are denoted by $u_p^n = u_h^n(x_p)$, then

$$\sum_{q\in N(p)} \int_{e_{pq}} \frac{1}{|\nabla u_h^{n-1}|}\frac{\partial u_h^n}{\partial \nu}\, ds = \sum_{q\in N(p)} \left(\sum_{T\in E_{pq}} \frac{c_{pq}^T}{|\nabla u_T^{n-1}|}\right)\frac{u_q^n - u_p^n}{d_{pq}}, \tag{3.19}$$

where $|\nabla u_T^{n-1}|$ denotes the constant value of the gradient of u_h^{n-1} in the simplex T. The complementary volume method approximates the left-hand side of (3.17) by

$$\frac{m(p)(u_p^n - u_p^{n-1})}{kg(|\nabla u_p^c|)|\nabla u_p^{n-1}|}, \tag{3.20}$$

where $|\nabla u_p^{n-1}|$, $|\nabla u_p^c|$ denote an approximation of the gradient in co-volume p. For that purpose we have chosen the average value of gradients in the co-volume [60]

$$|\nabla u_p| \approx \sum_{T\in S(p)} \frac{m(T\cap p)}{m(p)}|\nabla u_T|. \tag{3.21}$$

Figure 18: Processing of a noisy image by the adaptive finite element method for the anisotropic diffusion equation (2.1). Upper part: Original image and its noise-corrupted version; lower part: smoothed image after 8 discrete scale steps and corresponding 2D triangulation which is fine only along edges [5].

If we denote

$$b_p^{n-1} = \frac{m(p)}{g(|\nabla u_p^c|)|\nabla u_p^{n-1}|}, \tag{3.22}$$

$$a_{pq}^{n-1} = \frac{1}{d_{pq}} \sum_{T \in E_{pq}} \frac{c_{pq}^T}{|\nabla u_T^{n-1}|}, \tag{3.23}$$

we can write the

Linear fully discrete complementary volume scheme for solving equation (2.10) *For $n = 1, \ldots, N$ we look for u_p^n, $p = 1, \ldots, M$, satisfying the equation*

$$b_p^{n-1} u_p^n + k \sum_{q \in N(p)} a_{pq}^{n-1}(u_p^n - u_q^n) = b_p^{n-1} u_p^{n-1}. \tag{3.24}$$

Of course, before solving (3.24), we have to put $|\nabla u_p^c|$ into (3.22). The function $u_h^c \in V_h(\mathcal{T})$ is found by the same idea as given in (3.17)–(3.20) applied to (3.8); i.e., we solve (3.24) with u_p^n replaced by u_p^c, and with $b_p^{n-1} \equiv b_p = m(p)$, $a_{pq}^{n-1} \equiv a_{pq} = m(e_{pq})/d_{pq}$. Then $|\nabla u_p^c|$ is computed by (3.21) and put into the Perona-Malik function g. The system (3.24) gives a symmetric positive definite M-matrix with diagonal dominance, so standard or preconditioned linear solvers can be used.

Using the nonnegativeness of b_p^{n-1}, a_{pq}^{n-1}, we can get a uniform L_∞-stability estimate for the fully discrete scheme (3.24) in the form

$$\min_{p \in \mathcal{T}} u_p^0 \leq \min_{p \in \mathcal{T}} u_p^n \leq \max_{p \in \mathcal{T}} u_p^n \leq \max_{p \in \mathcal{T}} u_p^0, \quad 1 \leq n \leq N, \tag{3.25}$$

which means

$$\|u_h^n\|_{L_\infty(\Omega)} \leq \|u_h^0\|_{L_\infty(\Omega)}, \quad 1 \leq n \leq N. \tag{3.26}$$

To see (3.25), let us rewrite (3.24) in the form

$$u_p^n + \frac{k}{b_p^{n-1}} \sum_{q \in N(p)} a_{pq}^{n-1}(u_p^n - u_q^n) = u_p^{n-1}, \tag{3.27}$$

and let $\max u_h^n = \max_{r \in \mathcal{T}} u_r^n$ be achieved in the node p. Then the second term is nonnegative and thus $u_p^n \leq u_p^{n-1} \leq \max_{r \in \mathcal{T}} u_r^{n-1}$, which gives the result for max. The relation for min is derived in a similar way.

Already in the description of semidiscrete schemes we mentioned the use of the Evans-Spruck regularization [16], $|\nabla u|_\varepsilon = \sqrt{\varepsilon + |\nabla u|^2}$, instead of $|\nabla u|$ everywhere in the schemes to prevent possible zero gradients in the denominators. One can observe that the Evans-Spruck regularization is well suited for image smoothing purposes. For zero gradients such regularized discretization (3.24) reduces to a discretization of the linear heat equation. This is quite desirable at points with small change in intensity. On the other hand, the influence of ε can be neglected, and thus we have a curvature-driven flow of level lines, for large gradients. All results which were (formally) derived so far for fully discrete or semi-discrete schemes, namely the L_∞ and $W^{1,1}$ estimates (3.25) and (3.6), are valid also for the ε-regularization.

Moreover, they do not depend on the regularization parameter ε. Thus, we can pass to the limit and understand the solution of (3.24) in a generalized sense (see [23] for details). Concerning the computational point of view, if ε is very small the diagonal dominance of the system can be very weak. In [23] efficient linear solvers also for such a situation are studied. Fig. 19 gives an example of an application of the complementary volume scheme to a mammogram image (171×192 pixels).

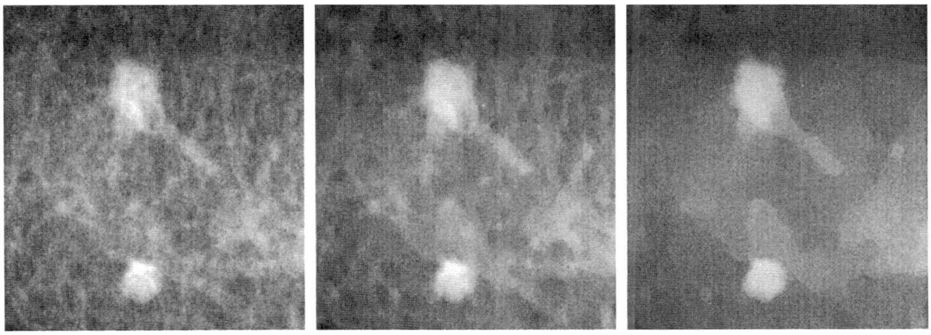

Figure 19: Initial image (left), result of multiscale analysis by equation (2.10) after 30 (middle) and 100 (right) discrete scale steps [23].

Using the integration (3.17) and the relation (3.19), one can also derive a complementary volume discretization of equation (2.1) or, more precisely, for the semi-discrete coupling (3.1), (3.8). In such a way, we get the system (3.24) with

$$b_p^{n-1} = m(p), \qquad a_{pq}^{n-1} = \frac{1}{d_{pq}} \sum_{T \in E_{pq}} c_{pq}^T g(|\nabla u_T^c|), \tag{3.28}$$

where u_h^c is computed in the same way as above. Let us note that such a discretization of (2.1) can be considered as a special mass lumping approximation in the finite element method.

3.6 Finite volume discretization

In this subsection we will consider the finite volume discretization of equation (2.1) only. In our approach, the finite volume grid corresponds to the dual mesh τ, and finite volumes correspond to co-volumes as they are identical with the pixel/voxel structure of the image. In general, the finite volume can be either a simplex of the triangulation or a co-volume of the dual mesh [17]; we use the second strategy. The main difference as compared to the complementary volume technique, is that the approximating functions in the finite volume method are not in $V_h(\mathcal{T})$, but they are just piecewise constant on finite volumes. Thus, we cannot work directly with gradients or normal derivatives since they are either zero (inside finite volumes) or infinite (on their boundaries). The derivation of the finite volume scheme follows the ideas of the last paragraph in the previous subsection. We integrate the equation (2.1) in every finite volume p. Then, by means of $\overline{u}_p$ representing an approximate value of the solution inside the finite volume $p, p \in \tau$, we approximate fluxes through the boundary of p (compare (3.19)). The value of diffusion coefficient along e_{pq} is approximated by its value at the point x_{pq}. We denote by $T_{pq} = m(e_{pq})/d_{pq}$ the transmissivity coefficient, and get the

Linear fully discrete finite volume scheme for solving equation (2.1) [41] *Let $N \in \mathbb{N}$, $k = T/N$ and $\sigma > 0$ be fixed numbers, $t_n = nk$, $n = 0, \dots, N$. For $n = 1, \dots, N$ we look for $\overline{u}_p^n, p \in \tau$, satisfying*

$$m(p)\overline{u}_p^n + k \sum_{q \in N(p)} g_{pq}^{\sigma,n-1} T_{pq} \left(\overline{u}_p^n - \overline{u}_q^n\right) = m(p)\overline{u}_p^{n-1}, \tag{3.29}$$

starting with

$$\overline{u}_p^0 = \frac{1}{m(p)} \int_p u^0(x)\, dx, \quad p \in \tau, \tag{3.30}$$

where

$$g_{pq}^{\sigma,n-1} = g(|\nabla G_\sigma * \tilde{u}_{h,k}(x_{pq}, t_{n-1})|) \tag{3.31}$$

and $\tilde{u}_{h,k}$ is an extension of the piecewise constant function $\overline{u}_{h,k}$ ($h = \max_{p \in \tau} \operatorname{diam}(p)$) defined as follows

$$\overline{u}_{h,k}(x,t) = \sum_{n=0}^{N} \sum_{p \in \tau} \overline{u}_p^n \chi_{\{x \in p\}} \chi_{\{t_{n-1} < t \leq t_n\}} \tag{3.32}$$

with the boolean function $\chi_{\{A\}} = \begin{cases} 1 & \textit{if } A \textit{ is true} \\ 0 & \textit{elsewhere.} \end{cases}$

The function $\overline{u}_{h,k}$, constructed using discrete values given by the scheme (3.29), is considered as the approximation of the solution and its convergence to a unique weak solution of (2.1) as $h, k \to 0$ can be proved ([41], see also the next subsection for the main ideas). It is clear that (3.29) gives a linear system with a symmetric, strictly diagonally dominant M-matrix, so there exists a unique discrete solution at every discrete scale step. Moreover, using the same trick as in (3.27), we get L_∞-stability of the scheme

$$\min_{p \in \tau} \overline{u}_p^0 \leq \min_{p \in \tau} \overline{u}_p^n \leq \max_{p \in \tau} \overline{u}_p^n \leq \max_{p \in \tau} \overline{u}_p^0, \quad 1 \leq n \leq N. \tag{3.33}$$

3.7 Convergence of the finite volume scheme

3.1 Definition A *weak solution* of the regularized Perona-Malik problem (2.1)–(2.3) is a function $u \in L_2(I, V)$ satisfying the identity

$$\int_0^T \int_\Omega u \frac{\partial \varphi}{\partial t}(x,t)\, dx\, dt + \int_\Omega u_0(x)\varphi(x,0)\, dx - \int_0^T \int_\Omega g(|\nabla G_\sigma * u|) \nabla u \nabla \varphi\, dx\, dt = 0 \tag{3.34}$$

for all $\varphi \in \Psi$, where Ψ is the space of smooth test functions

$$\Psi = \{\varphi \in C^{2,1}(\overline{\Omega} \times [0,T]), \nabla\varphi.\overrightarrow{n} = 0 \text{ on } \partial\Omega \times (0,T), \varphi(\cdot, T) = 0\}. \tag{3.35}$$

In [9] Catté, Lions, Morel and Coll proved that there exists a unique weak solution of (2.1)–(2.3) which is also the classical solution of the problem at the same time. To get the existence they used Schauder's fixed point theorem with iterations in entire parabolic equations. In [41] we look for that solution in a computationally natural and efficient way using a semi-implicit finite volume scheme. In this subsection we give the main ideas of the proof of the convergence of the finite volume solution to the weak solution of (2.1)–(2.3). The strategy is to prove some *a priori* estimates of the numerical solution, using the scheme (3.29). These estimates will lead to the so-called *space* and *time translate estimates* for the approximate solution $\overline{u}_{h,k}$. The space and time translate estimates are specific formulations of the equicontinuity in the following well-known criterion (see e.g. [35]):

Kolmogorov's relative compactness criterion in $L_2(Q_T)$ *The set $K \subset L_2(Q_T)$ is relatively compact if and only if*

(i) *K is bounded, i.e., there exists $C > 0$ such that $\|f\| \le C$ for every $f \in K$;*

(ii) *K is mean equicontinous, i.e., for every $\varepsilon > 0$ there exists $\delta > 0$ such that*

$$\int_{Q_T} (f(x+\gamma) - f(x))^2 \, dx < \varepsilon^2$$

for each $f \in K$ and γ with $|\gamma| < \delta$.

Proving relative compactness one gets that there exists a limit function $u \in L_2(Q_T)$ of the sequence $\overline{u}_{h,k}$ as $h, k \to 0$. Moreover, the space translates give that this limit is in $L_2(I, V)$, so it is a good candidate for the weak solution. It will be the last step of the convergence proof that u fulfills the weak identity (3.34).

3.2 Lemma (*a priori* estimates) *There exists a positive constant C, independent of h and k, such that*

(i) $\max_{0 \le n \le N} \sum_{p \in \tau} (\overline{u}_p^n)^2 m(p) \le C$;

(ii) $\sum_{n=1}^{N} k \sum_{(p,q) \in \mathcal{E}} T_{pq} (\overline{u}_p^n - \overline{u}_q^n)^2 \le C$;

(iii) $\sum_{n=1}^{N} \sum_{p \in \tau} (\overline{u}_p^n - \overline{u}_p^{n-1})^2 m(p) \le C$.

Proof Let us multiply the scheme (3.29) by $\overline{u}_p^n$ to obtain

$$(\overline{u}_p^n - \overline{u}_p^{n-1}) \overline{u}_p^n m(p) = k \sum_{q \in N(p)} g_{pq}^{\sigma, n-1} T_{pq} (\overline{u}_q^n - \overline{u}_p^n) \overline{u}_p^n. \tag{3.36}$$

Using (3.4) on the left-hand side of (3.36) and summing over $p \in \tau$, we have that

$$\begin{aligned} \frac{1}{2} \sum_{p \in \tau} (\overline{u}_p^n)^2 m(p) - \frac{1}{2} \sum_{p \in \tau} (\overline{u}_p^{n-1})^2 m(p) + \frac{1}{2} \sum_{p \in \tau} (\overline{u}_p^n - \overline{u}_p^{n-1})^2 m(p) \\ = k \sum_{p \in \tau} \sum_{q \in N(p)} g_{pq}^{\sigma, n-1} T_{pq} (\overline{u}_q^n - \overline{u}_p^n) \overline{u}_p^n. \end{aligned} \tag{3.37}$$

The following trick is often used in the finite volume technique for anti-symmetric $a_{pq} = -a_{qp}$:

$$\sum_{p\in\mathcal{T}}\sum_{q\in N(p)} a_{pq}b_p = \sum_{(p,q)\in\mathcal{E}} a_{pq}b_p = \sum_{(q,p)\in\mathcal{E}} a_{qp}b_q = -\sum_{(p,q)\in\mathcal{E}} a_{pq}b_q,$$

hence

$$2\sum_{p\in\mathcal{T}}\sum_{q\in N(p)} a_{pq}b_p = -\sum_{(p,q)\in\mathcal{E}} a_{pq}(b_q - b_p) \tag{3.38}$$

whence

$$\sum_{p\in\mathcal{T}}\sum_{q\in N(p)} g_{pq}^{\sigma,n-1}T_{pq}(\overline{u}_q^n - \overline{u}_p^n)\overline{u}_p^n = -\frac{1}{2}\sum_{(p,q)\in\mathcal{E}} g_{pq}^{\sigma,n-1}T_{pq}(\overline{u}_p^n - \overline{u}_q^n)^2. \tag{3.39}$$

Applying (3.39) in (3.37) and summing over $n = 1, \dots, m \le N$, we have

$$\frac{1}{2}\sum_{p\in\mathcal{T}}(\overline{u}_p^m)^2 m(p) + \frac{1}{2}\sum_{n=1}^{m}\sum_{p\in\mathcal{T}}(\overline{u}_p^n - \overline{u}_p^{n-1})^2 m(p) + \frac{1}{2}\sum_{n=1}^{m} k\sum_{(p,q)\in\mathcal{E}} g_{pq}^{\sigma,n-1}T_{pq}(\overline{u}_p^n - \overline{u}_q^n)^2$$
$$= \frac{1}{2}\sum_{p\in\mathcal{T}}(\overline{u}_p^0)^2 m(p).$$

Since $u^0 \in L_\infty(\Omega)$ we get the results (i)and (iii). We also have

$$\left|\frac{\partial}{\partial x_i}G_\sigma * \tilde{u}_{h,k}(x_{pq}, t_{n-1})\right| \le \int_{\mathbb{R}^d}\left|\frac{\partial}{\partial x_i}G_\sigma(x_{pq} - \xi)\tilde{u}(\xi, t_{n-1})\right| d\xi \le C_\sigma \|\tilde{u}(., t_{n-1})\|_{L_\infty(\Omega)}.$$

Because of (3.33) we have that $|\nabla G_\sigma * \tilde{u}_{h,k}(x_{pq}, t_{n-1})| < \infty$, which in turn implies that there exists a positive constant α such that $g_{pq}^{\sigma,n-1} > \alpha > 0$, from which one can deduce the assertion (ii) of the lemma. □

In order to apply Kolmogorov's compactness criterion we have to estimate the integral

$$\int_{Q_T}(\overline{u}_{h,k}(x+\xi, t+s) - \overline{u}_{h,k}(x,t))^2\, dx\, dt$$
$$\le 2\int_{Q_T}(\overline{u}_{h,k}(x+\xi, t+s) - \overline{u}_{h,k}(x, t+s))^2\, dx\, dt + 2\int_{Q_T}(\overline{u}_{h,k}(x, t+s) - \overline{u}_{h,k}(x,t))^2\, dx\, dt.$$

Using an extension of the solution it is sufficient to restrict the two integrals on the right-hand side to domains $\Omega_\xi \times (0,T)$ with $\Omega_\xi = \{x \in \Omega, [x, x+\xi] \in \Omega\}$ and $\Omega \times (0, T-s)$ with $s \in (0,T)$, respectively [17, 41].

3.3 Lemma (Space translate estimates) *For any vector $\xi \in \mathbb{R}^d$ there exists a positive constant C such that*

$$\int_{\Omega_\xi\times(0,T)}(\overline{u}_{h,k}(x+\xi, t) - \overline{u}_{h,k}(x,t))^2\, dx\, dt \le C|\xi|(|\xi| + 2h), \tag{3.40}$$

where $\Omega_\xi = \{x \in \Omega, [x, x+\xi] \in \Omega\}$.

Proof Let $\xi \in \mathbb{R}^d$ be a given vector. For all $(p,q) \in \mathcal{E}$, let us denote $\xi_{pq} = \xi/|\xi|.n_{pq}$. For all $x \in \Omega_\xi$, we denote by $E(x,p,q)$ the function defined as follows:

$$E(x,p,q) = \begin{cases} 1 & \text{if the segment } [x, x+\xi] \text{ intersects } e_{pq},\ p \text{ and } q, \text{ and } \xi_{pq} > 0 \\ 0 & \text{otherwise.} \end{cases}$$

For any $t \in (0,T)$ there exists $n \in \mathbb{N}$ such that $(n-1)k < t \leq nk$. Then for almost all $x \in \Omega_\xi$ we can see that

$$\overline{u}_{h,k}(x+\xi,t) - \overline{u}_{h,k}(x,t) = \overline{u}^n_{p(x+\xi)} - \overline{u}^n_{p(x)} = \sum_{(p,q)\in\mathcal{E}} E(x,p,q)(\overline{u}^n_q - \overline{u}^n_p),$$

where $p(x)$ is the volume of $p \in \tau$ and $x \in p$. By the Cauchy-Schwarz inequality we obtain

$$\begin{aligned}(\overline{u}_{h,k}(x+\xi,t) &- \overline{u}_{h,k}(x,t))^2 \\ &\leq \left(\sum_{(p,q)\in\mathcal{E}} E(x,p,q)\xi_{pq}d_{pq}\right)\left(\sum_{(p,q)\in\mathcal{E}} E(x,p,q)\frac{(\overline{u}^n_q - \overline{u}^n_p)^2}{\xi_{pq}d_{pq}}\right)\end{aligned} \tag{3.41}$$

and using the fact that $\xi_{pq}d_{pq} = \xi/|\xi|.n_{pq}d_{pq} = \xi/|\xi|.(x_q - x_p)$ we have

$$\sum_{(p,q)\in\mathcal{E}} E(x,p,q)\xi_{pq}d_{pq} = \frac{\xi}{|\xi|}.(x_{p(x+\xi)} - x_{p(x)}) \leq |x_{p(x+\xi)} - x_{p(x)}| \leq 2h + |\xi|.$$

Now, we integrate the relation (3.41) on $\Omega_\xi \times (0,T)$:

$$\begin{aligned}\int_{\Omega_\xi\times(0,T)} &(\overline{u}_{h,k}(x+\xi,t) - \overline{u}_{h,k}(x,t))^2\, dx\, dt \\ &\leq (2h+|\xi|)\sum_{n=1}^{N} k \sum_{(p,q)\in\mathcal{E}} \frac{(\overline{u}^n_q - \overline{u}^n_p)^2}{\xi_{pq}d_{pq}} \int_{\Omega_\xi} E(x,p,q)\, dx.\end{aligned} \tag{3.42}$$

Taking into account the area of a parallelogram we have

$$\int_{\Omega_\xi} E(x,p,q)\, dx \leq m(e_{pq})\xi.n_{pq} = m(e_{pq})\frac{\xi}{|\xi|}.n_{pq}|\xi| = m(e_{pq})|\xi|\xi_{pq},$$

and applying this result in (3.42) we obtain

$$\int_{\Omega_\xi\times(0,T)} (\overline{u}_{h,k}(x+\xi,t) - \overline{u}_{h,k}(x,t))^2\, dx\, dt \leq (2h+|\xi|)|\xi| \sum_{n=1}^{N} k \sum_{(p,q)\in\mathcal{E}} T_{pq}(\overline{u}^n_q - \overline{u}^n_p)^2. \tag{3.43}$$

Finally, using *a priori* estimate (ii) of Lemma 3.2 we complete the proof. □

In a more technical way, but basically using again the *a priori* estimate (ii) of Lemma 3.2 we get:

3.4 Lemma (Time translate estimate) *There exists a positive constant C such that*

$$\int_{\Omega\times(0,T-s)} (\overline{u}_{h,k}(x,t+s) - \overline{u}_{h,k}(x,t))^2\, dx\, dt \leq Cs$$

for all $s \in (0,T)$.

3.5 Lemma (Convergence of $\overline{u}_{h,k}$) *There exists $u \in L_2(Q_T)$ such that for some subsequence of $\overline{u}_{h,k}$*

$$\overline{u}_{h,k} \to u \quad in \quad L_2(Q_T)$$

as $h, k \to 0$. Moreover, this limit function is in $L_2(I, V)$.

Proof From the estimate (i) of Lemma 3.2 we have that $\|\overline{u}_{h,k}\|_{L_2(Q_T)} \le C$, so the space and time translate estimates allow us to use Kolmogorov's relative compactness criterion in $L_2(Q_T)$ and we have the first assertion of the lemma.

Let $\varphi \in C_0^\infty(Q_T)$, $\varepsilon > 0$ and $\varphi(x,t) = 0$ if $|x - \partial\Omega| < \varepsilon$. Let $0 < |\xi| < \varepsilon$. Then by the Cauchy-Schwarz inequality

$$\int_{\Omega\times(0,T)} \frac{\overline{u}_{h,k}(x+\xi,t) - \overline{u}_{h,k}(x,t)}{|\xi|}\varphi(x,t)\,dx\,dt \le \frac{\sqrt{C|\xi|(|\xi|+h)}}{|\xi|}\|\varphi\|_{L_2(Q_T)}.$$

For the limit function u we have

$$\int_{\Omega\times(0,T)} \frac{u(x+\xi,t) - u(x,t)}{|\xi|}\varphi(x,t)\,dx\,dt \le \sqrt{C}\|\varphi\|_{L_2(Q_T)}.$$

On the other hand, by a changing of the variables $y = x + \xi$ we get

$$\begin{aligned}
&\int_{\Omega\times(0,T)} \frac{u(x+\xi,t) - u(x,t)}{|\xi|}\varphi(x,t)\,dx\,dt \\
&= \int_{\Omega\times(0,T)} \frac{u(y,t)}{|\xi|}\varphi(y-\xi,t)\,dy\,dt - \int_{\Omega\times(0,T)} \frac{u(y,t)}{|\xi|}\varphi(y,t)\,dy\,dt \\
&= -\int_{\Omega\times(0,T)} \frac{\varphi(y,t) - \varphi(y-\xi,t)}{|\xi|}u(y,t)\,dy\,dt \\
&\le C\|\varphi\|_{L_2(Q_T)}.
\end{aligned}$$

Let $\xi = \omega e_i$, where e_i is i-th coordinate vector, and let $\omega \to 0$. Then

$$-\int_{\Omega\times(0,T)} \frac{\partial\varphi(x,t)}{\partial x_i}u(x,t)\,dx\,dt \le C\|\varphi\|_{L_2(Q_T)}, \quad \forall\varphi \in C_0^\infty(\Omega).$$

Thus u has generalized spatial derivatives in $L_2(Q_T)$, so it is in $L_2(I, V)$. □

The last step is to prove that u given in Lemma 3.5 fulfills the weak identity (3.34) from Definition 3.1, and thus is a weak solution of the regularized Perona-Malik problem. Since such a solution is unique [9], not only a subsequence of $\overline{u}_{h,k}$ but the whole sequence will converge to u. Let $\varphi \in \Psi$ be given. We obtain a discrete analogy of the weak solution identity multiplying the finite volume scheme

$$(\overline{u}_p^n - \overline{u}_p^{n-1})m(p) = k\sum_{q\in N(p)} g_{pq}^{\sigma,n-1}T_{pq}(\overline{u}_q^n - \overline{u}_p^n)$$

by $\varphi(x_p, t_{n-1})$ and summing the resulting identity over all $p \in \tau$ and $n = 1,\dots,N$:

$$\sum_{n=1}^{N} k\sum_{p\in\tau} \frac{(\overline{u}_p^n - \overline{u}_p^{n-1})}{k}\varphi(x_p,t_{n-1})m(p) = \sum_{n=1}^{N} k\sum_{p\in\tau}\sum_{q\in N(p)} g_{pq}^{\sigma,n-1}T_{pq}(\overline{u}_q^n - \overline{u}_p^n)\varphi(x_p,t_{n-1}).$$

Next we perform a discrete integration by parts

$$\sum_{n=1}^{N}(a^n - a^{n-1})b^{n-1} = a^N b^N - a^0 b^0 - \sum_{n=1}^{N} a^n(b^n - b^{n-1})$$

on the left-hand side, and a rearrangement of the sum

$$\sum_{p\in\mathcal{T}} \sum_{q\in N(p)} a_{pq} b_p = -\frac{1}{2} \sum_{(p,q)\in\mathcal{E}} a_{pq}(b_q - b_p)$$

on the right-hand side, which together with $\varphi(x_p, t_N) = 0$ gives

$$\sum_{n=1}^{N} k \sum_{p\in\mathcal{T}_h} \overline{u}_p^n \frac{\varphi(x_p, t_n) - \varphi(x_p, t_{n-1})}{k} m(p) + \sum_{p\in\mathcal{T}_h} \overline{u}_p^0 \varphi(x_p, 0) m(p)$$

$$- \frac{1}{2}\sum_{n=1}^{N} k \sum_{(p,q)\in\mathcal{E}} g_{pq}^{\sigma,n-1} T_{pq}(\overline{u}_q^n - \overline{u}_p^n)(\varphi(x_q, t_{n-1}) - \varphi(x_p, t_{n-1})) = 0.$$

The correspondence of the discrete terms of the previous equation and the continuous integral terms of the weak identity (3.34) can be clearly seen, and the convergence as $h, k \to 0$ can be proved (see [41] for details). Thus we get:

3.6 Theorem *The sequence $\overline{u}_{h,k}$ converges strongly in $L_2(Q_T)$ to the unique weak solution u of* (2.1)-(2.3) *as $h, k \to 0$.*

3.8 Adaptivity in the finite volume method

Similarly as in the finite element method we can accelerate the numerical solution of the finite volume scheme using adaptive (coarsened) grids. For the adaptive grid construction we choose an approach based on quadtrees, where the adaptive grid is represented by the leaves of a quadtree structure [33, 34]. We consider the coarsening criterion to merge quadruples of pixels if the difference in intensities is below a prescribed tolerance ε. The example of the adaptive grid is given in Fig. 20. One can see so-called hanging nodes in the grid, which can cause some troubles in the finite element approach, but can be handled naturally in the finite volume scheme. Our adjustment of the finite volume scheme (3.29) to adaptive grids is the following. In the definition of $g_{pq}^{\sigma,n-1}$ in (3.31), x_{pq} will be the midpoint of the common boundary of two neighboring cells (with possibly unequal areas). Then the difference will be that the sum in (3.9) can be evaluated over unequal control volumes. However, we again compute all possible coefficients of the sum in advance for every larger candidate cell on higher levels of the hierarchy. Let us mention that we allow only the so-called *balanced grids* where the ratio of the sides of neighbouring cells is $1:1$, $1:2$ or $2:1$. If the solution is piecewise constant on non-uniform cells, we can assume exchange of intensity between neighbours just in a strip of unit thickness along a cell boundary (assuming unit size of the cells in the case of a uniform rectangular grid). Then we can define

$$T_{pq} = \min\{l_p, l_q\}, \tag{3.44}$$

where l_p and l_q are the lengths of the sides of two adjacent cells p, q. With these adjustments the scheme (3.29) gives a linear system with a symmetric and strictly diagonally dominant M-matrix which guarantees existence of a unique solution. Moreover, L_∞-stability can be easily proved in every discrete scale step.

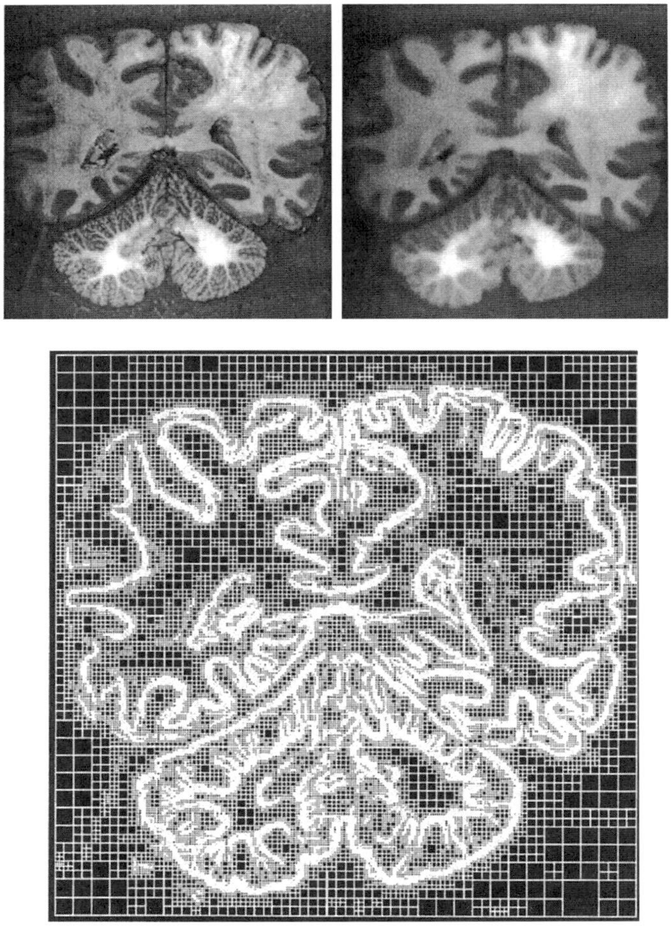

Figure 20: Processing of a medical image [33]; original (top left), multiscale analysis after 25 scale steps (top right), corresponding adaptive finite volume grid (bottom).

Acknowledgements The results presented in this paper were obtained together with Angela Handlovičová, Zuzana Krivá, Fiorella Sgallari, Eberhard Bänsch, Jozef Kačur, Claudio Lamberti, Narisoa Ramarosy, Alessandro Sarti and Daniel Ševčovič. The author is greatly indebted to all these colleagues for cooperation in computational methods for image processing and their applications. We would like to thank the TomTec company for collecting the sequence of 3D echocardiographic data used in our presentation. Our research was carried out with the support of the ICARUS and MINOS European projects at CINECA, Bologna, by SFB 256, IWR, University of Heidelberg, and with the support of the grant VEGA 1/7132/20 of the Scientific Grant Agency of the Slovak Republic. Thanks also go to A. Bourlioux, M. Gander, and G. Sabidussi from Montréal, the organizers of the NATO-ASI, where the lectures containing the material of this paper were presented.

References

[1] L. Alvarez, F. Guichard, P. L. Lions, and J.-M. Morel, Axioms and fundamental equations of image processing, *Arch. Rational Mech. Anal.* **123** (1993), 200–257.

[2] L. Alvarez and J.-M. Morel, Formalization and computational aspects of image analysis, Acta Numer.,Cambridge Univ. Press, 1994, 1–59.

[3] L. Alvarez, P. L. Lions, and J.-M. Morel, Image selective smoothing and edge detection by nonlinear diffusion II, *SIAM J. Numer. Anal.* **29** (1992), 845–866.

[4] E. Bänsch, Local mesh refinement in 2 and 3 dimensions, *IMPACT of Computing in Science and Engineering* **3** (1991), 181–191.

[5] E. Bänsch and K. Mikula, A coarsening finite element strategy in image selective smoothing, *Computing and Visualization in Science* **1** (1) (1997), 53–61.

[6] E. Bänsch and K. Mikula, Adaptivity in 3D image processing, *Computing and Visualization in Science* **4** (1) (2001), 21–30.

[7] M. Beneš, Mathematical and computational aspects of solidification of pure substances, *Acta Math. Univ. Comenian.* **70** (1) (2001), 124–152.

[8] V. Caselles, J.-M. Morel, and G. Sapiro (eds.), *IEEE Trans. Image Processing* **7** (1998).

[9] F. Catté, P. L. Lions, J.-M. Morel, and T. Coll, Image selective smoothing and edge detection by nonlinear diffusion, *SIAM J. Numer. Anal.* **29** (1992), 182–193.

[10] P. F. Ciarlet, *The Finite Element Method For Elliptic Problems*, North-Holland, Amsterdam, 1978.

[11] Y.-G. Chen, Y. Giga, and S. Goto, Uniqueness and existence of viscosity solutions of generalized mean curvature flow equation, *J. Differential Geom.* **33** (1991), 749–786.

[12] M. G. Crandall, H. Ishii, and P. L. Lions, User's guide to viscosity solutions of second order partial differential equations, *Bull. Amer. Math. Soc. (NS)* **27** (1992), 1–67.

[13] U. Diewald, T. Preusser, M. Rumpf, and R. Strzodka, Diffusion models and their accelerated solution in image and surface processing, *Acta Math. Univ. Comenian.* **70** (1) (2001), 15–34.

[14] G. Dziuk, Convergence of a semidiscrete scheme for the curve shortening flow, *Math. Models Methods Appl. Sci.* **4** (1994), 589–606.

[15] G. Dziuk, Algorithm for evolutionary surfaces, *Numer. Math.* **58** (1991), 603–611.

[16] L. C. Evans and J. Spruck, Motion of level sets by mean curvature I, *J. Differential Geom.* **33** (1991), 635–681.

[17] R. Eymard, T. Gallouet, and R. Herbin, The finite volume method, in: *Handbook for Numerical Analysis* **7** (Ph. Ciarlet, P. L. Lions, eds.), Elsevier, 2000.

[18] G. Golub and C. van Loan, *Matrix Computations*, 3rd ed., The Johns Hopkins Univ. Press, Baltimore, MD, 1996.

[19] F. Guichard, Axiomatisation des analyses multi-échelles d'images et de films, Ph.D. Thesis, Université Paris IX Dauphine (1994).

[20] A. Greenbaum, *Iterative Methods For Solving Linear Systems*, SIAM, Philadelphia, PA, 1999.

[21] I. Gustafsson, Modified incomplete Cholesky (MIC) methods, in: *Preconditioning Methods, Theory and Applications* (D. J. Evans, e.d), Gordon and Breach, New York, 1983, 265–293.

[22] A. Handlovičová, K. Mikula, and A. Sarti, Numerical solution of parabolic equations related to level set formulation of mean curvature flow, *Computing and Visualization in Science* **1** (2) (1999), 179–182.

[23] A. Handlovičová, K. Mikula, and F. Sgallari, Semi-implicit complementary volume scheme for solving level set like equations in image processing and curve evolution, *Numer. Math.* DOI 10.1007/s002110100374 (2001).

[24] A. Handlovičová, K. Mikula, and F. Sgallari, Variational numerical methods for solving nonlinear diffusion equations arising in image processing, *J. Visual Communication and Image Representation* (to appear).

[25] J. Kačur, *Method of Rothe in Evolution Equations*, Teubner-Texte Math. **80**, Leipzig, 1985.

[26] J. Kačur and K. Mikula, Solution of nonlinear diffusion appearing in image smoothing and edge detection, *Appl. Numer. Math.* **17** (1995), 47–59.

[27] J. Kačur and K. Mikula, Slow and fast diffusion effects in image processing, *Computing and Visualization in Science* **3** (4) (2001), 185–195.

[28] M. Kass, A. Witkin, and D. Terzopulos, Snakes: active contour models, *Internat. J. Computer Vision* **1** (1987), 321–331.

[29] S. Kichenassamy, The Perona-Malik paradox, *SIAM J. Appl. Math.* **57** (5) (1997), 1328–1342.

[30] R. Kimmel, R. Malladi, and N. Sochen, Images as embedded maps and minimal surfaces: movies, color, texture, and volumetric medical images, *Internat. J. Computer Vision* **39** (2) (2000), 111–129.

[31] M. Kimura, Numerical analysis for moving boundary problems using the boundary tracking method, *Japan J. Indust. Appl. Math.* **14** (1997), 373–398.

[32] J. J. Koenderink, The structure of images, *Biol. Cybern.* **50** (1984), 363–370.

[33] Z. Krivá and K. Mikula, An adaptive finite volume scheme for solving nonlinear diffusion equations in image processing, *J. Visual Communication and Image Representation* (to appear).

[34] Z. Krivá and K. Mikula, An adaptive finite volume scheme in processing of color images, in: *Proc. ALGORITMY 2000, Conf. on Scientific Computing, Podbanské, Slovakia (2000)*, 174–188.

[35] A. Kufner, O. John, and S. Fučík, *Function Spaces*, Academia, Prague, 1977.

[36] O. A. Ladyženskaja, N. N. Uralceva, *Linear and Quasilinear Elliptic Equations*, Academic Press, New York, NY, 1968.

[37] R. Malladi, J. A. Sethian, and B. Vemuri, Shape modeling with front propagation: a level set approach, *IEEE Trans. Pattern Anal. Machine Intelligence* **17** (2) (1995), 158–174.

[38] J. A. Meijerink and H. A. van der Vorst, An iterative solution method for systems of which the coefficient matrix is a symmetric M-matrix, *Math. Comp.* **31** (1977), 148–162.

[39] K. Mikula, Solution of nonlinear curvature driven evolution of plane convex curves, *Appl. Numer. Math.* **23** (1997), 347–360.

[40] K. Mikula and J. Kačur, Evolution of convex plane curves describing anisotropic motions of phase interfaces, *SIAM J. Sci. Comput.* **17** (1996), 1302–1327.

[41] K. Mikula and N. Ramarosy, Semi-implicit finite volume scheme for solving nonlinear diffusion equations in image processing, *Numer. Math.* **89** (3) (2001), 561–590.

[42] K. Mikula, A. Sarti, and C. Lamberti, Geometrical diffusion in 3D-echocardiography, in: *Proc. ALGORITMY'97 — Conf. on Scientific Computing* (J. Kačur and K. Mikula, eds.), Acta Math. Univ. Comenian. (N.S.) **67**, Comenius Univ. Press, Bratislava, 1998, 167–181.

[43] K. Mikula and D. Ševčovič, Solution of nonlinearly curvature driven evolution of plane curves, *Appl. Numer. Math.* **31** (1) (1999), 191–207.

[44] K. Mikula and D. Ševčovič, Evolution of plane curves driven by a nonlinear function of curvature and anisotropy, *SIAM J. Appl. Math.* **61** (5) (2001), 1473–1501.

[45] K. Mikula and D. Ševčovič, A direct method for solving an anisotropic mean curvature flow of planar curve with an external force, Preprint, Department of Mathematics and Descriptive Geometry, Slovak University of Technology, Bratislava, 2001.

[46] M. Nielsen, P. Johansen, O. F. Olsen, and J. Weickert (eds.), *Scale-Space Theories in Computer Vision*, Lecture Notes in Comput. Sci. **1682**, Springer, Berlin, 1999.

[47] M. Nitzberg and T. Shiota, Nonlinear image filtering with edge and corner enhancement, *IEEE Trans. Pattern Analysis and Machine Intelligence* **14** (8) (1992), 826–833.

[48] R. Nochetto, M. Paolini, and C. Verdi, Sharp error analysis for curvature dependent evolving fronts, *Math. Models Methods Appl. Sci.* **3** (6) (1993), 711–723.

[49] K. N. Nordström, Biased anisotropic diffusion — A unified approach to edge detection, Preprint, Department of Electrical Engineering and Computer Sciences, University of California, Berkeley, 1989.

[50] S. Osher and J. Sethian, Front propagating with curvature dependent speed: algorithms based on the Hamilton-Jacobi formulation, *J. Comput. Phys.* **79** (1988), 12–49.

[51] P. Perona and J. Malik, Scale space and edge detection using anisotropic diffusion, *Proc. IEEE Computer Society Workshop on Computer Vision* (1987).

[52] T. Preusser and M. Rumpf, An adaptive finite element method for large scale image processing, in: *Proceedings of ScaleSpace'99* (1999), 223–234.

[53] L. I. Rudin, S. Osher, and E. Fatemi, Nonlinear total variation based noise removal algorithms, *Physica D* **60** (1992), 259–268.

[54] Y. Saad, *Iterative Methods For Sparse Linear Systems*, PWS Publ. Comp., Boston-Toronto, 1996.

[55] G. Sapiro, *Geometric Partial Differential Equations and Image Processing*, Cambridge Univ. Press, 2001.

[56] A. Sarti, K. Mikula, and F. Sgallari, Nonlinear multiscale analysis of 3D echocardiographic sequences, *IEEE Trans. Medical Imaging* **18** (6) (1999), 453–466.

[57] A. Schmidt and K. G. Siebert, ALBERT — Software for scientific computations and applications, *Acta Math. Univ. Comenian.* **70** (1) (2001), 105–122.

[58] J. A. Sethian, Numerical algotithm for propagating interfaces: Hamilton-Jacobi equations and conservation laws, *J. Differential Geom.* **31** (1990), 131–161.

[59] J. A. Sethian, *Level Set Methods and Fast Marching Methods: Evolving Interfaces in Computational Geometry, Fluid Mechanics, Computer Vision, and Materials Science*, Cambridge Monogr. Appl. Comput. Math. **3**, Cambridge Univ. Press, 1999.

[60] N. J. Walkington, Algorithms for computing motion by mean curvature, *SIAM J. Numer. Anal.* **33** (6) (1996), 2215–2238.

[61] J. Weickert, *Anisotropic Diffusion in Computer Vision*, Teubner-Stuttgart, 1998.

[62] J. Weickert, B. M. t. H. Romeny, M. A. Viergever, Efficient and reliable schemes for nonlinear diffusion filtering, *IEEE Trans. Image Processing* **7** (1998), 398–410.

[63] J. Weickert, Coherence-enhancing diffusion of colour images, *Image and Vision Computing* **17** (1999), 201–212.

[64] R Whitacker and G. Gerig, Vector-valued diffusion, *Geometry Driven Diffusion in Computer Vision* (B. M. t. H. Romemy, ed.), Kluwer, Dodrecht, 1994.

[65] A. P. Witkin, Scale-space filtering, in: *Proc. Eight Internat. Conf. on Artificial Intelligence*, vol. 2 (1983), 1019–1022.

[66] S. Wolfram, *The Mathematica Book*, 3rd ed., Cambridge Univ. Press, 1996.

Interface connections in domain decomposition methods

Frédéric NATAF
CMAP, CNRS UMR7641
École Polytechnique
91128 Palaiseau
France

Abstract

Parallel computers are increasingly used in scientific computing. They enable one to perform large scale computations. New algorithms which are well suited to such architectures have to be designed. Domain decomposition methods are a very natural way to exploit the possibilities of multiprocessor computers, but such algorithms are very useful when used on monoprocessor computers as well.

The idea is to decompose the computational domain into smaller subdomains. The equations are solved on each subdomain. In order to enforce the matching of the local solutions, interface conditions have to be written on the boundary between subdomains. These conditions are imposed iteratively. The convergence rate is very sensitive to these interface conditions. Theoretical and numerical results are given.

1 Introduction

1.1 Why domain decomposition methods?

Domain decomposition is a tool introduced artificially to ease large scale computations or that is natural in some situations. These methods are well adapted to parallel computers and are very popular in parallel computing. They are also very efficient when used on sequential computers. This is due to the nonlinear cost of a simulation. Since the 50's, and long before the advent of parallel computers, these methods have been (and are still) used. They enable one to perform robust large scale computations even on a monoprocessor computer. In some sense, for domain decomposition methods, parallel computing is a secondary issue. Also, in some situations, the domain decomposition is natural from the physics of the problem: different physics in different subdomains, moving domains, strongly heterogeneous media.

Three dimensional numerical simulations are very demanding in CPU or/and in memory. When explicit schemes of the general following form

$$U(t + \Delta t, .) = U(t, .) + \Delta t F(U(t, .))$$

or Monte Carlo methods are used, the limitation comes mostly from the CPU time. The parallelization is clear. The correct balancing of tasks is then a crucial issue in order to use parallel computers efficiently. The use of a parallel computer is very profitable. But

A. Bourlioux and M.J. Gander (eds.), Modern Methods in Scientific Computing and Applications, 323–364.

very often, numerical simulations involve the solving very large linear systems arising from Poisson or Helmholtz equations or from the Jacobian matrix in a Newton's method. These computations are very demanding both in CPU time and in memory. Generally speaking, direct solvers are too costly and iterative solvers are not robust enough especially for problems with strongly heterogeneous media and ill conditioned matrices, see e.g. [Nep91]. There is a need for hybrid iterative/direct solvers: these are domain decomposition methods. Roughly speaking, the computational domain is decomposed into smaller subdomains. The equations in the subdomains are solved by a direct method and the matching of the solutions is imposed iteratively. Since the cost of a simulation is nonlinear w.r.t. the number of unknowns, breaking the initial problem into a set of smaller subproblems is profitable. Also, in some situations, the domain decomposition is natural from the physics of the problem: different physics in different subdomains (e.g. fluid/structure interaction), moving domains (e.g. rotor and stator in an electric motor), strongly heterogeneous media: sliding blocks along faults in subsurface modeling, see Fig. 1).

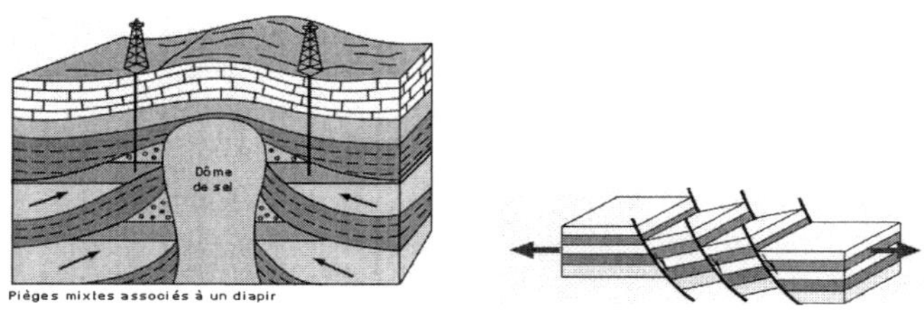

Figure 1: Subsurface modeling of a geological basin — extract from `http://www.ggl.ulaval.ca`.

In a slightly different context, the mesh generation is a complex and time-consuming task both for the user and the computer. In order to speed up this task, an increasingly popular possibility is to first generate a decomposition of the domain into large subdomains and then to mesh each subdomain concurrently. The mesh generation becomes a parallel task. The resulting mesh is of course non-conforming on the interfaces between the subdomains. Therefore new tools for handling non-conforming mesh are needed. In this case, "Domain connection" would be a more appropriate term, see Fig. 2 and references [BMP93, BD98, CDS99, AMW99, AK95, AKP95, Woh99].

1.2 The original Schwarz method (1870)

The first domain decomposition method was developed at the end of the 19th century by the mathematician H. A. Schwarz. His goal was to study the Laplace operator. At that time, the main tool for this purpose was Fourier analysis and more generally the use of special functions. Geometries of the domain were essentially restricted to simple configurations: rectangles and disks, see Fig. 3. His idea was to study the case of a domain that is the union

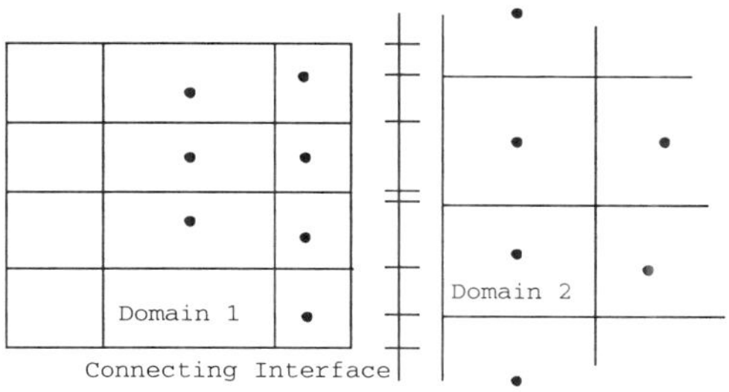

Figure 2: Non conforming mesh.

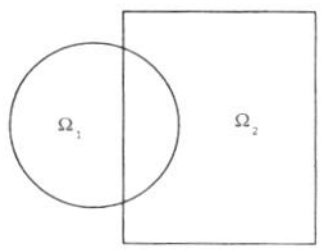

Figure 3: Overlapping domain decomposition.

of simple domains. For example, let $\Omega = \Omega_1 \cup \Omega_2$ with $\Omega_1 \cap \Omega_2 \neq \emptyset$. We want to solve

$$\begin{aligned} -\Delta(u) &= f \quad \text{in} \quad \Omega \\ u &= 0 \quad \text{on} \quad \partial\Omega. \end{aligned} \tag{1.1}$$

Schwarz proposed the following algorithm (Multiplicative Schwarz Method, **MSM**):

Let (u_1^n, u_2^n) be an approximation to $(u_{|\Omega_1}, u_{|\Omega_2})$ at step n of the algorithm, (u_1^{n+1}, u_2^{n+1}) is defined by

$$\begin{aligned} -\Delta(u_1^{n+1}) &= f \quad \text{in} \quad \Omega_1 \\ u_1^{n+1} &= 0 \quad \text{on} \quad \partial\Omega_1 \cap \partial\Omega \\ u_1^{n+1} &= u_2^n \quad \text{on} \quad \partial\Omega_1 \cap \overline{\Omega_2}. \end{aligned} \qquad \begin{aligned} -\Delta(u_2^{n+1}) &= f \quad \text{in} \quad \Omega_2 \\ u_2^{n+1} &= 0 \quad \text{on} \quad \partial\Omega_2 \cap \partial\Omega \\ u_2^{n+1} &= u_1^{n+1} \quad \text{on} \quad \partial\Omega_2 \cap \overline{\Omega_1}. \end{aligned}$$

Problem in domain Ω_1 has to be solved before problem in domain Ω_2. This algorithm is sequential.

A slight modification of the algorithm is the additive Schwarz method (**ASM**)

$$\begin{aligned} -\Delta(u_1^{n+1}) &= f \quad \text{in} \quad \Omega_1 \\ u_1^{n+1} &= 0 \quad \text{on} \quad \partial\Omega_1 \cap \partial\Omega \\ u_1^{n+1} &= u_2^n \quad \text{on} \quad \partial\Omega_1 \cap \overline{\Omega_2}. \end{aligned} \qquad \begin{aligned} -\Delta(u_2^{n+1}) &= f \quad \text{in} \quad \Omega_2 \\ u_2^{n+1} &= 0 \quad \text{on} \quad \partial\Omega_2 \cap \partial\Omega \\ u_2^{n+1} &= u_1^n \quad \text{on} \quad \partial\Omega_2 \cap \overline{\Omega_1}. \end{aligned} \tag{1.2}$$

Problems in domains Ω_1 and Ω_2 may be solved concurrently. The ASM is a parallel algorithm and is adapted to parallel computers. Schwarz proved the linear convergence of (u_1^n, u_2^n) to $(u_{|\Omega_1}, u_{|\Omega_2})$ as n tends to infinity.

The benefit of these algorithms is the saving in memory requirements. Indeed, if the problems are solved by direct methods, the cost of the storage is non-linear with respect to the number of unknowns. By dividing the original problem into smaller pieces the amount of storage can be significantly reduced. As far as CPU is concerned, the original algorithms ASM and MSM are very slow. Another weakness is the need of overlapping subdomains. Indeed, only the continuity of the solution is imposed and nothing is imposed on the matching of the fluxes. When there is no overlap convergence is thus impossible.

The slowness of the method and the need for overlapping subdomains are linked. Indeed, it can be proved that the convergence rate of the Schwarz method is a continuous function of the size of the overlap denoted δ. For small overlaps the convergence rate is close to one. Actually it can be proved that for small overlaps the convergence rate varies as $1 - C^t\delta$.

1.3 Towards faster methods: two families of methods

In order to remedy the drawbacks of the original Schwarz method, two families of methods have been developed. They both work in the non-overlapping case and consist of introducing the normal derivative of the solution, but in two very different ways:

- write a substructured formulation of the domain decomposition problem where the matching of the solution and of its normal derivative along the interface are imposed explicitly.

- Modify the original Schwarz method by replacing the Dirichlet interface conditions on $\partial\Omega_i \backslash \partial\Omega$, $i = 1, 2$, by Robin interface conditions ($\partial_{n_i} + \alpha$, where n is the outward normal to subdomain Ω_i), see [Lio90].

The first approach is explained in section 2 and the second in section 3.

More generally, a complete overview of various domain decomposition methods may be found in a few books [CM94, SBG96, QV99] or in the proceedings of various conferences on domain decomposition methods, see e.g. [CGPW89, PEBK97, LBCW98] and references therein.

2 Substructured formulation

We explain the basic ideas for a two domain decomposition first at the continuous level and then at the matrix level. Then we give some references on the general case.

2.1 The symmetric positive definite continuous case

2.1.1 Substructured formulation

We still consider equation (1.1). The domain Ω is decomposed into two non-overlapping subdomains Ω into Ω_1 and Ω_2. The interface $\partial\Omega_1 \cap \partial\Omega_2$ is denoted by Γ. A first formulation of (1.1) as a domain decomposition method consists of looking for $u_i = u|_{\Omega_i}$, $i = 1, 2$, which must satisfy

$$\begin{aligned}
-\Delta(u_i) &= f \quad \text{in} \quad \Omega_i, \quad i=1,2 \\
u_i &= 0 \quad \text{on} \quad \partial\Omega_i \setminus \Gamma, \quad i=1,2 \\
u_1 &= u_2 \quad \text{on} \quad \Gamma, \\
\left(\frac{\partial u_1}{\partial n_1} + \frac{\partial u_2}{\partial n_2}\right) &= 0 \quad \text{on} \quad \Gamma.
\end{aligned}$$

The substructured formulation consists in formulating the problem in terms of the common value of u_1 and u_2 on Γ, denoted by u_Γ. This is obtained by eliminating the internal unknowns via the solving of local subproblems. We introduce a Dirichlet BVP in each subdomain with $u|_\Gamma$ as a Dirichlet data:

$$\begin{aligned}
-\Delta(u_i) = f \quad \text{in} \quad \Omega_i, \\
u_i = u|_\Gamma \quad \text{on} \quad \Gamma, \qquad u_i = 0 \quad \text{on} \quad \partial\Omega_i \setminus \Gamma.
\end{aligned}$$

The jump of the normal derivative across the interface is a function of f and $u|_\Gamma$,

$$\mathcal{S}(f, u|_\Gamma) = \left(\frac{\partial u_1}{\partial n_1} + \frac{\partial u_2}{\partial n_2}\right)\bigg|_\Gamma . \tag{2.1}$$

The substructured interface problem reads: Find $u|_\Gamma$ such that

$$\mathcal{S}(0, u|_\Gamma) = -\mathcal{S}(f, 0). \tag{2.2}$$

The corresponding discretized operator $\mathcal{S}_h$ (the Schur complement of the matrix) is a matrix whose coefficients can be computed by solving a Dirichlet boundary value problem for each interface unknown (or d.o.f.). This task is long. Moreover, this matrix is full and the solving of the substructured problem by a direct method would be very costly. For these reasons, the discretized problem corresponding to (2.2) is solved by a Krylov type method such as CG, GMRES, BICGSTAB, QMR, ... , as only matrix-vector products are needed. In our case, it corresponds to solving a Dirichlet boundary value problem in each subdomain. Therefore, there is no need to build explicitly the substructured matrix $\mathcal{S}_h$. Moreover, this is a parallel task.

The convergence is enhanced when compared to a Krylov type method applied to the original problem. Indeed, the condition number of the discretized operator $\kappa(\mathcal{S}_h)$ is $O(1/h)$ whereas the condition number of a finite difference, finite volume or finite element discretization of the Laplace operator is $O(1/h^2)$. The size of the substructured problem is much smaller than the size of the overall problem. Moreover, as we shall see next, there are very good preconditioners $\mathcal{T}_h$ available for $\mathcal{S}_h$ which are almost optimal ($\kappa(\mathcal{T}_h\,\mathcal{S}_h) \simeq O(1)$). These remarks are even more relevant for problems with large jumps in the coefficients. In this case, there are no robust preconditioners for the original problem, whereas they exist for the substructured formulation, see [DSW96, LT94].

2.1.2 The basis for the Neumann-Neumann preconditioner

A very popular preconditioner for the operator $\mathcal{S}(0,.)$ has been proposed in [BGLTV]. The basis for the preconditioner comes from the special case where the domain Ω is decomposed along a symmetry axis, e.g. a rectangle decomposed into two half rectangles or the whole plane decomposed into two half planes. In that case we have

- the operator $\mathcal{S}(0,.)$ is naturally split into the sum of two operators $\mathcal{S}_1$ and $\mathcal{S}_2$ which are defined for each subdomain and such that $\mathcal{S}_1 = \mathcal{S}_2$.

- The inverses $\mathcal{S}_1^{-1}$ and $\mathcal{S}_2^{-1}$ are obtained via the solving of boundary value problems in each subdomain.

Then

$$\mathcal{T} := \frac{1}{4}(\mathcal{S}_1^{-1} + \mathcal{S}_2^{-1}) \tag{2.3}$$

is an exact preconditioner: $\mathcal{T}\mathcal{S}(0,.) = \mathrm{Id}$.

For simplicity, we explicitly give these operators for the following model problem:
Find $u : \mathbf{R}^2 \to \mathbf{R}$ such that

$$(\eta - \Delta)(u) = f,$$

where η is positive and f is a given function. The domain is decomposed into two non-overlapping half-planes $\Omega_1 = (-\infty, 0) \times \mathbf{R}$ and $\Omega_2 = (0, \infty) \times \mathbf{R}$, and the interface is $\Gamma = \{0\} \times \mathbf{R}$. We define the operators $\mathcal{S}_i$, $i = 1, 2$, which maps a function living on Γ to a function living on Γ via the solving of a Dirichlet boundary value problem

$$\mathcal{S}_i : v_\Gamma \to \left. \frac{\partial v}{\partial n_i} \right|_\Gamma,$$

where n_i is the outward normal to the domain Ω_i, and v_i is the solution to the following BVP:

$$\begin{cases} (\eta - \Delta)(v_i) = 0 & \text{in } \Omega_i, \\ v_i = v_\Gamma & \text{on } \Gamma. \end{cases}$$

The operator $\mathcal{S}_i$ is the Steklov-Poincaré a.k.a. the Dirichlet to Neumann (DtN) operator of domain Ω_i. By symmetry of the operator and of the decomposition, we have $\mathcal{S}_1 = \mathcal{S}_2$. It is clear that $\mathcal{S}(0,.) = \mathcal{S}_1 + \mathcal{S}_2$.

The inverse of $\mathcal{S}_i$ corresponds to the solving of a Neumann problem

$$\mathcal{S}_i^{-1} : g_\Gamma \to v_i|_\Gamma$$

with v_i the solution to the following BVP

$$\begin{cases} (\eta - \Delta)(v_i) = 0 & \text{in } \Omega_i, \\ \dfrac{\partial v_i}{\partial n_i} = g_\Gamma & \text{on } \Gamma. \end{cases}$$

The name "Neumann-Neumann" preconditioner (2.3) comes from the fact that applying it amounts to solving a Neumann BVP in each subdomain.

Only the symmetry of the operator and of the domain decomposition and the well-posedness of Dirichlet and Neumann BVP's are used. Therefore, this preconditioner can naturally be extended to elasticity or the Stokes system. It is also easily defined at the discrete level which eases its implementation.

2.1.3 The FETI or dual method

It is possible to consider the Neumann-Neumann preconditioner the other way round. The unknown is the flux at the interface between the subdomains. Let g be a function that lives on the interface Γ, and consider the Neumann BVP's ($i = 1, 2$):

$$\begin{cases} -\Delta v_i = f & \text{in } \Omega_i, \\ \dfrac{\partial v_i}{\partial n_i} = (-1)^i g_\Gamma & \text{on } \Gamma. \end{cases}$$

The jump of the solutions across the interface is a function of f and g:

$$\mathcal{T}(f, g_\Gamma) := v_2 - v_1.$$

The substructured formulation is
Find g_Γ such that

$$\mathcal{T}(0, g_\Gamma) := -\mathcal{T}(f, 0).$$

A good preconditioner is

$$\frac{1}{4}\mathcal{S}(0, .)$$

which involves the solving in parallel of a Dirichlet problem in each subdomain. This approach has been developed in [FR91, CMW95, KW01].

2.2 At the matrix level

When the problem (1.1) is discretized by a finite element method for instance, it yields a linear system of the form $AU = F$, where F is a given right-hand side and U is the set of unknowns. Corresponding to the domain decomposition, the set of unknowns U is decomposed into interior nodes of the subdomains U_1 and U_2, and to unknowns, U_Γ, associated to the interface Γ. This leads to a block decomposition of the linear system

$$\begin{pmatrix} A_{11} & A_{1\Gamma} & 0 \\ A_{\Gamma 1} & A_{\Gamma\Gamma} & A_{\Gamma 2} \\ 0 & A_{2\Gamma} & A_{22} \end{pmatrix} \begin{pmatrix} U_1 \\ U_\Gamma \\ U_2 \end{pmatrix} = \begin{pmatrix} F_1 \\ F_\Gamma \\ F_2 \end{pmatrix}. \tag{2.4}$$

The substructuring of the linear system corresponds to the Gauss elimination of the unknowns U_1 using the first line of (2.4), and of U_2 using the last line. These eliminations correspond to solving discretized Dirichlet boundary value problems. The resulting linear system reads:
Find U_Γ such that

$$\mathcal{S}_h(0, U_\Gamma) := (A_{\Gamma\Gamma} - A_{\Gamma 1}A_{11}^{-1}A_{1\Gamma} - A_{\Gamma 2}A_{22}^{-1}A_{2\Gamma})(U_\Gamma) = F_\Gamma - A_{\Gamma 1}A_{11}^{-1}F_1 - A_{\Gamma 2}A_{22}^{-1}F_2$$

The matrix $\mathcal{S}_h(0, .)$ (the Schur complement of the original matrix) is full, due to A_{ii}^{-1} in its expression.

We now describe the Neumann-Neumann preconditioner at the matrix level. In order to split the matrix $\mathcal{S}_h(0, .)$ into two matrices, we use the natural decomposition of $A_{\Gamma\Gamma}$ into its contribution coming from each subdomain $A_{\Gamma\Gamma} = A^1_{\Gamma\Gamma} + A^2_{\Gamma\Gamma}$. More precisely, for the Laplace operator, let k, l be the indices of two degrees of freedom on the interface associated with two basis functions ϕ^k and ϕ^l. The corresponding entry $a^{kl}_{\Gamma\Gamma}$ can be decomposed into a sum $a^{kl}_{\Gamma\Gamma} = a^{1,kl}_{\Gamma\Gamma} + a^{2,kl}_{\Gamma\Gamma}$, where

$$a^{i,kl}_{\Gamma\Gamma} = \int_{\Omega_i} \nabla\phi^k \nabla\phi^l, \quad i = 1,2.$$

Then we define the discrete counterparts of $\mathcal{S}_i$ $(i = 1,2)$:

$$\mathcal{S}_{i,h} : U_\Gamma \to A^i_{\Gamma\Gamma} U_\Gamma + A^i_{\Gamma i} U_i,$$

where $A_{11}U_1 = -A_{1\Gamma}U_\Gamma$ (Dirichlet problem). Its inverse reads

$$\mathcal{S}_{i,h}^{-1} : G \to V_\Gamma,$$

where

$$\begin{pmatrix} A_{ii} & A_{i\Gamma} \\ A_{i\Gamma} & A^i_{\Gamma\Gamma} \end{pmatrix} \begin{pmatrix} V_i \\ V_\Gamma \end{pmatrix} = \begin{pmatrix} 0 \\ G \end{pmatrix} \qquad \text{(Neumann problem)}.$$

We have $\mathcal{S}_{i,h}(0,.) = \mathcal{S}_{1,h} + \mathcal{S}_{2,h}$ and the Neumann-Neumann preconditioner is

$$\mathcal{T}_h = \frac{1}{4}(\mathcal{S}_{1,h}^{-1} + \mathcal{S}_{2,h}^{-1}).$$

For a symmetric decomposition of the domain, it is exact.

For a definition of the preconditioner for an arbitrary decomposition, see e.g. [LT94].

2.3 The non-symmetric continuous case: The Robin-Robin preconditioner

We shall see that the substructured formulation works as in the SPD case but that the Neumann-Neumann preconditioner has to be modified.

We consider the convection-diffusion equation arising from the time discretization by a backward Euler scheme of the time-dependent equation

$$\mathcal{L} := \frac{1}{\Delta t} + \boldsymbol{a}.\nabla - \nu\Delta,$$

where Δt is the time step, $\boldsymbol{a}$ is a given vector field and ν is the viscosity. The domain on which the equations are posed is the plane $\mathbf{R}^2$ decomposed into two half-planes as in section 2.1.2. By replacing $-\Delta$ by $\mathcal{L}$ in the definition of $\mathcal{S}$ (2.1), the substructured formulation is still given by (2.2). The Neumann-Neumann preconditioner is still given by

$$\mathcal{T} := \frac{1}{4}(\mathcal{S}_1^{-1} + \mathcal{S}_2^{-1}), \tag{2.5}$$

where

$$\mathcal{S}_i : v_\Gamma \to \left.\frac{\partial v_i}{\partial n_i}\right|_\Gamma \tag{2.6}$$

and

$$\begin{cases} \mathcal{L}(v_i) = 0 & \text{in } \Omega_i, \\ v_i = v_\Gamma & \text{on } \Gamma. \end{cases}$$

It is clear that $\mathcal{S}(0,.) = \mathcal{S}_1 + \mathcal{S}_2$. But, due to the non symmetry of the operator $\mathcal{L}$, the operators $\mathcal{S}_1$ and $\mathcal{S}_2$ are different in general. Thus, the Neumann-Neumann preconditioner $\mathcal{T}$ is no longer exact, $\mathcal{T}\mathcal{S}(0,.) \neq \text{Id}$. In order to see exactly how $\mathcal{S}_1$ and $\mathcal{S}_2$ differ, we perform a Fourier analysis. It will lead us to a new preconditioner adapted to the non-symmetric case: the Robin-Robin preconditioner.

2.3.1 A Fourier analysis

By using Fourier transform, we give an explicit form for the operators $\mathcal{S}_i$, $i = 1, 2$, see above (2.6). We denote the partial Fourier transform of $f(x, y) : \mathbf{R}^2 \to \mathbf{R}$ in the y variable by

$$\hat{f}(x,k) = \mathcal{F}_y(f)(x,k) := \int_{-\infty}^{\infty} e^{-Iky} f(x,y)\, dy$$

($I^2 = -1$), and the inverse Fourier transform of $\hat{f}(x, k)$ by

$$f(x,y) = \mathcal{F}_y^{-1}(\hat{f})(x,y) := \frac{1}{2\pi} \int_{-\infty}^{\infty} e^{Iky} \hat{f}(x,k)\, dk.$$

Our analysis will also involve the Fourier transform of a convolution operator with kernel $h(y)$,

$$\Lambda(u)(y) := \int_{-\infty}^{\infty} h(y-z) u(z)\, dz,$$

whose Fourier transform is given by

$$\mathcal{F}_y(\Lambda(u))(k) = \hat{h}(k)\hat{u}(k)$$

or equivalently with $\hat{\Lambda}(k) := \hat{h}(k)$,

$$\Lambda(u) = \mathcal{F}_y^{-1}(\hat{\Lambda}(k)\hat{u}(k)).$$

The function $\hat{\Lambda}(k)$ is called the symbol of the operator Λ. For example, the symbol of the operator $-\partial_{yy}$ is the polynomial k^2. More generally, the symbol of any constant coefficient differential operator is a polynomial in the Fourier variable k and conversely.

The vector field $\boldsymbol{a} = (a_1, a_2)^T$ and the coefficient ν are assumed to be constants and $a_1 > 0$. We take the partial Fourier transform of (2.6) in the y direction

$$\left(\frac{1}{\Delta t} + a_1 \partial_x + a_2 Ik - \nu \frac{\partial^2}{\partial x^2} + \nu k^2\right) (\hat{v}_i(x,k) = 0.$$

For a fixed k, this is an ordinary differential equation in x whose solution is sought in the form $\sum_\alpha c_\alpha exp(\lambda_\alpha x)$, so that λ_α is a root of the second order polynomial

$$\frac{1}{\Delta t} + a_1 \lambda_\alpha + a_2 Ik - \nu \lambda_\alpha^2 + \nu k^2.$$

We have two roots with opposite signs:

$$\lambda^{\pm} = \frac{a_1 \pm \sqrt{a_1^2 + 4\nu(\frac{1}{\Delta t} + Ika_2 + \nu k^2)}}{2\nu}. \tag{2.7}$$

The general form of $\hat{v}_i(x, k)$ is thus

$$\hat{v}_i(x,k) = c_i^+(k) e^{\lambda^+ x} + c_i^-(k) e^{\lambda^- x}.$$

The coefficients $c_i^{\pm}$ are computed from the boundary conditions. We consider first $\hat{v}_1$. The solution must be bounded as $x \to -\infty$ so that $c_i^-(k) \equiv 0$. From the Dirichlet boundary condition at $x = 0$ we get

$$\hat{v}_1(x,k) = \hat{v}_\Gamma(k) e^{\lambda^+ x}$$

and similarly,

$$\hat{v}_2(x,k) = \hat{v}_\Gamma(k) e^{\lambda^- x}.$$

The symbols of $\mathcal{S}_1$ and $\mathcal{S}_2$ are thus

$$\hat{\mathcal{S}}_1(k) = \lambda^+(k) \tag{2.8}$$

and (due to $n_2 = -\partial_x$)

$$\hat{\mathcal{S}}_2(k) = -\lambda^-(k). \tag{2.9}$$

The operators $\mathcal{S}_1$ and $\mathcal{S}_2$ differ: $\mathcal{S}_1(k) - \mathcal{S}_2(k) = a_1/\nu$.

2.3.2 Definition of the Robin-Robin preconditioner

Thanks to these formulas it is possible to split the operator $\mathcal{S}(0,.)$ into two equal contributions of each subdomain:

$$\mathcal{S}(0,.) = \tilde{\mathcal{S}}_1 + \tilde{\mathcal{S}}_2,$$

where

$$\hat{\tilde{\mathcal{S}}}_1 = \hat{\tilde{\mathcal{S}}}_2 = \hat{\mathcal{S}}_1 - \frac{a_1}{2\nu} = \hat{\mathcal{S}}_2 + \frac{a_1}{2\nu}.$$

Noticing that $a_1 = \boldsymbol{a}.n_1 = -\boldsymbol{a}.n_2$, it is possible to give an intrinsic definition of the operators $\tilde{\mathcal{S}}_i$, $i = 1, 2$:

$$\tilde{\mathcal{S}}_i : v_\Gamma \to \left.\frac{\partial v_i}{\partial n_i}\right|_\Gamma - \frac{\boldsymbol{a} \cdot n_i}{2\nu} v_i, \tag{2.10}$$

where

$$\begin{cases} \mathcal{L}(v_i) = 0 & \text{in } \Omega_i, \\ v_i = v_\Gamma & \text{on } \Gamma. \end{cases}$$

The inverse of $\tilde{\mathcal{S}}_i$, $i = 1, 2$, amounts to solving a Robin boundary value problem:

$$\tilde{\mathcal{S}}_i^{-1} : g_\Gamma \to w_i|_\Gamma, \tag{2.11}$$

where

$$\begin{cases} \mathcal{L}(w_i) = 0 & \text{in } \Omega_i, \\ \left(\left.\frac{\partial}{\partial n_i}\right|_\Gamma - \boldsymbol{a}.n_i/2\nu \right)(w_i) = g_\Gamma & \text{on } \Gamma. \end{cases}$$

The Robin-Robin preconditioner defined by

$$\tilde{\mathcal{T}} := \frac{1}{4}(\tilde{\mathcal{S}}_1^{-1} + \tilde{\mathcal{S}}_2^{-1}) \tag{2.12}$$

is exact, $\tilde{\mathcal{T}}\mathcal{S}(0,.) = \text{Id}$.

2.3.3 Numerical results

As examples we show 2D and 3D numerical results.

Consider a two dimensional flow with a velocity with a boundary layer near a wall

$$\begin{aligned} \boldsymbol{a} &= 3 - (300 * (x_2 - 0.1)^2)\boldsymbol{e}_1 \quad \text{if} \quad x_2 < 0.1 \\ \boldsymbol{a} &= 3\boldsymbol{e}_1 \quad \text{if} \quad x_2 \geq 0.1. \end{aligned} \tag{2.13}$$

The computational domain is the unit square. To capture the boundary layer, the mesh is refined in the x_2-direction, near the wall $x_2 = 0$ with a geometric progression of ratio 0.9. The advection-diffusion is discretized on a Cartesian grid by a Q1-streamline-diffusion method. The system for the nodal values at the interface is solved by a preconditioned GMRES algorithm, and the stopping criterion is to reduce the initial residual by a factor 10^{-10}. The preconditioners are either of the type Robin-Robin (R-R), Neumann-Neumann (N-N) or the identity (–).

Table 1: Iteration counts.

Partition		4×1	8×1	12×1	24×1	36×1
Grid		20×40	20×40	20×40	20×40	20×40
$\Delta t = 1$	R-R	11	18	25	39	51
$\nu = 0.001$	–	51	75	91	> 100	> 100
	N-N	49	> 100	> 100	> 100	> 100

In the three dimensional case, we solved the convection diffusion problem, with $\boldsymbol{a} = (y/2 - 0.5, -x/2 + 0.5, 0)$, $\Delta t = 10$ and $\nu = 1$ on the unstructured decomposition of Fig. 4. Here the unit cube contains 24576 tetrahedric second order finite elements and is split into 45 subdomains by an automatic mesh partitioner. This is why the boundaries between subdomains are less regular than for the other computations. For this decomposition the algorithm converges in 48 iterations with the R-R preconditioner.

2.4 Generalities

Except for numerical results, we have considered so far very simple geometries. Of course, the ideas presented above are used for arbitrary decompositions: see e.g. [LT94] for the SPD case and [ATNV00] for the convection-diffusion equation.

For arbitrary decompositions, the Neumann-Neumann or the Robin-Robin preconditioners are no longer exact. A general theory has been developed for SPD scalar problems. In the case of the scalar Laplace operator, the main result is that the condition number of the preconditioned system is $O(1/H^2(1 + \log(H/h)^2))$, where h is a typical mesh size and H is a typical diameter of a subdomain. The term $\log(H/h)^2$ comes from multiple intersection points. The more problematic term $1/H^2$ comes from the lack of global exchange mechanism in the preconditioner in order to capture the "average" value of the solution. By adding a coarse grid preconditioner, see [LT94, CMW95], it is possible to improve the condition

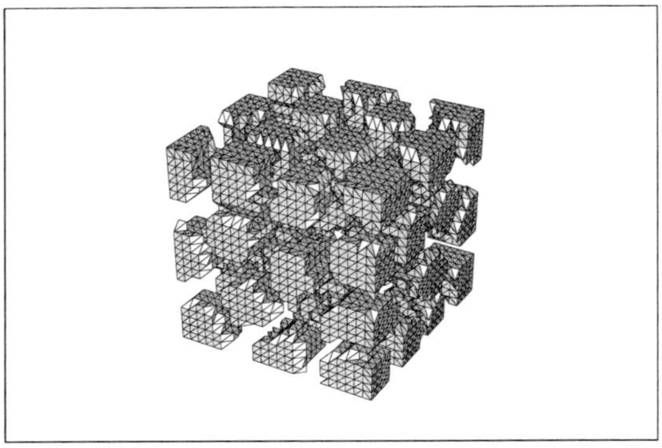

Figure 4: Three dimensional triangulation and automatic decomposition into 45 subdomains.

number of the preconditioned system. Roughly speaking, the coarse grid preconditioner consists in decomposing the solution in each subdomain into an average value and its variation. Solving a global problem for these average values improves the convergence rate so that it is $O((1+\log(H/h)^2))$. The iteration count is then almost mesh/decomposition independent.

3 Modified Schwarz method

The Additive Schwarz Method, (1.2), presents the drawback of needing overlapping subdomains in order to converge. In this chapter, we consider several improvements:

- replacement of the Dirichlet interface conditions by mixed interface conditions which yield convergence for non overlapping domain decompositions, see section 3.1;
- optimization of the interface conditions for faster convergence, see section 3.3;
- replacement of the fixed point iterative strategy of (1.2) by Krylov type methods, see section 3.4.1.

3.1 A general convergence result

A major improvement of the ASM method comes from the use of other interface conditions. It has first been proposed by P. L. Lions to replace the Dirichlet interface conditions by Robin interface conditions, see [Lio90]. Let α be a positive number; the modified algorithm reads:

$$\begin{aligned} -\Delta(u_1^{n+1}) &= f \quad \text{in} \quad \Omega_1, \\ u_1^{n+1} &= 0 \quad \text{on} \quad \partial\Omega_1 \cap \partial\Omega, \\ \left(\frac{\partial}{\partial n_1} + \alpha\right)(u_1^{n+1}) &= \left(-\frac{\partial}{\partial n_2} + \alpha\right)(u_2^n) \quad \text{on} \quad \partial\Omega_1 \cap \overline{\Omega_2} \end{aligned}$$

(n_1 and n_2 are the outward normals on the boundary of the subdomains),

$$\begin{aligned} -\Delta(u_2^{n+1}) &= f \quad \text{in} \quad \Omega_2, \\ u_2^{n+1} &= 0 \quad \text{on} \quad \partial\Omega_2 \cap \partial\Omega \\ \left(\frac{\partial}{\partial n_2} + \alpha\right)(u_2^{n+1}) &= \left(-\frac{\partial}{\partial n_1} + \alpha\right)(u_1^n) \quad \text{on} \quad \partial\Omega_2 \cap \overline{\Omega_1}. \end{aligned}$$

The convergence proof given by P. L. Lions in the elliptic case was extended by B. Després [Des93] to the Helmholtz equation. A general presentation is given in [CGJ00]. We treat here the elliptic case with second order tangential derivatives in the interface conditions.

Let Ω be an open set. We consider the following problem: Find u such that

$$\begin{aligned} \eta(\mathbf{x})u - \operatorname{div}(\kappa(\mathbf{x})\nabla u) &= f \quad \text{in} \quad \Omega, \\ u &= 0 \quad \text{on} \quad \partial\Omega, \end{aligned}$$

where the functions $x \mapsto \eta(\mathbf{x}), \kappa(\mathbf{x})$ are bounded from below by a positive constant.

The domain is decomposed into N nonoverlapping subdomains $(\Omega_i)_{1\le i\le N}$, $\bar{\Omega} = \bigcup_{i=1}^N \bar{\Omega}_i$ and $\bar{\Omega}_i \cap \bar{\Omega}_j = \emptyset$ for $i \neq j$. Let Γ_{ij} denote the interface $\Gamma_{ij} = \partial\Omega_i \cap \partial\Omega_j$, $i \neq j$. For two disjoints subdomains, $\Gamma_{ij} = \emptyset$.

For the sake of simplicity in writing the interface conditions, we consider the two dimensional case ($\Omega \subset \mathbf{R}^2$) although the proof is valid in arbitrary dimension. The interface conditions include second order tangential derivatives and have the form

$$\kappa(\mathbf{x})\frac{\partial}{\partial n_i} + \alpha_{ij}(\mathbf{x}) - \frac{\partial}{\partial \tau_i}\left(\beta_{ij}(\mathbf{x})\frac{\partial}{\partial \tau_i}\right),$$

where α_{ij} and β_{ij} are functions from Γ_{ij} into $\mathbf{R}$.

The algorithm reads:

$$\begin{aligned} \eta(\mathbf{x})u_i^{n+1} - \operatorname{div}(\kappa(\mathbf{x})\nabla u_i^{n+1}) &= f \quad \text{in} \quad \Omega_i, \\ u_i^{n+1} &= 0 \quad \text{on} \quad \partial\Omega \cap \partial\Omega_i \\ \kappa(\mathbf{x})\frac{\partial u_i^{n+1}}{\partial n_i} + \alpha_{ij}(\mathbf{x})u_i^{n+1} - \frac{\partial}{\partial \tau_i}\left(\beta_{ij}(\mathbf{x})\frac{\partial u_i^{n+1}}{\partial \tau_i}\right) & \\ = -\kappa(\mathbf{x})\frac{\partial u_j^n}{\partial n_j} + \alpha_{ij}(\mathbf{x})u_j^n - \frac{\partial}{\partial \tau_j}&\left(\beta_{ij}(\mathbf{x})\frac{\partial u_j^n}{\partial \tau_j}\right) \quad \text{on} \quad \Gamma_{ij}. \end{aligned} \tag{3.1}$$

We make the following assumptions on the coefficients of the interface conditions:

$$\begin{aligned} \alpha_{ij}(\mathbf{x}) &= \alpha_{ji}(\mathbf{x}) \ge \alpha_0 > 0, \\ \beta(\mathbf{x})_{ij} = \beta(\mathbf{x})_{ji} \ge 0 \quad &\text{and} \quad \beta_{ij}(\mathbf{x}) = 0 \quad \text{on} \quad \partial\Gamma_{ij} \end{aligned}$$

3.1 Theorem *With the above assumptions, algorithm* (3.1) *converges in* H^1, *i.e.*

$$\lim_{n\to\infty} \|u_i^n - u_{|\Omega_i}\|_{H^1(\Omega_i)}, \quad \textit{for} \quad i = 1, \ldots, N.$$

Proof Let us denote the operator

$$\Lambda_{ij} = \alpha_{ij}(\mathbf{x}) - \frac{\partial}{\partial \tau_i}\left(\beta_{ij}(\mathbf{x})\frac{\partial}{\partial \tau_i}\right), \quad \mathbf{x} \in \Gamma_{ij}.$$

From the assumptions of the theorem, we have the following properties of Λ_{ij}:

- $\Lambda_{ij} = \Lambda_{ji}$;
- Λ_{ij} is SPD (symmetric positive definite);
- Λ_{ij} is invertible.

Therefore, Λ_{ij} has an invertible SPD square root, denoted by $\Lambda_{ij}^{1/2}$, whose inverse is denoted by $\Lambda_{ij}^{-1/2}$. These operators are SPD as well.

The interface condition is rewritten as

$$\Lambda_{ij}^{-1/2}\left(\kappa(\mathbf{x})\frac{\partial u_i}{\partial n_i}\right) + \Lambda_{ij}^{1/2}(u_i) = -\Lambda_{ij}^{-1/2}\left(\kappa(\mathbf{x})\frac{\partial u_j}{\partial n_j}\right) + \Lambda_{ij}^{1/2}(u_j) \quad \text{on} \quad \Gamma_{ij}.$$

The proof follows the arguments given in [CGJ00] and is based on an energy estimate.

3.2 Lemma (Energy estimate) *Let* u *denote a function that satisfies*

$$\eta(\mathbf{x})u - \operatorname{div}(\kappa(\mathbf{x})\nabla u) = 0 \quad \textit{in} \quad \Omega_i$$
$$u = 0 \quad \textit{on} \quad \partial\Omega_i \cap \partial\Omega,$$

Then

$$\int_{\Omega_i} \eta(\mathbf{x})|u_i|^2 + \kappa(\mathbf{x})|\nabla u_i|^2 + \frac{1}{4}\sum_{j\neq i}\int_{\partial\Gamma_{ij}}\left(\Lambda_{ij}^{-1/2}\left[\kappa(\mathbf{x})\frac{\partial u_i}{\partial n_i} - \Lambda_{ij}(u_i)\right]\right)^2$$
$$= \frac{1}{4}\sum_{j\neq i}\int_{\partial\Gamma_{ij}}\left(\Lambda_{ij}^{-1/2}\left[\kappa(\mathbf{x})\frac{\partial u_i}{\partial n_i} + \Lambda_{ij}(u_i)\right]\right)^2.$$

Proof From

$$\eta(\mathbf{x})u_i - \operatorname{div}(\kappa(\mathbf{x})\nabla u_i) = 0 \quad \text{in} \quad \Omega_i,$$

we get

$$\begin{aligned}
\int_{\Omega_i} \eta(\mathbf{x})|u_i|^2 + \kappa(\mathbf{x})|\nabla u_i|^2 &= \int_{\partial\Omega_i} \kappa(\mathbf{x})\frac{\partial u_i}{\partial n_i}u_i \\
&= \sum_{j\neq i}\int_{\partial\Gamma_{ij}} \kappa(\mathbf{x})\frac{\partial u_i}{\partial n_i}u_i \\
&= \sum_{j\neq i}\int_{\partial\Gamma_{ij}} \kappa(\mathbf{x})\frac{\partial u_i}{\partial n_i}\Lambda_{ij}^{-1/2}\Lambda_{ij}^{1/2}(u_i) \\
&= \sum_{j\neq i}\int_{\partial\Gamma_{ij}} \Lambda_{ij}^{-1/2}\left(\kappa(\mathbf{x})\frac{\partial u_i}{\partial n_i}\right)\Lambda_{ij}^{1/2}(u_i).
\end{aligned}$$

From $ab = 1/4((a+b)^2 - (a-b)^2)$ we infer

$$\int_{\Omega_i} \eta(\mathbf{x})|u_i|^2 + \kappa(\mathbf{x})|\nabla u_i|^2 + \sum_{j\neq i} \frac{1}{4}\int_{\partial\Gamma_{ij}} \left(\Lambda_{ij}^{-1/2}\left(\kappa(\mathbf{x})\frac{\partial u_i}{\partial n_i}\right) - \Lambda_{ij}^{1/2}(u_i)\right)^2$$
$$= \sum_{j\neq i} \frac{1}{4}\int_{\partial\Gamma_{ij}} \left(\Lambda_{ij}^{-1/2}\left(\kappa(\mathbf{x})\frac{\partial u_i}{\partial n_i}\right) + \Lambda_{ij}^{1/2}(u_i)\right)^2$$

□

Proof of Theorem 3.1 We prove that $e_i^n = u_i^n - u_{\Omega_i}$ converges to zero. By the linearity of the equations and of the algorithm, it is clear that the error e_i^n satisfies

$$\eta(\mathbf{x})e_i^{n+1} - \operatorname{div}(\kappa(\mathbf{x})\nabla e_i^{n+1}) = 0 \quad \text{in} \quad \Omega_i,$$
$$e_i^{n+1} = 0 \quad \text{on} \quad \partial\Omega \cap \partial\Omega_i$$
$$\Lambda_{ij}^{-1/2}\left(\kappa(\mathbf{x})\frac{\partial e_i^{n+1}}{\partial n_i}\right) + \Lambda_{ij}^{1/2}(e_i^{n+1}) = -\Lambda_{ij}^{-1/2}\left(\kappa(\mathbf{x})\frac{\partial e_j^n}{\partial n_j}\right) + \Lambda_{ij}^{1/2}(e_j^n) \quad \text{on} \quad \Gamma_{ij}.$$

We apply the energy estimate to e_i^{n+1} and taking into account the interface condition (3.1) and noticing that by assumption we have $\Lambda_{ij} = \Lambda_{ji}$, we get

$$\int_{\Omega_i} \eta(\mathbf{x})|e_i^{n+1}|^2 + \kappa(\mathbf{x})|\nabla u_i^{n+1}|^2 = \sum_{j\neq i} \frac{1}{4}\int_{\partial\Gamma_{ij}} \left(\Lambda_{ji}^{-1/2}\left(-\kappa(\mathbf{x})\frac{\partial e_j^n}{\partial n_j}\right) + \Lambda_{ji}^{1/2}(e_j^n)\right)^2$$
$$- \left(\Lambda_{ij}^{-1/2}\left(\kappa(\mathbf{x})\frac{\partial e_i^{n+1}}{\partial n_i}\right) - \Lambda_{ij}^{1/2}(e_i^{n+1})\right)^2.$$

We introduce some notations:

$$\mathcal{E}_i^{n+1} := \int_{\Omega_i} \eta(\mathbf{x})|u_i^{n+1}|^2 + \kappa(\mathbf{x})|\nabla u_i^{n+1}|^2,$$

and

$$\mathcal{C}_{ij}^{n+1} := \frac{1}{4}\int_{\partial\Gamma_{ij}} \left(\Lambda_{ij}^{-1/2}\left(\kappa(\mathbf{x})\frac{\partial u_i^{n+1}}{\partial n_i}\right) - \Lambda_{ij}^{1/2}(u_i^{n+1})\right)^2.$$

The above estimate then reads:

$$\mathcal{E}_i^{n+1} + \sum_{j\neq i} \mathcal{C}_{ij}^{n+1} = \sum_{j\neq i} \mathcal{C}_{ji}^n.$$

After summation over the subdomains, we have

$$\sum_{i=1}^N \mathcal{E}_i^{n+1} + \sum_{\substack{i,j \\ (j\neq i)}} \mathcal{C}_{ij}^{n+1} = \sum_{\substack{i,j \\ (j\neq i)}} \mathcal{C}_{ji}^n = \sum_{\substack{i,j \\ (j\neq i)}} \mathcal{C}_{ij}^n.$$

We introduce the further notations: $\mathcal{E}^{n+1} = \sum_{i=1}^N \mathcal{E}_i^{n+1}$ and $\mathcal{C}^n = \sum_{i,j(j\neq i)} \mathcal{C}_{ji}^n$.

So far we have

$$\mathcal{E}^{n+1} + \mathcal{C}^{n+1} = \mathcal{C}^n.$$

Hence, by summation over n, we get

$$\sum_{n=0}^{\infty} \mathcal{E}^{n+1} \leq \mathcal{C}^0.$$

The strong convergence of the algorithm in H^1 is proved. □

The same kind of proof holds for the Maxwell system [DJR92] and the convection-diffusion equation [NR95].

3.2 Optimal interface conditions

In the preceding section, we have proved a general convergence result for interface conditions with second order tangential derivatives. Actually these conditions are not the most general. Rather than give the general conditions in an *a priori* form, we shall derive them in this section so as to have the fastest convergence. We establish the existence of interface conditions which are optimal in terms of iteration counts. The corresponding interface conditions are pseudo-differential and are not practical. Nevertheless, this result is a guide for the choice of partial differential interface conditions. Moreover, this result establishes a link between the optimal interface conditions and artificial boundary conditions. This is also a help when dealing with the design of interface conditions since it gives the possibility to use the numerous papers and books published on the subject of artificial boundary conditions, see e.g. [EM77, Giv92].

We consider a general linear second order elliptic partial differential operator $\mathcal{L}$ and the problem:

Find u such that $\mathcal{L}(u) = f$ in a domain Ω and $u = 0$ on $\partial\Omega$.

The domain Ω is decomposed into two subdomains Ω_1 and Ω_2. We suppose that the problem is regular so that $u_i := u|_{\Omega_i}$, $i = 1, 2$, is continuous and has continuous normal derivatives across the interface $\Gamma_i = \partial\Omega_i \cap \bar{\Omega}_j$, $i \neq j$.

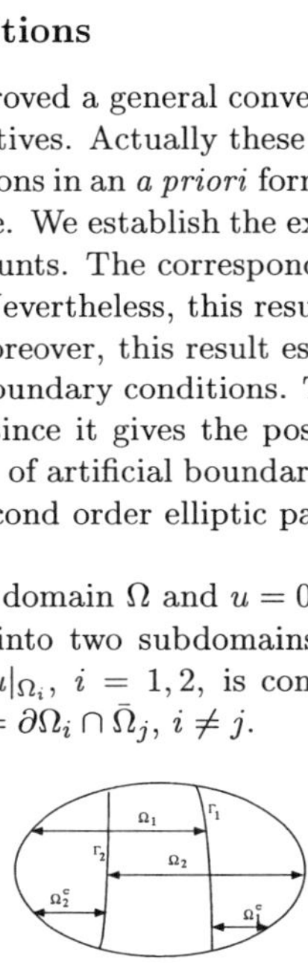

A modified Schwarz type method is considered.

$$\begin{array}{ll}
\mathcal{L}(u_1^{n+1}) = f \quad \text{in} \quad \Omega_1 & \mathcal{L}(u_2^{n+1}) = f \quad \text{in} \quad \Omega_2 \\
u_1^{n+1} = 0 \quad \text{on} \quad \partial\Omega_1 \cap \partial\Omega & u_2^{n+1} = 0 \quad \text{on} \quad \partial\Omega_2 \cap \partial\Omega \\
\mu_1 \nabla u_1^{n+1}.\boldsymbol{n_1} + \mathcal{B}_1(u_1^{n+1}) & \mu_2 \nabla u_2^{n+1}.\boldsymbol{n_2} + \mathcal{B}_2(u_2^{n+1}) \\
= -\mu_1 \nabla u_2^n.\boldsymbol{n_2} + \mathcal{B}_1(u_2^n) \quad \text{on} \quad \Gamma_1 & = -\mu_2 \nabla u_1^n.\boldsymbol{n_1} + \mathcal{B}_2(u_1^n) \quad \text{on} \quad \Gamma_2
\end{array} \tag{3.2}$$

where μ_1 and μ_2 are real-valued functions and $\mathcal{B}_1$ and $\mathcal{B}_2$ are operators acting along the interfaces Γ_1 and Γ_2. For instance, $\mu_1 = \mu_2 = 0$ and $\mathcal{B}_1 = \mathcal{B}_2 = \text{Id}$ correspond to the ASM algorithm (1.2); $\mu_1 = \mu_2 = 1$ and $\mathcal{B}_i = \alpha \in \mathbf{R}$, $i = 1, 2$, has been proposed in [Lio90] by P. L. Lions.

The question is:

> *Are there other possibilities in order to have convergence in a minimal number of steps?*

In order to answer this question, we note that by linearity, the error e satisfies ($\mu_1 = \mu_2 = 1$)

$$\begin{aligned} \mathcal{L}(e_1^{n+1}) &= 0 \quad \text{in} \quad \Omega_1 \\ e_1^{n+1} &= 0 \quad \text{on} \quad \partial\Omega_1 \cap \partial\Omega \\ \nabla e_1^{n+1}.\boldsymbol{n_1} &+ \mathcal{B}_1(e_1^{n+1}) \\ = -\nabla e_2^n.\boldsymbol{n_2} &+ \mathcal{B}_1(e_2^n) \quad \text{on} \quad \Gamma_1 \end{aligned} \qquad \begin{aligned} \mathcal{L}(e_2^{n+1}) &= 0 \quad \text{in} \quad \Omega_2 \\ e_2^{n+1} &= 0 \quad \text{on} \quad \partial\Omega_2 \cap \partial\Omega \\ \nabla e_2^{n+1}.\boldsymbol{n_2} &+ \mathcal{B}_2(e_2^{n+1}) \\ = -\nabla e_1^n.\boldsymbol{n_1} &+ \mathcal{B}_2(e_1^n) \quad \text{on} \quad \Gamma_2 \end{aligned}$$

The initial guess e_i^0 is arbitrary so that it is impossible to have convergence at step 1 of the algorithm. Convergence needs at least two iterations.

Having $e_1^2 \equiv 0$ requires

$$-\nabla e_2^1.\boldsymbol{n_2} + \mathcal{B}_1(e_2^1) \equiv 0.$$

The only meaningful information on e_2^1 is that

$$\mathcal{L}(e_2^1) = 0 \quad \text{in} \quad \Omega_2.$$

In order to use this information, we introduce the DtN (Dirichlet to Neumann) map (a.k.a. Steklov-Poincaré): Let

$$\begin{aligned} u_0 &: \Gamma_1 \to \mathbf{R} \\ \mathrm{DtN}_2(u_0) &:= \nabla v.n_{2|\partial\Omega_1 \cap \bar{\Omega}_2}, \end{aligned} \tag{3.3}$$

where n_2 is the outward normal to $\Omega_2 \setminus \bar{\Omega}_1$, and v satisfies the following boundary value problem:

$$\begin{aligned} \mathcal{L}(v) &= 0 \quad \text{in} \quad \Omega_2 \setminus \bar{\Omega}_1 \\ v &= 0 \quad \text{on} \quad \partial\Omega_2 \cap \partial\Omega \\ v &= u_0 \quad \text{on} \quad \partial\Omega_1 \cap \bar{\Omega}_2. \end{aligned}$$

We take

$$\mathcal{B}_1 := \mathrm{DtN}_2.$$

This choice is optimal since we have

$$-\nabla e_2^1.\boldsymbol{n_2} + \mathcal{B}_1(e_2^1) \equiv 0.$$

Indeed, in $\Omega_2 \setminus \bar{\Omega}_1 \subset \Omega_2$, e_2^1 satisfies

$$\mathcal{L}(e_2^1) = 0.$$

Hence,

$$\begin{aligned} \nabla e_2^1.n_2 &= \mathrm{DtN}_2(e_2^1) \\ \nabla e_2^1.\boldsymbol{n_2} &= \mathcal{B}_1(e_2^1) \quad (\mathcal{B}_1 = \mathrm{DtN}_2) \end{aligned}$$

We have formally proved

3.3 Result *The use of $\mathcal{B}_i = \mathrm{DtN}_j$ $(i \neq j)$ as interface conditions in* (3.2) *is optimal: we have (exact) convergence in two iterations.*

The two-domain case for an operator with constant coefficients has been first treated in [HTJ88]. The multidomain case for a variable coefficient operator with both positive results [NRdS94] and negative conjectures [Nie99] has been considered as well.

3.4 Remark The main feature of this result is to be very general since it does not depend on the exact form of the operator $\mathcal{L}$ and can be extended to systems or to coupled systems of equations as well with a proper care of the well posedness of the algorithm.

As an application, we take $\Omega = \mathbf{R}^2$ and $\Omega_1 =]-\infty, 0[\times\mathbf{R}$. Using the same Fourier technique that was presented in section 2.3.1, it is possible to give the explicit form of the DtN operator for a constant coefficient operator. If $\mathcal{L} = \eta - \Delta$, the DtN map is a pseudo-differential operator whose symbol is

$$B_{i,\mathrm{opt}}(k) = \sqrt{\eta + k^2},$$

i.e., $\mathcal{B}_{i,\mathrm{opt}}(u)(0,y) = \int_{\mathbf{R}} B_{i,\mathrm{opt}}(k)\hat{u}(0,k)e^{Iky}\,dk$.

If $\mathcal{L}$ is a convection-diffusion operator $\mathcal{L} := \eta + \boldsymbol{a}\nabla - \nu\Delta$, the symbol of the DtN map is

$$B_{i,\mathrm{opt}}(k) = \frac{-\boldsymbol{a}.\boldsymbol{n_i} + \sqrt{(\boldsymbol{a}.\boldsymbol{n_i})^2 + 4\nu(\eta + \boldsymbol{a}.\boldsymbol{\tau_i}k\nu + \nu^2k^2)}}{2\nu}.$$

These symbols are not polynomials in the Fourier variable k so that the operators and hence the optimal interface conditions are not a partial differential operator. They correspond to exact absorbing conditions, see the contribution of L. Halpern in this volume. These conditions are used on the artificial boundary resulting from the truncation of a computational domain. On this boundary, boundary conditions have to be imposed. The solution on the truncated domain depends on the choice of this artificial condition. We say that it is an exact absorbing boundary condition if the solution computed on the truncated domain is the restriction of the solution of the original problem. Surprisingly enough, the notions of exact absorbing conditions for domain truncation and that of optimal interface conditions in domain decomposition methods coincide.

As the above examples show, they are pseudodifferential. Therefore they are difficult to implement. Moreover, in the general case of a variable coefficient operator and/or a curved boundary, the exact form of these operators is not known, although they can be approximated by partial differential operators which are easier to implement. The approximation of the DtN has been addressed by many authors since the seminal paper [EM77] by Engquist and Majda on this question.

3.3 Optimized interface conditions

The results obtained so far are quite general. In section 3.1, we have proved convergence of the domain decomposition method with interface conditions of the type

$$\partial_n + \alpha - \partial_\tau \beta \partial_\tau \tag{3.4}$$

for a general but non overlapping domain decomposition. In section 3.2, we have exhibited interface conditions which are optimal in terms of iteration counts but are pseudodifferential operators difficult to use in practice.

These results are not sufficient for the design of effective boundary conditions which for the sake of simplicity must have the form (3.4). From section 3.2, we know that the parameters α and β must somehow be such that (3.4) approximates the optimal interface conditions

$$\frac{\partial}{\partial n_i} + \text{DtN}.$$

At first sight, it seems that the approximations proposed in the field of artificial boundary conditions are also relevant in the context of domain decomposition methods. Actually this is not the case, as was proved for the convection-diffusion equation, see [JN00, JNR01].

In order to clarify the situation, we need an estimate of the convergence rate as a function of the parameters α and β, the size of the overlap and the coefficients of the partial differential operator. In particular it will provide a means for choosing the interface conditions in an optimal way. This type of study is limited to a very simple situation: a constant coefficient operator and a whole space decomposed into two half-spaces. But, let us insist on the fact that these limitations concern only this theoretical study. The optimized values of the parameters of the interface conditions can be used with success in complex applications, see section 3.4. The robustness of the approach comes from the general convergence result of section 3.1 and from the replacement of the fixed point algorithm on the interface by a Krylov type method as explained in section 3.4.1. The efficiency comes from the study below which is made possible by the use of Fourier techniques similar to the ones used in artificial boundary conditions. The method is general and has also been applied to other types of equations; see [EZ98] for the Laplace equation, [Che98] for the Maxwell system, and [WFNS98] for porous flow media.

We shall consider here the example of a symmetric positive definite problem

$$(\eta - \Delta)(u) = f \quad \text{in} \quad \mathbf{R}^2,$$

$\eta = C^t > 0$. The domain is decomposed into to half-planes $\Omega_1 = (-\infty, \delta) \times \mathbf{R}$ and $\Omega_2 = (0, \infty) \times \mathbf{R}$. We introduce an optimization procedure which allows the choice of simplified interface conditions of the form $\partial_n + \alpha$ ($\beta = 0$) which are easy to implement and lead to a good convergence of the iterative method. We consider the Schwarz algorithm

$$\begin{aligned}
&(\eta - \Delta)(u_1^{n+1}) = f(x,y), \quad (x,y) \in \Omega_1 \\
&u_1^{n+1} \text{ is bounded at infinity} \\
&\left(\frac{\partial}{\partial n_1} + \alpha\right)(u_1^{n+1})(\delta, y) = \left(-\frac{\partial}{\partial n_2} + \alpha\right)(u_2^n)(\delta, y), \quad y \in \mathbf{R}
\end{aligned} \tag{3.5}$$

and

$$\begin{aligned}
&(\eta - \Delta)(u_2^{n+1}) = f(x,y), \quad (x,y) \in \Omega_2 \\
&u_2^{n+1} \text{is bounded at infinity} \\
&\left(\frac{\partial}{\partial n_2} + \alpha\right)(u_2^{n+1})(0, y) = \left(-\frac{\partial}{\partial n_1} + \alpha\right)(u_1^n)(0, y), \quad y \in \mathbf{R}
\end{aligned} \tag{3.6}$$

and compute its convergence rate.

Computation of the convergence rate

We introduce the errors $u_i^n - u|_{\Omega_i}$, $i = 1, 2$. By linearity, the errors satisfy the above algorithm with $f = 0$:

$$\begin{aligned} (\eta - \Delta)(e_1^{n+1}) = 0 \quad \text{in} \quad \Omega_1 \\ e_1^{n+1} \text{is bounded at infinity} \\ \left(\frac{\partial}{\partial n_1} + \alpha\right)(e_1^{n+1})(\delta, y) = \left(-\frac{\partial}{\partial n_2} + \alpha\right)(e_2^n)(\delta, y), \end{aligned} \tag{3.7}$$

and

$$\begin{aligned} (\eta - \Delta)(e_2^{n+1}) = 0 \quad \text{in} \quad \Omega_2 \\ e_2^{n+1} \text{is bounded at infinity} \\ \left(\frac{\partial}{\partial n_2} + \alpha\right)(e_2^{n+1})(0, y) = \left(-\frac{\partial}{\partial n_1} + \alpha\right)(e_1^n)(0, y). \end{aligned} \tag{3.8}$$

By taking the partial Fourier transform of the first line of (3.7) in the y direction we get:

$$\left(\eta - \frac{\partial^2}{\partial x^2} + k^2\right)(\hat{e}_1^{n+1}(x, k)) = 0 \quad \text{in} \quad \Omega_1.$$

For a given k, this is an ODE whose solution is sought in the form $\sum_j \gamma_j(k) \exp(\lambda_j(k)x)$. A simple calculation shows that there are two possible values for the lambdas:

$$\lambda^{\pm}(k) = \pm\sqrt{\eta + k^2}.$$

Therefore we have

$$\hat{e}_1^{n+1}(x, k) = \gamma_+^{n+1}(k) \exp(\lambda^+(k)x) + \gamma_-^{n+1}(k) \exp(\lambda^-(k)x).$$

From the second line of (3.7), the solution must be bounded at $x = -\infty$. This implies that $\gamma_-^{n+1}(k) \equiv 0$. Thus we have

$$\hat{e}_1^{n+1}(x, k) = \gamma_+^{n+1}(k) \exp(\lambda^+(k)x)$$

or equivalently, by changing the value of the coefficient γ_+,

$$\hat{e}_1^{n+1}(x, k) = \gamma_1^{n+1}(k) \exp(\lambda^+(k)(x - \delta))$$

and similarly,

$$\hat{e}_2^{n+1}(x, k) = \gamma_2^{n+1}(k) \exp(\lambda^+(k)x)$$

with $\gamma_{1,2}^{n+1}$ to be determined. From the interface conditions we get

$$\gamma_1^{n+1}(k)(\lambda^+ + \alpha) = \gamma_2^n(k)(\lambda^- + \alpha) \exp(\lambda^-(k)\delta)$$

and

$$\gamma_2^{n+1}(k)(-\lambda^- + \alpha) = \gamma_1^n(k)(-\lambda^+ + \alpha) \exp(-\lambda^+(k)\delta).$$

Combining these two and denoting $\lambda(k) = \lambda^+(k) = -\lambda^-(k)$, we get for $i = 1, 2$,

$$\gamma_i^{n+1}(k) = \rho(k; \alpha, \delta)^2 \gamma_i^{n-1}(k)$$

with

$$\rho(k; \alpha, \delta) = \left| \frac{\lambda(k) - \alpha}{\lambda(k) + \alpha} \right| \times \exp(-\lambda(k)\delta), \tag{3.9}$$

where $\lambda(k) = \sqrt{\eta + k^2}$ and $\alpha > 0$. This formula deserves a few remarks.

- For all $k \in \mathbf{R}$, $\rho(k) < 1$ so that $\gamma_i^n(k) \to 0$ as n goes to infinity.
- When domains overlap ($\delta > 0$), $\rho(k)$ is uniformly bounded from above by a constant smaller than one, $\rho(k; \alpha, \delta) < \exp(-\sqrt{\eta}\,\delta) < 1$ and $\rho \to 0$ as k tends to infinity.
- When there is no overlap ($\delta = 0$), $\rho \to 1$ as k tends to infinity.
- Let $\xi \in \mathbf{R}$. By taking $\alpha = \lambda(\xi)$, we have $\rho(\xi) = 0$.
- For the original Schwarz method (1.2), the convergence rate is $\exp(-\lambda(k)\delta)$. For $\delta = 0$ we see once again that there is no convergence. Replacing the Dirichlet interface conditions by Robin conditions enhances the convergence by a factor $|(\lambda(k) - \alpha)/(\lambda(k) + \alpha)|$.

Optimization of the interface condition

It is possible to optimize the choice of the parameter α in order to minimize the convergence rate in the physical space which is $\max_k \rho(k; \alpha, \delta)$.

When the subdomains overlap we have seen that the convergence rate is bounded from above by a positive constant so that it can be checked that the following min-max problem

$$\max_k \rho(k; \alpha_{\text{opt}}, \delta) = \min_\alpha \max_k \rho(k; \alpha, \delta)$$

admits a unique solution.

When the subdomains do not overlap, then for any choice of α we have $\max_k \rho(k; \alpha, 0) = 1$, so that the above min-max problem is ill-posed. Anyhow, the purpose of domain decomposition methods is not to solve partial differential equations. They are used to solve the corresponding linear systems arising from their discretizations. It is possible to study the convergence rate of the related domain decomposition methods at the discrete level based on the discretization scheme, see [Nat96]. Fourier transform is replaced by discrete Fourier series, i.e. the decomposition on the vectors $V_k = (e^{ij\Delta y\, k})_{j \in \mathbf{Z}}$, $k \in \pi/(\mathbf{Z}\Delta y)$ with Δy the mesh size in the y direction. The convergence rate depends as before on the parameters of the continuous problem but also on the discrete parameters: mesh size in x and y. The resulting formula is quite complex and would be very difficult to optimize.

Nevertheless, comparison with the continuous case and numerical experiments prove that a semi-continuous approach is sufficient for finding an optimal value for the parameter α. This of course due to the fact that as the discretization parameters go to zero, the discrete convergence rate tends to its continuous counterpart.

A semi continuous approach

For the sake of simplicity, we consider only the non-overlapping case, $\delta = 0$. We keep the formula of the convergence rate in the continuous case:

$$\rho(k;\alpha) := \left| \frac{\lambda(k) - \alpha}{\lambda(k) + \alpha} \right| \tag{3.10}$$

with $\lambda(k) = \sqrt{\eta + k^2}$. But we observe that the mesh induces a truncation in the frequency domain. We have $|k| < \pi/\Delta y := k_{\max}$. For a parameter α, the convergence rate is approximated by

$$\rho_h(\alpha) = \max_{|k|<\pi/\Delta y} \rho(k;\alpha).$$

The optimization problem reads:

Find $\alpha_{\text{opt}}^{\text{sc}}$ such that

$$\rho_h(\alpha_{\text{opt}}^{\text{sc}}) = \min_{\alpha} \max_{k<\pi/\Delta y} \rho(k;\alpha). \tag{3.11}$$

It is easy to check that the optimum is given by the relation $\rho(0;\alpha_{\text{opt}}^{\text{sc}}) = \rho(k_{\max};\alpha_{\text{opt}}^{\text{sc}})$. Let $\lambda_m = \lambda(0)$ and $\lambda_M = \lambda(k_{\max})$, we have

$$\alpha_{\text{opt}}^{\text{sc}} = \sqrt{\lambda_m \lambda_M}. \tag{3.12}$$

It can then easily be checked that in the limit of small Δy,

$$\rho_h(\alpha_{\text{opt}}^{\text{sc}}) \simeq 1 - 2\sqrt{\frac{\sqrt{\eta}\Delta y}{\pi}}$$

and

$$\alpha_{\text{opt}}^{\text{sc}} \simeq \eta^{1/4} \frac{\pi}{\Delta y}.$$

Whereas for α independent of Δy, we have

$$\rho_h(\alpha) \simeq 1 - 2\frac{\alpha \Delta y}{\pi}$$

for small Δy. Numerical tests on the model problem of a rectangle divided into two half-rectangles and a finite difference discretization shows a good agreement with the above formulas. In Table 2, the iteration counts are given for two possible choices of the parameter α, $\alpha = 1$ or $\alpha_{\text{opt}}^{\text{sc}}$ given by formula (3.12). The reduction error factor is 10^{-6}.

Table 2: Number of iterations for different values of the mesh size and two possible choices for α.

$1/\Delta y$	10	20	40	80
$\alpha_{\text{opt}}^{\text{sc}}$	6	7	10	16
$\alpha = 1$	27	51	104	231

3.4 A more complex example: optimized interface conditions for the Helmholtz equation

This study is joint work with M. Gander and F. Magoulès, see [GMN01] for a complete presentation.

We consider the Helmholtz equation

$$\mathcal{L}(u) := (-\omega^2 - \Delta)(u) = f(x, y), \quad x, y \in \Omega.$$

The difficulty comes from the negative sign of the term of order zero of the operator.

Although the following analysis could be carried out on rectangular domains as well, we prefer for simplicity to present the analysis in the domain $\Omega = \mathbf{R}^2$ with the Sommerfeld radiation condition at infinity,

$$\lim_{r=\infty} \sqrt{r} \left(\frac{\partial u}{\partial r} + i\omega u \right) = 0,$$

where $r = \sqrt{x^2 + y^2}$. We decompose the domain into two non-overlapping subdomains $\Omega_1 = (-\infty, 0] \times \mathbf{R}$ and $\Omega_2 = [0, \infty) \times \mathbf{R}$ and consider the Schwarz algorithm

$$\begin{aligned} -\Delta u_1^{n+1} - \omega^2 u_1^{n+1} &= f(x, y), \quad x, y \in \Omega_1 \\ \mathcal{B}_1(u_1^{n+1})(0) &= \mathcal{B}_1(u_2^n)(0) \end{aligned} \tag{3.13}$$

and

$$\begin{aligned} -\Delta u_2^{n+1} - \omega^2 u_2^{n+1} &= f(x, y), \quad x, y \in \Omega_2 \\ \mathcal{B}_2(u_2^{n+1})(0) &= \mathcal{B}_2(u_1^n)(0) \end{aligned} \tag{3.14}$$

where $\mathcal{B}_j$, $j = 1, 2$, are two linear operators. Note that for the classical Schwarz method $\mathcal{B}_j$ is the identity, $\mathcal{B}_j = I$ and without overlap the algorithm cannot converge. But even with overlap in the case of the Helmholtz equation, only the evanescent modes in the error are damped, while the propagating modes are unaffected by the Schwarz algorithm [GMN01]. One possible remedy is to use a relatively fine coarse grid [CW92] or Robin transmission conditions, see for example [Des93, CCEW98]. We consider here a new type of transmission conditions which lead to a convergent non-overlapping version of the Schwarz method. We assume that the linear operators $\mathcal{B}_j$ are of the form

$$\mathcal{B}_j := \partial_x + \mathcal{S}_j, \quad j = 1, 2,$$

for two linear operators $\mathcal{S}_1$ and $\mathcal{S}_2$ acting in the tangential direction on the interface. Our goal is to use these operators to optimize the convergence rate of the algorithm. For the analysis it suffices by linearity to consider the case $f(x, y) = 0$ and to analyze convergence to the zero solution. Taking a Fourier transform in the y direction we obtain

$$\begin{aligned} -\frac{\partial^2 \hat{u}_1^{n+1}}{\partial x^2} - (\omega^2 - k^2)\hat{u}_1^{n+1} &= 0, \\ (\partial_x + \sigma_1(k))(\hat{u}_1^{n+1})(0) &= (\partial_x + \sigma_1(k))(\hat{u}_2^n)(0) \end{aligned} \qquad x < 0, \ k \in \mathbf{R} \tag{3.15}$$

and

$$-\frac{\partial^2 \hat{u}_2^{n+1}}{\partial x^2} - (\omega^2 - k^2)\hat{u}_2^{n+1} = 0, \qquad x > 0,\ k \in \mathbf{R} \qquad (3.16)$$
$$(\partial_x + \sigma_2(k))(\hat{u}_2^{n+1})(0) = (\partial_x + \sigma_2(k))(\hat{u}_1^n)(0)$$

where $\sigma_j(k)$ denotes the symbol of the operator $\mathcal{S}_j$, and k is the Fourier variable, which we also call frequency. The general solutions of these ordinary differential equations are

$$\hat{u}_j^{n+1} = A_j e^{\lambda(k)x} + B_j e^{-\lambda(k)x}, \quad j = 1, 2,$$

where $\lambda(k)$ denotes the root of the characteristic equation $\lambda^2 + (\omega^2 - k^2) = 0$ with positive real or imaginary part,

$$\begin{aligned} \lambda(k) &= \sqrt{k^2 - \omega^2} \quad \text{for} \quad |k| \geq \omega, \\ \lambda(k) &= i\sqrt{\omega^2 - k^2} \quad \text{for} \quad |k| < \omega. \end{aligned} \qquad (3.17)$$

Since the Sommerfeld radiation condition excludes growing solutions as well as incoming modes at infinity, we obtain the solutions

$$\begin{aligned} \hat{u}_1^{n+1}(x,k) &= \hat{u}_1^{n+1}(0,k)e^{\lambda(k)x} \\ \hat{u}_2^{n+1}(x,k) &= \hat{u}_2^{n+1}(0,k)e^{-\lambda(k)x}. \end{aligned}$$

Using the transmission conditions and the fact that

$$\begin{aligned} \frac{\partial \hat{u}_1^{n+1}}{\partial x} &= \lambda(k)\hat{u}_1^{n+1} \\ \frac{\partial \hat{u}_2^{n+1}}{\partial x} &= -\lambda(k)\hat{u}_2^{n+1} \end{aligned}$$

we obtain over one step of the Schwarz iteration

$$\begin{aligned} \hat{u}_1^{n+1}(x,k) &= \frac{-\lambda(k) + \sigma_1(k)}{\lambda(k) + \sigma_1(k)} e^{\lambda(k)x} \hat{u}_2^n(0,k) \\ \hat{u}_2^{n+1}(x,k) &= \frac{\lambda(k) + \sigma_2(k)}{-\lambda(k) + \sigma_2(k)} e^{-\lambda(k)x} \hat{u}_1^n(0,k). \end{aligned}$$

Evaluating the second equation at $x = 0$ for iteration index n and inserting it into the first equation, we get after evaluating again at $x = 0$

$$\hat{u}_1^{n+1}(0,k) = \frac{-\lambda(k) + \sigma_1(k)}{\lambda(k) + \sigma_1(k)} \cdot \frac{\lambda(k) + \sigma_2(k)}{-\lambda(k) + \sigma_2(k)} \hat{u}_1^{n-1}(0,k).$$

Defining the convergence rate ρ by

$$\rho(k) := \frac{-\lambda(k) + \sigma_1(k)}{\lambda(k) + \sigma_1(k)} \cdot \frac{\lambda(k) + \sigma_2(k)}{-\lambda(k) + \sigma_2(k)} \qquad (3.18)$$

we find by induction that

$$\hat{u}_1^{2n}(0,k) = \rho(k)^n \hat{u}_1^0(0,k),$$

and by a similar calculation on the second subdomain,

$$\hat{u}_2^{2n}(0,k) = \rho(k)^n \hat{u}_2^0(0,k).$$

Choosing in the Fourier transformed domain

$$\sigma_1(k) := \lambda(k), \qquad \sigma_2(k) := -\lambda(k)$$

corresponds to using exact absorbing boundary conditions as interface conditions. So we get $\rho(k) \equiv 0$ and the algorithm converges in two steps independently of the initial guess. Unfortunately this choice becomes difficult to use in the real domain where computations take place, since the optimal choice of the symbols $\sigma_j(k)$ leads to non-local operators $\mathcal{S}_j$ in the real domain, caused by the square root in the symbols. We have to construct local approximations for the optimal transmission conditions.

In [EM77], the approximation valid for the truncation of an infinite computational domain is obtained via Taylor expansions of the symbol in the vicinity of $k = 0$:

$$\mathcal{S}_j^{\mathrm{app}} = \pm i \left(\omega - \frac{1}{2\omega} \partial_{\tau\tau} \right),$$

which leads to the zeroth or second order Taylor transmission conditions, depending on whether one keeps only the constant term or also the second order term. But these transmission conditions are only effective for the low frequency components of the error. This is sufficient for the truncation of a domain since there is an exponential decay of the high frequency part (large k) of the solution away from the artificial boundary.

But in domain decomposition, what is important is the convergence rate which is given by the maximum over k of $\rho(k)$. Since there is no overlap between the subdomains, it is not possible to profit from any decay. We present now an approximation procedure suited to domain decomposition methods. To avoid an increase in the bandwidth of the local subproblems, we take polynomials of degree at most 2, which leads to transmission operators $\mathcal{S}_j^{\mathrm{app}}$ which are at most second order partial differential operators acting along the interface. By symmetry of the Helmholtz equation there is no interest in a first order term. We therefore approximate the operators $\mathcal{S}_j$, $j = 1, 2$, in the form $\mathcal{S}_j^{\mathrm{app}} = \pm(a + b\partial_{\tau\tau})$ with $a, b \in \mathbf{C}$ and where τ denotes the tangent direction at the interface.

Optimized Robin interface conditions for the Helmholtz equation We approximate the optimal operators $\mathcal{S}_j$, $j = 1, 2$, in the form

$$\mathcal{S}_j^{\mathrm{app}} = \pm(p + qi), \quad p, q \in \mathbf{R}^+. \tag{3.19}$$

The non-negativity of p, q comes from the Shapiro-Lopatinski necessary condition for the well-posedness of the local subproblems (3.13)–(3.14). Inserting this approximation into the convergence rate (3.18) we find

$$\rho(p,q,k) = \begin{cases} \dfrac{p^2 + \left(q - \sqrt{\omega^2 - k^2}\right)^2}{p^2 + \left(q + \sqrt{\omega^2 - k^2}\right)^2}, & \omega^2 \geq k^2 \\[2ex] \dfrac{q^2 + \left(p - \sqrt{k^2 - \omega^2}\right)^2}{q^2 + \left(p + \sqrt{k^2 - \omega^2}\right)^2}, & \omega^2 < k^2. \end{cases} \tag{3.20}$$

First note that for $k^2 = \omega^2$ the convergence rate $\rho(p, q, \omega) = 1$, no matter what one chooses for the free parameters p and q. In the Helmholtz case one can not uniformly minimize the convergence rate over all relevant frequencies, as in the case of positive definite problems, see [Jap98, GMN01, JNR01]. The point $k = \omega$ represents however only one single mode in the spectrum, and a Krylov method will easily take care of this when the Schwarz method is used as a preconditioner, as our numerical experiments will show. We therefore consider the optimization problem

$$\min_{p,q\in\mathbf{R}^+}\left(\max_{k\in(k_{\min},\omega_-)\cup(\omega_+,k_{\max})}|\rho(p,q,k)|\right), \tag{3.21}$$

where ω_- and ω_+ are parameters to be chosen, and $k_{\min}$ denotes the smallest frequency relevant to the subdomain, and $k_{\max}$ denotes the largest frequency supported by the numerical grid. This largest frequency is of the order π/h. For example, if the domain Ω is a strip of height L with homogeneous Dirichlet conditions on top and bottom, the solution can be expanded in a Fourier series with the harmonics $\sin(j\pi y/L)$, $j \in \mathbf{N}$. Hence the relevant frequencies are $k = j\pi/L$. They are equally distributed with a spacing π/L and thus choosing $\omega_- = \omega - \pi/L$ and $\omega_+ = \omega + \pi/L$ leaves precisely one frequency $k = \omega$ for the Krylov method and treats all the others by the optimization. If ω falls in between the relevant frequencies, say $j\pi/L < \omega < (j+1)\pi/L$ then we can even get the iterative method to converge by choosing $\omega_- = j\pi/L$ and $\omega_+ = (j+1)\pi/L$, which will allow us to directly verify our asymptotic analysis numerically without the use of a Krylov method. How to choose the optimal parameters p and q is given by the following:

3.5 Theorem (Optimized Robin conditions) *Under the three assumptions*

$$2\omega^2 \le \omega_-^2 + \omega_+^2, \quad \omega_- < \omega \tag{3.22}$$

$$2\omega^2 > k_{\min}^2 + \omega_+^2, \tag{3.23}$$

$$2\omega^2 < k_{\min}^2 + k_{\max}^2, \tag{3.24}$$

the solution to the min-max problem (3.21) *is unique and the optimal parameters are given by*

$$p^* = q^* = \sqrt{\frac{\sqrt{\omega^2-\omega_-^2}\,\sqrt{k_{\max}^2-\omega^2}}{2}}. \tag{3.25}$$

The optimized convergence rate (3.21) *is then given by*

$$\max_{k\in(k_{\min},\omega_-)\cup(\omega_+,k_{\max})}\rho(p^*,q^*,k) = \frac{1-\sqrt{2}\left(\frac{\omega^2-\omega_-^2}{k_{\max}^2-\omega^2}\right)^{1/4}+\sqrt{\frac{\omega^2-\omega_-^2}{k_{\max}^2-\omega^2}}}{1+\sqrt{2}\left(\frac{\omega^2-\omega_-^2}{k_{\max}^2-\omega^2}\right)^{1/4}+\sqrt{\frac{\omega^2-\omega_-^2}{k_{\max}^2-\omega^2}}} \tag{3.26}$$

For the proof, see [GMN01].

Optimized second order interface conditions for the Helmholtz equation We seek interface conditions in the form $\mathcal{S}_j^{\text{app}} = \pm(a + b\partial_{\tau\tau})$ with $a, b \in \mathbf{C}$ and where τ denotes the tangent direction at the interface. The design of optimized second order interface conditions is simplified by the following:

3.6 Lemma *Let u_1 and u_2 be two functions which satisfy*

$$\mathcal{L}(u_j) \equiv (-\omega^2 - \Delta)(u) = f \quad in \quad \Omega_j, \quad j = 1, 2,$$

and the interface condition

$$\left(\frac{\partial}{\partial n_1} + \alpha\right)\left(\frac{\partial}{\partial n_1} + \beta\right)(u_1) = \left(-\frac{\partial}{\partial n_2} + \alpha\right)\left(-\frac{\partial}{\partial n_2} + \beta\right)(u_2) \tag{3.27}$$

with $\alpha, \beta \in \mathbf{C}$, $\alpha + \beta \neq 0$, and n_j denoting the unit outward normal to domain Ω_j. Then the following second order interface condition is satisfied as well:

$$\left(\frac{\partial}{\partial n_1} + \frac{\alpha\beta - \omega^2}{\alpha + \beta} - \frac{1}{\alpha + \beta}\frac{\partial^2}{\partial \tau_1^2}\right)(u_1) = \left(-\frac{\partial}{\partial n_2} + \frac{\alpha\beta - \omega^2}{\alpha + \beta} - \frac{1}{\alpha + \beta}\frac{\partial^2}{\partial \tau_2^2}\right)(u_2) \tag{3.28}$$

Proof Expanding the interface condition (3.27) yields

$$\left(\frac{\partial^2}{\partial n_1^2} + (\alpha + \beta)\frac{\partial}{\partial n_1} + \alpha\beta\right)(u_1) = \left(\frac{\partial^2}{\partial n_2^2} - (\alpha + \beta)\frac{\partial}{\partial n_2} + \alpha\beta\right)(u_2).$$

Now using the equation $\mathcal{L}(u_1) = f$, we can substitute $-(\partial^2/\partial\tau_1^2 + \omega^2)(u_1) - f$ for $\partial^2/\partial n_1^2(u_1)$, and similarly we can substitute $-(\partial^2/\partial\tau_2^2 + \omega^2)(u_2) - f$ for $\partial^2/\partial n_2^2(u_2)$. Hence we get

$$\left(-\frac{\partial^2}{\partial \tau_1^2} - \omega^2 + (\alpha + \beta)\frac{\partial}{\partial n_1} + \alpha\beta\right)(u_1) - f = \left(-\frac{\partial^2}{\partial \tau_2^2} - \omega^2 - (\alpha + \beta)\frac{\partial}{\partial n_2} + \alpha\beta\right)(u_2) - f.$$

Now the terms f on both sides cancel, and division by $\alpha + \beta$ yields (3.28). □

Note that Higdon has already proposed approximations to absorbing boundary conditions in factored form in [Hig86]. In our case, this special choice of approximating $\sigma_j(k)$ by

$$\sigma_j^{\text{app}}(k) := \pm\left(\frac{\alpha\beta - \omega^2}{\alpha + \beta} + \frac{1}{\alpha + \beta}k^2\right) \tag{3.29}$$

leads to a particularly elegant formula for the convergence rate. Inserting $\sigma_j^{\text{app}}(k)$ into the convergence rate (3.18) and simplifying, we obtain

$$\begin{aligned}\rho(k; \alpha, \beta) := \left(\frac{-\lambda(k) - \sigma_1}{\lambda(k) + \sigma_1}\right)^2 &= \left(\frac{-(\alpha + \beta)\lambda(k) + \alpha\beta + k^2 - \omega^2}{(\alpha + \beta)\lambda(k) + \alpha\beta + k^2 - \omega^2}\right)^2 \\ &= \left(\frac{\lambda(k)^2 - (\alpha + \beta)\lambda(k) + \alpha\beta}{\lambda(k)^2 + (\alpha + \beta)\lambda(k) + \alpha\beta}\right)^2 \\ &= \left(\frac{\lambda(k) - \alpha}{\lambda(k) + \alpha}\right)^2\left(\frac{\lambda(k) - \beta}{\lambda(k) + \beta}\right)^2\end{aligned} \tag{3.30}$$

where $\lambda(k)$ is defined in (3.17), and the two parameters $\alpha, \beta \in \mathbf{C}$ can be used to optimize the performance. By the symmetry of $\lambda(k)$ with respect to k, it suffices to consider only positive k to optimize performance. We thus need to solve the min-max problem

$$\min_{\alpha, \beta \in \mathbf{C}} \left(\max_{k \in (k_{\min}, \omega_-) \cup (\omega_+, k_{\max})} |\rho(k; \alpha, \beta)| \right), \tag{3.31}$$

where ω_- and ω_+ are again the parameters to exclude the frequency $k = \omega$ where the convergence rate equals 1, as in the zeroth order optimization problem. The convergence rate $\rho(k; \alpha, \beta)$ consists of two factors, and λ is real for vanishing modes and imaginary for propagative modes. If we chose $\alpha \in i\mathbf{R}$ and $\beta \in \mathbf{R}$ then for λ real the first factor is of modulus one and the second one can be optimized using β. If λ is imaginary, then the second factor is of modulus one and the first one can be optimized independently using α. Hence for this choice of α and β the min-max problem decouples. We therefore consider here the simpler min-max problem

$$\min_{\alpha \in i\mathbf{R},\ \beta \in \mathbf{R}} \left(\max_{k \in (k_{\min}, \omega_-) \cup (\omega_+, k_{\max})} |\rho(k; \alpha, \beta)| \right) \tag{3.32}$$

which has an elegant analytical solution. Note however that the original minimization problem (3.31) might have a solution with better convergence rate, an issue investigated in [GMN01].

3.7 Theorem (Optimized second order conditions) *The solution of the min-max problem* (3.32) *is unique and the optimal parameters are given by*

$$\alpha^* = i\left((\omega^2 - k_{\min}^2)(\omega^2 - \omega_-^2)\right)^{1/4} \in i\mathbf{R} \tag{3.33}$$

and

$$\beta^* = \left((k_{\max}^2 - \omega^2)(\omega_+^2 - \omega^2)\right)^{1/4} \in \mathbf{R}. \tag{3.34}$$

The convergence rate (3.32) *is then for the propagating modes given by*

$$\max_{k \in (k_{\min}, \omega_-)} |\rho(k, \alpha^*, \beta^*)| = \left(\frac{(\omega^2 - \omega_-^2)^{1/4} - (\omega^2 - k_{\min}^2)^{1/4}}{(\omega^2 - \omega_-^2)^{1/4} + (\omega^2 - k_{\min}^2)^{1/4}} \right)^2, \tag{3.35}$$

and for the evanescent modes it is

$$\max_{k \in (\omega_+, k_{\max})} \rho(k, \alpha^*, \beta^*) = \left(\frac{(k_{\max}^2 - \omega^2)^{1/4} - (\omega_+^2 - \omega^2)^{1/4}}{(k_{\max}^2 - \omega^2)^{1/4} + (\omega_+^2 - \omega^2)^{1/4}} \right)^2. \tag{3.36}$$

Proof For $k \in (k_{\min}, \omega_-)$ we have

$$\left| \frac{i\sqrt{\omega^2 - k^2} - \beta}{i\sqrt{\omega^2 - k^2} + \beta} \right| = 1$$

since $\beta \in \mathbf{R}$ and thus

$$|\rho(k; \alpha, \beta)| = \left| \frac{i\sqrt{\omega^2 - k^2} - \alpha}{i\sqrt{\omega^2 - k^2} + \alpha} \right|^2$$

depends only on α. Similarly, for $k \in (\omega_+, k_{\max})$ we have

$$\left|\frac{\sqrt{k^2-\omega^2}-\alpha}{\sqrt{k^2-\omega^2}+\alpha}\right| = 1$$

since $\alpha \in i\mathbf{R}$, and therefore

$$|\rho(k;\alpha,\beta)| = \left|\frac{\sqrt{k^2-\omega^2}-\beta}{\sqrt{k^2-\omega^2}+\beta}\right|^2$$

depends only on β. The solution (α, β) of the minimization problem (3.32) is thus given by the solution of the two independent minimization problems

$$\min_{\alpha\in i\mathbf{R}}\left(\max_{k\in(k_{\min},\omega_-)}\left|\frac{i\sqrt{\omega^2-k^2}-\alpha}{i\sqrt{\omega^2-k^2}+\alpha}\right|\right) \tag{3.37}$$

and

$$\min_{\beta\in\mathbf{R}}\left(\max_{k\in(\omega_+,k_{\max})}\left|\frac{\sqrt{k^2-\omega^2}-\beta}{\sqrt{k^2-\omega^2}+\beta}\right|\right). \tag{3.38}$$

We show the solution for the second problem (3.38) only, the solution for the first problem (3.37) is similar. First note that the maximum of

$$|\rho_\beta| := \left|\frac{\sqrt{k^2-\omega^2}-\beta}{\sqrt{k^2-\omega^2}+\beta}\right|$$

is attained on the boundary of the interval $[\omega_+, k_{\max}]$, because the function ρ_β (but not $|\rho_\beta|$) is monotone increasing with $k \in [\omega_+, k_{\max}]$. On the other hand, as a function of β, $|\rho_\beta(\omega_+)|$ grows monotonically with β while $|\rho_\beta(k_{\max})|$ decreases monotonically with β. The optimum is therefore reached when we balance the two values on the boundary, $\rho_\beta(\omega_+) = -\rho_\beta(k_{\max})$, which implies that the optimal β satisfies the equation

$$\frac{\sqrt{k_{\max}^2-\omega^2}-\beta}{\sqrt{k_{\max}^2-\omega^2}+\beta} = -\frac{\sqrt{\omega_+^2-\omega^2}-\beta}{\sqrt{\omega_+^2-\omega^2}+\beta} \tag{3.39}$$

whose solution is given in (3.34). □

The optimization problem (3.38) arises also for symmetric positive definite problems when an optimized Schwarz algorithm without overlap and Robin transmission conditions are used and the present solution can be found in [WFNS98].

Fig. 5 shows the convergence rate obtained for a model problem on the unit square with two subdomains, $\omega = 10\pi$ and $h = 1/50$. The optimal parameters were found to be $\alpha^* = 20.741i$ and $\beta^* = 47.071$, which gives a convergence rate $\rho = 0.0419$ for the propagating modes and $\rho = 0.2826$ for the evanescent modes. It is interesting to note that with the current practice in engineering of choosing about 10 grid points per wavelength, we have $h \approx \pi/(5\omega)$, and thus for the propagating modes the optimized Schwarz method presented here has an asymptotic convergence rate of

$$\rho_p = 1 - O(h^{1/4}).$$

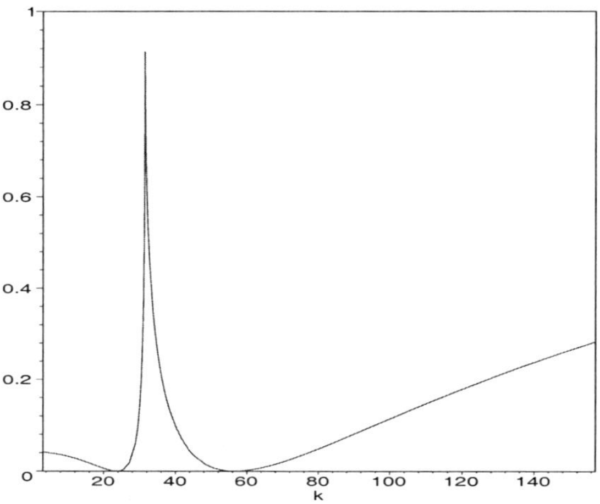

Figure 5: Convergence rate of the optimized Schwarz method with second order transmission conditions in Fourier space for $\omega = 10\pi$.

3.4.1 Numerical implementation and acceleration via Krylov type methods

This section is concerned with the Finite Element implementation of the interface conditions of Robin type and of the ones with second order tangential derivatives along the interface. We show that thanks to a reformulation of the algorithm they are as easy to implement as Neumann boundary conditions. We first treat the case of a decomposition into two subdomains and then an arbitrary decomposition of the domain.

Two-domain decomposition

We present the discretization scheme for a decomposition of a domain Ω into two subdomains Ω_1 and Ω_2 with interface Γ_{12}. So far, we have considered the optimized Schwarz algorithm at the continuous level,

$$\begin{aligned}
-\Delta u_1^{n+1} - \omega^2 u_1^{n+1} &= f_1 && \text{in} \quad \Omega_1 \\
\frac{\partial u_1^{n+1}}{\partial n_1} + \mathcal{S}_1^{\text{app}}(u_1^{n+1}) &= -\frac{\partial u_2^n}{\partial n_2} + \mathcal{S}_1^{\text{app}}(u_2^n) && \text{on} \quad \Gamma_{12} \\
-\Delta u_2^{n+1} - \omega^2 u_2^{n+1} &= f_2 && \text{in} \quad \Omega_2 \\
\frac{\partial u_2^{n+1}}{\partial n_2} + \mathcal{S}_2^{\text{app}}(u_2^{n+1}) &= -\frac{\partial u_1^n}{\partial n_1} + \mathcal{S}_2^{\text{app}}(u_1^n) && \text{on} \quad \Gamma_{12}.
\end{aligned} \tag{3.40}$$

A direct discretization would require the computation of the normal derivatives along the interfaces in order to evaluate the right hand sides in the transmission conditions of (3.40). This can be avoided by introducing two new variables,

$$\lambda_1^n = -\frac{\partial u_2^n}{\partial n_2} + \mathcal{S}_1^{\text{app}}(u_2^n) \qquad \text{and} \qquad \lambda_2^n = -\frac{\partial u_1^n}{\partial n_1} + \mathrm{S}_2^{\text{app}}(u_1^n).$$

The algorithm then becomes

$$
\begin{aligned}
-\Delta u_1^{n+1} - \omega^2 u_1^{n+1} &= f_1 \quad \text{in} \quad \Omega_1 \\
\frac{\partial u_1^{n+1}}{\partial n_1} + \mathcal{S}_1^{\text{app}}(u_1^{n+1}) &= \lambda_1^n \quad \text{on} \quad \Gamma_{12} \\
-\Delta u_2^{n+1} - \omega^2 u_2^{n+1} &= f_2 \quad \text{in} \quad \Omega_2 \\
\frac{\partial u_2^{n+1}}{\partial n_2} + \mathcal{S}_2^{\text{app}}(u_2^{n+1}) &= \lambda_2^n \quad \text{on} \quad \Gamma_{12} \\
\lambda_1^{n+1} &= -\lambda_2^n + (\mathcal{S}_1^{\text{app}} + \mathcal{S}_2^{\text{app}})(u_2^{n+1}) \\
\lambda_2^{n+1} &= -\lambda_1^n + (\mathcal{S}_1^{\text{app}} + \mathcal{S}_2^{\text{app}})(u_1^{n+1}).
\end{aligned}
\tag{3.41}
$$

We can interpret this new algorithm as a fixed point algorithm in the new variables λ_j, $j = 1, 2$, to solve the substructured problem

$$
\begin{aligned}
\lambda_1 &= -\lambda_2 + (\mathcal{S}_1^{\text{app}} + \mathcal{S}_2^{\text{app}})(u_2(\lambda_2, f_2)), \\
\lambda_2 &= -\lambda_1 + (\mathcal{S}_1^{\text{app}} + \mathcal{S}_2^{\text{app}})(u_1(\lambda_1, f_1)),
\end{aligned}
\tag{3.42}
$$

where $u_j = u_j(\lambda_j, f_j)$, $j = 1, 2$, are solutions of

$$
\begin{aligned}
-\Delta u_j - \omega^2 u_j &= f_j \quad \text{in} \quad \Omega_j, \\
\frac{\partial u_j}{\partial n_j} + \mathcal{S}_j^{\text{app}}(u_j) &= \lambda_j \quad \text{on} \quad \Gamma_{12}.
\end{aligned}
$$

Instead of solving the substructured problem (3.42) by the fixed point iteration (3.41), one usually uses a Krylov subspace method to solve the substructured problem directly. This corresponds to using the optimized Schwarz method as a preconditioner for the Krylov subspace method. A finite element discretization of the substructured problem (3.42) leads to the linear system

$$
\begin{aligned}
\lambda_1 &= -\lambda_2 + (S_1 + S_2) B_2 u_2 \\
\lambda_2 &= -\lambda_1 + (S_1 + S_2) B_1 u_1 \\
\widetilde{K}_1 u_1 &= f_1 + B_1^T \lambda_1 \\
\widetilde{K}_2 u_2 &= f_2 + B_2^T \lambda_2
\end{aligned}
\tag{3.43}
$$

where B_1 and B_2 are the trace operators of the domains Ω_1 and Ω_2 on the interface Γ_{12}, and we omit the superscript app in the discretization S_j of the continuous operators $\mathcal{S}_j^{\text{app}}$ to reduce the notation. If the two vectors u_1 and u_2 containing the degrees of freedom have their first components corresponding to the interior unknowns

$$
u_j = \begin{bmatrix} u_j^i \\ u_j^b \end{bmatrix}, \quad j = 1, 2,
\tag{3.44}
$$

where the indices i and b correspond to interior and interface degrees of freedom respectively for domain Ω_j, then the discrete trace operators B_1 and B_2 are just the boolean matrices corresponding to the decomposition (3.44) and they can be written as

$$
B_j = \begin{bmatrix} 0 & I \end{bmatrix}, \quad j = 1, 2,
\tag{3.45}
$$

where I denotes the identity matrix of appropriate size. For example, $B_1 u_1 = u_1^b$ and $B_2 u_2 = u_2^b$. The matrices $\widetilde{K}_1$ and $\widetilde{K}_2$ arise from the discretization of the local Helmholtz subproblems along with the interface conditions $\partial_n + a - b\partial_{\tau\tau}$,

$$\widetilde{K}_j = K_j - \omega^2 M_j + B_j^T(aM_{\Gamma_{12}} + bK_{\Gamma_{12}})B_j, \quad j = 1,2. \tag{3.46}$$

Here K_1 and K_2 are the stiffness matrices, M_1 and M_2 are the mass matrices, $M_{\Gamma_{12}}$ is the interface mass matrix, and $K_{\Gamma_{12}}$ is the interface stiffness matrix,

$$[M_{\Gamma_{12}}]_{nm} = \int_{\Gamma_{12}} \phi_n \phi_m \, d\xi \quad \text{and} \quad [K_{\Gamma_{12}}]_{nm} = \int_{\Gamma_{12}} \nabla_\tau \phi_n \nabla_\tau \phi_m \, d\xi. \tag{3.47}$$

The functions ϕ_n and ϕ_m are the basis functions associated with the degrees of freedom n and m on the interface Γ_{12}, and $\nabla_\tau \phi$ is the tangential component of $\nabla\phi$ on the interface. We have

$$S_j = aM_{\Gamma_{12}} + bK_{\Gamma_{12}}, \quad j = 1,2.$$

For given λ_1 and λ_2, the acoustic pressure u_1 and u_2 can be computed by solving the last two equations of (3.43). Eliminating u_1 and u_2 in the first two equations of (3.43) using the last two equations of (3.43), we obtain the substructured linear system

$$F\lambda = d, \tag{3.48}$$

where $\lambda = (\lambda_1, \lambda_2)$ and the matrix F and the right hand side d are given by

$$\begin{aligned} F &= \begin{pmatrix} I & I - (S_1+S_2)B_2\widetilde{K}_2^{-1}B_2^T \\ I - (S_1+S_2)B_1\widetilde{K}_1^{-1}B_1^T & I \end{pmatrix} \\ d &= \begin{pmatrix} (S_1+S_2)B_1\widetilde{K}_1^{-1}f_1 \\ (S_1+S_2)B_2\widetilde{K}_2^{-1}f_2. \end{pmatrix} \end{aligned} \tag{3.49}$$

The linear system (3.48) is solved by a Krylov subspace method. The matrix vector product amounts to solving a subproblem in each subdomain and to send interface data between subdomains. Note that the optimization of the interface conditions was performed for the convergence rate of the additive Schwarz method and not for a particular Krylov method applied to the substructured problem. In the positive definite case one can show that minimizing the convergence rate is equivalent to minimizing the condition number of the substructured problem [JN00]. Numerical experiments in the next section indicate that for the Helmholtz equation our optimization also leads to parameters close to the best ones for the preconditioned Krylov method.

The general case of a decomposition into an arbitrary number of subdomains is treated in [GMN01].

Numerical results

We present two sets of numerical experiments. The first set corresponds to the model problem analyzed in this paper and the results obtained illustrate the analysis and confirm the asymptotic convergence results. The second numerical experiment comes from industry and consists of analyzing the noise levels in the interior of a VOLVO S90.

Model problem

We study a two dimensional cavity on the unit square Ω with homogeneous Dirichlet conditions on top and bottom and on the left and right radiation conditions of Robin type. We thus have the Helmholtz problem

$$\begin{aligned} -\Delta u - \omega^2 u &= f & 0 < x, y < 1 \\ u &= 0 & 0 < x < 1,\ y = 0, 1 \\ \frac{\partial u}{\partial x} - i\omega u &= 0 & x = 0,\ 0 < y < 1 \\ -\frac{\partial u}{\partial x} - i\omega u &= 0 & x = 1,\ 0 < y < 1. \end{aligned} \tag{3.50}$$

We decompose the unit square into two subdomains of equal size, and we use a uniform rectangular mesh for the discretization. We perform all our experiments directly on the error equations, $f = 0$ and choose the initial guess of the Schwarz iteration so that all the frequencies are present in the error. We show two sets of experiments: The first one with $\omega = 9.5\pi$, thus excluding ω from the frequencies k relevant in this setting, $k = n\pi$, $n = 1, 2, \ldots$. This allows us to test directly the iterative Schwarz method, since with optimization parameters $\omega_- = 9\pi$ and $\omega_+ = 10\pi$ we obtain a convergence rate which is uniformly less than one for all k. Table 3 shows the number of iterations needed for different values of the mesh parameter h for both the zeroth and second order transmission conditions. The Taylor transmission

Table 3: Number of iterations for different transmission conditions and different mesh parameter for the model problem.

	Order Zero				Order Two			
	Iterative		Krylov		Iterative		Krylov	
h	Taylor	Optimized	Taylor	Optimized	Taylor	Optimized	Taylor	Optimized
1/50	-	457	26	16	-	22	28	9
1/100	-	126	34	21	-	26	33	10
1/200	-	153	44	26	-	36	40	13
1/400	-	215	57	34	-	50	50	15
1/800	-	308	72	43	-	71	61	19

conditions do not lead to a convergent iterative algorithm, because for all frequencies $k > \omega$, the convergence rate equals 1. However, with Krylov acceleration, GMRES in this case, the methods converge. Note however that the second order Taylor condition is only a little better than the zeroth order Taylor conditions. The optimized transmission conditions lead, in the case where ω lies between two frequencies, already to a convergent iterative algorithm. The iterative version even beats the Krylov accelerated Taylor conditions in the second order case. No wonder that the optimized conditions lead by far to the best algorithms when they are accelerated by a Krylov method, the second order optimized Schwarz method is more than a factor three faster than any Taylor method. Note that the only difference in cost of the various transmission conditions consists of different entries in the interface matrices, without enlarging the bandwidth of the matrices. Fig. 6 shows the asymptotic behavior of the methods considered, on the left for zeroth order conditions and on the right for second order conditions. Note that the scale on the right for the second order transmission conditions

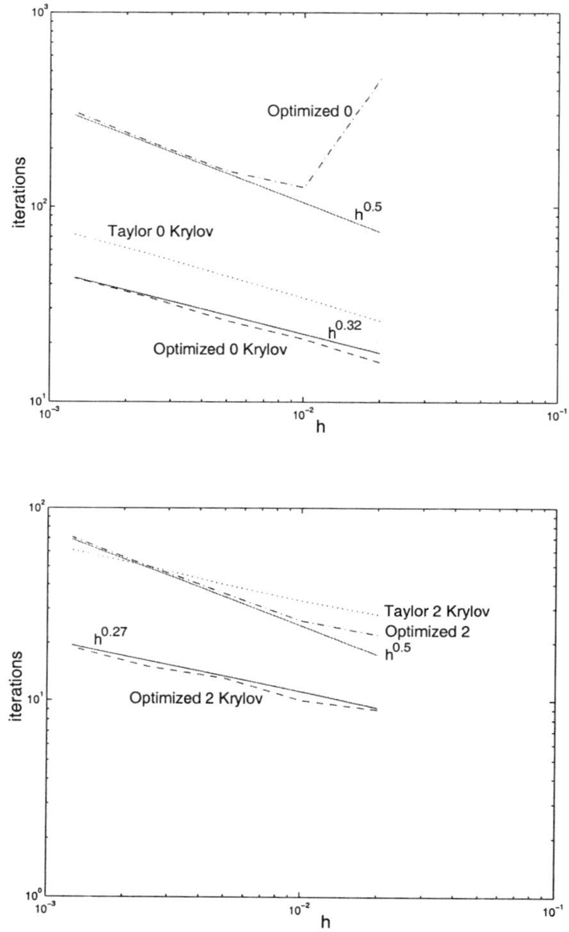

Figure 6: Asymptotic behavior for the zeroth order transmission conditions (above) and for the second order transmission conditions (below).

is different by an order of magnitude. In both cases the asymptotic analysis is confirmed for the iterative version of the optimized methods. In addition one can see that the Krylov method improves the asymptotic rate by almost an additional square root, as expected from the analysis in ideal situations. Note the outlier of the zeroth order optimized transmission condition for $h = 1/50$. It is due to the discrepancy between the spectrum of the continuous and the discrete operator: $\omega = 9.5\pi$ lies precisely in between two frequencies 9π and 10π at the continuous level, but for the discrete Laplacian with $h = 1/50$ this spectrum is shifted to 8.88π and 9.84π and thus the frequency 9.84π falls into the range $[9\pi, 10\pi]$ neglected by the optimization. Note however that this is of no importance when Krylov acceleration is used, so it is not worthwhile to consider this issue further. Now we put ω directly onto a frequency of the model problem, $\omega = 10\pi$, so that the iterative methods cannot be considered any more, since for that frequency the convergence rate equals one. The Krylov accelerated versions however are not affected by this, as one can see in Table 4. The number of iterations

Table 4: Number of iterations for different transmission conditions and different mesh parameter for the model problem when ω lies precisely on a frequency of the problem and thus Krylov acceleration is mandatory.

	Order Zero		Order Two	
h	Taylor	Optimized	Taylor	Optimized
1/50	24	15	27	9
1/100	35	21	35	11
1/200	44	26	41	13
1/400	56	33	52	16
1/800	73	43	65	20

does not differ from the case where ω was chosen to lie between two frequencies, which shows that with Krylov acceleration the method is robust for any values of ω. We finally tested for the smallest resolution of the model problem how well Fourier analysis predicts the optimal parameters to use. Since we want to test both the iterative and the Krylov versions, we need to put again the frequency ω in between two problem frequencies, and in this case it is important to be precise. We therefore choose ω to be exactly between two frequencies of the discrete problem, $\omega = 9.3596\pi$, and optimized using $\omega_- = 8.8806\pi$ and $\omega_+ = 9.8363\pi$. Fig. 7 shows the number of iterations the algorithm needs to achieve a residual of $10e-6$ as a function of the optimization parameters p and q of the zeroth order transmission conditions, on the left in the iterative version and on the right for the Krylov accelerated version. The Fourier analysis shows well where the optimal parameters lie and when a Krylov method is used, the optimized Schwarz method is very robust with respect to the choice of the optimization parameter. The same holds also for the second order transmission conditions, as Fig. 8 shows.

Noise levels in a VOLVO S90

We analyze the noise level distribution in the passenger cabin of a VOLVO S90. The vibrations are stemming from the part of the car called firewall. This example is representative for a large class of industrial problems where one tries to determine the acoustic response in the interior of a cavity caused by vibrating parts. We perform a two dimensional simulation

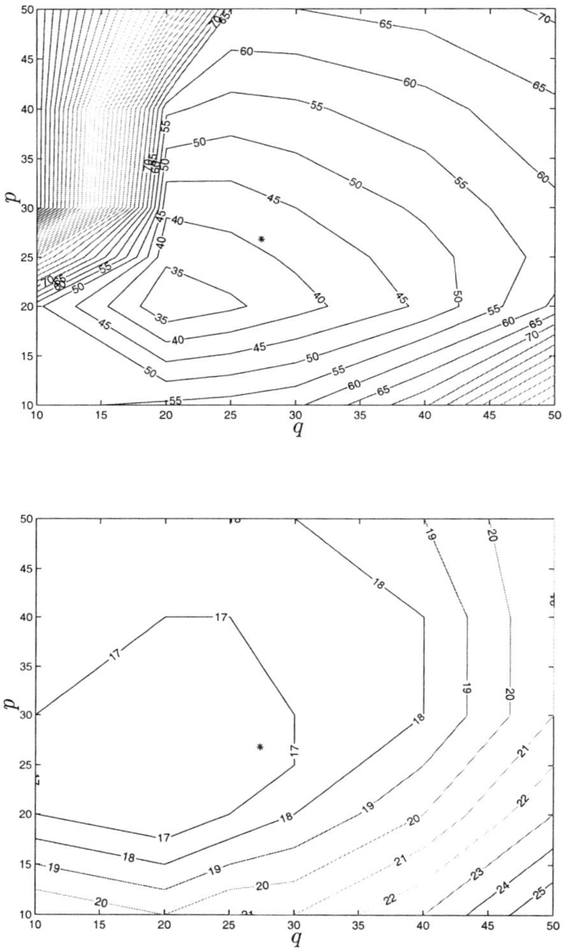

Figure 7: Number of iterations needed to achieve a certain precision as function of the optimization parameters p and q in the zeroth order transmission conditions, for the iterative algorithm (above) and for the Krylov accelerated algorithm (below). The star denotes the optimized parameters p^* and q^* found by our Fourier analysis.

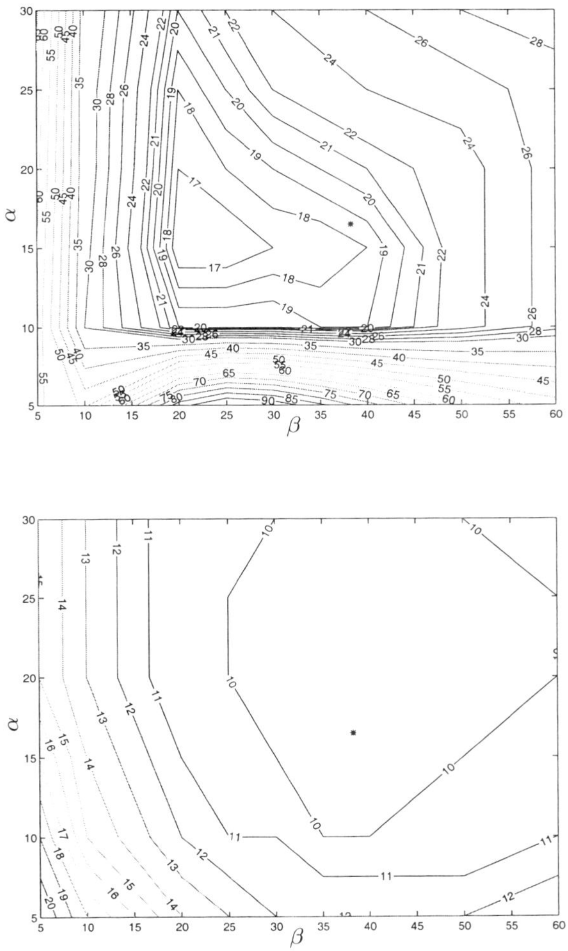

Figure 8: Number of iterations needed to achieve a certain precision as function of the optimization parameters α and β in the second order transmission conditions, for the iterative algorithm (above) and for the Krylov accelerated algorithm (below). The star denotes the optimized parameters α^* and β^* found by our Fourier analysis.

on a vertical cross section of the car. Fig. 9 shows the decomposition of the car into 16 subdomains. The computations were performed in parallel on a network of sun workstations with 4 processors.

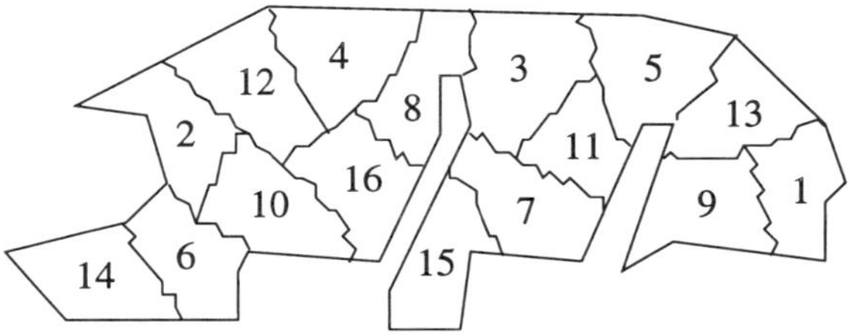

Figure 9: Decomposition of the passenger compartment into 16 subdomains.

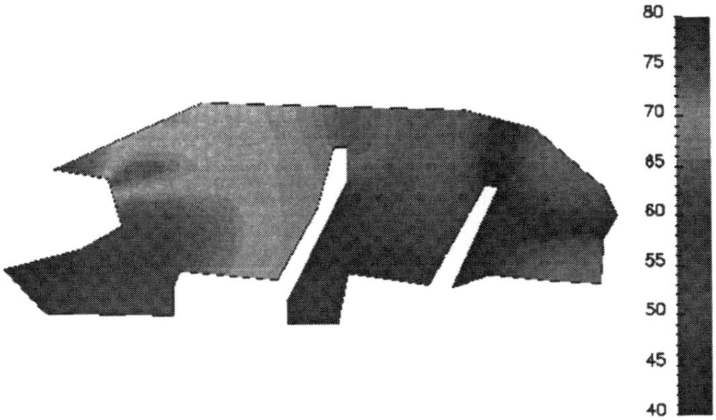

Figure 10: Acoustic field in the passenger compartment of the VOLVO S90.

The problem is characterized by $\omega a = 18.46$ which corresponds to a frequency of 1000 Hz in the car of length a. To solve the problem, the optimized Schwarz method was used as a preconditioner for the Krylov method ORTHODIR, and as convergence criterion we used

$$\|\widetilde{K}u - f\|_{L_2} \leq 10^{-6}\|f\|_{L_2}. \tag{3.51}$$

When using zeroth order Taylor conditions and a decomposition into 16 subdomains, the method needed 105 iterations to converge, whereas when using second order optimized transmission conditions, the method converged in 34 iterations, confirming that also in real applications the optimized Schwarz method is about a factor 3 faster, as we found for the model problem earlier. Fig. 10 shows the acoustic field obtained in the passenger compartment of the VOLVO S90.

4 Conclusion

In this presentation, we have stressed the influence of interface conditions in domain decomposition methods. By choosing them carefully, it is possible to achieve better and more reliable convergence behaviors for various types of scalar equations and the Stokes or Maxwell systems.

For complex systems of equations (e.g. multiphase flow, compressible Navier-Stokes equations) treated in a coupled way, the theory is still in development.

On the end-user point of view (physicists, engineers, ...) there is a lack of black-box routines working at the matrix level and implementing the latest methods. This could be achieved by a closer collaboration between numerical analysis and computer science.

Acknowledgement I wish to thank Victorita Dolean and Laurent Saas for their careful reading and useful suggestions.

References

[AK95] Y. Achdou and Y. A. Kuznetsov, Substructuring preconditioners for finite element methods on nonmatching grids, *East-West J. Numer. Math.* **3** (1) (1995), 1–28.

[AKP95] Y. Achdou, Y. Kuznetsov, and O. Pironneau, Substructuring preconditioners for the Q_1 mortar element method, *Numer. Math.* **71** (1995), 419–449.

[AMW99] Y. Achdou, Y. Maday, and O. B. Widlund, Iterative substructuring preconditioners for mortar element methods in two dimensions, *SIAM J. Numer. Anal.* **36** (2) (1999), 551–580.

[ATNV00] Y. Achdou, P. L. Tallec, F. Nataf, and M. Vidrascu, A domain decoposition preconditioner for an advection-diffusion problem, *Comput. Methods Appl. Mech. Engrg.* **184** (2000), 145–170.

[BD98] D. Braess and W. Dahmen, Stability estimates of the mortar finite element method for 3-dimensional problems, *East-West J. Numer. Math.* **6** (4) (1998), 249–264.

[BGLTV] J.-F. Bourgat, R. Glowinski, P. Le Tallec, and M. Vidrascu, Variational formulation and algorithm for trace operator in domain decomposition calculations, in: *Domain Decomposition Methods* (T. Chan et al., eds.), SIAM, Philadelphia, PA, 1989, 3–16.

[BMP93] C. Bernardi, Y. Maday, and A. T. Patera, Domain decomposition by the mortar element method, in: *Asymptotic and Numerical Methods for Partial Differential Equations with Critical Parameters* (H. G. Kaper and M. Garbey, eds.), NATO ASI Ser. C **384**, Kluwer, Dordrecht, 1993, 269–286.

[CCEW98] X.-C. Cai, M. A. Casarin, F. W. Elliott Jr., and O. B. Widlund, Overlapping Schwarz algorithms for solving Helmholtz's equation, in: *Domain Decomposition Methods 10* (J. Mandel et al., eds.), Contemp. Math. **218**, Amer. Math. Soc., Providence, RI, 1998, 391–399.

[CDS99] X.-C. Cai, M. Dryja, and M. V. Sarkis, Overlapping nonmatching grid mortar element methods for elliptic problems, *SIAM J. Numer. Anal.* **36** (1999), 581–606.

[CGJ00] F. Collino, S. Ghanemi, and P. Joly, Domain decomposition methods for harmonic wave propagation : a general presentation, *Comput. Methods Appl. Mech. Engrg.* (2–4) (2000), 171–211.

[CGPW89] T. Chan, R. Glowinski, J. Périaux, and O. Widlund, eds., *Domain Decomposition Methods*, SIAM, Philadelphia, PA, 1989.

[Che98] P. Chevalier, *Méthodes numériques pour les tubes hyperfréquences. Résolution par décomposition de domaine*, Ph.D. thesis, Université Paris VI, 1998.

[CM94] T. F. Chan and T. P. Mathew, Domain decomposition algorithms, in: *Acta Numerica 1994*, Cambridge University Press, 1994, 61–143.

[CMW95] L. C. Cowsar, J. Mandel, and M. F. Wheeler, Balancing domain decomposition for mixed finite elements, *Math. Comp.* **64** (211) (1995), 989–1015.

[CW92] X.-C. Cai and O. B. Widlund, Domain decomposition algorithms for indefinite elliptic problems, *SIAM J. Sci. Statist. Comput.* **13** (1) (1992), 243–258.

[Des93] B. Després, Domain decomposition method and the Helmholtz problem. II, in: *Second International Conference on Mathematical and Numerical Aspects of Wave Propagation (Newark, DE, 1993)*, SIAM, Philadelphia, PA, 1993, 197–206.

[DJR92] B. Després, P. Joly, and J. E. Roberts, A domain decomposition method for the harmonic Maxwell equations, in: *Iterative Methods in Linear Algebra (Brussels, 1991)*, North-Holland, Amsterdam, 1992, 475–484.

[DSW96] M. Dryja, M. V. Sarkis, and O. B. Widlund, Multilevel Schwarz methods for elliptic problems with discontinuous coefficients in three dimensions, *Numer. Math.* **72** (3) (1996), 313–348.

[EM77] B. Engquist and A. Majda, Absorbing boundary conditions for the numerical simulation of waves, *Math. Comp.* **31** (139) (1977), 629–651.

[EZ98] B. Engquist and H.-K. Zhao, Absorbing boundary conditions for domain decomposition, *Appl. Numer. Math.* **27** (4) (1998), 341–365.

[FR91] C. Farhat and F.-X. Roux, A method of finite element tearing and interconnecting and its parallel solution algorithm, *Internat. J. Numer. Methods Engrg.* **32** (1991), 1205–1227.

[Giv92] D. Givoli, *Numerical Methods for Problems in Infinite Domains*, Elsevier (1992).

[GMN01] M. J. Gander, F. Magoulès, and F. Nataf, Optimized Schwarz methods without overlap for the Helmholtz equation, *SIAM J. Numer. Anal.* (2001), to appear.

[Hig86] R. Higdon, Absorbing boundary conditions for difference approximations to the multi-dimensional wave equations, *Math. Comp.* **47** (176) (1986), 437–459.

[HTJ88] T. Hagstrom, R. P. Tewarson, and A. Jazcilevich, Numerical experiments on a domain decomposition algorithm for nonlinear elliptic boundary value problems, *Appl. Math. Lett.* **1** (3), 1988.

[Jap98] C. Japhet, Optimized Krylov-Ventcell method, Application to convection-diffusion problems, in: *Proc. 9th International Conference Domain Decomposition Methods* (P. E. Bjørstad et al., eds.), 1998, 382–389.

[JN00] C. Japhet and F. Nataf, The best interface conditions for domain decomposition methods: Absorbing boundary conditions, to appear in: *Artificial Boundary Conditions, with Applications to Computational Fluid Dynamics Problems* (L. Tourrette, ed.), Nova Science, 2000.

[JNR01] C. Japhet, F. Nataf, and F. Rogier, The optimized order 2 method. Application to convection-diffusion problems, *Future Generation Computer Systems FUTURE* **18**, 2001.

[KW01] A. Klawonn and O. B. Widlund, FETI and Neumann-Neumann iterative substructuring methods: Connections and new results, *Comm. Pure Appl. Math.* **54** (2001), 57–90.

[LBCW98] *Proc. Eleventh International Conference Domain Decomposition Methods, 1998* (C. H. Lai et al., eds.).

[Lio90] P.-L. Lions, On the Schwarz alternating method. III: A variant for nonoverlapping subdomains, in: *Third International Symposium on Domain Decomposition Methods for Partial Differential Equations* (T. F. Chan et al., eds.), SIAM, Philadelphia, PA, 1990.

[LT94] P. Le Tallec, Domain decomposition methods in computational mechanics, in: *Computational Mechanics Advances* vol. 1 (2) (J. Tinsley Oden, ed.), North-Holland, Amsterdam, 1994, 121–220.

[Nat96] F. Nataf, Absorbing boundary conditions in block Gauss-Seidel methods for convection problems, *Math. Models Methods Appl. Sci.* **6** (4) (1996), 481–502.

[Nep91] S. V. Nepomnyaschikh, Application of domain decomposition to elliptic problems on with discontinuous coefficients, in: *Fourth International Symposium on Domain Decomposition Methods for Partial Differential Equations* (R. Glowinski et al., eds.), SIAM, Philadelphia, PA, 1991, 242–251.

[Nie99] F. Nier, Remarques sur les algorithmes de décomposition de domaines, in: *Séminaire : Équations aux Dérivées Partielles, 1998–1999*, Exp. No. IX, 26, École Polytechnique, 1999.

[NR95] F. Nataf and F. Rogier, Factorization of the convection-diffusion operator and the Schwarz algorithm, *Math. Models Methods Appl. Sci.* **5** (1) (1995), 67–93.

[NRdS94] F. Nataf, F. Rogier, and E. de Sturler, Optimal interface conditions for domain decomposition methods, Technical report, CMAP (École Polytechnique), 1994.

[PEBK97] *Proc. Ninth International Conference on Domain Decomposition Methods, 1997*, (M. S. Espedal el al., eds.).

[QV99] A. Quarteroni and A. Valli, *Domain Decomposition Methods for Partial Differential Equations*, Oxford Sci. Publ., Oxford University Press, 1999.

[SBG96] B. F. Smith, P. E. Bjørstad, and W. Gropp, *Domain Decomposition: Parallel Multilevel Methods for Elliptic Partial Differential Equations*, Cambridge University Press, 1996.

[WFNS98] F. Willien, I. Faille, F. Nataf, and F. Schneider, Domain decomposition methods for fluid flow in porous medium, in: *6th European Conference on the Mathematics of Oil Recovery*, 1998.

[Woh99] B. Wohlmuth, Hierarchical a posteriori error estimators for mortar finite element methods with Lagrange multipliers, *SIAM J. Numer. Anal.* **36** (1999), 1636–1658.

A review of level set and fast marching methods for image processing

J. A. SETHIAN*

Department of Mathematics
University of California
Berkeley, CA 94720
USA

Abstract

In recent years, a considerable amount of work has been developed concerning a partial differential equations-based approach to image processing. This work has been focussed on the interplay between geometric motion and non-linear evolution equations, and its applicability to problems in image enhancement and denoising, image segmentation, and shape recognition and classification. In this review article, we provide an overview for some of these ideas.

1 Overview and introduction

A large number of computational problems and physical phenomena involve the motion of interfaces separating two or more regions. These can include problems in such areas as fluid mechanics, combustion, materials science, meteorology, and computer vision. In these problems, challenging issues often involve:

- Interfaces that change topology, break, and merge as they move.
- Formation of sharp corners, cusps, and singularities.
- Dependence of the interface motion on delicate geometric quantities such as curvature and normal direction.
- Complexities in three dimensions and higher.
- Subtle feedback between the physics and chemistry off the interface and the position/motion of the front itself.

One approach to formulating, modeling, and building computational techniques for some aspects of these problems is provided by level set methods, introduced by Osher and Sethian

*This work was supported in part by the Applied Mathematical Science subprogram of the Office of Energy Research, U.S. Department of Energy, under Contract Number DE–AC03–76SF00098, and the Division of Mathematical Sciences of the National Science Foundation.

A. Bourlioux and M.J. Gander (eds.), Modern Methods in Scientific Computing and Applications, 365–396.

[28]. These techniques work by embedding the propagating interface as the zero level set of a time-dependent, implicit function, and then solving the resulting equations of motion in a fixed grid Eulerian setting. They have been used with considerable success in a wide collection of settings, including fluid mechanics, crystal growth, combustion, medical imaging, A general overview of the theory, numerical approximation and range of applications may be found in [42].

Level set methods rely in part on the theory of curve and surface evolution given in [36] and on the link between front propagation and hyperbolic conservation laws discussed in [37]. They recast interface motion as a time-dependent Eulerian initial value partial differential equation, and rely on viscosity solutions to the appropriate differential equations to update the position of the front, using an interface velocity that is derived from the relevant physics both on and off the interface. These viscosity solutions are obtained by exploiting schemes from the numerical solution of hyperbolic conservation laws. Level set methods are specifically designed for problems involving topological change, dependence on curvature, formation of singularities, and a host of other issues that often appear in interface propagation techniques. Over the past few years, various aspects of these techniques have been refined to the point where a general computational approach to arbitrary front propagation problems can be developed. This general computational approach allows one to track the motion of very complex interfaces, with significant and delicate coupling between the relevant physics and the interface motion.

Level set methods cast interface propagation in terms of a time-dependent initial value problem. More recently, a set of numerical techniques, known as Fast Marching Methods [38], have been developed for solving the eikonal equation, which is a boundary value partial differential equation. These techniques rely on a marriage between the numerical technology for computing the solution to hyperbolic conservation laws and the causality relationships inherent in finite difference upwind schemes. Fast Marching Methods are Dijkstra-type methods, in that they are closely connected to Dijkstra's well-known network path algorithms [9], however they approximate the solution to the underlying eikonal equation in a consistent manner. While the eikonal equation itself describes some front propagation problems, the important link we shall emphasize in this review is that Fast Marching Methods provide a general, efficient, and accurate way to actually implement some aspects of level set methods.

Both sets of techniques, that is, level set methods and Fast Marching methods, require an adaptive methodology to obtain computational efficiency. In the case of level set methods, this leads to the Narrow Band level set method introduced by Adalsteinsson and Sethian in [1]. In the case of Fast Marching methods, adaptivity and speed stem from the causality relationship and the use of heap data structures.

In this set of lecture notes, we discuss the application of these techniques to problems in image processing. We will provide algorithms and applications to problems in image segmentation applied to tumor analysis, cardiac segmentation, denoising of various imaging modalities including MRIs, CT scans, Electron Beam Tomography, and ultrasound. The techniques can be use to both process video streams to extract data and images and perform recognition and identification tasks. We provide a large number of computational examples, and a framework for devising new applications.

We hasten to point out that these notes are by no means complete; a more detailed accounting of some of these ideas may be found in the book by Sethian [42], review articles [39, 43], and two recent books about these techniques specifically applied to image processing given in [27, 31]. These notes are meant only to give a flavor of the ideas pursued by one group of reseachers; computational details may be found in the above resources. Finally, we note a web site at `http://math.berkeley.edu/~sethian/level_set.html` which provides an overview of many of these ideas, replete with movies, Java Applets, and other interactive applications.

2 Goals

The goal of these lectures is to show how a family of interrelated PDEs has applications to Image Processing, and to provide motivation to think about the geometric aspects of various tasks of image processing, including

- Denoising and Enhancement;
- Segmentation and Representation;
- Recognition and Classification.

2.1 A hierarchy of equations

We begin by considering a hierarchy of partial differential equations which stand at the core of these ideas. Each will ultimately have relevance to a geometric view of image processing.

- **The distance equation**

  $$|\nabla\phi| = 1, \quad \phi(x, y) = 0 \quad \text{on} \quad \Gamma.$$

  The solution measures the distance from a curve or surface. It is easy to see that the characteristics are straight lines and orthogonal to level lines. The solution is non-differentiable, even for smooth curves, and a formal viscosity solution may be defined which is the regularized limit of the equation.

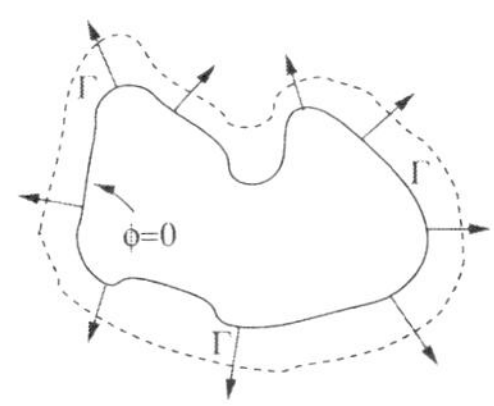

- **The eikonal equation**

  $$|\nabla\phi| = F(x, y), \quad \phi(x, y) = 0 \quad \text{on} \quad \Gamma.$$

  This is the distance function with a metric on the domain; we see that characteristics are not straight, but are still orthogonal to level lines of ϕ. Again, the solution is non-differentiable, even for smooth curves, and a viscosity formulation provides a unique solution.

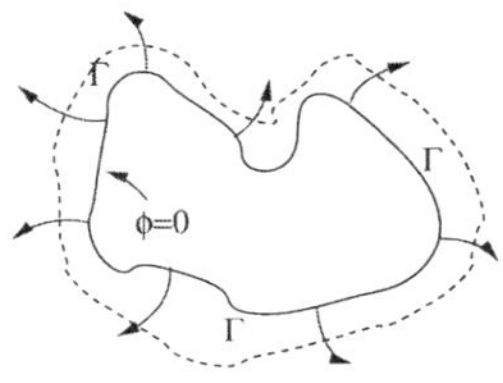

- **Static Hamilton-Jacobi equations and control theory**

$$|\nabla\phi| = F(x, y, \phi_x, \phi_y), \quad \phi(x, y) = 0 \quad \text{on} \quad \Gamma.$$

The general form of this static Hamilton-Jacobi equation is $H(u, Du) = 0$. Here, the characteristics are not orthogonal to level lines; once again, fast techniques to solve these equations are given by the class of Ordered Upwind Methods introduced by Sethian and Vladimirsky in [46, 47].

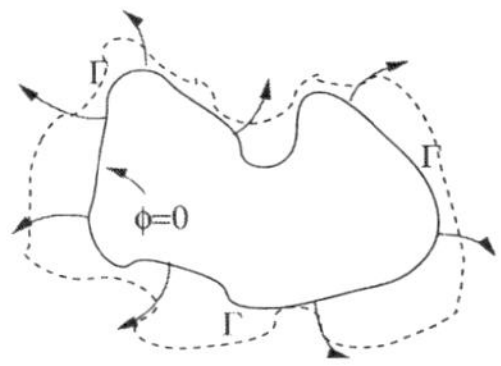

- **Level set equations for front propagation**

$$\phi_t + F|\nabla\phi| = 0, \quad \phi(x, y) = 0 \quad \text{on} \quad \Gamma.$$

This is a time-dependent partial differential equation which describes motion of a curve or surface normal to itself with arbitrary speed F. Algorithms based on this view were introduced by Osher and Sethian in [28].

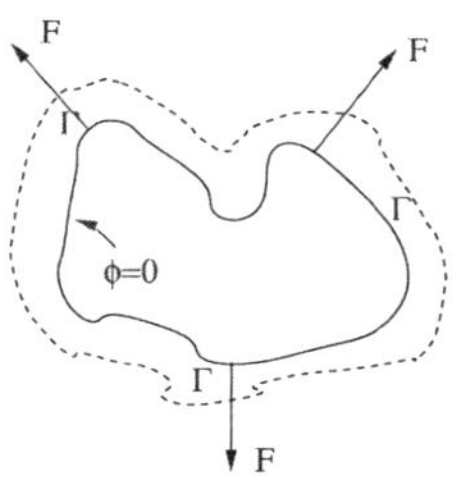

- **Curvature flow**

$$\phi_t - \kappa|\nabla\phi| = 0, \quad \phi(x, y) = 0.$$

This is a special case of level set equations, corresponding to anisotropic diffusion; it characterizes a surface evolving under its own curvature. Theoretical results obtained by Grayson show that in two dimensions, simple closed curves shrink to a point [12]; in three dimensions, this result is not true [13].

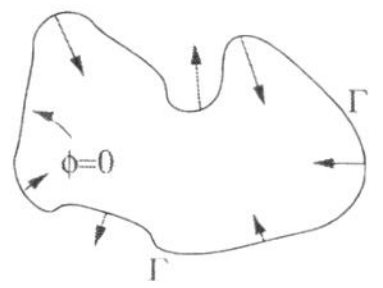

3 The chain of image processing

Typically, image processing is executed through a series of steps. Starting with a raw image, the first goal, during preprocessing, is to suitably prepare the image by removing artifacts which may occur due to errors in the collection device (such as background noise in an ultrasound device) and by enhancing certain desired features. The goal of this enhancement is to help delineate features for the next state, namely edge detection and segmentation. The goal in edge detection is to locate certain desired objects in the image field, such as cardiac boundaries. The segmentation step provides a reasonable representation of this extracted boundary; while chain-coded pixel descriptors are possible, a particularly effective representation is through a mathematical and/or geometric description. Finally, the recognition and classification stage is concerned with comparing the extracted object to a given library and identifying the result. The chain is schematically presented Fig. 1 below:

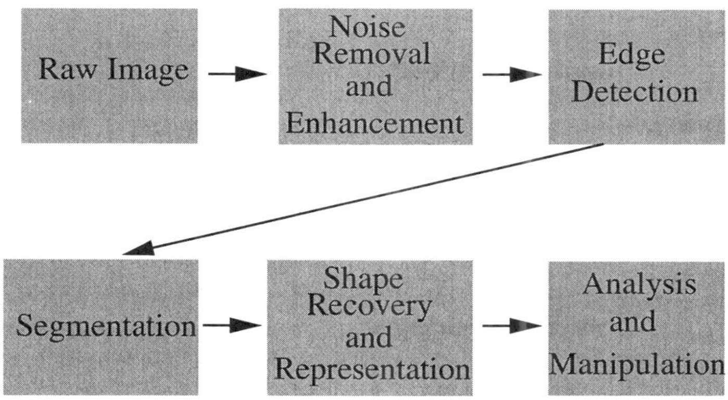

Figure 1: Chain of image processing.

We shall discuss the application of PDE-based techniques to these various steps, but not in order.

4 A PDE view of segmentation

4.1 Thresholding

As an introductory example, consider the goal of segmenting (or finding) the boundary of a white circle on a black background, as in Fig. 2.

Figure 2: White circle on black background.

One approach is provided by simple thresholding. The idea is to simply choose a grayscale value somewhere between 0, which corresponds to white, and 255, which corresponds to black; a well-chosen value can pick up the desired boundaries. However, a disadvantage with this approach soon becomes apparent: thresholding may find the boundaries of noise as unwanted edges, as in Fig. 3.

Figure 3: Thresholding.

4.2 An evolutionary approach

Instead, consider an approach in which we try to grow an object from a starting seed so that its motion attaches itself to the desired boundaries, as in Fig. 4.

In order for this approach to work, we will need several things:

- First, an ability to propagate an initial seed so that it can undergo topological changes as it moves.
- Second, an efficient technique that is robust, fast, and can handle sharp corners and cusps in the propagating front.
- Third, a stopping criterion that attaches the growing interface to the desired boundaries.

Such algorithms are provided by the combination of level set techniques and Fast Marching Methods discussed later in these notes. We note that we will want these techniques to work for a wide collection of images, including grayscale images, vector-valued (color) images (including (R, G, B) channels), volumetric images (MRI, CT), and time-varying sequences of images (movies), as well as images with texture information, such as fingerprint images.

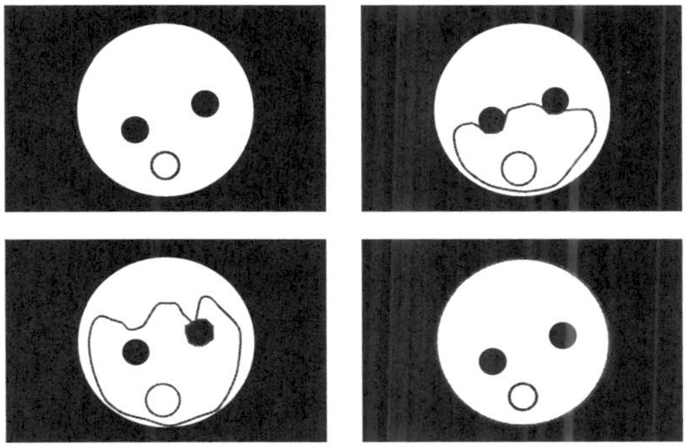

Figure 4: An evolutionary approach.

4.3 Equations of motion

The original idea of segmenting images through a growing seed was due to [15], and relied on Lagrangian marker particle formulation together with an energy function to attach the evolving interface to the desired boundary. This approach, known as "snakes", opened up a great deal of work in image segmentation. There are several drawbacks to this approach, including the fact that it does not allow the interface to change topology, requires adaptive addition and removal of points in the discrete interface representation, and has difficulty working in three dimensions and beyond. A different approach using level set methods, avoids these difficulties, and was introduced simultaneously by Malladi, Sethian, and Vemuri in [25] and Caselles, Catte, Coll, and Dibos [6]. A much more efficient technique, introduced by Malladi and Sethian [20], comes from combining narrow band level set methods [1] and Fast Marching Methods [38]. We now explain these approaches. Additional work using this approach may be found in [17, 18, 19, 20, 21, 22, 23, 24, 25, 26, 27].

Given a noisy image function $I(\mathbf{x})$, the objective is to construct mathematical descriptions of certain shapes contained in it. Shape recovery is possible if we imagine an initial curve propagating in the image domain with speed that is a decreasing function of the image gradient (see Fig. 5).

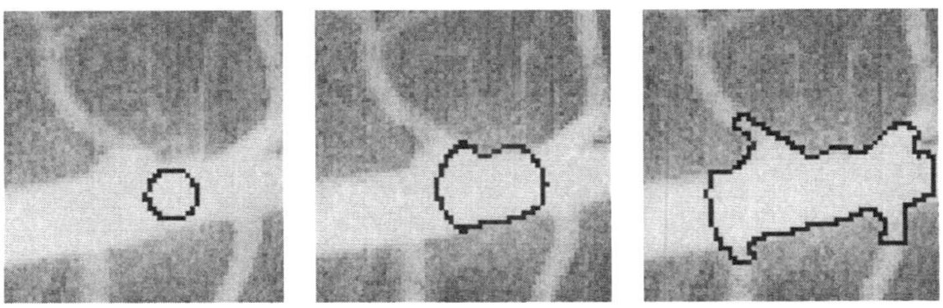

Figure 5: Propagating interface in segmentation.

We now formulate both the equations of motion and a stopping criterion. Let $I(\mathbf{x})$ be a noisy image function $I(\mathbf{x})$. Now, consider the eikonal equation

$$|\nabla\phi| = \frac{1}{k_I}, \tag{4.1}$$

where

$$k_I(\mathbf{x}) = e^{-\alpha|\nabla I(\mathbf{x})|}. \tag{4.2}$$

We note that the right-hand-side of the above eikonal equation gets large where the image gradient $|\nabla I|$ is large. Thus, since the solution to the eikonal equation corresponds to the arrival time of a interface starting at the seed, the boundary is segmented by chosing those points in the image field that are reached late in time.

A more sophisticated and robust method is realized by adding a smoothness constraint (regularization) and an additional forcing term, namely

$$\phi_t + k_I(1 - \epsilon K) \mid \nabla\phi \mid - \nabla k_I \cdot \nabla\phi = 0. \tag{4.3}$$

The various components here may be thought of as follows:

- $1 * |\nabla\phi|$: This corresponds to an inflationary force which drives the seed interface outwards.

- $-\epsilon K$: This corresponds to a curvature (smoothness) term which regularizes the evolving front and makes sure that it does not balloon or bleed through small gaps in the boundary.

- $\nabla k_I \cdot \nabla\phi$: This terms, introduced in [7], acts as a restoring force to retract the moving interface back to the boundary.

The fundamental strategy for solving the equations comes from a two step approach. In the first stage, using Fast Marching Methods for the eikonal equation [21], we consider a reduced system: $F = k_I(\mathbf{x})$. We then advance solution using these very fast techniques to get the interface close to the desired answer. Once this is done, in the second stage we employ narrow band level set methods [1] as follows: we use all the terms in the equation given by

$$\phi_t + k_I(1 - \epsilon K) \mid \nabla\phi \mid - \nabla k_I \cdot \nabla\phi = 0.$$

4.4 Examples of static segmentation

We begin with an example of cardiac segmentation taken from an UltraFast CT Scanner; this is part of a software package currently in use at Imatron Corporation. The segmentation results reveal the inner left and right ventricles (Fig. 6), as well as the outer ventricles.

Next, we show the extraction of a liver from three-dimensional data (Fig. 7). The data comes as a collection of stacked two-dimensional slices; here, nonetheless, the segmentation is directly performed in three dimensions.

Finally, we show the segmentation of the cortical stucture starting from an set of seed spheres (Fig. 8).

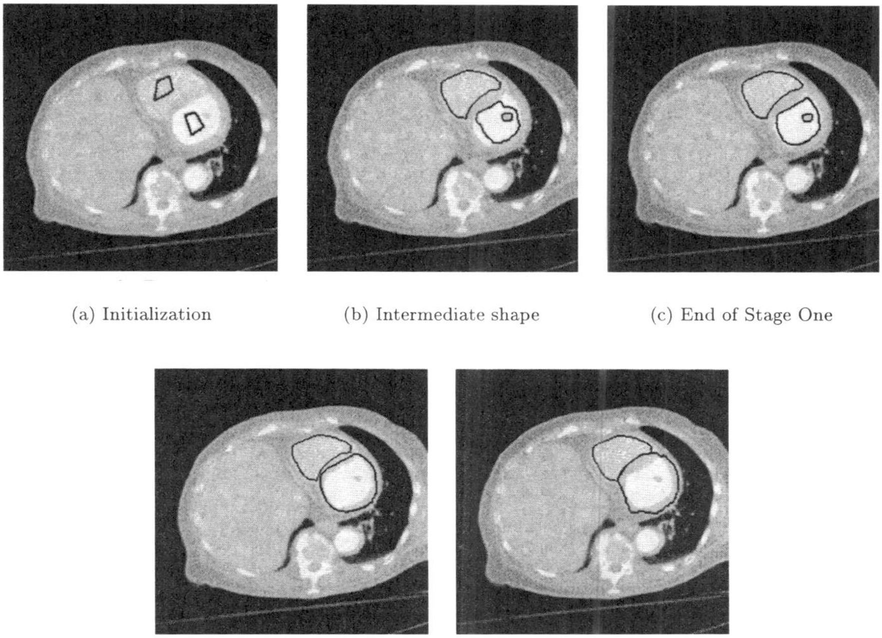

(a) Initialization (b) Intermediate shape (c) End of Stage One

(d) Intermediate Stage Two (e) End of Stage Two

Figure 6: Shape extraction from heart data.

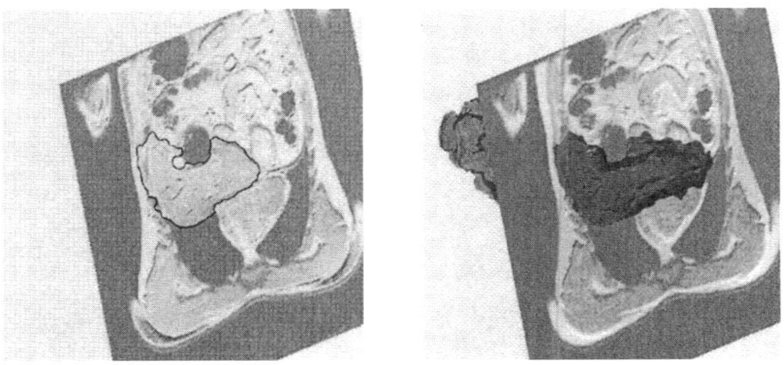

(a) Single two-dimensional contour (b) Three-dimensional reconstruction

Figure 7: Reconstruction of three-dimensional liver.

Figure 8: Segmentation of cortical structure.

4.5 Segmentation of Time-Dependent Data

The above techniques can be extended to four dimensions (three dimensions cross time), in order to deal with time-dependent data. In [32], a plot of volume fraction vs. time was developed using an extraction technique of the left ventricle. This is of great use in determining the performance and efficiency of the heart cycle. An inflationary full model was used to segment initial 2D shape, and was then taken as a starting guess for nearby shapes in space and time; this provides a more efficient technique rather than segmenting each still image from an initial seed. In Fig. 9 we show cross-sections of the extracted images, the full three-dimensional segmented result (Fig. 10), and finally (Fig. 11) a plot of the time-dependent volume fraction.

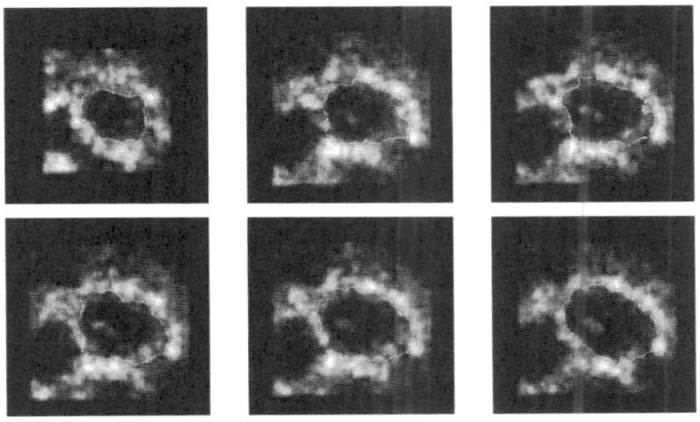

Figure 9: Two-dimensional cross sections of left ventricle.

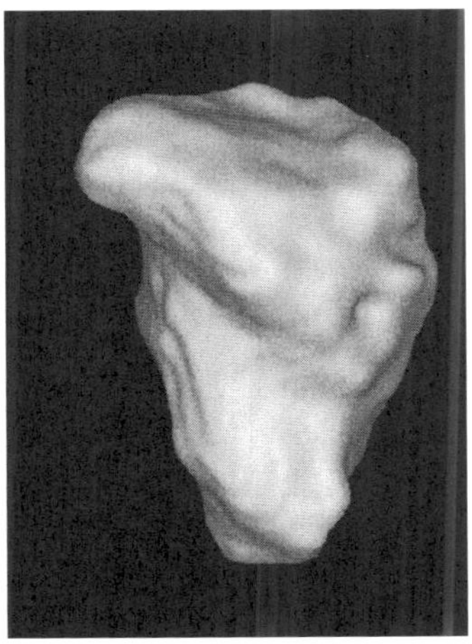

Figure 10: Three-dimensional time slice of left ventricle.

These techniques can also be used to isolate frames in movies. The basic idea is to build the stopping criterion in terms of the image gradient in both space and time; this isolates

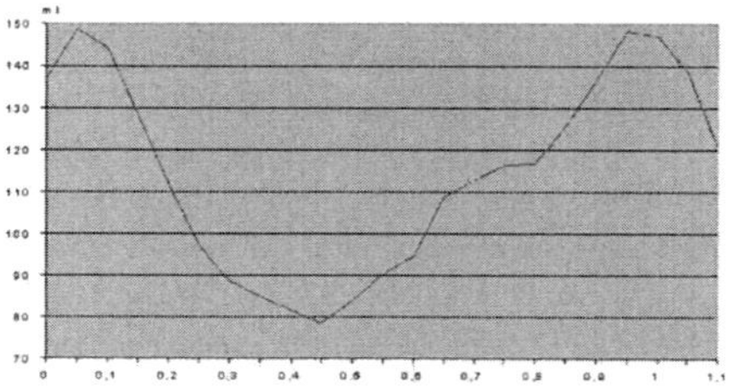

Figure 11: Time-volume plot.

moving objects from still backgrounds. In an image sequence taken from [11], a stationary video camera is fixed on a sidewalk; at some point in time, people (and a cat!) are filmed walking on the sidewalk. As postprocessing, we are able to extract the shapes by performing segmentation in three dimensions cross time (Fig. 12).

Figure 12: Extraction from video sequence.

4.6 Subjective surfaces

Up until now, we have tried to extract boundaries from information that is present in the image. However, there are circumstances in which one wants to extract boundaries from information that is missing, due to such factors as dropout in the collection procedure, boundary occlusion, etc. Here, we discuss work on "subjective surfaces", first presented in [33, 34].

Figure 13: Kanisza triangle.

Consider Fig. 13 above. The so-called Kanisza triangle is apparent from the wedges that are missing in the three disks. Of course, the triangle is not actually present; nonetheless, the eye extracts this boundary by amalgamating the separate pieces into one unified object. Our goal is to somehow "reconstruct" these boundaries. The basic approach is as follows. We will start with a user-defined point-of-view. Rather than flow a contour towards the boundary, we will flow a surface in such as way that existing parts of the image form boundary conditions for the evolving surface.

Why would one want to do such a thing? One reason is that certain images often suffer from dropout, such as in echocardiagrams. As an example, Fig. 14 shows an echocardiagram with considerable missing and blurry data; the goal is to extract the boundary which will require completion of missing parts.

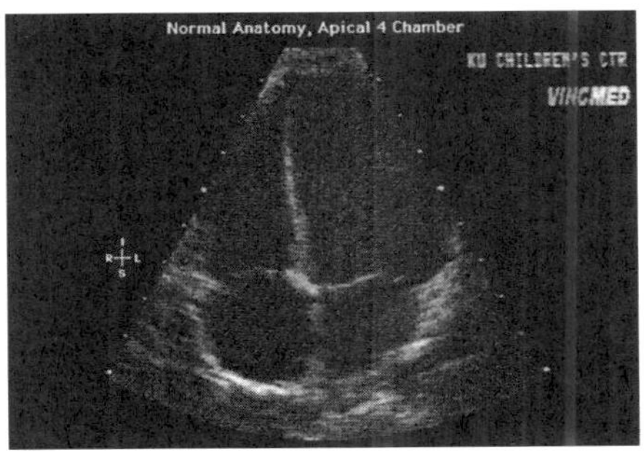

Figure 14: Echocardiagraphic data.

The approach to subjective segmentation taken in [33, 34] is to view the image map as a graph over the domain. Consider a surface height function $f(x, y)$, and the associated surface area given by

$$\text{Surface Area} = \int \sqrt{1 + f_x^2 + f_y^2}.$$

Suppose we mollify this surface area by a function which vanishes in the neighborhood of large image gradients, that is,

$$\text{Constrained Surface Area} = \int g(x,y) * \sqrt{1+f_x^2+f_y^2},$$
$$g(x,y) = \frac{1}{1+|\nabla I(x,y)|}.$$

Through an Euler-Lagrange formulation, this leads to a moving interface formulation for a level set function $\phi(x,y,t)$ given by

$$\frac{\partial \Phi}{\partial t} = g\frac{(1+\Phi_x^2)\Phi_{yy} - 2\Phi_x\Phi_y\Phi_{xy} + (1+\Phi_y^2)\Phi_{xx}}{1+\Phi_x^2+\Phi_y^2} + (g_x\Phi_x + g_y\Phi_y). \tag{4.4}$$

Several properties of this flow map are apparent:

- It flattens out the surface away from existing edges.
- The first term on the right corresponds to a parabolic term that evolves the surface in the normal direction under its mean curvature weighted by the edge indicator g. The surface motion is slowed down in the vicinity of edges ($g \to 0$).
- The second term on the right corresponds to a pure passive advection of the surface along the underlying velocity field $-\nabla g$ whose direction and strength depend on position. This term pushes/attracts the surface in the direction of the image edges.
- We additionally note that the term $g(I(x,y))$ is not a function of the third coordinate, therefore the vector field $-\nabla g$ lies entirely on the (x,y) plane.

The virtue of this flow is that it works even if part of the boundary is missing. Where information exists, the advection term drives the surface towards the edges: level sets of the surface also get attracted to the edge and accumulate. The spatial gradient increases and the surface begins to develop a discontinuity. Finally, we note that with spatial derivatives $\Phi_x, \Phi_y \gg 1$, the equation is approximated by

$$\frac{\partial \Phi}{\partial t} \approx g\frac{\Phi_x^2\Phi_{yy} - 2\Phi_x\Phi_y\Phi_{xy} + \Phi_y^2\Phi_{xx}}{\Phi_x^2+\Phi_y^2} + (g_x\Phi_x + g_y\Phi_y), \tag{4.5}$$

which is the level set flow for segmentation.

The region inside the objects where $g \to 1$, $|\nabla g| \to 0$ the surface is driven by the Euclidean mean curvature motion towards a flat surface. Here, $\Phi_x, \Phi_y \ll 1$ and the equation is approximated by

$$\frac{\partial \Phi}{\partial t} \approx g(\Phi_{xx} + \Phi_{yy}) + (g_x\Phi_x + g_y\Phi_y). \tag{4.6}$$

The equation is a non-uniform diffusion equation and denotes the steepest descent of the weighted $L2$ norm. Thus, subjective contours are, in fact, continuation of existing edge fragments. The equation is a combination of level set flow and pure diffusion. In a steady state solution, the points inside the objects are characterized by pure linear diffusion, while points on the boundary are characterized by the level set edge enhancing flow. Thus, we see that the flow continues missing boundaries as straight lines.

4.6.1 Examples of subjective segmentation

First, Fig. 15 shows the segmentation of the Kanisza Triangle given earlier. The technique extracts the perceived boundary with ease.

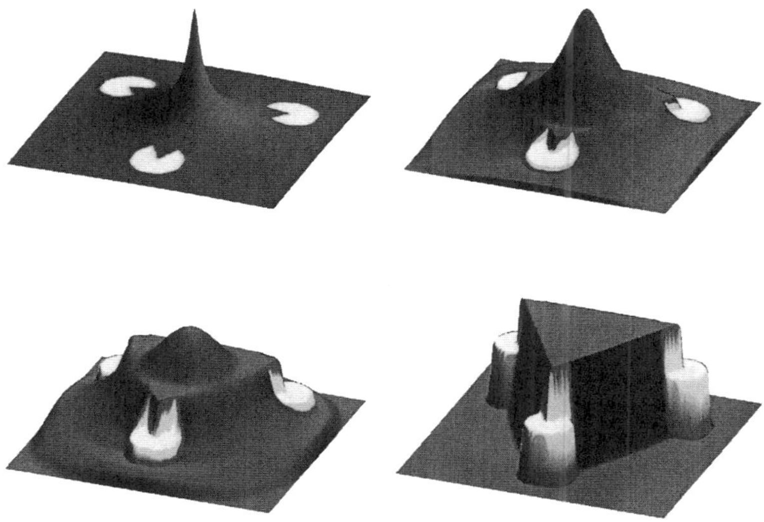

Figure 15: Subjective segmentation of Kanisza triangle.

We note that the algorithm also works in images without any aligned gradient, as shown Fig. 16 below.

We end this section of segmentation with the extraction of boundaries from echographic images (Fig. 17).

5 Shape processing, recognition, and classification

5.1 Shape smoothing

We now discuss some aspects of the next part of the image processing chain, namely shape processing, recognition and classification. To begin, suppose that one wants to smooth the extracted shape even further; this may be because of noise in the data, or the desire to compare the extracted shape against some generic smoothed library test boundaries. To do so, a surface smoothing algorithm was developed by Malladi and Sethian [19], based on the idea of a min/max flow.

The basic goal is to remove noise from image with automatic stopping criteria. The derived flow exploits curvature flow ideas, and makes use of a "min/max" switch. The resulting flow

Figure 16

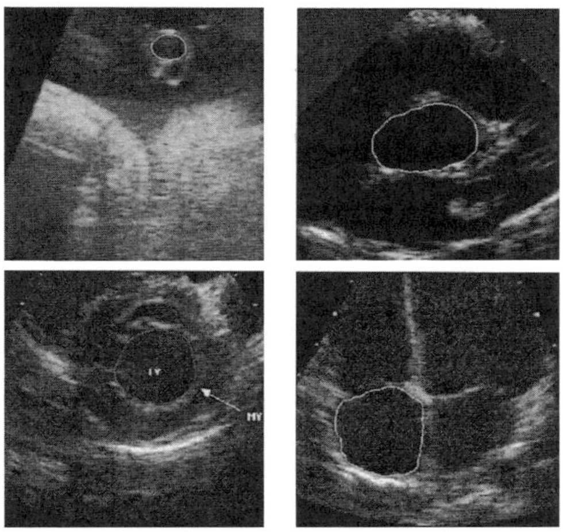

Figure 17

is applicable to shape smoothing, and has several virtues. First, it diffuses small-scale noise, second, it maintains larger, global properties, third, noise removal level is scale-dependent, fourth, edge definition is maintained, and finally, the flow algorithm stops automatically.

The basic idea of the min/max algorithm is as follows. We begin by noting that under the flow:

- $F = \min(\kappa, 0)$: concave indentations grow outwards, and
- $F = \max(\kappa, 0)$: protruding regions grow inwards.

as in Fig. 18.

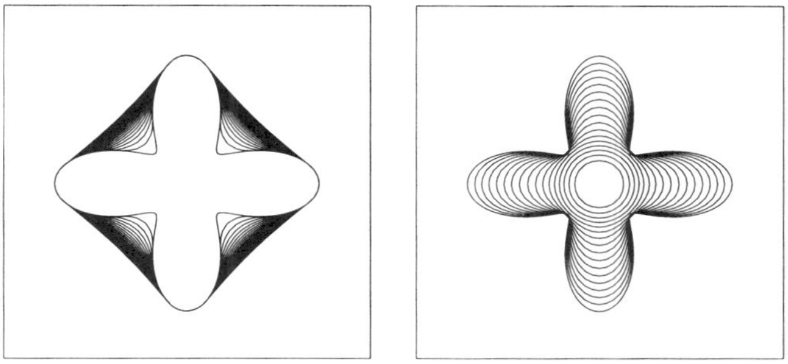

(a) $F = \min(\kappa, 0.0)$ (b) $F = \max(\kappa, 0.0)$

Figure 18: Motion of curve under min/max flow.

This suggests the following "switched flow":

$$\bar{F}_{\text{min}/\text{max}} = \begin{cases} \max(\kappa, 0), & \text{if } \text{Ave}_{\phi(x,y)}^{R=kh} < T_{\text{threshold}} \\ \min(\kappa, 0), & \text{otherwise} \end{cases} \tag{5.1}$$

We note that $T_{\text{threshold}}$ depends only on pixels in neighborhood of radius k, this flow automatically stops when noise is smaller than k pixels is removed, and that features bigger than k pixels are preserved.

We can apply this flow to shape smoothing as follows. Fig. 19 shows both mean curvature flow and the above min/max flow applied to a noisy surface. It is clear that mean curvature flow reduces the noise and smooths the surface, however, the cost is a significant decrease in the volume of the object. In contrast, the min/max flow smooths the surface while maintaining the underlying shape and volume.

As an application, in Fig. 20, we apply this flow to a three-dimensional ultrasound, obtain significant smoothing of the boundary while maintaining the essential geometric structure.

5.2 Shape recognition and classification

Once a shape is extracted and processed, a possible next goal is to perform some sort of feature recognition. One such example was done by Brewer and Malladi [5], applied to meteorological data and cloud recognition. Here, the goal is to identify and categorize meteorological data. First, neural net classifiers were used to label pixels according to type. The above segmentation techniques were used to extract pixel/shape boundaries. This was done by synthesizing an appropriate speed function/stopping criteria, based on multi-channel

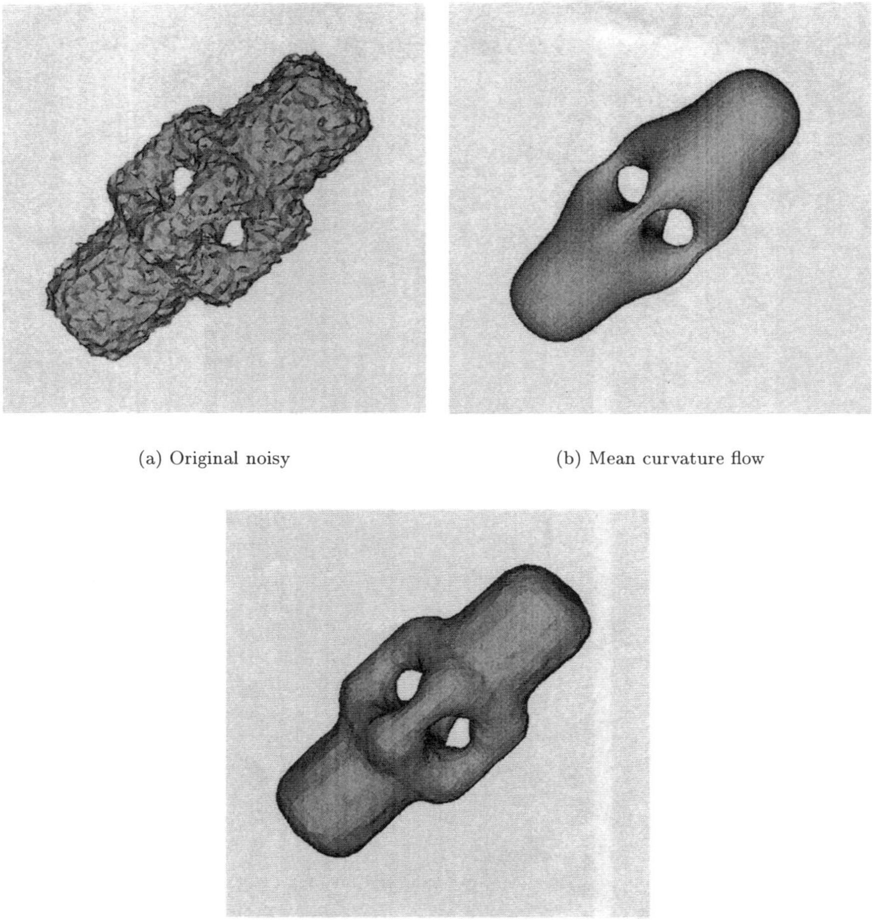

(a) Original noisy

(b) Mean curvature flow

(c) Min/Max flow

Figure 19: Two different shape smoothing flows.

data, including spectral features, texture features, and visible and infared reflections. The neural net classifier then picks the best match to a library of known shapes. The goal of this algorithm is to automatically extract deep convection and shallow convection structures. As an example of the flow steps, Fig. 21 shows the sequence of execution.

As examples, Fig. 22 shows the shape recovery of a deep convective cloud for two different smoothing parameter values, while Fig. 23 shows the shape recovery of deep convection and shallow convection overlaid onto the original visible image.

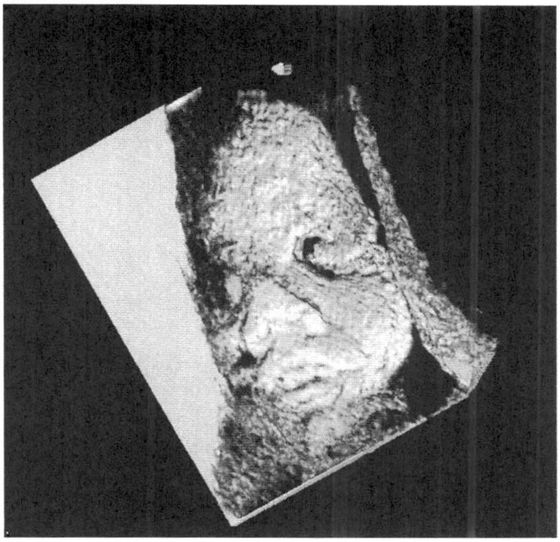

(a) Original ultrasound

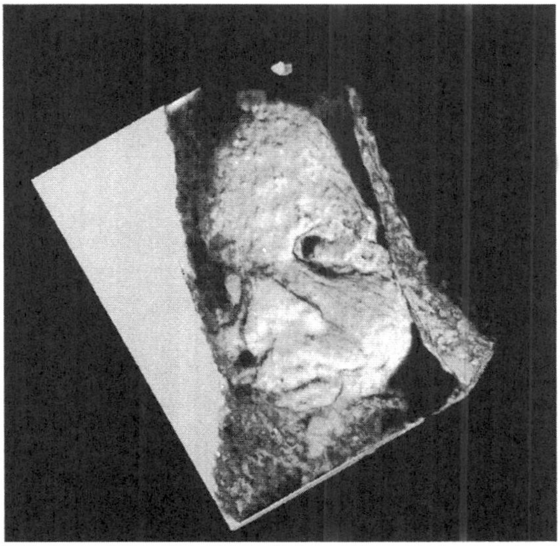

(b) Min/Max flow

Figure 20: Min/Max shape smoothing flow applied to ultrasound image.

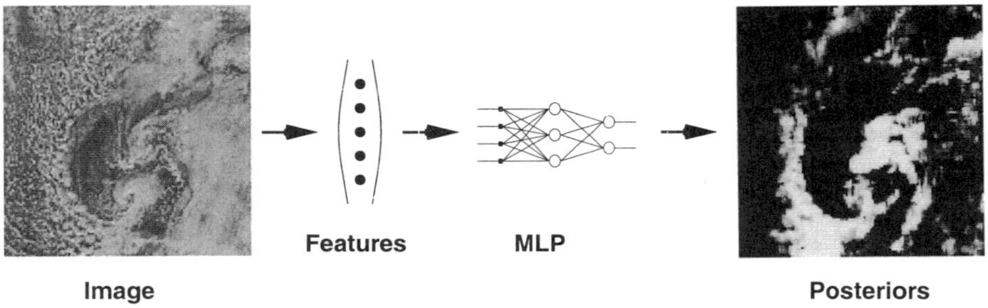

Figure 21: Classification of raw image using neural net (MLP) classifier.

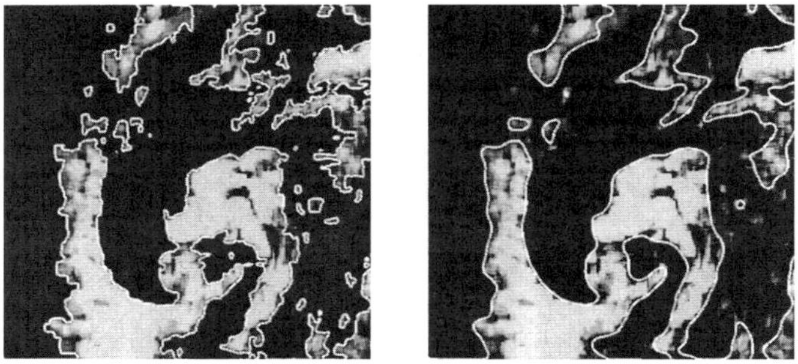

(a) $\epsilon = 0.0$.

(b) $\epsilon = 0.05$.

Figure 22: Shape recovery of a deep convective cloud for two different smoothing parameter values.

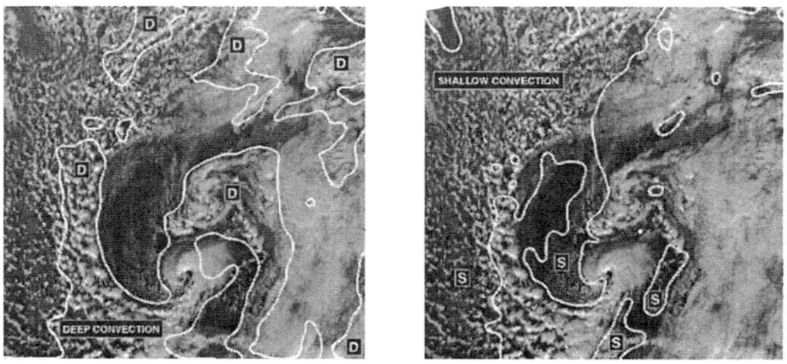

(a) Deep convection.

(b) Shallow convection.

Figure 23: Shape recovery of deep convection and shallow convection overlaid onto original visible image.

5.3 Character recognition

As a final example, developed by Malladi and Sethian [17], the above sequence of segmentation, smoothing, and neural net classification was applied to character recognition. Character recognition is typically either of one of two types:

- *Boundary-based*: Chain-code, Invariant features such as Fourier descriptors, etc.
- *Region-based*: "Skeleton" or the medial-axis transform.

Figure 24: Representations of the numeral 2.

The use of a signed-distance function of a given shape implicitly contains both these representations, since

- simple shapes, disconnected shapes, and shapes with holes can be embedded alike in the signed-distance function, and
- in this framework, we can perturb a shape in order to accentuate certain shape features, thereby increasing the chances of correct classification.

As an example, software was developed to perform handwritten numeral recognition. This required three stages: First, a neural net was trained on an implicit representation. Next, those elements that were not selected were perturbed in the acceptance set, by means of a chosen set of flow rules. As examples, the following characters (Fig. 25) were correctly identified using this software.

6 Image denoising and enhancement

A large amount of work has been performed in the area of image denoising and enhancement using partial differential equations based approaches. Here, the pioneering work was done by Malik and Perona [29] and Alvarez, Morel, Lions [3]. In brief, suppose we remove the constant speed term and then solve the equation with the noisy image, $I_0(\mathbf{x})$, as the initial condition

$$\phi_t + g(\mathbf{x})K \mid \nabla\phi \mid - \nabla g \cdot \nabla\phi = 0, \tag{6.1}$$

with $\phi(\mathbf{x}, t = 0) = I_0(\mathbf{x})$. Here, we note that $g(\mathbf{x})$ is a decreasing function of the image gradient, and that the first term smooths an image away from edge points and that the

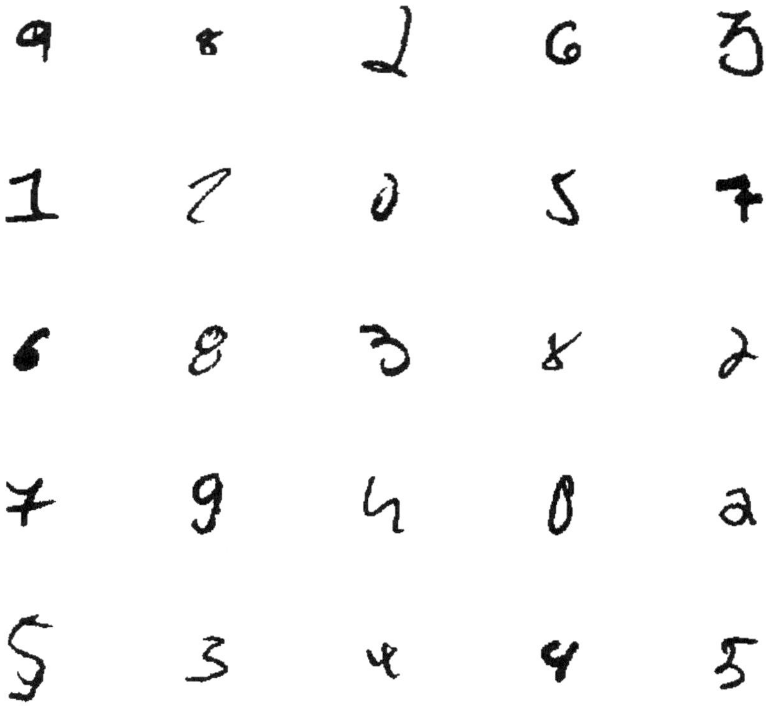

Figure 25: Hand-written NIST characters correctly identified.

second term attracts iso-intensity contours of image closer to "edges" as defined by the g function. As an option, g can either be kept constant through the evolution or recomputed every time step.

The above is a technique for dealing with gray scale images. One approach for tackling color images was presented by Kimmel, Malladi and Sochen [48]. There, the basic idea was to view an image as a surface: $\mathbb{R}^2 \to \mathbb{R}^3$, which, as an embedded map, becomes $\mathbb{R}^2 \to \mathbb{R}^5$. A unified approach for greyscale, color, volumetric images and movies was developed based on the ideas of a Beltrami Flow and an associated Polyakov Action. Again, the fundamental idea is to define a surface area which one then minimizes through some flow rules, leading to Euler-Lagrange equations. The choice of minimization depends on the choice of metrics, and produces different flows; each has benefits in terms of reducing the noise and variation from the smoothed state. The virtue of this approach is that it maintains sharp edges while removing noise. Since all three channels are processed simultaneously, one avoids the ringing effects and artificial colors that can result from processing each in a decoupled manner. We refer the reader to [48] for details.

7 Basic algorithms for interface advancement

Finally, we give an overview of how to construct algorithms for solving these basic partial differential equations.

In [35, 36], the equation for a curve propagating normal to itself with a given speed F and which remains a graph as it moves was studied. There, it was shown that curvature in front propagation plays a role similar to that of viscosity in hyperbolic conservation laws. An entropy condition, analogous to the one for shock dynamics, was formulated to extract a unique entropy-satisfying weak solution which corresponded to the physical boundary of a propagating front. In a later paper [37], this analogy was made more formal, and the strategy of attacking problems in interface motion by borrowing schemes from hyperbolic conservation laws was first introduced. This led to the level set method presented in [28], and was then followed by the adaptive narrow band method developed in [1] and the Fast Marching Method given in [38].

7.1 Level set methods

7.1.1 Equations of motion

Level set methods rely on two central embeddings; first the embedding of the interface as the zero level set of a higher dimensional function, and second, the embedding (or extension) of the interface's velocity to this higher dimensional level set function. More precisely, given a moving closed hypersurface $\Gamma(t)$, that is, $\Gamma(t=0) : [0, \infty) \to \mathbb{R}^N$, propagating with a speed F in its normal direction, we wish to produce an Eulerian formulation for the motion of the hypersurface propagating along its normal direction with speed F, where F can be a function of various arguments, including the curvature, normal direction, etc. Let $\pm d$ be the signed distance to the interface. If this propagating interface is embedded as the zero level set of a higher dimensional function ϕ, that is, let $\phi(x, t=0)$, where $x \in \mathbb{R}^N$ is defined by

$$\phi(x, t=0) = \pm d, \tag{7.1}$$

then an initial value partial differential equation can be obtained for the evolution of ϕ, namely

$$\phi_t + F|\nabla\phi| = 0, \tag{7.2}$$

$$\phi(x, t = 0) \quad \text{given.} \tag{7.3}$$

This is the implicit formulation of front propagation given in [28]. As discussed in [35, 36, 37], propagating fronts can develop shocks and rarefactions in the slope, corresponding to corners and fans in the evolving interface, and numerical techniques designed for hyperbolic conservation laws can be exploited to construct schemes which produce the correct, physically reasonable entropy solution.

There are certain advantages associated with this perspective. First, it is unchanged in higher dimensions; that is, for surfaces propagating in three dimensions and higher. Second, topological changes in the evolving front Γ are handled naturally; the position of the front at time t is given by the zero level set $\phi(x, y, t) = 0$ of the evolving level set function. This set need not be connected, and can break and merge as t advances. Third, terms in the speed function F involving geometric quantities such as the normal vector n and the curvature κ may be easily approximated through the use of derivative operators applied to the level set function, that is,

$$n = \frac{\nabla\phi}{|\nabla\phi|}, \qquad \kappa = \nabla \cdot \frac{\nabla\phi}{|\nabla\phi|}.$$

Fourth, the upwind finite difference technology for hyperbolic conservation laws may be used to approximate the gradient operators.

7.1.2 Approximation schemes

Entropy-satisfying upwind viscosity schemes for this initial value formulation were introduced in [28]. One of the simplest first order scheme is given as

$$\phi_{ijk}^{n+1} = \phi_{ijk}^{n} - \Delta t[\max(F_{ijk}, 0)\nabla^{+} + \min(F_{ijk}, 0)\nabla^{-}], \tag{7.4}$$

where

$$\nabla^{+} = \begin{bmatrix} \max(D_{ijk}^{-x}, 0)^2 + \min(D_{ijk}^{+x}, 0)^2 + \\ \max(D_{ijk}^{-y}, 0)^2 + \min(D_{ijk}^{+y}, 0)^2 + \\ \max(D_{ijk}^{-z}, 0)^2 + \min(D_{ijk}^{+z}, 0)^2 \end{bmatrix}^{1/2}$$

$$\nabla^{-} = \begin{bmatrix} \max(D_{ijk}^{+x}, 0)^2 + \min(D_{ijk}^{-x}, 0)^2 + \\ \max(D_{ijk}^{+y}, 0)^2 + \min(D_{ijk}^{-y}, 0)^2 + \\ \max(D_{ijk}^{+z}, 0)^2 + \min(D_{ijk}^{-z}, 0)^2 \end{bmatrix}^{1/2}$$

Higher order schemes are available, see [28].

The above formulation reveals two central embeddings.

(1) First, in the initialization step (Equ. (7.1)), the signed distance function is used to build a function ϕ which corresponds to the interface at the level set $\phi = 0$. This step is known as "initialization"; when performed at some later point in the calculation beyond $t = 0$, it is referred to as "re-initialization".

(2) Second, the construction of the initial value PDE given in Equ. (7.2) means that the velocity F is now defined for *all* the level sets, not just the zero level set corresponding to the interface itself. We can be more precise by rewriting the level set equation as

$$\phi_t + F^{\text{ext}}|\nabla\phi| = 0, \tag{7.5}$$

where F^{ext} is some velocity field which, at the zero level set, equals the given speed F. In other words,

$$F^{\text{ext}} = F \quad \text{on} \quad \phi = 0.$$

This new velocity field F^{ext} is known as the "extension velocity". Details on techniques for constructing extension velocities developed by Adalsteinsson and Sethian may be found in [2].

Both of these issues need to be confronted in order to efficiently apply level set methods to complex computational problems.

7.1.3 Adaptivity: the narrow band level set method

Equ. (7.4) is an explicit scheme, and hence can be solved directly. The time step requirement depends on the nature of the speed function F; for an F that depends only on position, the time step behaves like $\frac{\Delta t}{\Delta x}F \leq 1$. In the case when the speed function F depends on curvature terms (for example, $F = -\kappa$), the equation has a parabolic component, and hence the time step requirement resembles that of a non-linear heat equation; the time step depends roughly on $\frac{\Delta t}{\Delta x^2}$.

In the level set formulation, both the level set function and the speed are embedded into a higher dimension. This then implies computational labor through the entire grid, which is inefficient. A rough operation count for the original level set method assumes N grid points in each space dimension of a three-dimensional problem. For a simple problem of straightforward propagation with speed $F = 1$, assuming that it takes roughly N time steps for the front to propagate through the domain (here, the CFL condition is taken almost equal to unity), this produces an $O(N^4)$ method.

Considerable computational speedup in the level set method comes from the use of the "Narrow Band Level Set Method", introduced by Adalsteinsson and Sethian in [1]. It is clear that performing calculations over the entire computational domain is wasteful. Instead, an efficient modification is to perform work only in a neighborhood (or "narrow band") of the zero level set. This drops the operation count in three dimensions to $O(kN^3)$, where k is the number of cells in the narrow band. This is a significant cost reduction; it also means that extension velocities need only be constructed at points lying in the narrow band, as opposed to all points in the computational domain. This idea of limiting computation to a narrow band around the zero level set was introduced in Chopp [8], used in recovering shapes from images in Malladi, Sethian and Vemuri [25], and explored in depth in [1].

Use of narrow bands leads to level set front advancement algorithms that are computationally equivalent in terms of complexity to traditional marker methods and cell techniques, while maintaining the advantages of topological merger, accuracy, and easy extension to multi-dimensions. Typically, the speed associated with the Narrow Band Method is about ten times faster on a 160×160 grid than the full matrix method. Such a speed-up is substantial; in three-dimensional simulations, it can make the difference between computationally intensive problems and those that can be done with relative ease. Details on the accuracy, typical tube sizes, and number of times a tube must be rebuilt may be found in Adalsteinsson and Sethian [1].

Additional issues in the practical implementation of level set methods includes the construction of extension velocities and correct data structures; see [42] for complete details.

7.2 Fast Marching Methods for eikonal and related equations

Fast Marching Methods are finite difference techniques for solving the eikonal equation of the form

$$|\nabla T| F(x,y,z) = 1, \qquad T = 0 \quad \text{on} \quad \Gamma.$$

This can be thought of as a front propagation problem for a front initially located at Γ and propagating with speed $F(x,y,z) > 0$. We note that this is a *boundary value* partial differential equation as opposed to an initial value problem given by level set methods, even though it describes a moving interface. This eikonal equation describes a large number of physical phenomena, including those from optics, wave transport, seismology, photolithography and optimal path planning, and Fast Marching Methods have been used to solve these and a host of other problems. Here, we provide a brief overview of the techniques; we refer the reader to [41] and [42] for a large collection of applications based on this technique.

7.3 Basic algorithms

Consider the following upwind finite difference scheme for the eikonal equation, namely

$$\begin{bmatrix} \max(D^{-x}_{ijk}u, -D^{+x}_{ijk}u, 0)^2 + \\ \max(D^{-y}_{ijk}u, -D^{+y}_{ijk}u, 0)^2 + \\ \max(D^{-z}_{ijk}u, -D^{+z}_{ijk}u, 0)^2 \end{bmatrix}^{1/2} = F_{ijk}, \tag{7.6}$$

which was discussed by Rouy and Tourin [30]. One approach to solving this finite difference scheme (see [30]), is through iteration, which leads to an $O(N^4)$ algorithm in three dimensions, where N is the number of points in each direction. Instead, Fast Marching Methods take a different approach.

The Fast Marching Method [38] is connected to Huygens' principle. The viscosity solution to the eikonal equation $|\nabla u(x)| = F(x)$ can be interpreted through Huygens' principle in the following way: circular wavefronts are drawn at each point on the boundary, with the radius proportional to $F(x)$. The envelope of these wavefronts is then used to construct a new set of points, and the process is repeated; in the limit the eikonal solution is obtained. The Fast Marching Method mimics this construction; a computational grid is used to carry the solution u, and an upwind, viscosity-satisfying finite difference scheme is used to approximate this wavefront.

The order in which the grid values produced through these finite difference approximations are obtained is intimately connected to Dijkstra's method [9], which is a depth-search technique for computing shortest paths on a network. In that technique, the algorithm keeps track of the speed of propagation along the network links, and fans out along the network links to touch all the grid points. The Fast Marching Method exploits a similar idea in the context of a continuous finite difference approximation to the underlying partial differential equation, rather than discrete network links.

We note that Tsitsiklis [49] has developed a different Dijkstra-like algorithm which obtains the viscosity solution through a control-theoretic discretization which hinges on a causality relationship based on the optimality criterion.

In more detail, the Fast Marching Method is as follows. Suppose at some time the eikonal solution is known at a set of points (denoted *Accepted* points). For every not-yet accepted grid point such that it has an accepted neighbor, a trial solution to the above quadratic Equ. (7.6) is computed, using the given values for u at accepted points, and values of ∞ at all other points. Observe that the smallest of these trial solutions must be correct, since it depends only on accepted values which are themselves smaller. This " causality" relationship can be exploited to efficiently and systematically compute the solution as follows:

First, tag points in the initial conditions as *Accepted.* Then tag as *Considered* all points one grid point away and compute values at those points by solving Equ. (7.6). Finally, tag as *Far* all other grid points. Then the loop is :

(1) Begin Loop: Let *Trial* be the *Considered* point with smallest value of u.

(2) Tag as *Considered* all neighbors of *Trial* that are not *Accepted.* If the neighbor is in *Far*, remove it from that set and add it to the set *Considered.*

(3) Recompute the values of u at all *Considered* neighbors of *Trial* by solving the piecewise quadratic equation according to Equ. (7.6).

(4) Add point *Trial* to *Accepted*; remove from *Considered.*

(5) Return to top until the *Considered* set is empty.

This is the Fast Marching Method given by Sethian in [38]. The key to an efficient implementation of the above technique lies in a fast way of locating the grid point in the narrow band with the smallest value for u. An efficient scheme for doing so, discussed in detail in [42], can be devised using a min-heap structure, similar to what is done in Dijkstra's method. Given N elements in the heap, this allows one to change any element in the heap and re-order the heap in $O(\log N)$ steps. Thus, the computational efficiency of the Fast Marching Method for the mesh with N points is $O(N \log N)$: N steps to touch each mesh point with each step requiring $O(\log N)$, since the heap has to be re-ordered each time the values are changed.

The Fast Marching Method evolved in part from examining the limit of the Narrow Band level set method as the band was reduced to one grid cell. Fast Marching Methods, by taking the perspective of the large body of work on higher order upwind, finite difference approximants from hyperbolic conservation laws, allow for higher order versions on both structured and unstructured meshes. The Fast Marching Method has been extended to higher order finite difference approximations by Sethian in [41], first order unstructured meshes by Kimmel and Sethian [16], and higher order unstructured meshes by Sethian and Vladimirsky [45]. Some early applications include photolithography in [40], a comparison of a similar approach with volume-of-fluid techniques in [14], a fast algorithm for image segmentation in [21] and computation of seismic travel times by Sethian and Popovici [44]; again, see also [49] for a different Dijkstra-like algorithm for the eikonal equation.

The range of applications of these techniques is quite large, and extends considerably beyond image processing. In Fig. 26, we give a perspective on how some of these topics are related. There are many other contributors to the evolution of these ideas; the chart is meant to give perspective on how the theory, algorithms, and applications have evolved. The text and bibliography of [42] gives a somewhat more complete sense of the literature and the range of work underway.

Acknowledgements All calculations were performed at the University of California at Berkeley and the Lawrence Berkeley Laboratory. The detailed applications of Level Set and Fast Marching Methods applied to image processing discussed in this work are joint with R. Malladi, and A. Sarti.

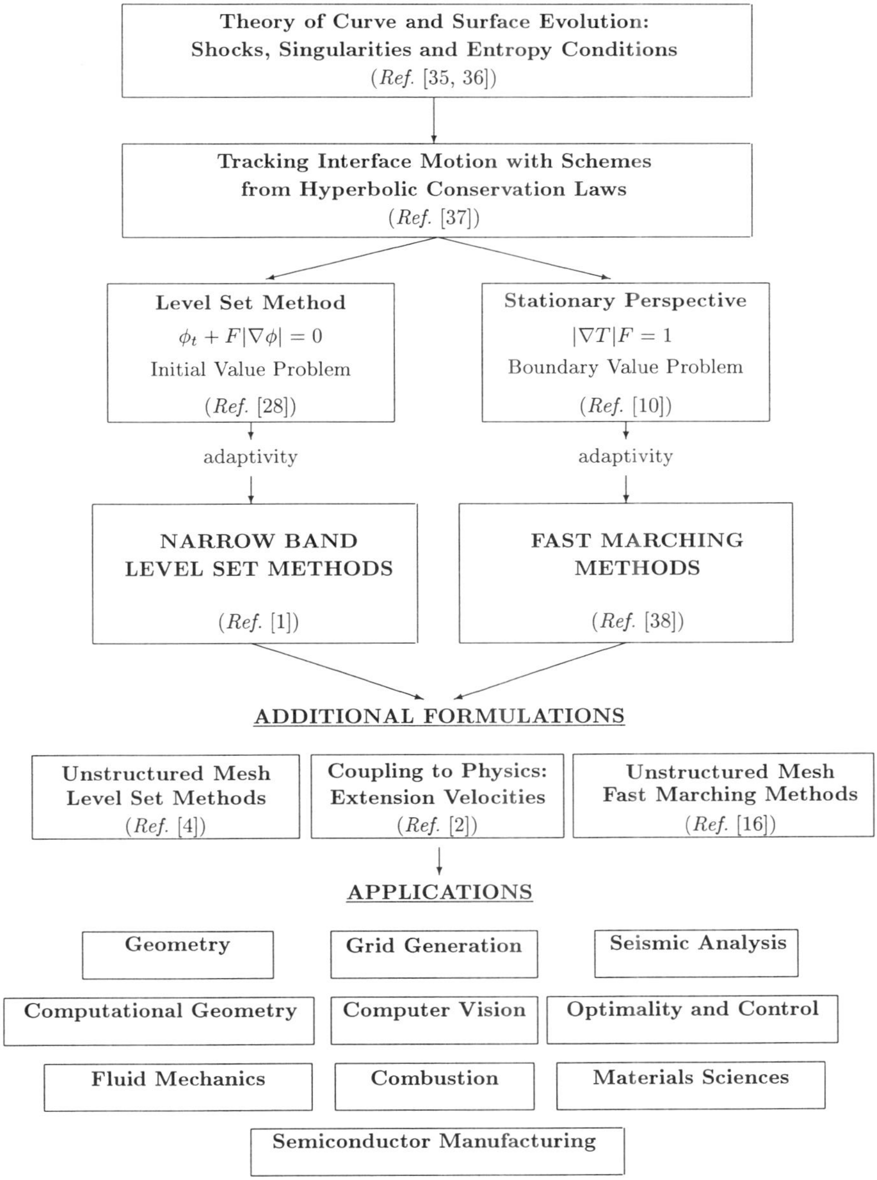

Figure 26: Algorithms and applications for interface propagation.

References

[1] D. Adalsteinsson and J. A. Sethian, A fast level set method for propagating interfaces, *J. Comp. Phys.* **118** (2) (1995), 269–277.

[2] D. Adalsteinsson and J.A. Sethian, The fast construction of extension velocities in level set methods, *J. Comp. Phys.* **148** (1999), 2–22.

[3] L. Alvarez, P. L. Lions, and M. Morel, Image selective smoothing and edge detection by nonlinear diffusion. II, *SIAM J. Numer. Anal.* **29** (3) (1992), 845–866.

[4] T. J. Barth and J. A. Sethian, Numerical schemes for the Hamilton-Jacobi and level set equations on triangulated domains, *J. Comp. Phys.* **145** (1) (1998), 1–40.

[5] M. R. Brewer, R. Malladi, G. Pankiewicz, B. Conway, and L. Tarassenko, Methods for large-scale segmentation of cloud images, in: *Proc. 1997 EUMETSAT Meteorological Satellite Data Users' Conference, Brussels.*

[6] V. Caselles, F. Catte, T. Coll, and F. Dibos, A geometric model for active contours in image processing recovery, *Numer. Math.* **66** (1) (1993), 1–31.

[7] V. Caselles, R. Kimmel, and G. Sapiro, Geodesic active contours, in: *Proc. Fifth IEEE Internat. Conf. on Computer Vision, ICCV '95, Cambridge, MA*, 694–699.

[8] D. L. Chopp, Computing minimal surfaces via level set curvature flow, *J. Comput. Phys.* **106** (1993), 77–91.

[9] E. W. Dijkstra, A note on two problems in connection with graphs, *Numer. Math.* **1** (1959), 269–271.

[10] L. C. Evans, *Partial Differential Equations*, Berkeley Math. Lecture Notes Ser. **3A**, **3B**, Center for Pure and Applied Mathematics, University of California, Berkeley, CA, 1994.

[11] R. Goldenberg, R. Kimmel, E. Rivlin, and M. Rudzsky, Fast geodesic active contours, *IEEE Trans. Image Processing* **10** (10) (2001), 1467–1475.

[12] M. Grayson, The heat equation shrinks embedded plane curves to round points, *J. Differential Geom.* **26** (1987), 285.

[13] M. Grayson, A short note on the evolution of surfaces via mean curvatures, *J. Differential Geom.* **58** (1989), 555.

[14] J. Helmsen, E. G. Puckett, P. Colella, and M. Dorr, Two new methods for simulating photolithography development, in: *Proc. SPIE 1996 Internat. Sympos. on Microlithography*, SPIE, 2726, June, 1996.

[15] M. Kass, A. Witkin, and D. Terzopoulos, Snakes: active contour models, *Internat. J. Computer Vision* **1** (1988), 321–331.

[16] R. Kimmel and J. A. Sethian, Fast marching methods on triangulated domains, *Proc. Nat. Acad. Sci.* **95** (1998), 8341–8435.

[17] R. Malladi and J. A. Sethian, A unified approach for shape segmentation, representation, and recognition, Center for Pure and Applied Mathematics, Report 614, University of California, Berkeley, 1994.

[18] R. Malladi and J. A. Sethian, Image processing via level set curvature flow, *Proc. Nat. Acad. Sci.* **92** (15) (1995), 7046–7050.

[19] R. Malladi and J. A. Sethian, Level set methods for curvature flow, image enhancement, and shape recovery in medical images, *Visualization and Mathematics* (H.-C. Hege and K. Polthier, eds.), Springer-Verlag, Berlin-Heidelberg-New York, 1997, 329–345.

[20] R. Malladi and J. A. Sethian, A unified approach to noise removal, image enhancement, and shape recovery, *IEEE Trans. Image Processing* **5** (11) (1996), 1554–1568.

[21] R. Malladi, and J. A. Sethian, An $O(N \log N)$ algorithm for shape modeling, *Proc. Nat. Acad. Sci. U.S.A.* **93** (1996), 9389–9392.

[22] R. Malladi and J. A. Sethian, Image processing: flows under min/max curvature and mean curvature, *Graphical Models and Image Processing* **58** (2) (1996), 127–141.

[23] R. Malladi and J. A. Sethian, Shape modeling in medical imaging with marching methods, Report LBNL–39541, LBNL, University of California, Berkeley, 1996.

[24] R. Malladi, and J. A. Sethian, Level set and fast marching methods in image processing and computer vision, in: *Proc. IEEE Internat. Conf. on Image Processing, Lausanne, 1996.*

[25] R. Malladi, J. A. Sethian, and B. Vemuri, A topology independent shape modeling scheme, in: *Proc. SPIE Conf. on Geometric Methods in Computer Vision II, Vol. 2031, San Diego, CA, 1993*, 246–258.

[26] R. Malladi, J. A. Sethian, and B. C. Vemuri, A fast level set based algorithm for topology-independent shape modeling, *J. Math. Imaging and Vision* **6** (2/3) (1996), 269–290.

[27] R. Malladi, *Geometric Methods in Biomedical Image Analysis* (R. Malladi, ed.), Springer Verlag, March 2002.

[28] S. Osher and J. A. Sethian, Fronts propagating with curvature-dependent speed: algorithms based on Hamilton-Jacobi Formulations, *J. Comput. Phys.* **79** (1988), 12–49.

[29] P. Perona and J. Malik, Scale-space and edge detection using anisotropic diffusion, *IEEE Trans. Pattern Anal. Machine Intelligence* **12** (7) (1990), 629–639.

[30] E. Rouy and A. Tourin, A viscosity solutions approach to shape-from-shading, *SIAM J. Numer. Anal* **29** (3) (1992), 867–884.

[31] G. Sapiro, *Geometric Partial Differential Equations and Image Analysis*, Cambridge University Press, 2001.

[32] A. Sarti and R. Malladi, A geometric level set model for ultrasound analysis, LBNL–44442, October 1999, to appear in: *Geometric Methods in Biomedical Image Analysis* (R. Malladi, ed.), Springer Verlag, March 2002.

[33] A. Sarti, R. Malladi, and J. A. Sethian, Subjective surfaces: a method for completing missing boundaries, *Proc. Nat. Acad. Sci. U.S.A.* **97** (2000), 6258–6263.

[34] A. Sarti, R. Malladi, and J. A. Sethian, Subjective surfaces: a geometric model for boundary completion, *IEEE Trans. Pattern Anal. Machine Intelligence* (2001).

[35] J. A. Sethian, An Analysis of Flame Propagation, Ph.D. Dissertation, Dept. of Mathematics, University of California, Berkeley, CA, 1982.

[36] J. A. Sethian, Curvature and the evolution of fronts, *Comm. Math. Phys.* **101** (1985), 487–499.

[37] J. A. Sethian, Numerical methods for propagating fronts, in: *Variational Methods for Free Surface Interfaces, Menlo Park, Calif., 1985* (P. Concus and R. Finn, eds.), Springer-Verlag, New York, 1987, 155–164.

[38] J. A. Sethian, A fast marching level set method for monotonically advancing fronts, *Proc. Nat. Acad. Sci. U.S.A.* **93** (4) (1996), 1591–1595.

[39] J. A. Sethian, A Review of the Theory, Algorithms, and Applications of Level Set Methods for Propagating Interfaces, *Acta Numerica*, Cambridge University Press, 1996.

[40] J. A. Sethian, Fast marching level set methods for three-dimensional photolithography development, in: *Proc. SPIE 1996 Internat. Sympos. on Microlithography, Santa Clara, CA.*

[41] J. A. Sethian, Fast marching methods, *SIAM Rev.* **41** (1999), 199–235.

[42] J. A. Sethian, *Level Set Methods and Fast Marching Methods*, Cambridge Monogr. Appl. Comput. Math. **3**, Cambridge University Press, 1999.

[43] J. A. Sethian, Evolution, implementation, and application of level set and fast marching methods for advancing fronts, *J. Comp. Phys.* **169** (2) (2001), 503-555.

[44] J. A. Sethian and M. Popovici, Fast marching methods applied to computation of seismic travel times, *Geophysics* **64** (2) (1999).

[45] J. A. Sethian and A. Vladimirsky, Fast methods for the eikonal and related Hamilton-Jacobi equations on unstructured meshes, *Proc. Nat. Acad. Sci. U.S.A.* **97** (11) (2000), 5699–5703.

[46] J. A. Sethian and A. Vladimirsky, Ordered upwind methods for static Hamilton-Jacobi equations, *Proc. Nat. Acad. Sci. U.S.A.* **98** (20) (2001), 11069–11074.

[47] J. A. Sethian and A. Vladimirsky, Ordered upwind methods for static Hamilton-Jacobi equations: theory and algorithms, Center for Pure and Applied Mathematics Technical Report PAM–792, University of California, Berkeley, 2001.

[48] N. Sochen, R. Kimmel, and R. Malladi, A general framework for low level vision, *IEEE Trans. Image Proc.* **7** (3) (1998), 310–318, 1998.

[49] J. N. Tsitsiklis, Efficient algorithms for globally optimal trajectories, *IEEE Trans. Automatic Control* **40** (1995), 1528–1538.

Recent developments in the theory of front propagation and its applications

Panagiotis E. SOUGANIDIS*
Department of Mathematics
The University of Texas at Austin
Austin, TX 78712-1082
USA

Abstract

We describe a new theory for studying the geometric motion of interfaces past the first times singularities develop. We also show how this theory allows one to justify and study the development of such interfaces in a wide range of applications, from the asymptotics of recation-diffusion equations and systems and hydrodynamic limits of particles systems in phase transitions to turbulent combustion.

1 Introduction

In this note I discuss in some detail, but without most of the cumbersome technicalities, a theory about propagating fronts, i.e., "hypersurfaces" moving with prescribed normal velocity, which may depend on the location of the front in space and time, its normal and its principal curvatures. I also discuss, in the same spirit, the way this theory allows for the rigorous study of the detailed asymptotics of a wide class of problems, like modeling in phase transitions, reaction-diffusion equations, interacting particle systems (stochastic Ising models), turbulent flame propagation and (premixed and non-premixed) combustion.

The study of the evolution of a hypersurface with presecribed normal velocity is a classical problem in geometry. Its connection with the applications comes up either by direct modeling, as in image processing and the phenomenological theory of sharp interfaces in continuum mechanics, or in the analysis of the asymptotic behavior of certain systems. This last claim can be loosely described as follows: An evolving order parameter, which, depending on the specific context, describes the different phases of a material or the total (averaged) magnetization of a stochastic system or the temperature of a reacting-diffusing system, etc., approaches the equilibrium states of the underlying system. Depending on the values of a threshold parameter, such systems have either a unique or multiple equilibrium states. The existence of multiple equilibrium states can be associated to phase transitions. When more than one equilibrium states exist, the evolving order parameter develops, for large times, interfaces. These are the boundaries of the regions where the order parameter converges to

*Partially supported by the NSF.

A. Bourlioux and M.J. Gander (eds.), Modern Methods in Scientific Computing and Applications, 397–449.

the different equilibria. The problem is then to justify the appearance of the interfaces and to understand in a qualitative way their dynamics, geometry, regularity, etc..

The main mathematical difficulty in the study of the evolution of hypersurfaces with prescribed normal velocity is the development of singularities in finite time, independently of the regularity of the initial surface. A great deal of work has been done during the last fifteen years in order to interpret the evolution past the singularities. The outcome has been the development of a weak notion of evolving fronts called generalized front propagation. The generalized evolution with prescribed normal velocity starting at a given surface is defined for all $t \geq 0$, although it may become extinct in finite time, and it agrees with the classical differential-geometric flow, as long as the latter exists. The generalized motion may, on the other hand, develop singularities, change topological type, etc..

In spite of these peculiarities, the generalized motion appears to be the right way to extend the classical motion past singularities. A strong indication in this direction is the fact that the generalized evolution governs the asymptotic behavior of solutions of semilinear reaction-diffusion equations and systems. Such equations are used in continuum mechanics to describe the time evolution of an order parameter determining the phases of a material (phase field theory). Similar equations, but with different reaction terms, are also used to model turbulent combustion.

Another recent striking application of the theory of generalized front propagation is the fact that it governs the macroscopic (hydrodynamic) behavior, for large times and in the context of grain coarsening, of interacting particle systems, like the stochastic Ising model, with long-range interactions and general spin flip dynamics. Such systems are standard Gibbsian models used in statistical mechanics to describe phase transitions. It turns out that the generalized front propagation not only describes the limiting behavior of such systems but also provides, in this context, a theoretical justification, from the microscopic point of view, of the several phenomenological sharp interface models in phase transitions.

The paper is organized as follows: Section 2 is about the generalized front propagation. It contains most of the definitions and results which are relevant for the study of the asymptotic problems. In Section 3, I present a general abstract theorem, which allows for the rigorous justification of the appearance of interfaces in a large class of asymptotic problems by checking a number of basic conditions and justifying formal asymptotics under arbitrary smoothness assumptions. Section 4 is about the asymptotics of reaction-diffusion equations with bistable nonlinearities. In Section 5, I discuss again reaction-diffusion equations but with KPP-type nonlinearities and I present some applications to turbulent flame propagation and combustion. Section 6 discusses the asymptotics of stochastic Ising models.

This paper is supposed to provide a general overview of the theory instead of a detailed account of all the results which have been obtained. I will therefore adopt a very informal style omitting, whenever possible, the technicalities. I will, however, provide precise references where all the details can be found.

The notion of viscosity solutions and their properties are fundamental for everything I am presenting here. The notes are written under the assumption that the reader is familiar with viscosity solutions. For a number, if not most, of the technical issues and statements, I refer to the "User's Guide" by Crandall, Ishii and Lions [CIL] as well as [Cr] and the references therein. Finally, in the text I will often say solution instead of viscosity solution.

2 Generalized front propagation

This section is about the evolution of fronts $\Gamma_t = \partial\Omega_t \subset \mathbb{R}^N$ starting at $\Gamma_0 = \partial\Omega_0$, where Ω_t is an open subset of $\mathbb{R}^N$, with normal velocity

$$V = v(Dn, n, x, t), \tag{2.1}$$

and n and Dn are the exterior normal vector to the front and its gradient, respectively.

Typical examples of motions governed by (2.1), which are of great interest in geometry but also appear in the kind of applications referred to in the Introduction, are the motion by mean curvature

$$V = -\operatorname{tr} Dn,$$

the motion by the Gaussian curvature

$$V = \kappa_1 \cdots \kappa_{N-1},$$

where $\kappa_1, \ldots, \kappa_{N-1}$ are the principal curvatures of the front, and the anisotropic law

$$V = -\operatorname{tr}[\theta(n, x, t)Dn] + c(n, x, t),$$

where θ is a nonnegative matrix and c is a scalar.

Such evolutions may develop singularities in finite time, independently of the smoothness of the initial surface. Classical examples in this direction are the evolution by mean curvature of "figure-eights", "crosses", "bar-bells" and "tori" in $\mathbb{R}^N$. For more discussion regarding this issue, I refer to Evans and Spruck [ESp], Ilmanen [Il1], Soner [Son1], Soner and Souganidis [SonS], Souganidis [Sou1], etc.

A great deal of work has been recently done in order to interpret the evolution (2.1) past the singularities. A number of a priori different approaches have been put forward leading to different definitions (e.g. level set, distance function, phase field, etc.). The task is then to show that all these approaches yield the same generalized front propagation, which can be accomplished assuming that there is "no-fattening", as it is explained later in this section. In these notes I use as the building block the geometric definition put forward by Barles and the author in [BS2] and I derive the other definitions as corollaries. In doing so I will also remark about the intrinsic merits of the other approaches and/or definitions. In what follows I discuss motions governed by (2.1) and not only mean curvature, for which one has, due to some additional geometric structure, more information and yet more ways to go about studying it.

Throughout this discussion I consider hypersurfaces $(\Gamma_t)_{t\in(a,b)}$ in $\mathbb{R}^N$, which are boundaries of open subsets $(\Omega_t)_{t\in(a,b)}$ of $\mathbb{R}^N$, and introduce the signed-distance function $d(x,t)$ from x to Γ_t defined by

$$d(x,t) = \begin{cases} d(x,\Gamma_t) & \text{if } x \in \Omega_t, \\ -d(x,\Gamma_t) & \text{otherwise,} \end{cases} \tag{2.2}$$

where $d(x,\Gamma_t)$ denotes the usual Euclidean distance from $x \in \mathbb{R}^N$ to Γ_t. If Γ_t is a smooth hypersurface, then d is a smooth function in a neighborhood of Γ_t and, for $x \in \Gamma_t$, $n(x,t) = -Dd(x,t)$ is the unit normal to Γ_t pointing away from Ω_t.

I begin considering smooth motions of smooth hypersurfaces $(\Gamma_t)_{t\in(a,b)}$ with a general prescribed normal velocity v. To this end, recall that the normal velocity $V(x,t)$ of such smoothly evolving hypersurfaces is defined at $x \in \Gamma_t$ by

$$V(x,t) = n(x,t) \cdot \dot{X}(t),$$

where $X : (a,b) \to \mathbb{R}^N$ is a C^∞-curve such that $X(s) \in \Gamma_s$ for all $s \in (a,b)$ and $X(t) = x$, and $p \cdot q$ denotes the usual inner product of the vectors p, q in $\mathbb{R}^N$. It is easy to check that such curves X exist and that the definition is independent of the particular choice of the curve.

The collection of smooth hypersurfaces $(\Gamma_t)_{t\in(a,b)}$ or of smooth open subsets $(\Omega_t)_{t\in(a,b)}$, is said to propagate with the prescribed normal velocity v if and only if, for all $x \in \Gamma_t$ and $t \in (a,b)$,

$$V(x,t) = v(x,t,\Gamma_t).$$

Similarly, it is said that the normal velocity is larger (resp. smaller) than v if, in the above equation, "=" is replaced by "≥" (resp. "≤").

A priori one may consider normal velocities v depending on (x,t) and on the different characteristics of Γ_t. Here, however, I focus attention on motions satisfying a number of assumptions, which are stated below.

To this end, let D^∞ be the set of C^∞-jets, and, if d is the signed distance to $(\Gamma_t)_{t\in(a,b)}$, define $[d] : \mathbb{R}^N \times (a,b) \to D^\infty$, the jet of the fundamental forms of Γ_t at (x,t), by

$$[d](x,t) = (Dd(x,t), D^2d(x,t), \dots, D^kd(x,t), \dots).$$

The first assumption is:

(A1) Locality and regularity *The normal velocity v at $x \in \Gamma_t$ depends only on x, t and the fundamental forms of Γ_t at x, i.e., there exists a function $v : \mathbb{R}^N \times (a,b) \times D^\infty \to \mathbb{R}$ such that the prescribed normal velocity is given by*

$$V(x,t) = v(x,t,[d](x,t)).$$

Moreover, v is a continuous function of x, t and Γ_t, in the sense that, if, as $n \to \infty$, $(x_n,t_n) \to (x,t)$ and $D^kd_n(x_n,t_n) \to D^kd(x,t)$ for all $k \in \mathbb{N}$, then

$$v(x_n,t_n,[d_n](x_n,t_n)) \to v(x,t,[d](x,t)).$$

Note that if, as $n \to \infty$, $D^kd_n \to D^kd$, uniformly in a neighborhood of Γ_t for all $k \in \mathbb{N}$, then $(x_n,t_n) \to (x,t)$ yields that $D^k_n(x_n,t_n) \to D^kd(x,t)$ for all $k \in \mathbb{N}$.

The second assumption is:

(A2) Monotonicity *If $(\Omega_t)_{t\in(a,b)}$ and $(\tilde{\Omega}_t)_{t\in(a,b)}$ are two collections of C^∞-open subsets of $\mathbb{R}^N$ such that $\Gamma_t = \partial\Omega_t$ and $\tilde{\Gamma}_t = \partial\tilde{\Omega}_t$ move smoothly with a normal velocity smaller and larger, respectively, than v and if $\Omega_t \subset \tilde{\Omega}_t$ for some $t \in (a,b)$, then*

$$\Omega_s \subset \tilde{\Omega}_s \quad \textit{for any} \quad s \in [t,b).$$

To state the definition of the generalized front propagation it is necessary to extend the domain of v to functions which are not necessarily distance functions. A natural way to understand this extension is to remark that, with the above notations, the set $\Omega_t = \{x : d(x,t) > 0\}$ can also be written as $\Omega_t = \{x : \phi(x,t) > 0\}$ for other suitable functions ϕ such that $|D\phi| \neq 0$ on $\partial\Omega_t$. Then, for all $x \in \partial\Omega_t$, one has

$$Dd(x,t) = \widehat{D\phi}(x,t), D^2d(x,t) = D(\widehat{D\phi}(x,t)), \ldots, D^kd(x,t) = D^{k-1}(\widehat{D\phi}(x,t)), \ldots,$$

where, for $p \in \mathbb{R}^N \setminus \{0\}$,

$$\hat{p} = |p|^{-1}p.$$

Define

$$[\phi](x,t) = (\widehat{D\phi}(x,t), D(\widehat{D\phi}(x,t)), \ldots, D^{k-1}(\widehat{D\phi}(x,t)), \ldots),$$

and consider the extension $\bar{v}$ of v given by

$$\bar{v}(x,t,[\phi](x,t)) = |D\phi(x,t)|v(x,t,[\phi](x,t)).$$

Notice that since v is defined on $\mathbb{R}^N \times (a,b) \times D^\infty$, the right hand side of this last equality is well defined at each point where $|D\phi| \neq 0$. Moreover, if the function ϕ is fixed, then $\bar{v}$ can also be seen as a continuous function of (x,t), which is defined on the open subset $\{|D\phi| \neq 0\}$.

In order to simplify the presentation below, I am going to assume throughout the paper that, for each fixed smooth function ϕ, $\bar{v}$ is locally bounded in $\mathbb{R}^N \times (a,b)$ in the following sense. For the general case, I refer to [BS2].

(A3) Local Boundedness *For any compact subset K of $\mathbb{R}^N \times (a,b)$, there exists a constant $C(K)$ such that, for all C^∞-functions ϕ and for all $(x,t) \in \{|D\phi| \neq 0\} \cap K$,*

$$|\bar{v}(x,t,[\phi](x,t))| \leq C(K).$$

In what follows it is also necessary to consider the upper- and lower-semicontinuous envelopes of the locally bounded function $\bar{v}$ considered as a function of x and t, which are denoted by $\bar{v}^*$ and $\bar{v}_*$, respectively. Recall that for a locally bounded function $f : A \to \mathbb{R}$, where A is a subset of some $\mathbb{R}^k$, the upper- and lower-semicontinuous envelopes f^* and f_* of f are given by

$$f^*(y) = \limsup_{z \to y} f(z) \quad \text{and} \quad f_*(y) = \liminf_{z \to y} f(z).$$

2.1 Definition A family $(\Omega_t)_{t\in(a,b)}$ of open subsets of $\mathbb{R}^N$ is called a *generalized super-flow* (resp. *sub-flow*) with normal velocity v if and only if for all $x_0 \in \mathbb{R}^N$, $t \in (a,b)$, $r > 0$, $\alpha > 0$ and for all smooth functions $\phi : \mathbb{R}^N \to \mathbb{R}$ such that $\{\phi \geq 0\} \subset \Omega_t \cap B_r(x_0)$, (resp. $\{\phi \leq 0\} \subset \overline{\Omega}_t^c \cap B_r(x_0)$), with $|D\phi| \neq 0$ on $\{\phi = 0\}$, there exists $h_0 > 0$ depending only on α and the C^∞-function ϕ through the properties of $[\phi]$ in $\overline{B}_r(x_0)$ such that, for all $h \in (0,h_0)$,

$$\{x \in \mathbb{R}^N : \phi(x) + h[\bar{v}_*(x,t,[\phi](x)) - \alpha] > 0\} \cap \overline{B}_r(x_0) \subset \Omega_{t+h}$$

(resp.

$$\{x \in \mathbb{R}^N : \phi(x) + h[\bar{v}^*(x,t,[\phi](x)) + \alpha] < 0\} \cap \overline{B}_r(x_0) \subset \overline{\Omega}_{t+h}^c).$$

A family $(\Omega_t)_{t\in(a,b)}$ of open subsets of $\mathbb{R}^N$ is called a *generalized flow* with normal velocity v if it is both a sub- and super-flow.

The following remark is essential in order to understand the definitions of sub- and super-flow.

2.2 Remark For smooth classical flows, if the orientation is changed, i.e., if d is replaced by $\tilde{d} = -d$, which, in other words, means that the motion of the family $(\overline{\Omega}_t^c)_{t\in(a,b)}$ is considered instead of $(\Omega_t)_{t\in(a,b)}$, then the prescribed normal velocity has to be changed into $-v(x,t,[-\tilde{d}])$. This elementary fact is used in the above definition. Indeed one easily sees that $(\Omega_t)_{t\in(a,b)}$ is a generalized sub-flow with normal velocity $v(x,t,[d])$ if and only if $(\overline{\Omega}_t^c)_{t\in(a,b)}$ is a generalized super-flow with normal velocity $-v(x,t,[-\tilde{d}])$.

From now I always assume that the assumptions (A1)–(A3) are satisfied. It follows that (i) if $(\Omega_t)_{t\in(a,b)}$ is a collection of open subsets which depends smoothly on t, then $(\Omega_t)_{t\in(a,b)}$ is a generalized super-flow (resp. sub-flow) with normal velocity v iff the collection $(\Gamma_t)_{t\in(a,b)}$, where $\Gamma_t = \partial\Omega_t$ for $t \in (a,b)$, propagates with normal velocity larger (resp. smaller) than v, (ii) the monotonicity assumption (A2) yields that there exists a continuous function G such that $\bar{v}(x,t,[\phi])$ is necessarily of the form

$$\bar{v}(x,t,[\phi](x,t)) = -|D\phi(x,t)|G(D(\widehat{D\phi})(x,t),\widehat{D\phi}(x,t),x,t), \tag{2.3}$$

and (iii) G is nonincreasing, i.e., degenerate elliptic, in $D^2\phi$.

To formulate these results and to emphasize the dependence of G in $D^2\phi$ and $D\phi$, it is necessary to introduce the function $F : \mathcal{S}^N \times \mathbb{R}^N \setminus \{0\} \times \mathbb{R}^N \times [0,\infty) \to \mathbb{R}$, $\mathcal{S}^N$ being the space of $N \times N$ symmetric matrices, given by

$$F(X,p,x,t) = |p|G(|p|^{-1}(X - X\hat{p}\otimes\hat{p}),\hat{p},x,t). \tag{2.4}$$

The monotonicity of G is expressed by saying that for all $X, Y \in \mathcal{S}^N$ and $(p,x,t) \in \mathbb{R}^N \setminus \{0\} \times \mathbb{R}^N \times (0,\infty)$, F satisfies the ellipticity condition

$$F(X,p,x,t) \le F(Y,p,x,t) \quad \text{if} \quad X \ge Y. \tag{2.5}$$

Notice that (2.4) and the local boundedness assumption (A3) yield that F is itself locally bounded. As it was done for $\bar{v}$, the locally bounded function F is extended to $p=0$ by taking upper or lower semi-continuous envelopes. In the rest of the paper the symbols v, $\bar{v}$ and F are used to denote the prescribed normal velocity.

The results which are presented here, are strongly inspired by the work of Alvarez, Guichard, Lions and Morel [AGLM] in image processing. It is, however, worth mentioning that the geometric formulation presented here does not allow for a direct application of [AGLM].

The main result is:

2.3 Theorem *Assume that* (A1)–(A3) *hold and that* $(\Omega_t)_{t\in(a,b)}$ *is a collection of smooth, open subsets of* $\mathbb{R}^N$ *which depend smoothly on* t. *Then:*

(i) *The collection* $(\Omega_t)_{t\in(a,b)}$ *is a generalized super-flow (resp. sub-flow) with normal velocity* v *if and only if the collection* $(\Gamma_t)_{t\in(a,b)}$, *where* $\Gamma_t = \partial\Omega_t$ *for all* $t \in (a,b)$, *propagates with a normal velocity larger (resp. smaller) than* v.

(ii) *There exists a continuous function G such that, for all $(x,t) \in \mathbb{R}^N \times [0,\infty)$ and for all smooth functions ϕ,* (2.4) *holds on* $\{|D\phi| \neq 0\}$.

(iii) *The function F defined by* (2.4) *satisfies the ellipticity condition* (2.5).

It turns out that it is occasionally more convenient to restate Definition 2.1 using F in place of v.

2.4 Definition A family $(\Omega_t)_{t\in(a,b)}$ of open subsets of $\mathbb{R}^N$ is called a *generalized super-flow* (resp. *sub-flow*) with normal velocity $-F(D^2d, Dd, x, t)$ if and only if, for all $x_0 \in \mathbb{R}^N$, $t \in (a,b)$, $r > 0$, $\alpha > 0$ and for all smooth functions $\phi : \mathbb{R}^N \to \mathbb{R}$ such that $\{\phi \geq 0\} \subset \Omega_t \cap B_r(x_0)$, (resp. $\{\phi \leq 0\} \subset \overline{\Omega}_t^c \cap B_r(x_0)$), with $|D\phi| \neq 0$ on $\{\phi = 0\}$, there exists $h_0 > 0$ depending only on α and ϕ through its C^4-norm in $\overline{B}_r(x_0)$ such that, for all $h \in (0, h_0)$,

$$\{x \in \mathbb{R}^N : \phi(x) - h[F^*(D^2\phi(x), D\phi(x), x, t) + \alpha] > 0\} \cap \overline{B}_r(x_0) \subset \Omega_{t+h},$$

(resp.

$$\{x \in \mathbb{R}^N : \phi(x) - h[F_*(D^2\phi(x), D\phi(x), x, t) - \alpha] < 0\} \cap \overline{B}_r(x_0) \subset \overline{\Omega}_{t+h}^c).$$

A family $(\Omega_t)_{t\in(a,b)}$ of open subsets of $\mathbb{R}^N$ is called a *generalized flow* with normal velocity $-F(D^2d, Dd, x, t)$ if it is both a sub- and super-flow.

The next remark emphasizes an observation which plays an important role in this theory, stemming from our discussion above.

2.5 Remark In view of Theorem 2.3, any motion with prescribed normal velocity satisfying (A1)–(A3), reduces to a generalized evolution as in Definition 2.4, i.e., the normal velocity must depend only on (x,t), n and Dn and must satisfy (2.5).

Such equations, with F of the special form considered here, arise in the so-called level-set approach to define weak motions of hypersurfaces with normal velocity $-F$. I briefly present the level-set approach below and show the connections with the above described approach. Here I just want to point out that Definition 2.1 mixes ideas coming from the level-set approach and also from a more geometric point of view, as it incorporates properties like (A2), inspired by what is known as avoidance and inclusion properties in the literature. I refer to Ilmanen [Il1] where properties of this type were remarked for the motion by mean curvature without, however, being put forward as a possible definition for such motions. The above defined notion of generalized flow is closely related to the notion of barriers introduced by De Giorgi [D], who proposed the use of some geometric maximum principle-type ideas to define the propagation of manifolds in $\mathbb{R}^N$—see Bellettini and Paolini [BP2] which expands on these ideas.

The level-set approach was introduced for numerical calculations by Osher and Sethian [OsS]. I also refer to Ohta, Jasnow and Kawasahi [OJK] in the physics literature and to Barles [Ba] for a first-order model for flame propagation. The classical derivation of this approach is as follows:

Let $(\Gamma_t)_{t\in(a,b)}$ be a collection of smooth hypersurfaces moving with normal velocity $-F$ and let $(D_t)_{t\in(a,b)}$ be a collection of smooth open subsets of $\mathbb{R}^N$ such that $\Gamma_t = \partial D_t$. Assume that $u : \mathbb{R}^N \times (a,b) \to \mathbb{R}$ is a C^∞-function such that

$$D_t = \{x \in \mathbb{R}^N : u(x,t) > 0\},\ \Gamma_t = \{x \in \mathbb{R}^N : u(x,t) = 0\} \text{ and } |Du| \neq 0 \text{ on } \bigcup_{t\in(a,b)} \Gamma_t \times \{t\}.$$

A straightforward computation—see for example [ESp]—yields, under the additional assumption that all the smooth level sets of u move with the same normal velocity, that u must satisfy the partial differential equation

$$u_t + F(D^2u, Du, x, t) = 0 \quad \text{in} \quad \mathbb{R}^N \times (a,b),$$

where $F : \mathcal{S}^N \times (\mathbb{R}^N \setminus \{0\}) \times \mathbb{R}^N \times \mathbb{R} \to \mathbb{R}$ is related to the normal velocity by (2.4).

To justify and extend this approach to the case of non-smooth motions, one has to use the notion of viscosity solutions for fully nonlinear elliptic and parabolic pde's, which provides the existence and uniqueness of viscosity solutions of fully nonlinear second- and first-order pde's under rather general assumptions. I refer to [ESp, CGG, BSS, IS, G] for such results and to [Cr] and the "User's Guide" by Crandall, Ishii and Lions [CIL] for a general overview of the theory of viscosity solutions.

The level-set approach can then be described in the following way. Let $\mathcal{E}$ be the collection of triplets (Γ, D^+, D^-) of mutually disjoint subsets of $\mathbb{R}^N$ such that Γ is closed and $D^\pm$ is open and $\mathbb{R}^N = \Gamma \cup D^+ \cup D^-$. For any $(\Gamma_0, D_0^+, D_0^-) \in \mathcal{E}$, first choose $u_0 \in UC(\mathbb{R}^N)$, the space of uniformly continuous functions defined on $\mathbb{R}^N$, so that

$$D_0^+ = \{u_0 > 0\}, \quad D_0^- = \{u_0 < 0\} \quad \text{and} \quad \Gamma_0 = \{u_0 = 0\},$$

and then consider the initial value problem

$$\begin{cases} \text{(i)} & u_t + F(D^2u, Du, x, t) = 0 & \text{in } \mathbb{R}^N \times [0,\infty), \\ \text{(ii)} & u = u_0 & \text{on } \mathbb{R}^N \times \{0\}, \end{cases} \tag{2.6}$$

where F is defined by (2.4). In view of the existence and uniqueness theory of [CGG, ESp, BSS, IS, Go], etc., under general assumptions on F, there exists, for every $u_0 \in UC(\mathbb{R}^N)$, a unique viscosity solution u of (2.6) in $UC(\mathbb{R}^N \times [0,T])$ for all $T > 0$.

Finally set

$$\Gamma_t = \{x \in \mathbb{R}^N : u(x,t) = 0\},\ D_t^+ = \{x \in \mathbb{R}^N : u(x,t) > 0\},\ D_t^- = \{x \in \mathbb{R}^N : u(x,t) < 0\}.$$

Since F is geometric, i.e., it satisfies, for all $(X,p,x,t) \in \mathcal{S}^N \times (\mathbb{R}^N \setminus \{0\}) \times \mathbb{R}^N \times (0,\infty)$, $\mu \in \mathbb{R}$ and $\lambda > 0$,

$$F(\lambda X + \mu p \otimes p, \lambda p, x, t) = \lambda F(X, p, x, t), \tag{2.7}$$

the collection $\{(\Gamma_t, D_t^+, D_t^-)\}_{t\geq 0} \subset \mathcal{E}$ is uniquely determined, independently of the choice of u_0, by the initial triplet (Γ_0, D_0^+, D_0^-). Recall that the main consequence of the property (2.7) is that the pde in (2.6) is invariant under the changes $u \to \varphi(u)$, for all nondecreasing functions $\varphi : \mathbb{R} \to \mathbb{R}$.

For each $t > 0$ define the mapping $E_t : \mathcal{E} \to \mathcal{E}$ by

$$E_t(\Gamma_0, D_0^+, D_0^-) = (\Gamma_t, D_t^+, D_t^-),$$

and notice that $\{E_t\}_{t\geq 0}$ satisfies the properties $E_0 = \mathrm{id}_{\mathcal{E}}$ and $E_{t+s} = E_t \circ E_s$ for all $t, s \geq 0$—see, for example, [ESp, CGG, IS, Go].

2.6 Definition (i) The collection $\{E_t\}_{t\geq 0}$ is called the *generalized level set evolution* with normal velocity $-F$.

(ii) Given $(\Gamma_0, D_0^+, D_0^-) \in \mathcal{E}$, the collection $\{\Gamma_t\}_{t\geq 0}$ of closed sets is called the *generalized level set front propagation* of Γ_0 with normal velocity $-F$.

Notice that the level set propagation is determined not only by Γ_0 but by the choice of the triplet (Γ_0, D_0^+, D_0^-). This corresponds to fixing an orientation for the normal to Γ_0. In particular, the evolution differs, in general, if D_0^+ and D_0^- are interchanged.

The properties of the generalized level evolution have been the object of extended study. One of the most intriguing issues is whether the *no interior condition* holds, i.e., whether the set Γ_t does not develop an interior. We say that the no interior condition holds if and only if

$$\{(x,t) : u(x,t) = 0\} = \partial\{(x,t) : u(x,t) > 0\} = \partial\{(x,t) : u(x,t) < 0\}. \tag{2.8}$$

It turns out that there are general geometric conditions on Γ_0 yielding (2.8)—see [BSS] for such conditions as well as examples where (2.8) fails. [BSS] also considered the issue of the existence and uniqueness of discontinuous solutions to (2.8) with $u_0 = \mathbb{I}_{D_0^+} - \mathbb{I}_{D_0^-}$, where, if A is a subset of some $\mathbb{R}^k$, $\mathbb{I}_A$ denotes the characteristic function of A, i.e., $\mathbb{I}_A(x) = 1$ if $x \in A$ and $\mathbb{I}_A(x) = 0$ if $x \in A^c$.

Notice that in view of (2.7) one can expect that any such solution only takes the values $+1$ and -1. In fact this is true if and only if (2.8) holds (see [BSS]). Since this plays some role in our analysis below, I state the relevant result of [BSS] in the following proposition and sketch its proof, which gives a good indication of the "power" of the theory of viscosity solutions.

2.7 Proposition (i) *The initial value problem* (2.6) *with* $u_0 = \mathbb{I}_{D_0^+} - \mathbb{I}_{D_0^-}$ *has a unique discontinuous solution if and only if the no interior condition* (2.8) *holds.*

(ii) *If* (2.8) *fails and* $u : \mathbb{R}^N \times [0,\infty) \to \mathbb{R}$ *is a discontinuous solution of* (2.6), *then, for all* $t > 0$,

$$D_t^\pm \subset \{x \in \mathbb{R}^N : u(x,t) = \pm 1\} \subset D_t^\pm \cup \Gamma_t,$$

where $(\Gamma_t, D_t^+, D_t^-) = E_t(\Gamma_0, D_0^+, D_0^-)$.

Proof (1) Let $u \in UC(\mathbb{R}^N \times [0,\infty))$ be a solution of (2.6)(i) such that $u = d_0$ on $\mathbb{R}^N \times \{0\}$, where d_0 is the signed distance to Γ_0. It follows from (2.7) that, for any $\epsilon > 0$, the function $u^\epsilon = \tanh(\epsilon^{-1}u)$ is also a solution of (2.6)(i).

(2) The stability properties of discontinuous viscosity solutions imply that the function $u_\infty = \lim_{\epsilon\to 0} u^\epsilon$ is also a solution of (2.6)(i). Moreover, the properties of tanh yield that

$$u_\infty = 1 \text{ in } \{u > 0\}, \quad u_\infty = -1 \text{ in } \{u < 0\}, \quad \text{and} \quad u_\infty = 0 \text{ in } \operatorname{int}\{u = 0\}.$$

For the rest of the points, the value of u_∞ depends on the lsc or usc envelope one considers in the definition of the discontinuous solutions.

(3) Fix $\alpha \in (0,1)$ and set

$$\overline{u}_\infty = \lim_{\epsilon\to 0} \tanh(\epsilon^{-1}(u_\infty + \alpha)) \quad \text{and} \quad \underline{u}_\infty = \lim_{\epsilon\to 0} \tanh(\epsilon^{-1}(u_\infty - \alpha)).$$

The functions $\overline{u}_\infty$ and $\underline{u}_\infty$ are again solutions of (2.6) (i). Moreover,

$$\overline{u}_\infty = \begin{cases} 1 & \text{in } \{u \geq 0\}, \\ -1 & \text{in } \{u < 0\}, \end{cases} \quad \text{and} \quad \underline{u}_\infty = \begin{cases} 1 & \text{in } \{u > 0\}, \\ -1 & \text{in } \{u \leq 0\}. \end{cases}$$

If $\bigcup_{t>0}(\Gamma_t \times \{t\})$ has non-empty interior, $\overline{u}_\infty$ and $\underline{u}_\infty$ are two different discontinuous solutions of (2.6)(i) with initial datum $\mathbb{I}_{\Omega_0} - \mathbb{I}_{\mathbb{R}^N \setminus \overline{\Omega}_0}$.

(4) Let w be a solution of (2.6)(i) with $w = \mathbb{I}_{\Omega_0} - \mathbb{I}_{\mathbb{R}^N \setminus \overline{\Omega}_0}$ on $\mathbb{R}^N \times \{0\}$ and choose a sequence $(\phi_n)_n$ of smooth functions such that

$$-1 \leq \phi_n \leq 1, \quad \phi_n \equiv 1 \text{ on } [0, +\infty), \quad \inf_n \phi_n = -1 \text{ on } (-\infty, 0] \quad \text{and} \quad \phi_n' \geq 0.$$

Since $w^* \leq \phi_n(d_0)$ on $\mathbb{R}^N \times \{0\}$, the comparison property of viscosity solutions yields $w^* \leq \phi_n(u)$ in $\mathbb{R}^N \times (0, \infty)$, and, in view of the properties of ϕ_n,

$$w^* \leq -1 = \inf_n \phi_n(u) \quad \text{on} \quad \{u < 0\}.$$

On the other hand (2.7) implies that $+1$ and -1 are respectively super- and sub-solutions of (2.6). Therefore

$$-1 \leq w_* \leq w^* \leq 1 \quad \text{on} \quad \mathbb{R}^N \times [0, \infty)$$

and, finally, $w^* = -1$ in $\{u < 0\}$, if the set $\bigcup_{t>0}(\Gamma_t \times \{t\})$ has empty interior.

(5) The same line of argument shows that $w_* = 1$ in $\{u > 0\}$, which in view of the non-empty interior assumption, implies that w is identified uniquely. □

The next theorem describes the relationship between the flows defined by Definitions 2.1 and 2.6. I refer to [BS2] for the proofs.

2.8 Theorem (i) *Assume* (A1)–(A3). *A family* $\{\Omega_t\}_{t\in[0,T]}$ *of open subsets of* $\mathbb{R}^N$ *is a generalized flow (resp. super-flow, resp. sub-flow) with normal velocity* v *if and only if the function* $\chi = \mathbb{I}_{\Omega_t} - \mathbb{I}_{\overline{\Omega}_t^c}$ *is a viscosity solution (resp. super-solution, resp. sub-solution) of* (2.6)(i) *in* $\mathbb{R}^N \times (0, \infty)$.

(ii) *Let* $\{\Omega_t\}_{t\in[0,T]}$ *be a generalized flow with normal velocity* v, *let* $(\Gamma_t, D_t^+, D_t^-)_{t\in[0,T]}$ *be the generalized level-set evolution of* (Γ_0, D_0^+, D_0^-), *where* $D_0^+ = \Omega_0$ *and* $D_0^- = \overline{\Omega}_0^c$, *and assume that the no interior condition* (2.8) *holds. Then, for all* $t > 0$,

$$\Omega_t = D_t^+.$$

(iii) *If the no interior condition* (2.8) *fails then, for all* $t > 0$,

$$D_t^+ \subset \Omega_t \subset D_t^+ \cup \Gamma_t.$$

I continue with a discussion of the properties of the signed distance d to a front $\Gamma_t = \partial\Omega_t$ evolving with a given normal velocity, which is defined by (2.2). One may use the signed distance to the front to give another formulation for the weak propagation. This was introduced by Soner [Son1] for the case of (x,t)-independent normal velocities and was extended in [BBS] to the general case. The point of view taken here, however, is not to give another definition but rather to develop some tools which play a fundamental role in the study of the generation of fronts as asymptotic limits of problems like the ones discussed in theIntroduction, i.e. phase field theory, stochastic systems, etc. It should be noted, however, that it is very natural to try to describe the front in terms of the signed distance function, since in the smooth case one has, of course, that

$$V = d_t, \quad n = -Dd \quad \text{and} \quad Dn = -D^2 d, \quad \text{and} \quad d_t = F(D^2 d, Dd, x, t) \text{ on } \Gamma_t.$$

As usual I begin with an open set $\Omega_0 \subset \mathbb{R}^N$ and consider the level set evolution $(\Gamma_t, \Omega_t^+, \Omega_t^-)$ of $(\Gamma_0, \Omega_0, \mathbb{R}^N \setminus \overline{\Omega}_0)$ and the signed distance function to Γ_t given by (2.2).

To state the main result we define the extinction time $t^* \in (0, +\infty]$ for Γ_t by

$$t^* = \sup\{t > 0 \text{ such that } \Gamma_t \neq \phi\}.$$

Notice that depending on the evolution under consideration it is possible to have $t^* < \infty$. This is for example the case of a bounded set Γ_0 moving by mean curvature. On the other hand, for general velocities it is possible to have $t_* = \infty$.

2.9 Theorem *Assume that Γ_t has empty interior for all $t > 0$. Then the functions $\underline{d} = d \wedge 0$ and $\overline{d} = d \vee 0$ satisfy respectively in $\mathbb{R}^n \times (0, t^*)$*

$$\text{(i) } \underline{d}_t \leq F(x - \underline{d}D\underline{d}, t, D\underline{d}, D^2 d) \quad \textit{and} \quad \text{(ii) } \overline{d}_t \geq F(x - \overline{d}D\overline{d}, t, D\overline{d}, D^2\overline{d}) \ . \tag{2.9}$$

Moreover,

$$\text{(i) } -(D^2\underline{d}D\underline{d} \mid D\underline{d}) \leq 0 \textit{ in } \{\underline{d} < 0\} \quad \textit{and} \quad \text{(ii) } -(D^2\overline{d}D\overline{d} \mid D\overline{d}) \geq 0 \textit{ in } \{\overline{d} > 0\}. \tag{2.10}$$

The assumption that Γ_t has empty interior was made only to simplify the presentation. In fact one can show that (2.9) and (2.10) still hold when Γ_t has empty interior but for different solutions. Indeed let $\overline{\Gamma}_t = \partial\{x : \overline{u}_\infty(x,t) = 1\}$ and $\underline{\Gamma}_t = \partial\{x : \underline{u}_\infty(x,t) = 1\}$ where $\underline{u}_\infty$ and $\overline{u}_\infty$ are defined as in the proof of Proposition 2.7. Then (2.9)(i), (2.10)(i) and (2.9)(ii), (2.10)(ii) hold true for $d(x, \overline{\Gamma}_t)$ and $d(x, \underline{\Gamma}_t)$. This again is related to the connections between the non-empty difficulty and the possible nonuniqueness in the geometric formulation of the motion in terms of discontinuous solutions to (2.6). For a detailed discussion I refer to [Son1, BSS].

Another important point is that the assertions of Theorem 2.9 are sharp independently of the smoothness of the distance function, which in the context of the weak evolution is only Lipschitz continuous with respect to x and discontinuous with respect to t. To see this, assume for the moment that F is (x,t)-independent. It is then a fact that even when the signed distance function is smooth it does not satisfy $d_t = F(D^2 d, Dd)$ away from the front as can be seen, for example, for the motion by mean curvature by an explicit computation. Indeed following, for example, Gilbarg and Trüdinger [GT], one sees that

$$d_t - \operatorname{tr}\left(I - \frac{Dd \otimes Dd}{|Dd|^2}\right) D^2 d \begin{cases} \geq 0 & \text{if } d > 0, \\ \geq 0 & \text{if } d < 0. \end{cases}$$

In the case of (x,t)-dependent F's the meaning of the term $x - dDd$, with the bars at the appropriate place, is that if $x \notin \Gamma_t$ one has to move distance d in the Dd direction to return on the front.

Finally, before I continue with the proof of Theorem 2.9, I state a corollary which follows immediately from the above and which re-emphasizes the point about the sharpness of Theorem 2.9.

2.10 Corollary *Assume that* $\Gamma_0 = \partial\Omega_0 = \partial(\mathbb{R}^N \setminus \overline{\Omega}_0)$ *and consider the mean curvature evolution* $(\Gamma_t, \Omega_t^+, \Omega_t^-)$ *of* $(\Gamma_0, \Omega_0, \mathbb{R}^N \setminus \overline{\Omega}_0)$*. Then the signed distance function to* Γ_t *satisfies*

$$d_t - \Delta d \begin{cases} \leq 0 & \text{in } \{d < 0\} \cap \mathbb{R}^N \times (0, t_*), \\ \geq 0 & \text{in } \{d > 0\} \cap \mathbb{R}^N \times (0, t_*). \end{cases}$$

Proof of Theorem 2.9 (1) I only prove (2.9)(i) and (2.10)(i). The proof of the other two claims is similar.

(2) Observe that the function

$$w = \begin{cases} 0 & \text{if } u_\infty = 1, \\ -\infty & \text{if } u_\infty = -1, \end{cases}$$

is a solution of (2.6)(i), where u_∞ is defined in the proof of Proposition 2.7.

(3) Let the function $\overline{w} : \mathbb{R}^N \times [0, T] \to \mathbb{R}$ be defined by

$$\overline{w}(x,t) = \sup_{y \in \mathbb{R}^N} \{w(y,t) - |x - y|\}.$$

An easy calculation yields that

$$\overline{w}(x,t) = d \wedge 0.$$

On the other hand, standard arguments from the theory of viscosity solutions (see, for example, Jensen, Lions and Souganidis [JLS, CIL, Cr], etc.) imply that $\overline{w}$ is a subsolution of (2.6)(i). The inequalities (2.9)(i) and (2.10)(ii) then follow easily. □

It has become, hopefully, clear by now that understanding the empty interior condition is important, since it may lead to some rather unintuitive situations. Unfortunately, if no conditions are imposed on Γ_0, interior may be created for $t > 0$. See for example [ESp] and Soner [Son1] for some simple examples in this direction for motion by mean curvature starting with non-smooth surfaces like “crosses”, i.e. intersection of planes, and “figure-eights”, i.e. two spheres touching at a point. It can, however, be argued that the interior in the examples of [ESp] and [Son1] is due mainly to the fact that the initial data are not smooth, which, in turn, yields that the normal direction is somehow not well defined. This, of course, raises the question of finding some necessary and sufficient conditions of Γ_0 so that no interior is created. I will address this question below for the case of first-order and second-order motions with geometric pde's of the form

$$u_t = v(x,t)|Du| \quad \text{in} \quad \mathbb{R}^N \times (0, \infty) \tag{2.11}$$

and

$$u_t = F(D^2u, Du) \quad \text{in} \quad \mathbb{R}^N \times (0, \infty). \tag{2.12}$$

Throughout this discussion I assume that

$$\Gamma_0 = \partial\Omega_0 = \partial(\mathbb{R}^N \setminus \overline{\Omega}_0),$$

with Ω_0 an open subset of $\subset \mathbb{R}^N$, which, in particular, implies that Γ_0 has no interior. The following theorem is proved in [BSS], to which I refer for the details.

2.11 Theorem *Assume $v \in W^{1,\infty}(\mathbb{R}^N \times (0,T))$ for all $T > 0$ and that either* (i) *v does not change sign in $\mathbb{R}^N \times (0,+\infty)$ or,* (ii) *v is independent of t. Then $\Gamma_t = \{x : u(x,t) = 0\}$ does not develop interior, where $u \in UC(\mathbb{R}^N \times (0,\infty))$ is the solution of* (2.6) *with F as in* (2.11).

This theorem does not include the case $v \equiv v(x,\hat{p})$ which was studied by Soravia [Sor]. On the other hand, the theorem is sharp in its context because there is an explicit counterexample in the case where α depends on t and changes sign. This example, which is discussed in detail in [BSS], says that there exists an open interval $I \subset \mathbb{R}$ such that the equation

$$\begin{cases} u_t = (t-x)|u_x| & \text{in } \mathbb{R} \times (0,\infty), \\ u = \mathbb{I}_I - \mathbb{I}_{\mathbb{R}\setminus I} & \text{on } \mathbb{R} \times \{0\}, \end{cases}$$

has more than one discontinuous solutions. Proposition 2.7 then yields that the evolution of I with normal velocity $t - x$ develops interior.

I next turn to the case of the motion governed by (2.12), the typical example here being motion by mean curvature. To state the result one needs the following additional condition on F, which is satisfied in the mean curvature case, that yields that the equation is invariant under rotation and dilation:

$$F(\mu Q^t p, \mu^2 Q^t X Q) = \mu^m F(p, X) \tag{2.13}$$

for some $m \in \mathbb{R}$ and all $\mu > 0$, $p \in \mathbb{R}^N$, $X \in \mathcal{S}^N$ and $Q \in O(N)$, where Q^t is the adjoint of Q and $O(N)$ is the group of $N \times N$ orthogonal matrices, i.e. $Q^t = Q^{-1}$. The following theorem is proved in [BSS].

2.12 Theorem *Assume that* (2.13) *holds. Let $(\Gamma_t, \Omega_t^+, \Omega_t^-)$ be the generalized evolution of $(\Gamma_0, \Omega^+_{,}\Omega_0^-)$ governed by* (2.12). *Assume that Γ_0 is of class C^2, compact and that there exist nonnegative constants c_i $(i = 1,2,3)$, a skew symmetric matrix H and $x_0 \in \mathbb{R}^N$ such that*

$$c_1(x - x_0) \cdot Dd(x) + c_2 H(x - x_0) \cdot Dd(x) + c_3 F(Dd(x), D^2 d(x)) \neq 0 \quad \text{on} \quad \Gamma_0, \tag{2.14}$$

where d is the signed distance to Γ_0. Then the set $\bigcup_{t>0}(\Gamma_t \times \{t\})$ has empty interior in $\mathbb{R}^N \times (0,+\infty)$.

The left-hand side of (2.14) is the generator of rotations, dilations and translations in (x,t) evaluated at $t = 0$ on Γ_0. Condition (2.14), which was introduced in [BSS] and includes, as special cases, results of Ilmanen [Il1] and Soner [Son1] for motion by mean curvature, is not necessary. Indeed the work of Soner and the author [SonS] for bodies of rotation moving by mean curvature shows that there exist smooth Γ_0's which do not satisfy (2.14), but their evolution never develops interior. It is proved however in [SonS] that (2.14) holds near the singularities of Γ_t. A related observation is that if (2.14) holds at a later time, this again yields no interior. For the case of mean curvature, [ESp] also showed that, under some assumptions on Γ_0, almost every level set of the solution to the mean curvature pde does not develop interior; I refer to [E2] for more details. More recently, Angenent, Chopp and Ilmanen [ACI] and Angenent, Ilmanen and Velazquez [AIV] constructed specific examples of smooth Γ_0 with the property that their mean curvature evolution develops interior for $t > 0$.

Since the proof of Theorem 2.12 is rather simple and intuitive, I present it below.

Proof of Theorem 2.12 (1) Let $u \in UC(\mathbb{R}^N \times (0,\infty))$ be the unique solution of (2.6) with F as in (2.12) and $u = d_0$ on $\mathbb{R}^N \times \{0\}$. For $h > 0$, define the function

$$u_h(x,t) = \Phi(u((1+c_1h)e^{c_2hH}(x-x_0)+x_0,(1+c_1h)t+c_3h)),$$

where Φ is a nondecreasing smooth function with $\Phi(0) = 0$ to be chosen later. Since $Q = e^{c_2hH} \in O(N)$, (2.13), u_h is also a solution of (2.6). Moreover, if h is small enough, there exists some $\eta > 0$ such that

$$|u(\cdot,0) - u_h(\cdot,0)| \geq \eta h \quad \text{on} \quad \mathbb{R}^N. \tag{2.15}$$

(2) Assuming for the moment the last estimate, one obtains from the comparison of viscosity solution that either

$$u_h \leq u - \eta h \quad \text{or} \quad u_h \geq u + \eta h \quad \text{in} \quad \mathbb{R}^N \times (0,\infty).$$

If $\bigcup_{t>0}(\Gamma \times \{t\})$ has interior, either of the above inequalities, however, yields a contradiction. Indeed if u vanishes in some neighborhood of a point (x_0,t_0), then so does u_h for h sufficiently small.

(3) To prove (2.15) first observe that Φ may be chosen so that one only needs to work in a small neighborhood of Γ_0. Since, for a suitable choice of such a neighborhood, u is smooth, a simple Taylor's expansion yields

$$\begin{aligned} &u((1+c_1h)e^{c_2hH}(x-x_0)+x_0,c_3h) \\ &\quad = u(x,0) + h(c_1(x-x_0)\cdot Du(x,0) + c_2H(x-x_0)\cdot Du(x,0) + c_3u_t(x,0) + o(h). \end{aligned}$$

Using (2.14) and the fact that $u = d$ on $\mathbb{R}^N \times \{0\}$ one concludes. □

To conclude the discussion about the non-empty interior difficulty, I remark that one can not expect to have a general theorem guaranteeing no interior without severe restrictions on the velocity if the latter happens to be dependent on t. The intuitive reason behind this claim is that, in principle, all motions have some "pathological" configuration leading to the development of interior. One can take any such motion, perturb its velocity by a time dependent forcing term so as to drive the front to the "pathological" configuration and then simply "turn off" the forcing term. Examples of such situations were given by Barles, Soner and Souganidis [BSS] for time-dependent forcing term and Belletini and Paolini [BP1] for constant forcing term.

As mentioned earlier, one can have a weak formulation of the propagation of a front in terms of whether the signed distance to the front satisfies the inequalities of Theorem 2.9. A natural question to ask is whether these inequalities are enough to identify the distance function uniquely, i.e. if z satisfies (2.9), (2.10) and $z(x,0) = d(x,\Gamma_0)$, is it true that $z \equiv d$? In addition to being a natural mathematical question to ask, having such information plays an important role in some of the analysis of the asymptotic problems to be discussed later in these notes.

Below I consider, as an example, the initial value problem

$$\begin{cases} \text{(i)} \quad u_t = \theta\left(\Delta u - \dfrac{(D^2uDu \mid Du)}{|Du|^2}\right) + \alpha(x,t)|Du| & \text{in } \mathbb{R}^N \times (0,\infty) \\ \text{(ii)} \quad u = d & \text{on } \mathbb{R}^N \times \{0\}, \end{cases} \tag{2.16}$$

with $\theta \geq 0$ and $\alpha \in W^{1,\infty}(\mathbb{R}^N \times (0,\infty))$. Some of the arguments and the conclusions hold if $\theta = \theta(x,t)$ (under suitable assumptions) as well as for anisotropic motions. This is discussed in detail in [BSS].

As usual $\Gamma_t = \{x : u(x,t) = 0\}$, where u solves (2.6)(i). Theorem 2.9 and the discussion following it say that the functions $d_1 = d(x, \overline{\Gamma}_t)$ and $d_2 = d(x, \underline{\Gamma}_t)$, where $\overline{\Gamma}_t = \partial\{x : u(x,t) > 0\}$ and $\underline{\Gamma}_t = \partial\{x : u(x,t) \geq 0\}$ satisfy the inequalities

$$z_t - \theta\Delta z - \alpha(x - zDz, t) \leq 0, \quad 1 - |Dz| = 0 \quad \text{in} \quad \{z < 0\}, \tag{2.17}$$

and

$$z_t - \theta\Delta z - \alpha(x - zDz, t) \geq 0, \quad |Dz| - 1 = 0 \quad \text{in} \quad \{z > 0\}. \tag{2.18}$$

Of course, if the no-interior condition holds for every $t > 0$, the above are satisfied by $z = d$.

Next I consider the question whether (2.17) and (2.18) are enough to identify z as the signed distance function. The following theorem which is proved in [BSS] gives a positive answer to this question.

2.13 Theorem *If the usc (resp. lsc) function z satisfies* (2.17)(ii), $|Dz| \leq 1$ *and* $z \wedge 0$ *is continuous from below (resp.* (2.18)(ii), $|Dz| \leq 1$ *and* $z \vee 0$ *is continuous from above), then*

$$z \leq d_2 \quad \textit{in} \quad \{z < 0\} \supset \{d_2 < 0\}, \quad (\textit{resp. } z \geq d_1 \textit{ in } \{z > 0\} \supset \{d_1 > 0\}).$$

If z satisfies (2.17), (2.18), $|Dz| \leq 1$, $z \wedge 0$ *and* $z \vee 0$ *are continuous from below and above respectively, and* Γ_t *does not develop interior for all* $t > 0$, *then* $z = d$ *in* $\mathbb{R}^N \times [0,\infty)$.

3 The abstract method

In most of the asymptotic problems which will be discussed in this paper, one is given a family $(u_\epsilon)_{\epsilon>0}$ of functions in $\mathbb{R}^N \times (0,T)$, uniformly bounded in ϵ, which are, typically, the solutions of a reaction-diffusion equation with a small parameter ϵ or the total magnetization of a stochastic system with interaction range ϵ^{-1}, etc. The goal is to prove the existence of a generalized flow $(\Omega_t)_{t\geq 0}$ with normal velocity v determined by the problem such that, as $\epsilon \to 0$,

$$u_\epsilon(x,t) \to b \text{ if } (x,t) \in \Omega = \bigcup_{t\in(0,T)} \Omega_t \times \{t\}, \quad \text{and} \quad u_\epsilon(x,t) \to a \text{ if } (x,t) \in \overline{\Omega}^c,$$

where $a, b \in \mathbb{R}$ are equilibrium states of this system.

Here I present an abstract formulation of a new general method for proving this kind of results. In the next sections, I discuss several concrete examples, which are analyzed by this method.

I assume that the family $(u_\epsilon)_{\epsilon>0}$ satisfies, for all $\epsilon > 0$, $t \geq 0$ and $h > 0$, the following properties, where $B(\mathbb{R}^N)$ denotes the set of real valued bounded functions on $\mathbb{R}^N$:

(H1) Causality *There exists a family of maps* $S^\epsilon_{t,t+h} : B(\mathbb{R}^N) \to B(\mathbb{R}^N)$, *such that*

$$u_\epsilon(\cdot, t+h) = S^\epsilon_{t,t+h} u_\epsilon(\cdot, t) \quad \textit{in} \quad \mathbb{R}^N.$$

(H2) Monotonicity *For all functions* $u, v \in B(\mathbb{R}^N)$,

$$\text{if} \quad u \le v \text{ in } \mathbb{R}^N, \quad \text{then} \quad S^{\epsilon}_{t,t+h}u \le S^{\epsilon}_{t,t+h}v.$$

(H3) Existence of equilibria *There exists* $a_\epsilon, b_\epsilon, a, b \in \mathbb{R}$ *such that*

$$a < b, \quad S^{\epsilon}_{t,t+h}a_\epsilon = a_\epsilon, \quad S^{\epsilon}_{t,t+h}b_\epsilon = b_\epsilon, \quad a_\epsilon \le u_\epsilon \le b_\epsilon \quad \text{in} \quad \mathbb{R}^N \times \{0\},$$

and, as $\epsilon \to 0$,

$$a_\epsilon \to a \quad \text{and} \quad b_\epsilon \to b.$$

(H4) Consistency *There exists a locally bounded function* $F : \mathcal{S}^N \times \mathbb{R}^N \setminus \{0\} \times \mathbb{R}^N \times [0, \infty) \to \mathbb{R}$ *such that:*

(i) *For all* $(x_0, t) \in \mathbb{R}^N \times [0, T)$, $r > 0$, $\alpha > 0$ *and for all smooth functions* $\phi : \mathbb{R}^N \to \mathbb{R}$ *such that* $\{\phi \ge 0\} \subset B_r(x_0)$ *and* $|D\phi| \ne 0$ *on* $\{\phi = 0\}$, *there exists* $\bar{\delta} > 0$ *and* $h_0 > 0$, h_0 *depending only on* α *and* ϕ *through its* C^4*-norm in* $\overline{B}_r(x_0)$, *such that, if* $\delta \le \bar{\delta}$, $x \in B_r(x_0)$ *and* $\phi(x) - h[F^*(D^2\phi(x), D\phi(x), x, t) + \alpha] > 0$ *for* $h \le h_0$, *then*

$$\liminf{}_*(S^{\epsilon}_{t,t+h}[(b_\epsilon - \delta)\mathbb{I}_{\{\phi \ge 0\}} + a_\epsilon \mathbb{I}_{\{\phi < 0\}}])(x) = b.$$

(ii) *For all* $(x_0, t) \in \mathbb{R}^N \times [0, T)$, $r > 0$, $\alpha > 0$ *and for all smooth functions* $\phi : \mathbb{R}^N \to \mathbb{R}$ *such that* $\{\phi \le 0\} \subset B_r(x_0)$ *and* $|D\phi| \ne 0$ *on* $\{\phi = 0\}$, *there exists* $\bar{\delta} > 0$ *and* $h_0 > 0$, h_0 *depending only on* α *and* ϕ *through its* C^4*-norm in* $\overline{B}_r(x_0)$ *such that, if* $\delta \le \bar{\delta}$, $x \in B_r(x_0)$ *and* $\phi(x) - h[F_*(D^2\phi(x), D\phi(x), x, t) - \alpha] < 0$ *for* $h \le h_0$, *then*

$$\limsup{}^*(S^{\epsilon}_{t,t+h}[b\mathbb{I}_{\{\phi > 0\}} + (a_\epsilon + \delta)\mathbb{I}_{\{\phi \le 0\}}])(x) = a.$$

Recall that if $f^\epsilon : A \to \mathbb{R}$ is a family of uniformly bounded functions, then

$$\liminf{}_* f^\epsilon(y) = \liminf_{\substack{z \to y \\ \epsilon \to 0}} f^\epsilon(z) \quad \text{and} \quad \limsup{}^* f^\epsilon(y) = \limsup_{\substack{z \to y \\ \epsilon \to 0}} f^\epsilon(z).$$

The main result is the following theorem.

3.1 Theorem *Assume that, for all* $\epsilon > 0$, $t \ge 0$ *and* $h > 0$, (H1)–(H4) *hold. For* $t > 0$, *set*

$$\Omega^1_t = \{x \in \mathbb{R}^N : \liminf{}_* u_\epsilon(x, t) = b\} \quad \text{and} \quad \Omega^2_t = \{x \in \mathbb{R}^N : \limsup{}^* u_\epsilon(x, t) = a\},$$

and define Ω^1_0 *and* Ω^2_0 *by*

$$\Omega^1_0 = \bigcap_{t > 0} \left(\bigcup_{0 < h \le t} \Omega^1_t \right) \quad \text{and} \quad \Omega^2_0 = \bigcap_{t > 0} \left(\bigcup_{0 < h \le t} \Omega^2_t \right).$$

If Ω^1_0 *(resp.* Ω^2_0*) is not empty, then* Ω^1_t *(resp.* Ω^2_t*) is not empty for sufficiently small* $t \ge 0$, *and the family* $(\Omega^1_t)_{t \ge 0}$ *(resp.* $((\overline{\Omega}^2_t)^c)_{t \ge 0}$*) is a generalized super-flow (resp. sub-flow) with normal velocity* $-F$. *Moreover,* F *is degenerate elliptic, i.e., it satisfies* (2.5).

An immediate consequence of Theorem 3.1 is the following corollary:

3.2 Corollary *Assume the hypotheses of Theorem* 3.1 *and, in addition, that* $\Omega_0^1 = (\overline{\Omega}_0^2)^c$. *Let* $(\Gamma_t, \Omega_t^+, \Omega_t^-)_{t\geq 0}$ *be the generalized level set evolution of* $(\partial\Omega_0^1, \Omega_0^1, \Omega_0^2)$ *with normal velocity* $-F$. *Then:*

(i) *For all* $t \geq 0$,

$$\Omega_t^+ \subset \Omega_t^1 \subset \Omega_t^+ \cup \Gamma_t \quad \textit{and} \quad \Omega_t^- \subset \Omega_t^2 \subset \Omega_t^- \cup \Gamma_t.$$

(ii) *If the no interior condition* (2.8) *holds, then, for all* $t \geq 0$,

$$\Omega_t^1 = \Omega_t^+ \quad \textit{and} \quad \Omega_t^2 = \Omega_t^-.$$

Before I present any proofs I discuss the meaning of (H1)–(H4) as well as what is involved in checking them. In most of the examples, (H1) is satisfied by definition as a consequence of a semigroup-type property, while (H2) follows from a maximum principle-type property and (H3) from the structure of the problem. Checking (H4) is in general the only difficult step.

Indeed, verifying (H4) consists of proving a similar result to the one to be derived for u_ϵ but only for smooth data, for compact smooth fronts and for small time. The exact regularity of the length of time is irrelevant. It is clear that these properties are a priori far easier to obtain—although by no means trivial—than the general case for u_ϵ. This reduction to an easier case is the main contribution of this new method.

The additional assumption that Ω_0^1 (resp. Ω_0^2) is not empty, which is used to initialize the moving front, is checked, in general, exactly in the same way as (H4), i.e., by studying the small time behavior of solutions which, locally in time, generate a smooth front. Finally, the reason for giving different definitions for Ω_0^1 and Ω_0^2 is to take into account the boundary layer which may occur at $t = 0$.

The intuitive idea here is that the map S^ϵ defines some approximation to a certain flow on sets and that (H4) amounts to finding the generator of this flow. It is in this context that this approach resembles, as mentioned in the Introduction, the formulation introduced in [BS3] to study convergence of numerical schemes.

Next I sketch the proof.

Proof of Theorem 3.1 (1) The fact that F satisfies (2.8) follows, once the rest of the theorem is proved. The key point here is that the fact that the validity of (H2) yields for the evolving fronts a monotonicity property like (A2).

(2) I only prove the result for $(\Omega_t^1)_{t\geq 0}$, the one for $(\Omega_t^2)_{t\geq 0}$ following similarly.

(3) Let $(x_0, t) \in \mathbb{R}^N \times (0, T)$, $r > 0$ and a smooth function $\phi : \mathbb{R}^N \to \mathbb{R}$ be such that $\{\phi \geq 0\} \subset \Omega_t^1 \cap B_r(x_0)$ and $|D\phi| \neq 0$ on $\{\phi = 0\}$. One needs to show that, for all sufficiently small $\alpha > 0$, there exists $h_0 > 0$ depending only on α, r and the C^4-norm of ϕ in $\overline{B}_r(x_0)$ such that, for all $0 < h < h_0$ and for α small enough,

$$\{x : \phi(x) - h[F^*(D^2\phi(x), D\phi(x), x, t) + \alpha] > 0\} \cap \overline{B}_r(x_0) \subset \Omega_{t+h}.$$

Since

$$a_\epsilon = S_{0,t}^\epsilon a_\epsilon \leq S_{0,t}^\epsilon u_\epsilon(\cdot, 0) = u_\epsilon(\cdot, t) \leq S_{0,t}^\epsilon b_\epsilon = b_\epsilon \quad \text{in} \quad \mathbb{R}^N,$$

assumption (H2) yields that

$$a_\epsilon \le u_\epsilon \le b_\epsilon \quad \text{in} \quad \mathbb{R}^N \times [0,T],$$

and hence

$$\limsup{}^* u_\epsilon \le b \quad \text{in} \quad \mathbb{R}^N \times [0,T].$$

(4) Classical arguments from the theory of viscosity solutions yield that, as $\epsilon \to 0$, $u_\epsilon \to b$, locally uniformly in $\bigcup_{t>0} \Omega_t^1 \times \{t\}$. Therefore, for ϵ small enough and δ as in (H4), we have

$$u_\epsilon(\cdot,t) \ge b_\epsilon - \delta \quad \text{on} \quad \{\phi \ge 0\} \subset \Omega_t^1.$$

Hence

$$u_\epsilon(\cdot,t) \ge (b_\epsilon - \delta)\mathbb{I}_{\{\phi\ge 0\}} + a_\epsilon \mathbb{I}_{\{\phi<0\}} \quad \text{in} \quad \mathbb{R}^N,$$

and, by (H2),

$$S^\epsilon_{t,t+h} u_\epsilon(\cdot,t) \ge S^\epsilon_{t,t+h}[(b_\epsilon - \delta)\mathbb{I}_{\{\phi\ge 0\}} + a_\epsilon \mathbb{I}_{\{\phi<0\}}] \quad \text{in} \quad \mathbb{R}^N.$$

(5) Choose $\alpha > 0$ and apply (H4). There exists $\bar\delta > 0$ and $h_0 > 0$ such that, if $\delta \le \bar\delta$, $0 < h \le h_0$, $x \in B_r(x_0)$ and

$$\phi(x) - h[F^*(D^2\phi(x), D\phi(x), x, t) + \alpha] > 0,$$

then

$$\begin{aligned} \liminf{}_* u_\epsilon(x,t+h) &= \liminf{}_* S^\epsilon_{t,t+h} u_\epsilon(\cdot,t)(x) \\ &\ge \liminf{}_* (S^\epsilon_{t,t+h}[(b_\epsilon - \delta)\mathbb{I}_{\{\phi\ge 0\}} + a_\epsilon \mathbb{I}_{\{\phi<0\}}])(x) = b. \end{aligned}$$

This yields that, for any $0 < h \le h_0$, where h_0 depends only on α and ϕ through its C^4-norm on $\overline{B}_r(x_0)$,

$$\{\phi(x) - h[F^*(D^2\phi(x), D\phi(x), x, t) + \alpha] > 0\} \cap \overline{B}_r(x_0) \subset \Omega^1_{t+h}.$$

The proof is now complete. □

4 Asymptotics of reaction-diffusion equations of bistable type — Phase Field Theory

Reaction-diffusion equations of the form

$$u_t - \Delta u + W'(u) = 0, \tag{4.1}$$

set in, for example, in $\mathbb{R}^N \times (0,\infty)$, where W is a double-well potential as well as their generalizations with a general second-order elliptic operator instead of Δ and W depending also on (x,t), arise in many areas of applications like phase transitions, flame propagation, pattern formation, chemical kinetics, etc. For example, (4.1) was put forward by Allen and Cahn [AC] to describe the evolution of an order parameter identifying the phases of a polycrystalline material.

It is expected that, as $t \to \infty$ the solution u will converge to the different equilibrium solutions of (4.1), which turn out to be the local extrema of W. An interface therefore develops, as $t \to \infty$, as the boundary of the regions where the solution u of (4.1) converges to the different equilibria. A great deal of work has been done over the last years towards understanding these interfaces. Some of the most striking results in this direction will be discussed in this section.

Formal arguments, some of which are explained below, yield that the front, which develops for $t \gg 1$, moves with normal velocity V which can be expanded in terms of t^{-1} like

$$V = \alpha + \kappa t^{-1} + O(t^{-2}) \quad (t \gg 1),$$

where κ denotes the mean curvature.

A convenient and by now classical way to study the fronts and understand the above expansion is to scale (4.1) in space and time, the goal being to reproduce, in finite time and in bounded space regions, the long-time behavior of (4.1). The correct scaling to find the first term in the expansion is, of course, the hyperbolic scaling $(x,t) \mapsto (\epsilon^{-1}x, \epsilon^{-1}t)$. Defining

$$u^\epsilon(x,t) = u(\epsilon^{-1}x, \epsilon^{-1}t),$$

where u solves (4.1), one finds that u^ϵ solves

$$u_t^\epsilon - \epsilon \Delta u^\epsilon + \epsilon^{-1} W'(u^\epsilon) = 0. \tag{4.2}$$

The problem then becomes the study of the asymptotics of (4.2), as $\epsilon \to 0$, with given initial conditions.

When $\alpha = 0$, the appropriate scaling is the parabolic one $(x,t) \mapsto (\epsilon^{-1}, \epsilon^{-2}t)$. Writing

$$u^\epsilon(x,t) = u(\epsilon^{-1}, \epsilon^{-2}t),$$

one is lead to the study of the asymptotics of

$$u_t^\epsilon - \Delta u^\epsilon + \epsilon^{-2} W'(u^\epsilon) = 0. \tag{4.3}$$

This section is divided into two parts as follows: In Part 1, I present results about (4.2) and (4.3) with some proofs and discuss some extensions. Part 2 is about new results regarding more complicated versions of (4.1) with anisotropies, bounded domains, etc., which are based on the theory presented in Section 3 of these notes.

4.1 The asymptotics of (4.1)

I begin by recalling some of the main properties of the potential W. These are:

$$\begin{cases} W : \mathbb{R} \to \mathbb{R} \text{ is a double-well potential with minima} \\ \text{at } \pm 1 \text{ and a local maximum at } \mu \in (-1,1). \end{cases} \tag{4.4}$$

It follows, see for example Aronson and Weinberger [AW], Fife and McLeod [FM], etc., that

$$\begin{cases} \text{there exist unique } \alpha \in \mathbb{R} \text{ and } q : \mathbb{R} \to \mathbb{R} \text{ such that} \\ \alpha \dot{q} + \ddot{q} = W'(q) \text{ on } \mathbb{R},\ q(\pm\infty) = \pm 1,\ q(\mu) = 0 \text{ and } \dot{q} > 0 \text{ in } \mathbb{R}, \text{ and} \\ \alpha = \left(\displaystyle\int_{-\infty}^{\infty} \dot{q}^2(\xi)\, d\xi \right)^{-1} (W(1) - W(-1)). \end{cases} \tag{4.5}$$

The results regarding the asymptotic behavior of (4.2) and (4.3) are:

4.1 Theorem *Assume* (4.4) *and let* u^ϵ *be the solution of* (4.2) *with initial datum* u_0 *on* $\mathbb{R}^N \times \{0\}$. *Define* $\Omega_0 = \{u_0 > 0\}$, $\Gamma_0 = \{u_0 = 0\}$ *and assume that* $\Gamma_0 = \partial\Omega_0 = \partial\mathbb{R}^N \setminus \overline{\Omega}_0$. *If* $(\Gamma_t, \Omega_t^+, \Omega_t^-)$ *is the level set evolution of* $(\Gamma_0, \Omega_0, \mathbb{R}^N \setminus \overline{\Omega}_0)$ *with normal velocity* $-\alpha$, *then, as* $\epsilon \to 0$, *locally uniformly,*

$$u^\epsilon \to 1 \text{ in } \mathcal{O} = \bigcup_{t>0}(\Omega_t^+ \times \{t\}), \quad \text{and} \quad u^\epsilon \to -1 \text{ in } \mathcal{O}' = \bigcup_{t>0}(\Omega_t^- \times \{t\}).$$

4.2 Theorem *Assume* (4.4) *and that* $W(-1) = W(+1) = 0$. *If* u^ϵ *is the solution of* (4.3) *with initial datum* u_0 *on* $\mathbb{R}^N \times \{0\}$, *define* $\Omega_0 = \{u_0 > 0\}$, $\Gamma_0 = \{: u_0 = 0\}$, *and assume that* $\Gamma_0 = \partial\Omega_0 = \partial(\mathbb{R}^N \setminus \overline{\Omega}_0)$. *If* $(\Gamma_t, \Omega_t^+, \Omega_t^-)$ *is the level set evolution by mean curvature of* $(\Gamma_0, \Omega_0, \mathbb{R}^N \setminus \overline{\Omega}_0)$, *then, as* $\epsilon \to 0$, *locally uniformly,*

$$u^\epsilon \to 1 \text{ in } \mathcal{O} = \bigcup_{t>0}(\Omega_t^+ \cup \{t\}), \quad \text{and} \quad u^\epsilon \to -1 \text{ in } \mathcal{O}' = \bigcup_{t>0}(\Omega_t^- \times \{t\}).$$

Some discussion about the history of Theorems 4.1 and 4.2 is now in order. Theorem 4.1 was proved, without this particular characterization of the sets $\mathcal{O}$ on $\mathcal{O}'$, by Gärtner [Ga] using a combination of probabilistic and pde techniques and by Barles, Bronsard and Souganidis [BBS] based on arguments related to viscosity solutions. In the form stated here and in a more general setting, it was proved in [BSS].

Theorem 4.2 has a considerably longer history. As mentioned earlier, this behavior of (4.3) was conjectured by Allen and Cahn [AC] and was justified formally in [Fi, Ca1, Ca2, RSK]. Rigorous results under the assumption that the mean curvature motion is smooth, i.e. short-time type results, were obtained by DeMottoni and Schatzman [DS], Chen [Ch] and later, for different but related W's and with sharp rates of convergence, by Nochetto, Paolini and Verdi [NPV]. Also Bronsard and Kohn [BrK] proved a global in time result but for radial solutions. It should also be noted that DeGiorgi [D] formulated an alternative approach to mean curvature evolution using the asymptotics of (4.3) as a definition. The first global in time result for (4.3), under no special assumptions, was obtained by Evans, Soner and Souganidis [ESS]. Nontrivial extensions and generalizations of Theorem 4.2 for more general pde's and motions, some of which will be presented at the end of this part of this section, were obtained in [BSS]. The particular case of the asymptotics of the Allen-Cahn equation in Theorem 4.2 as stated here has been the object of even more study. As mentioned earlier in this note there is another way to study mean curvature evolution, due to Brakke [Br], which is, however, very restricted to this kind of motion and is based on varifolds—see [E2] for details. Based on this approach to mean curvature, Ilmanen [Il2] gave another proof, after [ESS], for Theorem 4.2 with information about what happens when interior develops. Later Soner [Son2] reconciled some of the arguments of [Il2] with the ones from viscosity solutions and extended the results of [Il2].

Here I will present three different proofs for Theorem 4.2 and I will refer to [BSS] for the proof of Theorem 4.1. The reason for giving these different proofs is that each one of them contains different ingredients, which may be useful in different situations, and also to indicate the scope and strength of the theory of viscosity solutions.

For the first two proofs of Theorem 4.2, I will make the additional assumption that

$$u^\epsilon = q(\epsilon^{-1}d_0) \quad \text{on} \quad \mathbb{R}^N \times \{0\}, \tag{4.6}$$

where, as usual, d_0 is the signed distance function to Γ_0 and q is the traveling wave, in this case, standing wave solution associated with W'. The aforementioned extra condition on u^ϵ at $t = 0$ is essential for the proof I am presenting below. It can, however, be removed at the end by a short time analysis, carried out for (4.3) by [Ch], which implies that in time $t = O(\epsilon^2 |\ell n\ \epsilon|)$ the solution of u^ϵ of (4.3) can be put between two functions which look like $q^\epsilon(\epsilon^{-1} d_0)$ for some approximate q^ϵ's.

First proof of Theorem 4.1 (1) Let $z^\epsilon : \mathbb{R}^N \times [0, \infty) \to \mathbb{R}$ be defined by

$$u^\epsilon = q(\epsilon^{-1} z^\epsilon).$$

A straightforward computation yields that z^ϵ solves the following initial value problem:

$$\begin{cases} z_t^\epsilon - \Delta z^\epsilon + \epsilon^{-1} Q(\epsilon^{-1} z^\epsilon)(|Dz^\epsilon|^2 - 1) = 0 & \text{in } \mathbb{R}^N \times (0, \infty), \\ z^\epsilon = d_0 & \text{on } \mathbb{R}^N \times \{0\}, \end{cases} \tag{4.7}$$

where

$$Q(\xi) = -(\dot{q}(\xi))^{-1} \ddot{q}(\xi).$$

(2) Since, in view of (4.6),

$$\operatorname{sgn} Q(\xi) = \operatorname{sgn}(-\ddot{q}(\xi)) = \operatorname{sgn} q(\xi), \tag{4.8}$$

it is immediate that

$$z_t^\epsilon - \Delta z^\epsilon \begin{cases} \geq 0 & \text{in } \{z^\epsilon > 0\}, \\ \leq 0 & \text{in } \{z^\epsilon < 0\}. \end{cases} \tag{4.9}$$

Since $|Dz^\epsilon| \leq 1$ on $\mathbb{R}^N \times \{0\}$, it also follows that

$$|Dz^\epsilon| \leq 1 \quad \text{in} \quad \mathbb{R}^N \times (0, \infty). \tag{4.10}$$

This last estimate can be either obtained by standard maximum principle type estimates or directly by arguing about what happens at the maximum in $\mathbb{R}^N \times \mathbb{R}^N \times (0, \infty)$ of the function

$$(x, t) \mapsto z^\epsilon(x, t) - z^\epsilon(y, t) - |x - y|.$$

(3) Let t_* be the extinction time of the mean curvature evolution of $\{d_0 > 0\}$ and assume that, for any compact subset K of $\mathbb{R}^N \times [0, T]$,

$$\sup_{\epsilon > 0} \|z^\epsilon\|_{L^\infty(K)} < \infty;$$

refer to [BSS] for the arguments yielding this estimate.

(4) Consider the upper- and lower-semicontinuous functions z^* and $z_* : \mathbb{R}^N \times [0, t_*) \to \mathbb{R}$. It follows that

$$|Dz^*| \leq 1 \quad \text{and} \quad |Dz_*| \leq 1 \quad \text{in} \quad \mathbb{R}^N \times [0, t_*). \tag{4.11}$$

(5) Using once more (4.7), (4.8), (4.10) and standard arguments from the theory of viscosity solutions, one gets that z^* (resp. z_*) is a viscosity solution of

$$1 - |Dz| = 0 \text{ in } \{z < 0\}\ , \quad (\text{resp. } |Dz| - 1 \text{ in } \{z > 0\}).$$

Finally, letting $\epsilon \to 0$ in (4.9) also yields that

$$z_t^* - \Delta z^* \leq 0 \text{ in } \{z^* < 0\} \quad \text{and} \quad z_{*t} - \Delta z_* \geq 0 \text{ in } \{z_* > 0\}.$$

The uniqueness results of Theorem 2.11 now yield that

$$z^* \leq d \text{ in } \{d < 0\} \quad \text{and} \quad z_* \geq d \text{ in } \{d > 0\},$$

where d is the signed distance to Γ_t. In particular, if the evolution does not develop interior, it follows that

$$z^* = z_* = d.$$

(6) The extension of the result in $\mathbb{R}^N \times [t_*, \infty)$ is now more or less immediate—I refer to [ESS, BSS] for the precise argument. □

The above proof is rather simple. It can also be extended to more general situations where some of the estimates mentioned, for example the gradient bound etc., are not so obvious. One then has to use one more step involving changing the z^ϵ's and hence z^* and z_* in such a way as to produce an appropriate gradient bound. This argument was introduced in [BSS] and was employed rather successfully in [J] and [KS2].

The second proof of Theorem 4.2 follows along the lines of the original argument of [ESS] and its modification in [BSS] to cover more general problems. The basic idea is to use the signed distance to the evolving front to construct super- and sub-solutions to (4.3).

In preparation for the proof it is necessary to discuss the existence and the properties of traveling waves for nonlinearities of the form $u \mapsto W'(u) - \epsilon a$, where W is a double-well potential and ϵ and a are sufficiently small so that the equation

$$W'(u) - \epsilon a = 0$$

has exactly three zeroes $h_-^\epsilon < h_0^\epsilon < h_+^\epsilon$, which, of course, in the limit $\epsilon \to 0$, converge to the zeros of $W'(u)$, which for definiteness I call here $-1, 0$ and 1. Under some mild technical assumptions on W (see, for example, [AW, FM, BSS]) it turns out that for ϵ and a sufficiently small there exists a unique pair $(c^\epsilon(a,\epsilon), q^{a,\epsilon})$ where $c^\epsilon(a,\epsilon)$ is a constant and $q^{a,\epsilon} : \mathbb{R} \to \mathbb{R}$ is such that

$$\begin{cases} c(a,\epsilon)\dot{q}^{a,\epsilon} + \ddot{q}^{a,\epsilon} = W'(q^{a,\epsilon}) - \epsilon a \text{ in } \mathbb{R}, \text{ and} \\ q^{a,\epsilon}(\pm\infty) = h_\pm^{a,\epsilon},\ q^{a,\epsilon}(0) = h_0^{a,\epsilon} \text{ and } \dot{q}^{a,\epsilon} > 0 \text{ in } \mathbb{R}, \end{cases} \tag{4.12}$$

and, since W is assumed to have wells of the same depth,

$$c(a,\epsilon) \to 0 \quad \text{and} \quad \epsilon^{-1}c(a,\epsilon) \to c(a) \quad \text{as} \quad \epsilon \to 0 \quad \text{and} \quad c(a) \to 0 \quad \text{as} \quad a \to 0. \tag{4.13}$$

For fixed $\delta, a > 0$, let $u^{\delta,a}$ be the solution of

$$\begin{cases} u_t^{\delta,a} = \operatorname{tr}\left[\left(I - \dfrac{Du^{\delta,a} \otimes Du^{\delta,a}}{\widehat{Du}^{\delta,a}}\right) D^2u^{\delta,a}\right] - c(a)|Du^{\delta,a}| & \text{in } \mathbb{R}^N \times (0,\infty), \\ u^{\delta,a} = d_0 + \delta & \text{in } \mathbb{R}^N \times \{0\}, \end{cases} \tag{4.14}$$

where $c(a)$ is given in (4.13). Recall that the level sets of (4.14) move with normal velocity

$$V = \text{ mean curvature } - c(a),$$

and that, if $d^{\delta,a}$ is the signed distance from the set $\{x : u^{\delta,a}(x,t) = 0\}$, then

$$d_t^{\delta,a} \geq \Delta d^{\delta,a} - c(a)|Dd^{\delta,a}| \quad \text{and} \quad |Dd^{\delta,a}| = 1 \quad \text{in} \quad \{d^{\delta,a} > 0\} \tag{4.15}$$

Next it is necessary to introduce an auxiliary function

$$w^{\delta,a} = \eta_\delta(d^{\delta,a}) \quad \text{in} \quad \mathbb{R}^N \times (0,\infty), \tag{4.16}$$

where $\eta_\delta : \mathbb{R} \to \mathbb{R}$ is a smooth function satisfying, for some constant C independent of δ,

$$\begin{cases} \eta_\delta(z) = -\delta \text{ if } z \leq \delta/4, & \eta_\delta(z) = z - \delta \text{ if } z \geq \delta/2, \\ \eta_\delta(z) \leq \delta/2 \text{ if } z \leq \delta/2, & 0 \leq \eta_\delta'(z) \leq C \text{ and } |\eta_\delta''| \leq C\delta^{-1} \text{ on } \mathbb{R}. \end{cases} \tag{4.17}$$

The following lemma follows from a straightforward combination of (4.15) and (4.17)—see, for example Lemma 3.1 of [ESS].

4.3 Lemma *There exists a constant C, independent of δ and a, such that*

$$\begin{cases} \text{(i)} \quad w_t^{\delta,a} - \Delta w^{\delta,a} + c(a)|Dw^{\delta,a}| \geq -\delta^{-1}C & \text{in } \mathbb{R}^N \times (0,t^*), \\ \text{(ii)} \quad w_t^{\delta,a} - \Delta w^{\delta,a} + c(a)|Dw^{\delta,a}| \geq 0 & \text{in } \{d^{\delta,a} > \delta/2\}, \end{cases} \tag{4.18}$$

and

$$|Dw^{\delta,a}| = 1 \quad \text{in} \quad \{d^{\delta,a} > \delta/2\}, \tag{4.19}$$

where t^* is the extinction time of $\{x : u^{\delta,a}(x,t) = 0\}$.

Finally define

$$\Phi^\epsilon = q^{a,\epsilon}(\epsilon^{-1}w^{\delta,a}) \quad \text{on} \quad \mathbb{R}^N \times [0,\infty). \tag{4.20}$$

4.4 Proposition *For every $a > 0$, Φ^ϵ is a supersolution of (4.3), if $\epsilon \leq \epsilon_0(\delta,a)$ and $\delta \leq \delta_0(a)$.*

The proof of this proposition follows along the lines of Theorem 3.2 of [ESS] and Proposition 10.2 of [BSS] to which I refer for more details as well as more general statements. In the proof I present below, I will argue as if $w^{\delta,a}$ has actual derivatives, keeping in mind that actually everything has to be checked in the viscosity sense.

Proof of Proposition 4.4 (1) It follows from (4.12) and (4.20) that

$$\Phi_t^\epsilon - \Delta\Phi^\epsilon + \frac{1}{\epsilon^2}W'(\Phi^\epsilon) = \epsilon^{-1}\dot{q}^{a,\epsilon}(w_t^{\delta,a} - \Delta w^{\delta,a} + \epsilon^{-1}c(a,\epsilon)) - \ddot{\epsilon}^{-1}q^{a,\epsilon}(|Dw^{\delta,a}| - 1)^2 + \epsilon^{-1}a,$$

where $\dot{q}^{a,\epsilon}$ and $\ddot{q}^{a,\epsilon}$ are evaluated at $w^{\delta,a}/\epsilon$.

(2) If $\delta/2 < d^{\delta,a} < 2\delta$, (4.12), (4.18)(i), (4.19) and (4.17) yield, for sufficiently small ϵ, that

$$\Phi_t^\epsilon - \Delta\Phi^\epsilon + \epsilon^{-2}W'(\Phi^\epsilon) = \epsilon^{-1}\dot{q}^{a,\epsilon}(\epsilon^{-1}c(a,\epsilon) - c(a)) \geq 0.$$

(3) If $d^{\delta,a} \leq \delta/2$, the choice of η_δ yields that $w^{\delta,a} \leq -\delta/2$. Consequently the properties of the traveling waves and, in particular, the exponential decay of $\dot{q}$ and $\ddot{q}$ to 0 as $|\xi| \to \infty$, imply that there exists $K > 0$ such that

$$\epsilon^{-1}\dot{q}^{a,\epsilon}(\epsilon^{-1}w^{\delta,a}) + \epsilon^{-2}|\ddot{q}^{a,\epsilon}(\epsilon^{-1}w^{\delta,a})| \leq Ke^{-K\delta/\epsilon}.$$

Using the fact that $|Dw^{\delta,a}| \leq C$, which follows from (4.19) and $|Dd^{\delta,a}| \leq 1$, and (4.18)(i) give, in the limit $\epsilon \to 0$,

$$\Phi_t^\epsilon - \Delta\Phi^\epsilon + \epsilon^{-2}W'(\Phi^\epsilon) \geq Ke^{-K\delta/\epsilon}(-\delta^{-1} - c + o(1)) + \epsilon^{-1}a = o(1).$$

(4) If $d^{\delta,a} > 2\delta$, then $w^{\delta,a} > \delta$ and the argument is similar. □

All the ground has now been prepared for the:

Second proof of Theorem 4.2 (1) In view of (4.6), Proposition 4.4 and the maximum principle, it follows that, for ϵ sufficiently small, given δ and a,

$$u^\epsilon \leq \Phi^\epsilon \quad \text{in} \quad \mathbb{R}^N \times [0, t_*).$$

(2) Let $(x_0, t_0) \in \mathbb{R}^N \times [0, t_*)$ be such that $u(x_0, t_0) = \beta < 0$, where u is the solution of the mean curvature pde

$$\begin{cases} u_t = \operatorname{tr}(I - \widehat{Du} \otimes \widehat{Du})D^2u & \text{in } \mathbb{R}^N \times (0, \infty), \\ u = d_0 & \text{on } \mathbb{R}^N \times \{0\}. \end{cases}$$

The stability properties of viscosity solution yield the $u^{\delta,a} \to u$, as $\delta, a \to 0$, uniformly in (x,t). Hence, for sufficiently small a and δ, one has

$$u^{\delta,a}(x_0, t_0) < -\beta < 0.$$

But then $d^{\delta,a}(x_0, t_0) < 0$, hence

$$\limsup_{\epsilon \to 0} u^\epsilon(x_0, t_0) \leq \limsup_{\epsilon \to 0} \Phi^\epsilon(x_0, t_0) = -1.$$

Since, again by the maximum principle, $u^\epsilon \geq -1$, it follows that

$$u^\epsilon \to -1 \quad \text{in} \quad \bigcup_{t_* > t > 0} \{x : u(x,t) < 0\}.$$

(3) A similar argument, which requires the analogue of Proposition 4.4 but for subsolutions, which follows along the lines of the aforementioned propositions, yields that

$$u^\epsilon \to 1 \quad \text{in} \quad \bigcup_{0 < t < t_*} \{x : u(x,t) > 0\}.$$

(4) For the fact that the result can be extended to $\mathbb{R}^N \times (0,\infty)$ instead of $\mathbb{R}^N \times (0,t_*)$, I refer to the related discussion in the first proof of Theorem 4.2. □

The third proof of Theorem 4.2 relies on the abstract theory presented in Section 3, i.e., Theorem 3.1. It is necessary, however, to settle at this point the issue of the small time behavior. For the particular problem under consideration this was studied in [Ch]. The arguments of [Ch] can then be extended to apply to a more general situation (see [BS2]). Below I only remark what follows from [Ch] and I instead refer to the actual paper for the precise statement.

To this end assume that Γ_0 is a smooth hypersurface in $\mathbb{R}^N$. It follows that Γ_0 propagates smoothly by mean curvature for small time t. Denoting, as usual, by d the signed distance to the front, [Ch] proves that under suitable regularity assumptions on u_0, there exists some $M>0$ and $\tau>0$ such that (see also [BS2])

$$\begin{cases} u_\epsilon(x,t) \geq 1-\epsilon & \text{if } d(x,t) > M\epsilon|\ln\epsilon| \text{ and } \tau\epsilon^2|\ln\epsilon| \leq t, \\ \text{and} & \\ u_\epsilon(x,t) \leq -1+\epsilon & \text{if } d(x,t) < -M\epsilon|\ln\epsilon| \text{ and } \tau\epsilon^2|\ln\epsilon| \leq t. \end{cases} \tag{4.21}$$

Third proof of Theorem 4.2 (1) Assumptions (H1)–(H3) of Theorem 3.1 are obviously satisfied in view of the choice of W and the maximum principle.

(2) If $g = (1-\delta)\mathbb{I}_{\{\phi\geq 0\}} - \mathbb{I}_{\{\phi<0\}}$ in $\mathbb{R}^N$ is the initial datum for (4.3), where $0<\delta<1$ and ϕ is as in (H4), then (4.21) combined with the arguments in the second proof of Theorem 4.2 in the case where the motion is smooth, immediately yield that, for h sufficiently small and $d(x,h)>0$,

$$\liminf_{*} S^\epsilon(h)[(1-\delta)\mathbb{I}_{\{\phi\geq 0\}} - \mathbb{I}_{\{\phi>0\}}](x) = 1.$$

(3) Actually the function ϕ in (H4) does not satisfy the assumptions needed for the argument presented earlier. This can, however, be worked out by an easy sub-and supersolution argument using again (4.21) and observing that this ϕ can be put between two others that have the appropriate form, using $q^{a,\epsilon}$ and $q^{-a,\epsilon}$ for a and ϵ sufficiently small. □

The above ideas can be extended to treat more general reaction-diffusion equations of the form

$$u_t^\epsilon - \epsilon L u^\epsilon + \epsilon^{-1} W_\epsilon'(u^\epsilon,x,t) = 0 \quad \text{in} \quad \mathbb{R}^N \times (0,\infty),$$

and

$$u_t^\epsilon - L u^\epsilon + \epsilon^{-2} W_\epsilon'(u^\epsilon,x,t) = 0 \quad \text{in} \quad \mathbb{R}^N \times (0,\infty),$$

where L is now a general second-order elliptic operator with coefficients depending on (x,t), and $W_\epsilon(u,x,t)$ is a double-well potential for each (x,t), provided that the assumptions needed for the existence of traveling waves hold uniformly in (x,t) and the behavior, as $\epsilon\to 0$, of W_ϵ is well understood. The aforementioned problems are studied in [BSS].

They also lead to fronts moving with stochastic normal velocities of the form

$$V = \text{ mean curvature } + dB,$$

where B is the standard Brownian motion. This result follows from the theory of stochastic viscosity solutions which is being developed by P.-L. Lions and the author (see [LS1, LS2, LS3, LS4] to which I refer for a discussion).

The arguments presented here can also be used to study the asymptotics of certain systems of reaction-diffusion equations like the FitzHugh-Nagumo-type systems. This was studied by Soravia and Souganidis [SorS], using ideas related to the second proof of Theorem 4.2, and by Kim [K], using the abstract method of Section 3.

4.2 Some new results

Here I state without proof and, as usual, without all the technical assumptions some new results regarding the appearance of moving interfaces. These results are studied in [BS2] using the method described in Section 3.

The first result is about the asymptotics of

$$\begin{cases} u_t^\epsilon - \operatorname{tr}[A(\widehat{Du^\epsilon})D^2u^\epsilon] + \epsilon^{-2}W'(u^\epsilon) = 0 & \text{in } \mathbb{R}^N \times (0,\infty), \\ u^\epsilon = u_0 & \text{on } \mathbb{R}^N \times \{0\}, \end{cases}$$

where $A : \mathbb{R}^N \to \mathbb{R}^N$ is uniformly elliptic and W is a double-well potential with two wells of equal depth. As usual, the results extend to the more general W's.

The theorem is:

4.5 Theorem *Assume that* $\Gamma_0 = \partial\{x : u_0(x) > 0\} = \partial\{x : u_0(x) < 0\}$. *Let* $(\Gamma_t, \Omega_t^+, \Omega_t^-)$ *be the level set evolution of* $(\Gamma_0, \{u_0 > 0\}, \{u_0 < 0\})$ *with*

$$F(X,p) = \operatorname{tr}[A(\hat{p})X(I(A(\hat{p})p\cdot p)^{-1}A(\hat{p})p\otimes p) + 2(A(\hat{p})p\cdot p)X(I-\hat{p}\otimes\hat{p})D_pA(\hat{p})p\cdot p]$$

Then, as $\epsilon \to 0$, *and locally uniformly in*

$$u_\epsilon \to 1 \ in \ \bigcup_{t>0} \Omega_t^+ \times \{t\}, \quad and \quad u^\epsilon \to -1 \ in \ \bigcup_{t>0} (\Omega_t^- \times \{t\}.$$

Formal results corresponding to the above theorem were obtained in [BP3].

The next result is about the asymptotics of problems like,

$$\begin{cases} u^\epsilon - \epsilon\Delta u^\epsilon + b(\epsilon^{-1}x)\cdot Du^\epsilon + \epsilon^{-1}W'(u^\epsilon) = 0 & \text{in } \mathbb{R}^N \times (0,\infty), \\ u^\epsilon = u_0 & \text{on } \mathbb{R}^N \times (0,\infty), \end{cases}$$

where b is a periodic function in Q and, as usual, W is a double-well potential. It was proved by Xin [X] that under certain assumptions on b and W, the equation

$$u_t - \Delta u + b(x)\cdot Du + W'(u) = 0 \quad \text{in} \quad \mathbb{R}^N \times (0,\infty)$$

admits multidimensional traveling wave solutions, i.e., special solutions of the form

$$u(x,t) = q(x\cdot k - c(k)t),$$

where $k \in S^{N-1}$, the unit sphere in $\mathbb{R}^N$. More precisely, [X] shows that, for each direction $k \in S^{N-1}$, there exists a unique pair $(c(k), q)$—the uniqueness of q being understood always up to translations—where $q : \mathbb{R}\times\mathbb{R}^N \to \mathbb{R}$ solves the problem

$$\begin{cases} c(k)\dot{q} + \Delta_x q + 2k\Delta_x\dot{q} + |k|^2\ddot{q} = W'(q) \text{ in } \mathbb{R}\times\mathbb{R}^N \\ q(\pm\infty, x) = \pm 1, q(\xi,\cdot) \text{ periodic in } Q, \text{ and } \dot{q}(\xi,\cdot) > 0 \text{ in } \mathbb{R}\times Q. \end{cases}$$

The following theorem is a special case of a more general result proved in [BS2], to which I refer for the details.

4.6 Theorem *Assume that* $c \equiv 0$ *and let* $(\Gamma_t, \Omega_t^+, \Omega_t^-)$ *be the level set evolution of* $(\Gamma_0, \{u_0 > 0\}, \{u_0 < 0\})$ *with normal velocity* F *defined in* [BS2]. *Then, as* $\epsilon \to 0$, *and locally uniformly*

$$u^\epsilon \to 1 \textit{ in } \bigcup_{t>0} \Omega_t^+ \times \{t\} \quad \textit{and} \quad u^2 \to -1 \textit{ in } \bigcup_{t>0} (\Omega_t^- \times \{t\}.$$

The last example I discuss here is about the asymptotics of the following fully nonlinear equation

$$u_t^\epsilon - \epsilon^{-2}(J^\epsilon * u^\epsilon - u^\epsilon) + \epsilon^{-2} W'(u^\epsilon) = 0 \quad \text{in} \quad \mathbb{R}^N \times (0, \infty),$$

where W is the usual double-well potential and

$$J^\epsilon(x) = \epsilon^{-N} J(\epsilon^{-1} x),$$

J being nonnegative, even, i.e., $J(z) = J(-z)$, and satisfying some integrability conditions. Such equations arise at the mesoscopic limit of stochastic Ising models with long-range interactions and Kac potentials, which are discussed in detail in Section 6 below.

To state the result I make the assumption that for each $n \in S^{N-1}$ there exists a unique standing wave solution $u(x,t) = q(x \cdot n, n)$ of

$$u_t + [u - J * u] + W'(u) = 0 \quad \text{in} \quad \mathbb{R}^N \times (0, \infty),$$

i.e., a function $q : \mathbb{R} \times S^{N-1} \to \mathbb{R}$ such that

$$\begin{cases} \displaystyle\int J(y) q(\xi + n \cdot y, n)\, d\xi - q(\xi, n) = W'(q(\xi, n)) & \text{in } \mathbb{R}, \\ q(\pm\infty, n) = \pm 1 \quad \text{and} \quad \dot q(\xi, n) > 0 & \text{on } \mathbb{R}. \end{cases}$$

Notice that since J is not assumed to be isotropic, q depends in a nonlinear way on the direction n. For a discussion about the existence of such q's I refer to [BFRW, KS3].

Next for each $n \in S^{N-1}$ consider the matrix $\theta(n)$ defined by

$$\theta(n) = \frac{1}{2} \left(\int \dot q^2(\xi, n)\, d\xi \right)^{-1}$$
$$\times \iint J(y) \dot q(\xi, n) [\dot q(\xi + n \cdot y, n) y \otimes y + D_n q(\xi + n \cdot y, n) \otimes y + y \otimes D_n q(\xi + n \cdot y, n)]\, dy d\xi.$$

The following is proved in [BS2].

4.7 Theorem *Assume that* $\Gamma_0 = \partial\{x : u_0(x) > 0\} = \partial\{x : u_0(x) > 0\}$. *If* $(\Gamma_t, \Omega_t^+, \Omega_t^-)$ *is the level set evolution of* $(\Gamma_0, \{u_0 > 0\}, \{u_0 < 0\})$ *with normal velocity* $V = -\operatorname{tr}[\theta(n) Dn]$, *then, as* $\epsilon \to 0$, $u^\epsilon \to 1$ *locally uniformly in* $\bigcup_{t>0} \Omega_t^+ \times \{t\}$, *and* $u^\epsilon \to -1$ *locally uniformly in* $\bigcup_{t>0} (\Omega_t^-) \times \{t\}$.

At this point it is worth remarking about the qualitative difference between (4.3) and this last equation. The asymptotic behavior of (4.3) is always governed by mean curvature, as long as W is a double-well potential with wells of equal depth, independently of the particular W. On the other hand, this is not true anymore for the above model since the normal velocity depends on q which in turn depends nontrivially on the specific form of W.

5 Asymptotics of reaction diffusion equations and systems of KPP-type

Reaction-diffusion equations with KPP- (i.e., Kolmogorov, Petrovskii and Piskunov) type nonlinearties, a simple example being

$$u_t - \Delta u + u(u-1) = 0, \tag{5.1}$$

arise in applications similar to the ones for equations of bistable-type, the main difference being in the physical assumptions regarding the chemistry. The study of the long time and large space asymptotics of (5.1) and its generalizations, which after the usual (hyperbolic) scaling leads to asymptotic problems of the form

$$u_t^\epsilon - \epsilon\Delta u^\epsilon + \epsilon^{-1}u^\epsilon(u^\epsilon - 1) = 0, \tag{5.2}$$

have been the object of a great deal of research, which began before the one for equations of bistable type and has been recently revisited in some work regarding turbulent flame propagation and combustion.

This section is divided into four parts. The first part is about the asymptotics, as $\epsilon \to 0$, of problems like (5.2), but with both more general nonlinearities and second-order operators. It also contains a discussion about the validity of Huygens' principle and the connection with the asymptotics discussed in the previous section. In the second part, I discuss some results about KPP-type systems. In the third part, I present some more recent results regarding asymptotics of problems like (5.2) but with oscillatory coefficients and different scalings. This is related to turbulent flame propagation and combustion. Finally in the fourth part, I discuss some results about a system modelling non-premixed combustion.

5.1 General theory

Here I consider the behavior, as $\epsilon \to 0$, of

$$\begin{cases} \text{(i)} \quad u_t^\epsilon - a_{ij}(x)u^\epsilon_{x_ix_j} + b(x)\cdot Du^\epsilon + \epsilon^{-1}f(x,u^\epsilon) = 0 & \text{in } \mathbb{R}^N \times (0,\infty), \\ \text{(ii)} \quad u^\epsilon = g & \text{on } \mathbb{R}^N \times \{0\}, \end{cases} \tag{5.3}$$

where the matrix $A = (a_{ij})$ is uniformly elliptic, the nonlinearity $f : \mathbb{R}^N \times \mathbb{R} \to \mathbb{R}$ is of KPP-type, i.e.,

$$\begin{cases} f(x,u) < 0 \text{ for } u \in (0,1),\ f(x,u) > 0 \text{ for } u \in (-\infty,0)\cup(1,\infty), \\ \text{and} \\ c(x) = \dfrac{\partial f}{\partial u}(x,0) = \inf\limits_{0<u\le 1} \dfrac{f(x,u)}{u} < 0 \quad (x \in \mathbb{R}^N), \end{cases} \tag{5.4}$$

and the initial datum $g : \mathbb{R}^N \to \mathbb{R}$ is such that

$$g \geq 0 \quad \text{and} \quad \Omega_0 = \operatorname{spt}(g) \text{ is bounded.} \tag{5.5}$$

It is necessary, of course, to make a number of technical assumptions on the dependence and regularity on x of the coefficients of (5.3) for which I refer to [ESo2, MS1], etc. Moreover, the choice of the equilibrium value is only made to simplify the presentation.

The asymptotic behavior of (5.3) was first studied, under some conditions, using probabilistic ideas from thetheory of large deviations by Freidlin [Fr1, Fr2], and later by Evans and Souganidis [ESo1], using ideas from the theory of viscosity solutions, without the restrictions of [Fr1, Fr2]. More recently, Barles and Souganidis [BS3] revisited this issue and studied the rate of convergence as $\epsilon \to 0$ of the solution of u^ϵ to the equilibria.

The main result is:

5.1 Theorem *Assume* (5.4) *and* (5.5) *and let* $Z : \mathbb{R}^N \times [0,\infty) \to \mathbb{R}$ *be the unique viscosity solution of the variational inequality*

$$\begin{cases} \min[Z_t - a_{ij}(x)Z_{x_i}Z_{x_j} + b(x)\cdot DZ + c(x), Z] = 0 \quad in \quad \mathbb{R}^N \times (0,\infty), \\ Z = \begin{cases} 0 & in\ \Omega_0 \times \{0\}, \\ -\infty & in\ (\mathbb{R}^N \setminus \overline{\Omega}_0) \times \{0\}. \end{cases} \end{cases} \tag{5.6}$$

Then, as $\epsilon \to 0$,

$$u^\epsilon \to 0 \ in \ \{Z<0\}\ , \qquad u^\epsilon \to 1 \ in \ \mathrm{int}\{Z=0\},$$

locally uniform with exponential rate.

I continue with a sketch of the proof of the theorem and then discuss a number of interesting points and examples. In the proof below I make all the necessary assumptions to simplify the presentation and refer to [ESo1, BS3] for the details.

Proof of Theorem 5.1 (1) Assume, for simplicity, that $a_{ij} = \delta_{ij}$ and $b \equiv 0$. The general case follows in exactly the same way.

(2) Assume, again for simplicity, that $0 \le g \le 1$. Then the maximum principle yields that

$$0 < u < 1 \quad \text{in} \quad \mathbb{R}^N \times [0,\infty).$$

(3) Define $Z^\epsilon : \mathbb{R}^N \times [0,\infty) \to \mathbb{R}$ by

$$Z^\epsilon(x,t) = \epsilon \ln u^\epsilon(x,t).$$

A simple computation yields that

$$\begin{cases} Z^\epsilon_t - \epsilon\Delta Z^\epsilon - |DZ^\epsilon|^2 + (u^\epsilon)^{-1}f(u^\epsilon) = 0 \text{ in } \mathbb{R}^N \times [0,\infty), \\ Z^\epsilon = \begin{cases} \epsilon \ln g & \text{in } \Omega_0 \times \{0\}, \\ -\infty & \text{in } (\mathbb{R}^N \setminus \overline{\Omega}_0) \times \{0\}. \end{cases} \end{cases}$$

(4) Assume that for each compact subset K of $(\mathbb{R}^N \times (0,\infty)) \cup (\Omega_0 \times [0,\infty))$, there exists $C = C(K) > 0$ such that

$$\sup_{\epsilon>0} \|Z^\epsilon\|_{L^\infty(K)} \le C.$$

Then

$$Z^*(x,t) = \limsup_{\substack{\epsilon\to 0 \\ (y,s)\to(x,t)}} Z^\epsilon(y,s) \quad \text{and} \quad Z_*(x,t) = \limsup_{\substack{\epsilon\to 0 \\ (y,s)\to(x,t)}} Z^\epsilon(y,s)$$

are well defined upper- and lower-semicontinuous functions respectively and, in view of the bound in Step (2) above,

$$Z_* \le Z^* \le 0 \quad \text{in} \quad \mathbb{R}^N \times [0,\infty).$$

(5) Using (5.4) in the equation satisfied by Z^ϵ yields

$$Z_t^\epsilon - \epsilon\Delta Z^\epsilon - |DZ^\epsilon|^2 + c(x) \leq 0 \quad \text{in} \quad \mathbb{R}^N \times [0,\infty).$$

Passing now to the limit $\epsilon \to 0$ yields

$$\begin{cases} Z_t^* - |DZ^\epsilon|^2 + c(x) \leq 0 \text{ in } \mathbb{R}^N \times [0,\infty), \\ Z^* = \begin{cases} 0 & \text{in } \Omega_0 \times \{0\}, \\ -\infty & \text{in } (\mathbb{R}^N \setminus \overline{\Omega}_0) \times \{0\}. \end{cases} \end{cases}$$

(6) It also turns out that Z_* is a supersolution of (5.6). Indeed, since $Z_* \leq 0$, it suffices to check that

$$Z_{*t} - |DZ_*|^2 + c(x) \geq 0 \quad \text{in} \quad \{Z_* < 0\}.$$

But the relationship between Z^ϵ and u^ϵ and the KPP-assumption (5.4) on f immediately yield

$$\lim_{\epsilon\to 0}(u^\epsilon)^{-1} f(x,u^\epsilon) = c(x) \quad \text{in} \quad \{Z_* < 0\},$$

hence the claim.

(7) Since (5.6) admits a unique viscosity solution Z, it now follows that

$$Z^* = Z_* = Z \quad \text{in} \quad \mathbb{R}^N \times [0,\infty).$$

The issue of existence and uniqueness of solutions to a large class of parabolic pde's with unbounded initial datum, (5.6) being one of them, was studied by Crandall, Lions and Souganidis [CLS], to which I refer to for the details.

(8) Since

$$u^\epsilon = e^{Z^\epsilon/\epsilon},$$

it is immediate that

$$u^\epsilon \to 0 \quad \text{exponentially fast in} \quad \{Z < 0\}.$$

(9) The local uniform convergence of u^ϵ to 1 in $\operatorname{int}\{Z = 0\}$ is a bit more technical, hence I refer to [ESo1].

(10) If $w^\epsilon : \mathbb{R}^N \times [0,\infty) \to \mathbb{R}$ is defined by

$$u^\epsilon = 1 - e^{w^\epsilon/\epsilon},$$

then w^ϵ solves

$$w_t^\epsilon - \epsilon\Delta w^\epsilon - |Dw|^2 + \frac{f(u^\epsilon, x)}{u^\epsilon - 1} = 0 \quad \text{in} \quad \mathbb{R}^N \times (0,\infty).$$

It follows that, as $\epsilon \to 0$,

$$w^\epsilon \to w \quad \text{locally uniformly in} \quad \operatorname{int}\{Z = 0\},$$

where w is the unique viscosity solution of

$$\begin{cases} w_t - |Dw|^2 + d(x) = 0 \quad \text{in} \quad \operatorname{int}\{Z = 0\}, \\ w = \begin{cases} 0 & \text{on } \partial\{Z = 0\} \cap \{t > 0\}, \\ 0 & \text{in } \Omega_1 \times \{0\}, \\ -\infty & \text{in } (\mathbb{R}^N \setminus \Omega_1) \times \{0\}, \end{cases} \end{cases}$$

where

$$\Omega_1 = \{x \in \mathbb{R}^N : g(x) = 1\} \quad \text{and} \quad d(x) = \frac{\partial f}{\partial u}(u, 1) > 0;$$

notice that Ω_1 may be empty.

The uniqueness of solutions of the above initial boundary value problem, which in turn yields the convergence of the w^ϵ's, is proved in [BS3].

(11) Since $W < 0$ in $\operatorname{int}\{Z = 0\}$, the proof is complete. □

It turns out (see [ESo1]) that there is an explicit but rather complicated formula for Z as the value function of a zero sum, two player differential game with stopping times. In some cases, however, this formula can be simplified as is indicated below, without any proof. (See [ESo1] for details.)

5.2 Proposition *Assume that* $b \equiv 0$ *and* $c(x) = c$ *in* (5.3). *Then*

$$Z = \overline{Z} \wedge 0 \quad \text{in} \quad \mathbb{R}^N \times (0, \infty),$$

where $\overline{Z}$ *is the solution of*

$$\begin{cases} \overline{Z}_t - a_{ij}(x)\overline{Z}_{x_i}\overline{Z}_{x_j} + c = 0 & \text{in} \quad \mathbb{R}^N \times (0, \infty), \\ \overline{Z} = \begin{cases} 0 & \text{in } \Omega_0 \times \{0\}, \\ -\infty & \text{in } (\mathbb{R}^N \setminus \overline{\Omega}_0) \times \{0\}, \end{cases} \end{cases}$$

which is given by the formula

$$\overline{Z}(x, t) = -\frac{d^2(x, \Omega_0)}{4t} - ct \quad ((x, t) \in \mathbb{R}^N \times (0, \infty)).$$

Here $d(x, \Omega_0)$ *denotes the distance from* x *to* Ω_0 *in the Riemannian metric determined by* A^{-1}.

An immediate corollary is:

5.3 Corollary *Under the hypotheses of Proposition* 5.2, *as* $\epsilon \to 0$,

$$u^\epsilon(x, t) \to \begin{cases} 0 \\ \\ 1 \end{cases} \quad \text{exponentially fast and locally uniformly in } d(x, \Omega_0) < 2(-c)^{1/2}t,$$

Another interesting point is that under some relatively mild conditions on A and b, the front $\Gamma_t = \partial\{Z(\cdot, t) < 0\}$ is a Lipschitz continuous surface. This was proved by Freidlin and Lee [FL] using the formulae obtained in [ESo1]. (See also [MS1] for a direct pde proof.)

5.4 Proposition *Assume that* $c(x) = c$ *and, for simplicity,* $A = I$. *Then, if* $\|b\|_\infty < 2\sqrt{-c}$, $t \mapsto \Gamma_t$ *is a Lipschitz continuous surface in* $\mathbb{R}^N \times (0, \infty)$.

A natural question at this point is why is it that Theorem 5.1 cannot be proved in the same way as the analogous theorems in Section 3. To answer this question it is necessary to recall (see [AW] for example) that nonlinearities of KPP-type do not admit a unique pair

(α, q) with $\alpha \in \mathbb{R}$ and $q : \mathbb{R} \to \mathbb{R}$ (unique up to translations) such that—here for simplicity I neglect the x-dependence of f,

$$\alpha \dot{q} + \ddot{q} = f(q) \text{ in } \mathbb{R}, \qquad q(-\infty) = 0,\ q(+\infty) = 1 \text{ and } \dot{q} > 0 \text{ in } \mathbb{R},$$

in contrast to the behavior of nonlinearities of bistable type – but rather a whole continuum of velocities and, therefore, traveling waves! This rather intriguing fact, which is, of course, related to the different nature of the equilibria 0 and 1 for f's which are KPP-stable or unstable, respectively (as opposed to bistable f's, which are KPP-stable), is the main reason about the difference in the analysis of the asymptotics of (4.2) and (5.3).

As a matter of fact this is also the source of some qualitative differences in the asymptotic behavior of (4.2) and (5.3), which goes well beyond the technical issue of how they are studied. This is related to the issue of whether some kind of Huygens-principle is valid for the fronts obtained by Theorem 5.1, which, in this context, is quantified as whether there is a geometric pde governing their evolution. The issue of whether there is a Huygens-principle was discussed at length in [Fr1, Fr2], which contain a number of related examples. This problem was revisited by Majda and Souganidis [MS1], who gave some further examples and discussed the issue of existence or non-existence of a geometric pde. The lack of such equations exhibited in [MS1] came as the real surprise to a number of researchers in the combustion community where such equations, known as G-equations, were derived formally using nonlinearities which were not of KPP-type but "close" to that. Below I present some of the arguments and examples of [MS1]; for simplicity, however, I take

$$A = I.$$

It is immediate from (5.6) that Z satisfies

$$\min[Z_t + b(x) \cdot DZ - 2(-c)^{1/2}|DZ|, Z] \geq 0 \quad \text{in} \quad \mathbb{R}^N \times [0, \infty),$$

which yields, using the theory a generalized front propagation of Section 2, that

$$\{x : Z(x,t) < 0\} \subset \{x : W(x,t) < 0\},$$

where W solves the geometric pde

$$\begin{cases} W_t + b(x) \cdot DW = 2(-c(x))^{1/2}|DW| & \text{in} \quad \mathbb{R}^N \times (0, \infty), \\ W = \begin{cases} d(\cdot, \Omega_0) & \text{in } \Omega_0, \\ -d(\cdot, \Omega_0) & \text{in } \mathbb{R}^N \setminus \overline{\Omega}_0. \end{cases} \end{cases}$$

The question now is whether $\Gamma_t = \partial\{Z(\cdot, t) < 0\}$ moves with normal velocity

$$V = 2(-c(x))^{1/2} + b(x) \cdot n.$$

It turns out (see [MS1]) that this is indeed the case if $b(x) \equiv b$ and $c(x) \equiv c$ are x-independent. Actually, $A = I$ is not needed for this fact. This is, of course, related to Proposition 5.2 and Corollary 5.3. On the other hand, if either c or b depend on x it turns out that there is no geometric pde for Γ_t. This actually can be exhibited by the following two examples.

The first example shows that the velocity of Γ_t depends nontrivially on the location of Γ_0, which means that its motion is not geometric motion in the sense discussed in Section 2.

Indeed, consider the asymptotic problem

$$\begin{cases} u_t^\epsilon - \epsilon u_{xx}^\epsilon + x u_x^\epsilon + \epsilon^{-1} u^\epsilon (u^\epsilon - 1) = 0 & \text{in} \quad \mathbb{R} \times (0, \infty), \\ u^\epsilon = \chi_{(-\infty, a]} \quad \text{on} \quad \mathbb{R} \times \{0\}. \end{cases}$$

Since the solution Z of

$$\begin{cases} \min(Z_t - (Z_x)^2 + x Z_x - 1, Z) = 0 \quad \text{in} \quad \mathbb{R} \times (0, \infty), \\ Z = \begin{cases} 0 & \text{in } (-\infty, a] \times \{0\}, \\ -\infty & \text{in } (a, \infty) \times \{0\}, \end{cases} \end{cases}$$

is given by

$$Z(x,t) = \left(t - \frac{(x - e^t a)_+^2}{2(e^{2t} - 1)} \right) \wedge 0 \quad ((x,t) \in \mathbb{R} \times (0, \infty)),$$

where $x_+ = \max(x, 0)$, as one can verify by direct computation, it follows that the front Γ_t is the curve

$$x(t) = e^t a + (2t(e^{2t} - 1))^{1/2},$$

which clearly does not satisfy the ode

$$\dot{\tilde{x}} = 2 + \tilde{x},$$

the geometric motion derived earlier. Moreover, the velocity $\dot{x}(t)$ of $x(t)$ depends on the location a of Γ_0 and it is not universal.

The second example shows that although the front, in view of Proposition 5.4, can be characterized as a Lipschitz continuous function $t : \mathbb{R} \to \mathbb{R}$, i.e., $t = t(x)$, nevertheless there exist b such that the map $x \mapsto t(x)$ is not a monotone function of x. This yields that the front does not propagate according to Huygens' principle, since the lack of monotonicity of $t(x)$ implies that at some points ahead of the front "new sources" appear. This example is rather tedious. Here I only state it and refer to [MS1] for the details.

Consider again the one-dimensional problem

$$\begin{cases} u_t^\epsilon - \epsilon u_{xx}^\epsilon + b(x) u_x^\epsilon + \epsilon^{-1} u^\epsilon (u^\epsilon - 1) = 0 & \text{in} \quad \mathbb{R} \times (0, \infty), \\ u^\epsilon = \chi_{(-\infty, 0]} \quad \text{on} \quad \mathbb{R} \times \{0\}, \end{cases}$$

where now $b : \mathbb{R} \to \mathbb{R}$ is smooth and decreasing. It follows that, if b is within a small distance from the piecewise constant function

$$\bar{b}(x) = \begin{cases} \lambda b & \text{if } x < \overline{x}, \\ b & \text{if } x > \overline{x}, \end{cases}$$

for some $\overline{x} > 0$, where λ and b are chosen so that

$$1 < \lambda, \quad 0 < b < 2, \quad 0 < \lambda b < 2 \quad \text{and} \quad 1 < \left(\lambda^2 - \frac{1}{2} - \lambda \sqrt{\lambda^2 - 1} \right) b,$$

then $x \mapsto t(x)$ is not a monotone function of x.

5.2 Some results about systems

Here I present results regarding the asymptotics of KPP-type systems of reaction-diffusion equations coupled only through the zero-order terms. Partial results in this direction were first obtained in [Fr1, Fr2]. Evans and Souganidis [ESo2] obtained some related results for linear systems. The first general result about the asymptotics of KPP-systems were obtained by Barles, Evans and Souganidis [BES]. Extensions of these results were later obtained in [FL].

The problem considered in [BES] is the scaled reaction-diffusion system

$$\begin{cases} u^\epsilon_{k,t} - \epsilon d_k \Delta u^\epsilon_k + \epsilon^{-1} f_k(u^\epsilon) = 0 & \text{in} \quad \mathbb{R}^N \times (0,\infty), \\ u^\epsilon_k = g_k \quad \text{on} \quad \mathbb{R}^N \times \{0\} & (k = 1, \dots, m). \end{cases} \tag{5.7}$$

Here the positive constants $d_k > 0$ are given and the smooth, bounded functions $g : \mathbb{R}^N \to \mathbb{R}^m$ are such that

$$g_k \geq 0 \quad \text{and} \quad \Omega_0 = \{g_k > 0\} \text{ is bounded} \quad (k = 1, \dots, m). \tag{5.8}$$

As for the smooth and bounded vector field f: $\mathbb{R}^N \to \mathbb{R}^m$, the essential assumptions are:

$$f(0) = 0 \tag{5.9}$$

and, if

$$C \equiv Df(0) = \left(\left(\frac{\partial f_k}{\partial u_\ell}(0) \right) \right),$$

then

$$\begin{cases} \text{(i)} \quad c_{u\ell} < 0 & (1 \leq k, \ \ell \leq m), \\ \text{(ii)} \quad f_k(u) \geq c_{u\ell} u_\ell & (k = 1, \dots, m). \end{cases} \tag{5.10}$$

I refer to [BES] for one more structural assumption on f which is necessary to guarantee that some uniform L^∞-bounds on the $u^\epsilon : \mathbb{R}^N \to \mathbb{R}^m$ exist.

Given $p \in \mathbb{R}^N$ define now the $m \times m$-matrix

$$B(p) \equiv \operatorname{diag}(\dots, -d_k|p|^2, \dots)$$

and set

$$A(p) \equiv B(p) + C.$$

Since the matrix $A(p)$ has negative entries, the Perron-Frobenius theory asserts that $A(p)$ possesses a simple, real eigenvalue $\lambda^0 = \lambda^0(A(p))$ satisfying $\operatorname{Re}\lambda > \lambda^0$ for all other eigenvalues λ of $A(p)$. Define the Hamiltonian

$$H(p) = \lambda^0(A(p)) \quad (p \in \mathbb{R}^N),$$

and consider the unique viscosity solution Z of

$$\begin{cases} \min[Z_t + H(DZ), Z] = 0 \quad \text{in} \quad \mathbb{R}^N \times (0,\infty), \\ Z = \begin{cases} 0 & \text{on } \Omega_0 \times \{0\}, \\ -\infty & \text{in } \operatorname{int}(\mathbb{R}^N \setminus \Omega_0) \times \{0\}. \end{cases} \end{cases}$$

5.5 Theorem *Assume* (5.8), (5.9) *and* (5.10). *Then, as* $\epsilon \to 0$, $u^\epsilon \to 0$ *locally uniformly in* $\{Z > 0\}$ *and* $\liminf_{\epsilon \to 0} u^\epsilon_k > 0$ *uniformly on compact subsets of* $\{Z < 0\}$.

5.3 Front dynamics for turbulent reaction-diffusion equations

Recently Majda and Souganidis [MS1] developed simplified effective equations for the large scale front propagation of turbulent reaction-diffusion equations in the simplest prototypical situation involving advection for turbulent velocity fields with two separated scales. In addition to developing a rigorous theory, [MS1] studied the issue of the validity of the Huygens principle, discussed earlier in this section, and provided elementary but also sharp bounds on the velocity of the propagating front. In subsequent papers, Embid, Majda and Souganidis [EMS1, EMS2], studied the special case of steady incompressible velocity fields consisting of a mean flow plus a small-scale periodic shear both analytically and computationally. The results of [EMS1] give information about, among others, the dependence of the turbulent flame speed on the turbulence intensity. [EMS2] considers the comparison between the results of [MS1] and the theory of the G-equation and shows that although in some regimes of the ratio between the transverse component of the mean flow and the shear the agreement is excellent, in some other regimes the G-theory underestimates the flame speeds computed through the nonlinear averaging theory of [MS1]. This is, of course, related to the issue of failure of the Huygens principle.

Here I discuss the following asymptotic problem, which is similar to the one considered in [MS1]:

$$\begin{cases} u_t^\epsilon - \epsilon\Delta u^\epsilon + V(x,\delta^{-1}x)\cdot Du^\epsilon + \epsilon^{-1}u^\epsilon(u^\epsilon - 1) = 0 & \text{in} \quad \mathbb{R}^N \times (0,\infty), \\ u^\epsilon = g \quad \text{on} \quad \mathbb{R}^N \times \{0\}. \end{cases} \tag{5.11}$$

The main assumption on V is that

$$(x,y) \mapsto V(x,y) \text{ is periodic with respect to } y \text{ in } Q. \tag{5.12}$$

As before, the assumption on g is that

$$0 \le g \quad \text{and} \quad \Omega_0 = \{g > 0\} \text{ is bounded.} \tag{5.13}$$

Given $p, x \in \mathbb{R}^N$ and $a \in [0,\infty)$ consider next the issue of solving the "cell-problem"

$$-a\Delta w - |p + Dw|^2 + V(x,y)\cdot(p + Dw) = -H_a(p,x) \quad \text{in} \quad Q, \tag{5.14}$$

i.e., the problem of finding a unique $H_a(p,x)$ and a unique up to constant Q-periodic $w : \mathbb{R}^N \to \mathbb{R}$ solving (5.14). It turns out that the homogenization theory for Hamilton-Jacobi equations developed by Lions, Papanicolaou and Varadhan [LPV] (see also Evans [E1]) applies to this case and yields the existence of such a unique $H_a(p,x)$. Notice that when $a \neq 0$, then $-H_a(p,x)$ is the principal eigenvalue of the non-divergence form operator

$$-a\Delta - (2p - V(x,y))\cdot D - |p|^2 + V(x,y)\cdot p \quad \text{in} \quad Q.$$

Finally, consider the variational inequality

$$\begin{cases} \min[Z_t^a - H_a(DZ^a,x) - 1, Z^a] = 0 \quad \text{in} \quad \mathbb{R}^N \times (0,\infty), \\ Z^a = \begin{cases} 0 & \text{in } \Omega_0 \times \{0\}, \\ -\infty & \text{in } \operatorname{int}(\mathbb{R}^N \setminus \overline{\Omega}_0) \times \{0\}. \end{cases} \end{cases} \tag{5.15}$$

The asymptotic behavior of (5.11) is characterized by the following theorem, which in [MS1] is studied when $\delta = \epsilon^\alpha$ with $\alpha \in (0,1]$ and for general f's of KPP-type.

5.6 Theorem *Assume* (5.12) *and* (5.13) *and let* Z^a *be the solution of* (5.15).
If $\delta^{-1}\epsilon \to a \in [0,\infty)$, *then*

$$\lim_{\substack{\epsilon\to 0\\ \delta\to 0}} u^\epsilon(x,t) = \begin{cases} 0 & \text{in } \{Z^a < 0\},\\ 1 & \text{in } \operatorname{int}\{Z^a = 0\}. \end{cases}$$

The proof follows along the lines of the one of Theorem 5.5 appropriately modified to take care of the nonlinear averaging, which can be accomplished using the "perturbed-test function" method of Evans [E1] . Instead of writing all the technicalities, below I only present a formal argument and refer to [MS1] for the details.

To this end observe that the usual transformation

$$u^{\epsilon,\delta} = \exp \epsilon^{-1} Z^{\epsilon,\delta}$$

yields the equation

$$Z_t^{\epsilon,\delta} - \epsilon\Delta Z^{\epsilon,\delta} - |DZ^{\epsilon,\delta}|^2 + V(x,\delta^{-1}x)\cdot DZ^{\epsilon,\delta} + u^{\epsilon,\delta} - 1 = 0. \tag{5.16}$$

For simplicity neglect the $u^{\epsilon,\delta}$ term above, which is of course the source of the variational inequality, and consider the following expansion for $Z^{\epsilon,\delta}$:

$$Z^{\epsilon,\delta} = Z^0(x,t) + \delta Z^1(\delta^{-1}x) + O(\delta^2).$$

A simple computation yields that the following equation should hold:

$$Z_t^0 - \delta^{-1}\epsilon\Delta Z^0 - |DZ^0 + DZ^1|^2 + V(x,\delta^{-1}x)\cdot(DZ^0 + DZ^1) - 1 = 0$$

which, of course, explains the role played by $H^a(p,x)$ in Theorem 5.6.

I present next a specific example from the homogenization theory of Hamilton-Jacobi equations, which yields a formula for the solution $H_0(p,x)$ of the cell problem. Having such a formula is important in the context of turbulent flame propagation, because it provides a model which permits to validate numerical algorithms to compute the effective flame velocity.

To this end, consider the case $N = 2$, $a = 0$, $V(x,y) = \overline{v}(x) + (v(y_2),0)$ and $\langle v\rangle = 0$. In view of the uniqueness of $H_0(p,x)$, the cell problem reduces to finding $H_0(p,x)$ and a $Q\cap\mathbb{R}$-periodic $w:\mathbb{R}\to\mathbb{R}$ such that

$$\left|w' + \left(p_2 - \frac{1}{2}\overline{v}_2\right)\right|^2 = H_0(p,x) - p_1^2 + \frac{1}{4}(\overline{v}_2)^2 + (\overline{v}_1 + v(y))p_1 \quad \text{in} \quad \mathbb{R},$$

where, $/$ denotes differentiation with respect to $y\in\mathbb{R}$.

A simple argument (see [LPV, MS1]), which is based on the observation that, if w is a $Q\cap\mathbb{R}$-periodic solution of $|w'+\xi|^2 = F(y)$ and $F(y) > 0$ in $Q\cap\mathbb{R}$, then

$$|\xi| = \langle F(y)^{1/2}\rangle,$$

yields that

$$H_0(p,x) = \begin{cases} p_1^2 - \frac{1}{4}\overline{v}_2^2(x) - \overline{v}_1(x)p_1 + M & \text{if } \left|p_2 - \frac{\overline{v}_2(x)}{2}\right| \le \langle (M + v(y_2)p_1)^{1/2}\rangle \\ p_1^2 - \frac{1}{4}\overline{v}_2^2(x) - \overline{v}_1(x)p_1 + M + A & \text{if } \left|p_2 - \frac{\overline{v}_2(x)}{2}\right| = \langle (M + v(y_2)p_1 + A)^{1/2}\rangle, \end{cases}$$

where

$$M = \max_Q(-v(y)p_1).$$

The above results were extended recently by the author in [Sou2] to vector fields which are stationary random fields.

5.4 Isothermal non-premixed turbulent combustion

Isothermal non-premixed turbulent combustion is characterized by two variables Y_F and Y_θ representing the mass fraction of fuel and oxidizer, respectively. In non-premixed combustion, the fuel and oxidizer occupy different regions of space initially and are mixed to react together through advection by the fluid velocity field which has both large scale and turbulent components. With the assumptions that the fuel and oxidizer react through a simple single-step kinetics (see [W, LW]), the mass fractions satisfy the coupled reaction diffusion equations

$$\begin{cases} D_t Y_F = \mathcal{K}\Delta Y_F - K Y_F Y_\theta \\ D_t Y_\theta = \mathcal{K}\Delta Y_\theta - K Y_F Y_\theta \end{cases} \quad \text{in} \quad \mathbb{R}^N \times (0,\infty), \tag{5.17}$$

where $\mathcal{K} \geq 0$ is the diffusion coefficient and $K > 0$ is the reaction rate. The symbol D_t denotes the convective derivative

$$D_t = \partial_t + \vec{V} \cdot D$$

with imposed velocity $\vec{V}$.

In non-premixed combustion, the fuel and oxidizer initially occupy disjoint regions of space, i.e., there are two disjoint sets Ω_+ and Ω_- with $\Omega_+ \cup \Omega_- = \mathbb{R}^N$ such that

$$Y_F = \begin{cases} Y_F^0 > 0 & \text{on } \Omega_+ \times \{0\}, \\ Y_F^0 = 0 & \text{on } \Omega_- \times \{0\}, \end{cases} \quad \text{and} \quad Y_\theta = \begin{cases} Y_\theta^0 = 0 & \text{on } \Omega_+ \times \{0\}, \\ Y_\theta^0 > 0 & \text{on } \Omega_- \times \{0\}. \end{cases} \tag{5.18}$$

By introducing the variables

$$Y = Y_F - Y_\theta \quad \text{and} \quad Z = Y_F, \tag{5.19}$$

the equations in (5.17) can be rewritten as

$$\begin{cases} D_t Z = \mathcal{K}\Delta Z + KZ(Y - Z) \\ D_t Y = \mathcal{K}\Delta Y \end{cases} \quad \text{in} \quad \mathbb{R}^N \times (0,\infty), \tag{5.20}$$

with the initial data such that

$$Z_0 = \begin{cases} Y_F^0 & \text{in } \Omega_+ \times \{0\}, \\ 0 & \text{in } \Omega_- \times \{0\}, \end{cases} \quad \text{and} \quad Y_0 = \begin{cases} Y_F^0 & \text{in } \Omega_+ \times \{0\}, \\ -Y_\theta^0 & \text{in } \Omega_- \times \{0\}. \end{cases} \tag{5.21}$$

The crucial feature for the initial data (5.21) in non-premixed turbulent combustion is the assumption that

$$\begin{cases} \text{the inital data describe different stable equilibria} \\ \text{for the reaction in the regions } \Omega_+ \text{ and } \Omega_-, \end{cases} \tag{5.22}$$

since we have

$$\left.\frac{d}{dZ}\right|_{(Y_0^\pm, Z_0^\pm)} (Z(Y-Z)) < 0.$$

In contrast, for premixed turbulent combustion for the equations in (5.20), we have the "stoichiometric" balance condition $Y \equiv Y_0 > 0$ with Y_0 constant and the KPP equation

$$D_t Z = \mathcal{K}\Delta Z + KZ(Y_0 - Z) \quad \text{in} \quad \mathbb{R}^N \times (0,\infty), \tag{5.23}$$

with initial data

$$Z = \begin{cases} Y_0 & \text{in } \Omega_+ \times \{0\}, \\ 0 & \text{in } \Omega_- \times \{0\}. \end{cases} \tag{5.24}$$

In this situation the equilibrium in Ω_+ is stable, while the one in Ω_- is unstable, leading to turbulence enhanced front propagation in the fast reaction/slow diffusion limit ([Fr1, Fr2, ESp, MS1, MS2, MS3, EMS1, EMS2], etc.). We will see below that the phenomena for (5.20) and (5.21) are very different.

To study the effects of turbulence on mixing for the problem in (5.20), (5.21), we assume throughout this paper that the velocity field has the form

$$\vec{V}(x,t) = \bar{V}(x,t) + v(\varepsilon^{-\alpha}x, \varepsilon^{-\alpha}t), \tag{5.25}$$

where $\bar{V}$ is the large scale flow and is a zero mean space-time periodic velocity representing the effects of small scale turbulence with $\varepsilon \ll 1$ a small parameter and the parameter α fixed with $0 < \alpha \le 1$. Thus, the velocity field satisfies the same hypotheses as in [MS1].

(i) The effect of turbulence on mixing with fast reaction and slow diffusion

In many applications of turbulent reaction diffusion equations, it is natural to assume that the chemical reaction time is much shorter than the large scale eddy turnover time, while the large scale eddy turnover time is much shorter than the large scale diffusion time. If the units of space and time are determined by the large scale velocity field, the above requirements are satisfied by setting $K = \varepsilon^{-1}K'$ and $\mathcal{K} = \varepsilon\mathcal{K}'$ in (5.20). Thus in this fast reaction-slow diffusion regime, the turbulent reaction diffusion equations in (5.20) assume the form

$$\begin{cases} D_t^\epsilon Y^\varepsilon = \varepsilon\mathcal{K}\Delta Y^\varepsilon \\ D_t^\epsilon Z^\varepsilon = \varepsilon\mathcal{K}\Delta Z^\varepsilon + \varepsilon^{-1}KZ^\varepsilon(Y^\varepsilon - Z^\varepsilon) \end{cases} \quad \text{in} \quad \mathbb{R}^N \times (0,\infty), \tag{5.26}$$

where the primes have been dropped in (5.26) for notational simplicity, and we write D_t^ϵ instead of D_t to signify the dependence on ϵ of the fector field. The structure for the small scale velocity field in (5.25) represented by $v(\varepsilon^{-\alpha}x, \varepsilon^{-\alpha}t)$ has the following interpretation: As in actual turbulent flow fields, these velocities vary on smaller length scales and evolve on faster time scales than the mean flow for any fixed α with $0 < \alpha \le 1$, and yet vary on at least as large a spatial scale as $O(\varepsilon)$, which can be regarded as the natural turbulent dissipation length scale in the model. These are the same scalings utilized in [MS1] to study turbulence enhanced front propagation.

In contrast with [MS1], the following result is proved in [MS4]:

5.7 Theorem *Consider the solution $(Y^\varepsilon, Z^\varepsilon)$ of the reaction diffusion system (5.20) with smooth initial data converging to the initial data (Y_0, Z_0) defined in (5.21) and associated with two distinct stable equilibria, and assume that the velocity field v is incompressible. Assume $\alpha \in (1/2, 1]$ in (5.25). Then, as $\varepsilon \to 0$, $Y^\varepsilon \to Y^0$, locally uniformly in $\mathbb{R}^N \times [0, \infty)$, and $Z^\varepsilon \to \max(Y^0, 0)$, locally uniformly in $\mathbb{R}^N \times (0, \infty)$, where Y^0 satisfies the large scale advection equation*

$$Y_t^0 + \bar{V} \cdot DY^0 = 0 \quad in \quad \mathbb{R}^N \times (0, \infty),$$

with the initial data $Y^0|_{t=0} = Y_0$. In particular, the small scale velocity field does not affect the mixing zone in the limit and the mixing interface evolves with the large scale flow.

As mentioned earlier, this result contrasts strongly with the results of [MS1, MS2, MS3, EMS1, EMS2, EMS3] for the situation in (5.23) where one region has an unstable equilibrium, where it is established that small scale turbulence leads to enhanced effective large scale front propagation. For the proof of Theorem 5.7 and other more complicated models I refer to [MS4]. Theorem 5.7 provides a rigorous justification for numerical algorithms that are based on the fast reaction limit and are utilized in numerical modelling of condensation-evaporation phenomena in cloud physics [GS]. The theorem also provides a rigorous justification of the equilibrium chemistry approximation in the fast reaction limit for turbulent diffusion flames [W, LW] with the surprising additional fact that only the mean flow contributes to the location of the mixing zone in this limit.

(ii) The effect of turbulence on mixing with fast reaction and zero diffusion

The result in Theorem 5.7 is subtle. If one considers the limiting behavior as $\varepsilon \to 0$ of (5.25), with the initial data in (5.21) but with zero diffusion so that $\mathcal{K} = 0$ for all ε, which is the fast reaction/zero diffusion limit, the following is true:

5.8 Theorem *There are explicit examples of small scale incompressible velocity fields v, such that the fast reaction/zero diffusion limit necessarily has non-trivial mixing by the turbulence, and the mixing zone in the limit is not determined by the large scale flow.*

Thus Theorems 5.7 and 5.8 establish that turbulent mixing of a reactive scalar is very different in the fast reaction/slow diffusion and the fast reaction/zero diffusion limits.

(iii) Turbulent mixing with slow reaction and slow diffusion

Another relevant scaling is required in the case of slow reaction and slow diffusion, where the reaction and diffusion terms in (5.20) both satisfy $\mathcal{K} = \epsilon\mathcal{K}'$, $K = \varepsilon K'$ and $\alpha = 1$ for the turbulent velocity field in (5.25). For technical reasons, it is also necessary to require that the mean flow $\bar{V}$ is relatively weak, i.e., that it has the form

$$\bar{V} = \varepsilon \bar{V}'(x, \varepsilon t).$$

Clearly very little chemical reaction occurs in this slow reaction/slow diffusion limit for order $O(1)$ times. With these assumptions, it is natural to look for mixing in the reaction diffusion equations in (5.20) over the longer time scale $t' = \varepsilon t$, where the reaction rates are not negligible. By introducing the primed variables in (5.20) and then dropping (for notational simplicity) the prime notation in the result, one obtains the following reaction diffusion system

$$\begin{cases} Y_t^\varepsilon + (\bar{V}(x,t) + \epsilon^{-1}v(\epsilon^{-1}x, \epsilon^{-2}t)) \cdot DY^\epsilon = \mathcal{K}\Delta Y^\epsilon \\ Z_t^\epsilon + (\bar{V}(x,t) + \epsilon^{-1}v(\epsilon^{-1}x, \epsilon^{-2}t) \cdot DZ^\epsilon) \\ \qquad = \mathcal{K}\Delta Z^\epsilon + K(Z^\epsilon(Y^\epsilon - Z^\epsilon)). \end{cases} \quad \text{in} \quad \mathbb{R}^N \times (0,\infty), \tag{5.27}$$

Notice from (5.27) that the small-scale turbulence strongly stirs the fluid before significant chemical reaction occurs, so we can expect strong mixing effects of turbulence in the limit. The following result is presented in [MS4]:

5.9 Theorem *Assume that the velocity field v is incompressible. Then for any bounded initial data (Y_0, Z_0) and any $T > 0$, the solution (Y^ϵ, Z^ϵ) of (5.27) converges locally uniformly in $\mathbb{R}^N \times (0,T)$, as $\epsilon \to 0$, to the homogenized limit (Y, Z) which satisfies the equations,*

$$\begin{cases} Y_t + \bar{V}(x,t) \cdot DY = \mathcal{K}\Delta Y + D \cdot (\mathcal{K}^* DY) \\ Z_t + \bar{V}(x,t) \cdot DZ = \mathcal{K}\Delta Z + D \cdot (\mathcal{K}^* DZ) + KZ(Y - Z) \end{cases} \quad in \quad \mathbb{R}^N \times (0,\infty),$$

with the same initial data (Y_0, Z_0).

6 Macroscopic (hydrodynamic) limits of stochastic Ising models with long range interactions

One of the most striking applications of the theory of the generalized front propagation and viscosity solutions has been the development of a rigorous theory about macroscopic limits of stochastic Ising models and the rigorous justification of the appearance of moving interfaces in the limit. Such results are important both from the mathematical and statistical mechanics point of view. They also provide a theoretical justification, from the microscopic point of view, for the phenomenological theories of phase transitions, which are obtained with arguments from continuum mechanics, as well as for the numerical computations performed by physicists to compute evolving fronts.

In the first part of this section, I present very briefly some of the key facts from the equilibrium theory of Ising models. In the second part, I discuss the dynamic theory and derive the appropriate mean field equations. Finally in the third part, I present the results about the macroscopic limits.

6.1 Equilibrium theory

Stochastic Ising models are the canonical Gibbsian models used in statistical mechanics to model phase transitions. Describing in detail such models is well beyond the scope of these notes. Instead below, abusing, if needed, the mathematical rigor as well as the actual meaning of the terms I use, I present a brief summary of these models and refer to the monograph of DeMasi and Presutti [DP], the books by Spohn [Sp1] and Liggett [Lig], the papers by DeMasi, Orlandi, Presutti and Triolo [DOPT1, DOPT2, DOPT3] and the references therein for the complete and rigorous theory.

To this end, one considers the lattice $\mathbb{Z}^N$, the spin $\sigma(x) = \pm 1$ at $x \in \mathbb{Z}^N$, the configuration (sample) space $\Sigma = \{-1,1\}^{\mathbb{Z}^N}$ and the Gibbs (equilibrium) measures μ^β on Σ which depend on the inverse temperature $\beta > 0$ and the Hamiltonian (energy) function H, which is given by

$$H(\sigma) = -\sum_{x \neq y} J(x,y)\sigma(x)\sigma(y) - h\sum_{x} \sigma(x).$$

Here $J \geq 0$ is the interaction potential – the sign condition on J means that one deals with ferromagnetic models – and h is the external magnetization field. It should be noted that in the way that H is defined above, the sums on the right-hand side may diverge. This is corrected in the rigorous setting (see, for example, [DP]) by considering the model in some finite set of size Λ and then letting $\Lambda \to \infty$.

It turns out (see, for example, [Lig]) that for any $\beta > 0$, as long as $h \neq 0$, there exists a unique Gibbs measure. On the other hand, if $h = 0$ there exists β_c such that for $\beta < \beta_c$ there still exists a unique Gibbs measure, but for $\beta > \beta_c$ there exist at least two probability measures $\mu_\pm$ on Σ such that any linear combination $\alpha\mu_- + (1-\alpha)\mu_+$, where $\alpha \in [0,1]$, is also a Gibbs measure. The appearance of many Gibbs measures is a manifestation that phase transition occurs for $\beta > \beta_c$.

6.2 Dynamic (nonequilibrium) theory-mesoscopic limit

Studying the phase transitions from the dynamic point of view for $\beta > \beta_c$ amounts to introducing some dynamics, i.e., a Markov process on Σ, which have the Gibbs measures as invariant measures and to analyzing the way this process evolves any initial distribution (measure) to the equilibria measures. Convenient quantities (order parameters) to analyze in this context are the moments of the evolving measure, the first one being the total magnetization m, which, of course, will develop an interface for large times. The shape and evolution of this interface is of great interest both theoretically and in the applications.

A very general example of such dynamics which have the Gibbs measures as invariant measures is the spin-flip dynamics, which, loosely speaking, is a sequence of flips σ^x defined by

$$\sigma^x(y) = \begin{cases} \sigma(y) & \text{if } y \neq x, \\ -\sigma(y) & \text{if } y = x, \end{cases}$$

with rate

$$c(x,\sigma) = \Psi(\Delta_x H),$$

for an appropriate Ψ, where $\Delta_x H$ is the energy difference due to a spin flip at x. Besides some obvious continuity assumptions, the only restriction on Ψ, which is related to the requirement that the Gibbs measures are invariant for the dynamics, is that it satisfies the balance law

$$\Psi(r) = \Psi(-r)e^{-r}.$$

Classical choices of Ψ are $(1 + e^{-2r})$, which corresponds to the Glauber dynamics and $e^{-r/2}$, which is the Metropolis dynamics.

More precisely, the spin-flip dynamics is a Markov jump process on Σ with generator given by

$$Lf(\sigma) = \sum_{x} c(x,\sigma)[f(\sigma^x) - f(\sigma)],$$

acting on cylindrical functions on Σ—see [Lig] for the precise definition. The solution of the equation

$$\frac{d}{dt}f_t = L_\gamma f_t, \qquad f_0 = f$$

is then given by

$$f_t(\sigma) = \sum_{\sigma'} e^{L_\gamma t}(\sigma, \sigma') f(\sigma'),$$

where $e^{L_\gamma t}(\sigma, \sigma')$ are the transition probabilities of the process.

The full-stochastic jump process σ_t is constructed as follows: The initial configurations σ^0 are randomly distributed according to some measure μ on Σ. Given σ^0, $\sigma_t = \sigma^0$ for an exponentially distributed waiting time with rate $\sum_y c(y, \sigma^0)$. Then σ_t jumps to a new configuration $\sigma^1 = \sigma^x$ with probability $c(x, \sigma^0)/\sum_y c(y, \sigma^0)$. Then $\sigma_t = \sigma^1$ for another exponentially distributed waiting time with rate $\sum_y c(y, \sigma^1)$, etc. Notice that, in view of the positivity of J, the probability of a spin flip at x is higher when the spin at x is different from that of most of its neighbors than it is when the spin agrees with most of its neighbors. Thus the system prefers configurations in which the spins tend to be aligned with one another. This property, in the language of statistical mechanics, is referred to as ferromagnetism.

In view of the previous discussion one is interested in the behavior of the system as $t \to \infty$. Another classical limit, known as the Lebowitz-Penrose limit [LP], is to study the behavior of the system also as the interaction range tends to infinity. In this limit, known in the physics literature as grain coarsening, there is a law of large numbers effect that dampens the oscillations and causes the whole collection to evolve deterministically. An important question is whether these limits commute and, if not, whether there is a particular scaling or scalings for which one can study both.

These issues were addressed by Katsoulakis and Souganidis [KS3] for the general dynamics described above with long-range interactions, with rate

$$c_\gamma(x, \sigma) = \Psi(\Delta_x H_\gamma),$$

where γ^{-1} is the interaction range,

$$H_\gamma(x) = -\sum_{y \neq x} J_\gamma(x, y)\sigma(x)\sigma(y)$$

and

$$J_\gamma(x, y) = \gamma^N J(\gamma(x - y)).$$

Throughout the remainder of this section it is assumed that

$$\begin{cases} J : \mathbb{R}^N \to [0, \infty) \text{ satisfies some integrability conditions} \\ \text{and is even, i.e., } J(z) = J(-z). \end{cases} \tag{6.1}$$

The behavior of the model under the mesoscopic scaling $(x, t) \to (\gamma x, t)$ in the limit $\gamma \to 0$ was studied in [DOPT1] for isotropic J's and in [KS3] in the general setting of (6.1).

To state the result it is necessary to introduce for each $n \in \mathbb{Z}^+$ the notation

$$\mathbb{Z}_n^N = \{\underline{x} = (x_1, \ldots, x_n) \in (\mathbb{Z}^N)^n : x_1 \neq \cdots \neq x_n\},$$

and to consider the fully nonlinear integral differential equation

$$m_t + \Phi(\beta(J * m))[m - \tanh \beta J * m] = 0 \quad \text{in} \quad \mathbb{R}^N \times (0, \infty), \tag{6.2}$$

where $J * m$ denotes the usual convolution and Φ is given by

$$\Phi(r) = \Psi(-2r)(1 + e^{-2r})^{-1}. \tag{6.3}$$

The result, which is proved in [DOPT1] and [KS3] under some simplified assumption on the initial measure μ^γ, is:

6.1 Theorem *Assume that that intial measure μ^γ is a product measure such that*

$$\mathbb{E}_{\mu^\gamma}(\sigma(x)) = m_0(\gamma x) \quad (x \in \mathbb{Z}^N).$$

Then, for each $n \in \mathbb{Z}^+$,

$$\lim_{\gamma \to 0} \sup_{\underline{x} \in \mathbb{Z}^N_n} \left| \mathbb{E}_{\mu^\gamma} \left(\prod_{i=1}^n \sigma_t(x_i) \right) - \prod_{i=1}^n m(\gamma x_i, t) \right| = 0,$$

where m is the unique solution of (6.3) *with initial datum m_0.*

6.3 Macroscopic behavior of the mean field equation

Here I discuss the asymptotic behavior for large r and t of the solution of (6.2). To avoid confusion I use x to denote points on the lattice and r to denote points in $\mathbb{R}^N$.

Assuming that

$$\beta \hat{J} = \beta \int J(y)\, dy > 1, \tag{6.4}$$

it is easy to see that (6.2) admits three steady solutions $-m_\beta$, 0 and m_β such that $-m_\beta < 0 < m_\beta$. It then becomes a natural question, as is was for the reaction-diffusion problems, to try to understand the interface generated when, as $t \to \infty$, m approaches $\pm m_\beta$.

The appropriate scaling to consider is, as for the reaction-diffusion equation with a double-well potential with wells of the same depth, $(r, t) \mapsto (\lambda r, \lambda^2 t)$, which leads to the singular problem

$$m_t^\lambda + \lambda^{-2}\Phi(\beta J^\lambda * m^\lambda)[m^\lambda - \tanh \beta J^\lambda * m^\lambda] = 0 \quad \text{in} \quad \mathbb{R}^N \times (0, \infty), \tag{6.5}$$

where

$$m^\lambda(r, t) = m(\lambda^{-1} r, \lambda^{-2} t) \quad \text{and} \quad J^\lambda(r) = \lambda^{-N} J(\lambda^{-1} r).$$

An important step in the analysis to follow is the existence of traveling, in this case standing, wave solutions for (6.2), i.e., solutions of the form

$$m(r, t) = q(r \cdot n, n) \quad (n \in S^{N-1}), \quad \text{such that} \quad q(\pm\infty, n) = \pm m_\beta.$$

It turns out that q solves the equation

$$q(\xi, n) = \tanh \beta \int J(y) q(\xi + n \cdot y, n)\, dy. \tag{6.6}$$

Notice that this equation is independent of Φ and hence Ψ, i.e., the equation is independent of the dynamics assumed for the underlying spin models. Moreover, since J is not assumed to be isotropic, q must depend in a nontrivial way on the direction n. For the existence and properties of such q's I refer to [KS3]. In what follows I will use $\dot{q}$ and $D_n q$ to denote the derivative of q with respect to ξ and n respectively.

For each $n \in S^{N-1}$ consider the symmetric matrix

$$\begin{aligned}\theta(n) = &\left[\int \frac{\dot{q}^2(\xi,n)}{\Phi(\beta(J*q))(1-q^2(\xi,n))}\,d\xi\right]^{-1} \times \\ &\times \frac{\beta}{2}\left[\iint J(y)\dot{q}(\xi,n)[\dot{q}(\xi+n\cdot y,n)y\otimes y + D_n q(\xi+n\cdot y,n)\otimes y\right. \\ &\left. + y\otimes D_n q(\xi+n\cdot y,n)]\,dyd\xi\right].\end{aligned} \tag{6.7}$$

The result about the asymptotic behavior, in the limit $\lambda \to 0$, of the solution m^λ of (6.5) is stated, without all the technical conditions, in the following theorem.

6.2 Theorem *Let $\Omega_0 = \{x : m^\lambda(x,0) > 0\}$, assume that $\Gamma_0 = \partial\Omega_0 = \partial\{x : m^\lambda(x,0) < 0\}$ and consider the evolution $(\Gamma_t, \Omega_t^+, \Omega_t^-)$ of $(\Gamma_0, \Omega_0, \mathbb{R}^N \setminus \overline{\Omega}_0)$ with normal velocity $V = -\operatorname{tr}[\theta(n)Dn]$, with $\theta(n)$ given by (6.7). Then, as $\lambda \to 0$, $m^\lambda \to +m_\beta$ locally uniformly in $\bigcup_{t>0} \Omega_t^+ \times \{t\}$, and $m^\lambda \to -m_\beta$ locally uniformly in $\bigcup_{t>0} \Omega_t^- \times \{t\}$.*

This theorem, whose proof is based on the ideas developed in Section 2, is proved in [KS3]. Earlier related results were obtained by Katsoulakis and Souganidis [KS2], who studied (6.5) when J is isotropic, i.e., $J(r) = J(|r|)$ and for Glauber dynamics, and Jerrard [J] who studied a local version of (6.5) again in the isotropic case. Finally [DOPT1] studied again the isotropic problem and Glauber dynamics under the assumption that the moving interface is smooth.

I present next a formal argument to explain why the asymptotics of (6.5) are governed by the motion $V = -\operatorname{tr}[\theta(n)Dn]$. To this end, let $d(r,t)$ be the signed distance to the evolving front Γ_t and try to find an asymptotic expansion for m^λ of the form

$$m^\lambda(r,t) = q(\lambda^{-1}d(r,t), Dd(r,t)) + \lambda Q(\lambda^{-1}d(r,t), Dd(r,t)) + O(\lambda^2), \tag{6.8}$$

where Q is to be chosen.

Next insert (6.8) in (6.5) to find

$$\begin{aligned}&\lambda^{-1}\dot{q}(\xi,n)d_t + \lambda^{-2}\Phi(\beta J^\lambda * m^\lambda)[q(\xi,n) + \lambda Q(\xi,n) \\ &- \tanh\left[\beta\int J(y)\left[q\left(\frac{d(r+y)}{\lambda}, Dd(r+\lambda y)\right) + \lambda Q\left(\frac{d(r+\lambda y)}{\lambda}, Dd(r+\lambda y)\right)dy\right] + O(\lambda^2)\right] \\ &= O(1),\end{aligned}$$

where

$$\xi = \lambda^{-1}d(r,t) \quad \text{and} \quad n = Dd(r,t).$$

Expanding now the expression involving the tanh around $\tanh\beta\int J(y)q(\xi+n\cdot y,n)dy$, using (6.6) as well as the fact that $(\tanh u)' = 1-(\tanh u)^2$, one finds that, in order for the coefficient of λ^{-1} to be zero, Q must satisfy the equation

$$\begin{aligned}&\frac{1}{1-q^2(\xi,n)}Q(\xi,n)-\int J(y)Q(\xi+n\cdot y,n)\\&=\frac{\dot q(\xi,n)}{\Phi(\beta(J*q))}\,d_t-\frac{\beta}{2}\int J(y)[\dot q(\xi+y\cdot n,n)(D^2dy,y)+(D_nq(\xi+y\cdot n,n),D^2dy)]\,dy\end{aligned}\tag{6.9}$$

where d_t and D^2d are evaluated at (r,t).

If the operator L is defined by

$$LQ=\frac{1}{1-q^2}\,Q-\int J(y)Q(\xi+n\cdot y,n)\,dy,$$

differentiating (6.6) with respect to ξ yields that

$$L\dot q=0.$$

Then, by a Fredholm alternative type of argument, one finds that in order to have a Q satisfying (6.9), it is necessary for the right-hand side of (6.9) to be orthogonal to $\dot q(\xi,e)$, i.e., it is necessary to have the compatibility condition

$$\begin{aligned}&\left(\int\frac{\dot q^2(\xi,n)}{\Phi(\beta(J*q))(1-q^2(\xi,n))}\,d\xi\right)d_t\\&\qquad=\frac{\beta}{2}\iint J(y)\dot q(\xi,n)[\dot q(\xi+y\cdot n,n)(D^2dy,y)+D_nq(\xi+y\cdot n,n)D^2dy]\,dyd\xi.\end{aligned}$$

After some elementary manipulations this last expression yields

$$d_t=\operatorname{tr}(\theta(Dd)D^2d),$$

with $\theta(n)$ as in (6.7). This concludes the formal argument about the validity of Theorem 6.2.

I conclude the discussion about the macroscopic behavior of (6.5) with a few remarks about the form of $\theta(n)$. Indeed observe that $\theta(n)$ is given in (6.7) as a product of a constant, which can be identified as the inverse of the mobility, and a matrix which can be identified as the surface tension of the interface. Notice that the surface tension depends on J and the mobility depends on the dynamics in terms of Φ and the energy via J.

6.4 Macroscopic (hydrodynamic) limit for the stochastic Ising model

The results presented in the previous parts of this section suggest that if one could perhaps scale the particle system first $(x,t)\to(\gamma x,t)$ and then $(\gamma x,t)\to(\lambda\gamma x,\lambda^2 t)$, then the moving interface given by Theorem 6.2 will also govern the asymptotic behavior of the spin system. This is of course not obvious because it is not clear that the two aforementioned limits commute. Moreover, one does not have and will not have, in view of the lack of regularity for the moving front, enough smoothness to perform a good enough asymptotic expansion to study this issue. On the other hand, these two limits suggest that it may be possible to scale the particle system like $(x,t)\mapsto(\gamma\lambda(\gamma),\lambda^2(\gamma)t)$ where $\lambda(\gamma)\to 0$, as $\gamma\to 0$, so that, in the limit $\gamma\to 0$, one obtains a moving interface. This actually happens provided that one chooses $\lambda(\gamma)$ to converge to 0 relatively slowly.

The result, again stated in a rather informal way, is contained in the following theorem. The statement is however admittedly cumbersome due to the fact that the particle system and the evolving front are, of course, defined at a different scale.

6.3 Theorem *There exists $\rho^* > 0$ such that for any $\lambda(\gamma) \to 0$ slower than γ^{ρ^*} (i.e., $\lambda(\gamma)\gamma^{-\rho^*} \to +\infty$ as $\gamma \to 0$) the following hold: Let Ω_0 be an open subset of $\mathbb{R}^N$ such that $\Gamma_0 = \partial\Omega_0 = \partial(\mathbb{R}^N \setminus \overline{\Omega}_0)$ and assume that*

$$\mathbb{E}_{\mu^\gamma}\sigma(x) > 0 \quad \textit{iff} \quad \gamma\lambda(\gamma)x \in \Omega_0 \quad \textit{and} \quad \mathbb{E}_{\mu_\gamma}\sigma(x) < 0 \quad \textit{iff} \quad \gamma\lambda(\gamma)x \in \mathbb{R}^N \setminus \overline{\Omega}_0.$$

If $(\Gamma_t, \Omega_t^+, \Omega_t^-)$ is the evolution of $(\Gamma_0, \Omega_0, \mathbb{R}^N \setminus \overline{\Omega}_0)$ with normal velocity $V = -\operatorname{tr}[\theta(n)Dn]$, where $\theta(n)$ is given by (6.7), then

$$\sup_{\underline{x}\in\mathbb{Z}_n^N} \left| \mathbb{E}_{\mu^\gamma} \prod_{i=1}^{n} \sigma_{t\lambda(\gamma)^{-2}}(x_i) - m_\beta^n \prod_{x_i \in \mathcal{N}_t^\gamma} (-1) \right| = O(\lambda),$$

with the limit locally uniform in t, where $N_t^\gamma = \{x \in \mathbb{Z}^N : (\gamma\lambda(\gamma)x, t) \in \Omega_t^-\}$.

Theorem 6.3 is proved in [KS3]. A similar theorem but for the case of isentropic interactions and Glauber dynamics was obtained earlier in [KS2] for all times and in [DOPT1] as long as the resulting motion is smooth and for a particular choice of $\lambda(\gamma)$. An analogous result, but for nearest neighbor interaction Ising models in two dimensions, where the curvature motion is always smooth, was obtained by Spohn [Sp2]. When the dynamics of nearest neighbor models are coupled with Kawasaki dynamics at infinite temperature, which delocalize the interaction, the macroscopic limit, which turns out to be motion by mean curvature, was studied by Bonaventura [Bo] for smooth motions, i.e., for small time and a particular $\lambda(\gamma)$, and by Katsoulakis and Souganidis [KS1] in the generality of Theorem 6.3, even when fattening occurs. It should be noted that the mean field equation corresponding to the mesoscopic limit of Ising models with nearest neighbor spin flip Glauber dynamics and simple exchange (stirring) Kawasaki dynamics at infinite termpature is the Allen-Cahn equation (4.1) for an appropriate double-well potential W, as was proved by DeMasi, Ferrari and Lebowitz [DFL].

As mentioned earlier, in addition to its theoretical interest, Theorem 6.3 provides a justification, from microscopic considerations, to phenomenological theories of phase transitions like sharp-interface models derived by thermodynamic arguments ([Gu]) and sometimes by scaling of a Landau-Ginzburg model, i.e., reaction-diffusion equations. The theorem can also be thought of as providing a theoretical justification of the validity of some Monte Carlo-type methods, which have long been implemented in the physics literature, to approximate evolving fronts at any time. Notice that in this context the stochastic spin dynamics are unaffected by the possible appearance of singularities in the flow. A final remark about the importance of Theorem 6.3 is related to the discussion at the end of Part 3 of this Section and the form of $\theta(n)$. Indeed this result provides by microscopic considerations the "correct" form for the mobility of the interface, which depends, of course, on the dynamics but also on the equilibrium considerations via J, something which is occasionally neglected in the phenomenological theories, where the choice of the mobility appears to be sometimes arbitrary.

I conclude the discussion about Theorem 6.3 with a few words about its proof. At first glance it may appear that Theorem 6.3 follows by straightforward combinations of Theorems 6.1 and 6.2. This is far from being true since Theorem 6.1 does not provide any information at the appropriate time scale. The needed argument can be described as follows:

A careful study of the particle system involving a version of the local central limit theorem yields that the evolving spins are close, with respect to some γ-dependent seminorm and for short times, to the solution of (6.2), the error being, however, algebraic in γ. Unfortunately, this last observation eliminates the possibility of a straightforward discretization in time and then application of Theorem 6.2, since the errors add up! Instead one needs to consider two different fronts which move with velocities slightly larger and smaller than the desired ones and then to use these fronts as appropriate barriers to control the errors at each time step. It is exactly this choice of fronts and the need to control the errors that leads to the restriction on the scale $\lambda(\gamma)$.

References

[AC] S. M. Allen and J. W. Cahn, A macroscopic theory for antiphase boundary motion and its application to antiphase domain coarsening, *Acta Metallica* **27** (1979), 1085–1095.

[AGLM] L. Alvarez, F. Guichard, P.-L. Lions, and J.-M. Morel, Axioms and fundamental equations of image processing, *Arch. Rational Mech. Anal.* **123** (1992), 199-257.

[ACI] S. Angenent, D. L. Chopp, and T. Ilmanen, A computed example of nonuniqueness of mean curvature flow is $\mathbb{R}^3$, *Comm. Partial Differential Equations* **20** (1995), 1937–1958.

[AIV] S. Angenent, T. Ilmanen, and J. L. Velazquez, Nonuniqueness in geometric heat flows, preprint.

[AW] D. G. Aronson and H. Weinberger, Multidimensional nonlinear diffusion arising in population genetics, *Adv. Math.* **30** (1978), 33–76.

[Ba] G. Barles, Remark on a flame propagation model, Rapport INRIA 464 (1985).

[BBS] G. Barles, L. Bronsard, and P. E. Souganidis, Front propagation for reaction-diffusion equations of bistable type, *Anal. Nonlin.* **9** (1992), 479–506.

[BES] G. Barles, L. C. Evans, and P. E. Souganidis, Wavefront propagation for reaction-diffusion systems of PDE, *Duke Math. J.* **61** (1990), 835–858.

[BSS] G. Barles, H. M. Soner, and P. E. Souganidis, Front propagation and phase field theory, *SIAM J. Control Optim.* **31** (1993), 439–469.

[BS1] G. Barles and P. E. Souganidis, Convergence of approximation schemes for full nonlinear equations, *Asymptotic Anal.* **4** (1989), 271–283.

[BS2] G. Barles and P. E. Souganidis, A new approach to front propagation: Theory and applications, *Arch. Rational Mech. Anal.* **141** (1998), 237–296.

[BS3] G. Barles and P. E. Souganidis, A remark on the asymptotic behavior of the solution of the KPP equation, *C. R. Acad. Sci. Paris Sér. I Math.* **319** (1994), 679–684.

[BJ] E. N. Barron and R. Jensen, Semicontinuous viscosity solutions for Hamilton-Jacobi Equations with convex Hamiltonians, to appear in *Comm. Partial Differential Equations*.

[BFRW] P. Bates, P. Fife, X. Ren, and X. Wang, Traveling waves in a convolution model for phase transitions, preprint.

[BP1] G. Bellettini and M. Paolini, Two examples of fattening for the mean curvature flow with a driving force, *Math. Appl.* **5** (1994), 229–236.

[BP2] G. Bellettini and M. Paolini, Some results on minimal barriers in the sense of DeGiorgi to driven motion by mean curvature, *Rend. Accad. Naz. Sci. XL Mem. Mat. Appl. (5)* **19** (1995), 43–67.

[BP3] G. Bellettini and M. Paolini, Anisotropic motion by mean curvature in the context of Finsler geometry, *Hokkaido Math. J.* **25** (1996), 537–566.

[Bo] L. Bonaventura, Motion by curvature in an interacting spin system, preprint.

[Br] K. A. Brakke, *The Motion of a Surface by Its Mean Curvature*, Princeton University Press, Princeton, NJ, 1978.

[BrK] L. Bronsard and R. Kohn, Motion by mean curvature as the singular limit of Ginzburg-Landau model, *J. Differential Equations* **90** (1991), 211–237.

[Ca1] G. Caginalp, An analysis of a phase field model of a free boundary, *Arch. Rational Mech.* **92** (1986), 205–245.

[Ca2] G. Caginalp, Mathematical models of phase boundaries, in: *Material Instabilities in Continuum Mechanics and Related Mathematical Problems* (J. Ball, ed.), Clarendon Press, Oxford, 1988, 35–52.

[Ch] X.-Y. Chen, Generation and propagation of the interface for reaction-diffusion equation, *J. Differential Equations* **96** (1992), 116–141.

[CGG] Y.-G. Chen, Y. Giga, and S. Goto, Uniqueness and existence of viscosity solutions of generalized mean curvature flow equations, *J. Differential Geom.* **33** (1991), 749–786.

[Cr] M. G. Crandall, Viscosity solutions: a primer, in: *Viscosity Solutions and Applications* (I. Capuzzo Dolcetta and P.-L. Lions, eds.), Lecture Notes in Math. **1660**, Springer-Verlag, Berlin, 1997, 1–43.

[CIL] M. G. Crandall, H. Ishii, and P.-L. Lions, User's guide to viscosity solutions of second order partial differential equations, *Bull. Amer. Math. Soc.* **27** (1992), 1–67.

[CLS] M. G. Crandall, P.-L. Lions, and P. E. Souganidis, Universal bounds and maximal solutions for certain evolution equations, *Arch. Rational Mech. Anal.* **105** (1989), 163–190.

[D] E. DeGiorgi, Some conjectures on flow by mean curvature, *Proc. Capri Workshop 1990* (Benevento, Bruno, and Sbardone, eds.).

[DFL] A. DeMasi, P. Ferrari, and J. Lebowitz, Reaction-diffusion equations for interacting particle systems, *J. Statist. Phys.* **44** (1986), 589–644.

[DOPT1] A. DeMasi, E. Orlandi, E. Presutti, and L. Triolo, Glauber evolution with Kač potentials: I. Mesoscopic and macroscopic limits, interface dynamics, *Nonlinearity* **7** (1994), 633–696.

[DOPT2] A. DeMasi, E. Orlandi, E. Presutti, and L. Triolo, Motion by curvature by scaling non-local evolution equations, *J. Statist. Phys.* **73** (1993), 543–570.

[DOPT3] A. DeMasi, E. Orlandi, E. Presutti, and L. Triolo, Stability of the interface in a model of phase separation, *Proc. Royal Soc. Edinburgh Sect. A* **124** (1994), 1013–1022.

[DP] A. De Masi and E. Presutti, *Mathematical Methods for Hydrodynamic Limits*, Lecture Notes in Math. **1501**, Springer-Verlag, Berlin, 1991.

[DS] P. DeMottoni and M. Schatzman, Development of interfaces in $\mathbb{R}^N$, *Proc. Royal Soc. Edinburgh Sect. A* **116** (1990), 207–220.

[EMS1] P. F. Embid, A. Majda, and P. E. Souganidis, Effective geometric front dynamics for premixed turbulent combustion with separated velocity scales, *Combustion Sci. Techn.* **103** (1994), 85–115.

[EMS2] P. F. Embid, A. Majda and P. E. Souganidis, Comparison of turbulent flame speeds from complete averaging and the G-equation, *Phys. Fluids* **7** (1995), 2052–2060.

[EMS3] P. F. Embid, A. Majda, and P. E. Souganidis, Examples and counterexamples for Huygens Principle in premixed combustion, *Combustion Sci. Techn.* **120** (1996), 273–303.

[E1] L. C. Evans, The perturbed test function method for viscosity solutions of nonlinear PDE, *Proc. Royal Soc. Edinburgh Sect. A* **111** (1989), 359–375.

[E2] L. C. Evans, Regularity for fully nonlinear elliptic equations and motion by mean curvature, in: *Viscosity Solutions and Applications* (I. Capuzzo Dolcetta and P.-L. Lions, eds.), Lecture Notes in Math. **1660**, Springer-Verlag, Berlin, 1997, 98–133.

[ESS] L. C. Evans, H. M. Soner, and P. E. Souganidis, Phase transitions and generalized motion by mean curvature, *Comm. Pure Appl. Math.* **45** (1992), 1097–1123.

[ESo1] L. C. Evans and P. E. Souganidis, A PDE approach to geometric optics for certain reaction-diffusion equations, *Indiana Univ. Math. J.* **38** (1989), 141–172.

[ESo2] L. C. Evans and P. E. Souganidis, A PDE approach to certain large deviation problems for systems of parabolic equations, *Ann. Inst. H. Poincaré Anal. Non Linéaire* **6** *(Suppl.)* (1994), 229–258.

[ESp] L. C. Evans and J. Spruck, Motion of level sets by mean curvature I, *J. Differential Geom.* **33** (1991), 635–681.

[Fi] P. C. Fife, Nonlinear diffusive waves, CMBS Conf., Utah 1987, CMBS Conf. Series (1989).

[FM] P. C. Fife and B. McLeod, The approach of solutions of nonlinear diffusion equations to travelling solutions, *Arch. Rational Mech. Anal.* **65** (1977), 335–361.

[Fr1] M. Freidlin, *Functional Integration and Partial Differential Equations*, Ann. of Math. Stud. **109**, Princeton University Press, Princeton, NJ, 1985.

[Fr2] M. Freidlin, Limit theorems for large deviations of reaction-diffusion equations, *Ann. Probab.* **13** (1985), 639–675.

[FL] M. Freidlin and Y. T. Lee, Wave front propagation for a class of space nonhomogeneous reaction-diffusion systems, preprint.

[Ga] J. Gärtner, Bistable reaction-diffusion equations and excitable media, *Math. Nachr.* **112** (1983), 125–152.

[GS] Y. Giga and M.-H. Sato, Generalized interface evolution with Neumann boundary condition, *Proc. Japan Acad. Ser. A Math. Sci.* **67** (1991), 263–266.

[GT] D. Gilbarg and N. S. Trudinger, *Elliptic Partial Differential Equations of Second Order*, Grundlehren Math. Wiss. **224**, Springer-Verlag, New York, 1983.

[Go] S. Goto, Generalized motion of hypersurfaces whose growth speed depends superlinearly on curvature tensor, *J. Differential Integral Equations* **7** (1994), 323–343.

[G] W. Grabowski, Cumulus entrainment, fine scale mixing, and bouyancy reversal, *Quart. J. Roy. Meteorol. Soc.* **119** (1993), 935–956.

[GSm] W. Grabowski and P. Smolarkiewicz, Two-time level semi-Lagrangian modeling of precipitating clouds, *Monthly Weather Rev.* **124** (1996), 487–497.

[Gu] M. E. Gurtin, Multiphase thermodynamics with interfacial structure. 1. Heat conduction and the capillary balance law, *Arch. Rational Mech. Anal.* **104** (1988), 185–221.

[GSS] M. E. Gurtin, H. M. Soner, and P. E. Souganidis, Anisotropic motion of an interface relaxed by the formation of infinitesimal wrinkles, *J. Differential Equations* **119** (1995), 54–108.

[Il1] T. Ilmanen, The level-set flow on a manifold, in: *Differential Geometry: Partial Differential Equations on Manifolds* (R. Greene and S.-T. Yau, eds.), Proc. Sympos. Pure Math. **54**, Amer. Math. Soc., Providence, RI, 1993, 193–204.

[Il2] T. Ilmanen, Convergence of the Allen-Cahn equation to Brakke's motion by mean curvature, *J. Differential Geom.* **38** (1993), 417–461.

[Is1] H. Ishii, Hamilton-Jacobi equations with discontinuous Hamiltonians on arbitrary open sets, *Bull. Fac. Sci. Engrg. Chuo Univ. Ser. I Math.* **26** (1985), 5–24.

[Is2] H. Ishii, Parabolic pde with discontinuities and evolution of interfaces, *Adv. Differential Equations* **1** (1996), 51–72.

[IPS] H. Ishii, G. Pires, and P. E. Souganidis, Threshold dynamics type approximation schemes for propagating fronts, *J. Math. Soc. Japan* **51** (1999), 267–308.

[IS] H. Ishii and P. E. Souganidis, Generalized motion of noncompact hypersurfaces with velocity having arbitrary growth on the curvature tensor, *Tohoku Math. J.* **47** (1995), 227–250.

[JLS] R. Jensen, P.-L. Lions, and P. E. Souganidis, A uniqueness result for viscosity solutions of second-order fully nonlinear pde's, *Proc. Amer. Math. Soc.* **102** (1988), 975–978.

[J] R. Jerrard, Fully nonlinear phase transitions and generalized mean curvature motions, *Comm. Partial Differential Equations* **20** (1995), 223–265.

[KKR] M. Katsoulakis, G. Kossioris, and F. Reitich, Generalized motion by mean curvature with Neumann conditions and the Allen-Cahn model for phase transitions, to appear in *J. Geom. Anal.*.

[KS1] M. Katsoulakis and P. E. Souganidis, Interacting particle systems and generalized mean curvature evolution, *Arch. Rational Mech. Anal.* **127** (1994), 133–157.

[KS2] M. Katsoulakis and P. E. Souganidis, Generalized motion by mean curvature as a macroscopic limit of stochastic Ising models with long range interactions and Glauber dynamics, *Comm. Math. Phys.* **169** (1995), 61–97.

[KS3] M. Katsoulakis and P. E. Souganidis, Stochastic Ising models and anisotropic front propagation, to appear in *J. Statist. Phys.*.

[K] I. C. Kim, Singular limits of chemotaxis — growth model, *Nonlinear Anal.* **46** (2001), 817–834.

[LW] A. Linan and F. Williams, *Fundamental Aspects of Combustion*, Oxford University Press, 1993, Chapters 3 and 5.

[LP] J. Lebowitz and O. Penrose, Rigorous treatment of the Van der Waals Maxwell theory of the liquid vapour transition, *J. Math. Phys.* **98** (1966), 98–113.

[Lig] T. Liggett, *Interacting Particle Systems*, Grundlehren Math. Wiss. **276**, Springer-Verlag, New York, 1985.

[LPV] P.-L. Lions, G. Papanicolaou, and S. R. S. Varadhan, Homogenization of the Hamilton-Jacobi equations, preprint.

[LS1] P.-L. Lions and P. E. Souganidis, Viscosity solutions of fully nonlinear stochastic partial differential equations, *C. R. Acad. Sci. Paris Sér. I Math.* **326** (1998), 1085–1092.

[LS2] P.-L. Lions and P. E. Souganidis, Fully nonlinear stochastic partial differential equations: Nonsmooth equations and applications, *C. R. Acad. Sci. Paris Sér. I Math.* **327** (1998), 735–741.

[LS3] P.-L. Lions and P. E. Souganidis, Uniqueness of weak solutions for fully nonlinear stochastic partial differential equations, *C. R. Acad. Sci. Paris Sér. I Math.* **331** (2000), 783–790.

[LS4] P.-L. Lions and P. E. Souganidis, Fully nonlinear stochastic partial differential equations with semilinear stochastic dependence, *C. R. Acad. Sci. Paris Sér. I Math.* **331** (2000), 617–624.

[MS1] A. Majda and P. E. Souganidis, Large scale front dynamics for turbulent reaction-diffusion equations with separated velocity scales, *Nonlinearity* **7** (1994), 1–30.

[MS2] A. Majda and P. E. Souganidis, Bounds on enhanced turbulent flame speeds for combustion with fractal velocity fields, *J. Statist. Phys.* **83** (1996), 933–954.

[MS3] A. Majda and P. E. Souganidis, Flame fronts in a turbulent combustion model with fractal velocity fields, *Comm. Pure Appl. Math.* **51** (1998), 1337–1348.

[MS4] A. Majda and P. E. Souganidis, The effect of turbulence on mixing in prototype reaction diffusion systems, *Comm. Pure Appl. Math.* **53** (2000), 1284–1304.

[NPV] R. H. Nochetto, M. Paolini, and C. Verdi, Optimal interface error estimates for the mean curvature flow, *Ann. Scuola Norm. Sup. Pisa Cl. Sci. (4)* **21** (1994), 193–212.

[OJK] T. Ohta, D. Jasnow, and K. Kawasaki, Universal scaling in the motion of random intervaces, *Phys. Rev. Lett.* **49** (1982), 1223-1226.

[OsS] S. Osher and J. Sethian, Fronts moving with curvature dependent speed: Algorithms based on Hamilton-Jacobi equations, *J. Comput. Phys.* **79** (1988), 12–49.

[RSK] J. Rubinstein, P. Sternberg, and J. B. Keller, Fast reaction, slow diffusion and curve shortening, *SIAM J. Appl. Math.* **49** (1989), 116–133.

[Son1] H. M. Soner, Motion of a set by the curvature of its boundary, *J. Differential Equations* **101** (1993), 313–372.

[Son2] H. M. Soner, Ginzburg-Landau equation and motion by mean curvature, I: Convergence, *J. Geom. Anal.* **7** (1997), 437–475, II to appear *ibid.*

[SonS] H. M. Soner and P. E. Souganidis, Uniqueness and singularities of rotationally symmetric domains moving by mean curvature, *Comm. Partial Differential Equations* **18** (1993), 859–894.

[Sor] P. Soravia, Generalized motion of a front along its normal direction: A differential games approach, *Nonlinear Anal.* **22** (1994), 1242–1262.

[SorS] P. Soravia and P. E. Souganidis, Phase field theory for FitzHugh-Nagumo type systems, *SIAM J. Math. Anal.* **42** (1996), 1341–1359.

[Sou1] P. E. Souganidis, Front propagation: theory and application, in: *Viscosity Solutions and Applications* (I. Capuzzo Dolcetta and P.-L. Lions, eds.), Lecture Notes in Math. **1660**, Springer, Berlin, 1997, 186–242.

[Sou2] P. E. Souganidis, Stochastic Homogenization for Hamilton-Jacobi equations and applications, *Asymptotic Anal.* **20** (1999), 1–11.

[Sp1] H. Spohn, *Large Scale Dynamics of Interacting Particles*, Springer-Verlag, New York, 1991.

[Sp2] H. Spohn, Interface motion in models with stochastic dynamics, *J. Statist. Phys.* **71** (1993), 1081–1132.

[W] F. Williams, *Combustion Theory*, 2nd ed., Addison-Wesley.

[X] J. X. Xin, Existence and nonexistence of traveling waves and reaction-diffusion front propagation in periodic media, *J. Statist. Phys.* **73** (1993), 893–926.

Computing finite-time singularities in interfacial flows

Thomas P. WITELSKI
Department of Mathematics
Duke University
Durham, NC 27708-0320
USA

Abstract

Finite-time singularities occurring in mathematical models of free-surface flows indicate that important qualitative changes are taking place; for problems in solid and fluid mechanics this includes topological transitions—blow-up and pinch-off. For many problems, the dynamics leading to the formation of such singularities are described by self-similar solutions of the governing nonlinear partial differential equations. We present an analytical and numerical study of these similarity solutions and discuss their stability.

1 Introduction

Free-surface flows are among the most challenging areas of research in the physical sciences. They are important for a large number of industrial and technological processes such as coating flows, film drainage, fluid jetting, and droplet formation. Direct computation of the full governing equations for these problems is intractable in most cases, and even when possible, large-scale numerical simulations do not necessarily provide understanding of the fundamental physical mechanisms at work. In contrast, for limiting cases of these problems, asymptotic models have yielded valuable new insights. These mathematical models retain much of the rich dynamics of the physical systems, but take more tractable forms, spurring advances in analysis and computation.

One of the most dramatic behaviors that these models can exhibit is the formation of finite-time singularities. The occurrence of finite-time singularities in mathematical models indicates important qualitative changes taking place in the physical systems; for problems in solid and fluid dynamics this includes changes in topology: blow-up and pinch-off. Such fundamental structural changes are often just the first stage in the development of more rich and complex behavior in these physical systems. However, for the mathematical models blow-up (divergence of the solution, $h \to \infty$), or other singular limits, like pinch-off (extinction of the solution, $h \to 0$) may signal the break-down of the model. Analysis and computation of finite-time singularities has become a major area of research in pure and applied mathematics.

Singularities take place in a wide variety of areas in continuum mechanics [59, 91] including problems in fluid dynamics [44, 70], phase transitions, nonlinear optics [29], population dynamics, and combustion. In many cases, the formation of singularities is connected to

A. Bourlioux and M.J. Gander (eds.), Modern Methods in Scientific Computing and Applications, 451–487.

fundamental instabilities in the problems, such as the Rayleigh-Taylor or Kelvin-Helmholtz instabilities [41]. Earlier studies of these problems have yielded extensive results on complex behavior in the near-equilibrium regime: pattern formation, stability, bifurcation, and weakly nonlinear dynamics [38, 41]. The analysis of finite-time singularities is one of the new challenges in studying strongly nonlinear behavior for these problems. Surface tension driven Rayleigh-Benard convection,as studied in experiments by VanHook et al. [93, 94], is an interesting example of a new problem where both pattern formation and finite-time rupture singularities can co-exist.

In many problems the formation of singularities occurs as a result of self-similar dynamics driving the system to a singular limit [3, 59]. Such behavior has been studied thoroughly in model problems for blow-up singularities in reaction-diffusion equations [48, 50, 51, 69]. We build on this earlier body of research and extend it to higher order nonlinear problems connected with surface tension driven flows [9, 34, 43, 44, 75, 79, 82, 88] and other emerging problems in applied mathematics [63].

In Section 2 we introduce two model problems that are used as examples for our discussion of self-similar solutions for finite-time singularities. Section 3 begins with the analysis of instabilities in these problems in the near-equilibrium regime. Section 4 gives a thorough presentation of self-similar solutions leading to rupture singularities. Finally, Sections 5 and 6 conclude with a discussion of multi-dimensional problems and other extensions. The common theme herein is the development of numerical methods that build upon a fundamental implicit time-stepping scheme to carry out computations for all aspects of the problem [92].

2 Two case studies

Our study is directed at problems that have finite-time singularities and are governed by nonlinear parabolic evolution equations of the form

$$\frac{\partial h}{\partial t} = \mathcal{R}(h), \tag{2.1}$$

where $\mathcal{R}(h)$ is a nonlinear spatial operator, including appropriate boundary conditions. To make the presentation concrete, we focus on the formation of finite-time rupture singularities ($h \to 0$) in two model problems: one in solid mechanics, the other in fluid dynamics. The solid mechanics problem, pinch-off in surface diffusion [9], is a fourth-order extension of second-order geometric evolution equations for motion by mean curvature [31, 47, 56]. Similarly, the fluid dynamics problem, rupture in unstable lubrication flows, is a higher-order analogue to blow-up in second-order reaction-diffusion equations [48, 50, 51, 69, 85]. Both of these classes of second-order problems are classical models in the study of singularity formation. We investigate the dynamics of the more challenging fourth-order problems using a combination of numerical, analytical and asymptotic techniques. In studying these two problems side by side, we hope to present a unified description of self-similar singularities applicable to both of these fundamental fields.

2.1 Pinch-off of axisymmetric surfaces via surface diffusion

Surface diffusion is a process for mass transfer in solids that is driven by the shape of the solid's surface. This process plays a key role in describing problems such as the formation of pores in crystals [64], viscous sintering, and thermal grooving of materials [73]. The detailed physics of mass transport on surfaces are reviewed in [30, 74]. The mathematical formulation of surface diffusion is given by a geometric evolution equation, that is, the evolution of the solid is determined by geometric properties its surface [31]. Specifically, in surface diffusion the normal velocity of points on the smooth surface $\vec{\Gamma}(t)$ is proportional to the Laplacian of the surface curvature,

$$\frac{d\vec{\Gamma}}{dt} = -\hat{n}\nabla_s^2\mathcal{H}, \tag{2.2}$$

where $\mathcal{H}$ is the sum of the principal curvatures of the surface and is twice the local mean curvature [9]. Surface diffusion preserves the volume of the solid bounded by Γ while reducing its surface area,

$$A = \int_\Gamma dS, \qquad \frac{dA}{dt} = -\int_\Gamma |\nabla_s\mathcal{H}|^2\, dS \le 0. \tag{2.3}$$

Thus, the surface area A plays the role of an energy functional that is monotonically dissipated throughout the evolution. The study of the dynamics given by (2.2) for general surfaces is a formidable challenge. We will separate out many of the technical difficulties associated with studying intricate surfaces [54] by restricting our attention to surface diffusion on axisymmetric surfaces, a problem first studied by Nichols and Mullins [76] in connection with a number of applications. We will focus on the fact that equation (2.2) can form singularities in finite-time, see [9] and references therein. Singularities occur when the surface loses smoothness and the curvature can not be defined. This occurs when an axisymmetric surface pinches off; the surfaces necks down to a single point while the curvature diverges (see Fig. 1a). Indeed, this singularity is connected to a topology transition of the surface, as it changes from a single surface to two disconnected surfaces.

For axisymmetric bodies, the radial position of the bounding surface gives a direct measure of the thickness of the body. Consider a surface of revolution with the radius given as a function of the axial position on a finite interval $0 \le x \le L$, $r = h(x,t)$; the total curvature $\mathcal{H}$ for such a surface is given by

$$\mathcal{H} \equiv \kappa_1 + \kappa_2, \qquad \kappa_1 = \frac{1}{h\sqrt{1+h_x^2}}, \qquad \kappa_2 = -\frac{h_{xx}}{(1+h_x^2)^{3/2}}, \tag{2.4}$$

where κ_1 is the "hoop" or azimuthal curvature and κ_2 is the longitudinal curvature. Then (2.2) becomes a fourth-order nonlinear parabolic PDE,

$$\frac{\partial h}{\partial t} = \frac{1}{h}\frac{\partial}{\partial x}\left(\frac{h}{\sqrt{1+h_x^2}}\frac{\partial}{\partial x}\left\{\frac{1}{h\sqrt{1+h_x^2}} - \frac{h_{xx}}{(1+h_x^2)^{3/2}}\right\}\right). \tag{2.5}$$

For this simplified problem, the dissipation of surface area (2.3) takes the form,

$$A = 2\pi\int_0^L h\sqrt{1+h_x^2}\, dx, \qquad \frac{dA}{dt} = -2\pi\int_0^L \mathcal{H}_x^2 \frac{h\, dx}{\sqrt{1+h_x^2}} \le 0. \tag{2.6}$$

Using numerical simulations of (2.5) subject to periodic boundary conditions, Coleman, Falk, and Moakher [32, 33] demonstrated that for appropriate initial data, a pinch-off singularity can occur in finite time. Moreover, Wong, Miksis, Voorhees, and Davis [101] focused attention on the nature of the dynamics of pinch-off by obtaining a self-similar solution leading to the singularity. Here we review these results and study the finite-time similarity solutions of this problem further [9].

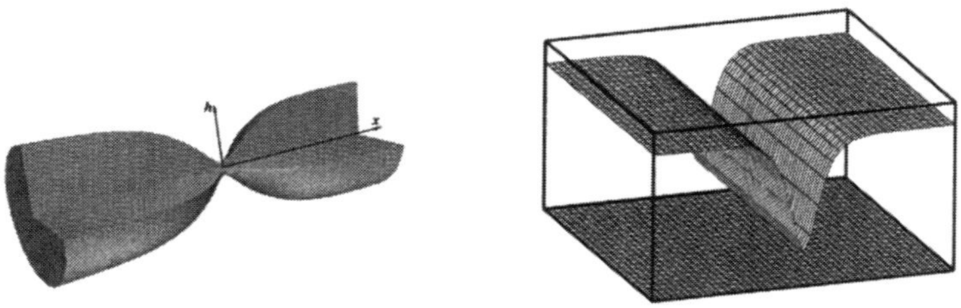

Figure 1: Illustrations of early stages in the development of finite-time pinch-off occurring in axisymmetric surface diffusion (left) and line rupture for van der Waals driven thin film rupture (right).

2.2 Van der Waals driven rupture of fluid films

Instabilities occurring in thin layers of fluid can lead to the formation of finite-time rupture singularities that change the flow topology and produce holes or "dry spots" in the fluid film [18, 55, 84, 102]. This behavior is of great significance in industrial coating processes, where non-uniformities are very undesirable [72, 79, 83]. For low Reynolds number flow of a thin, slowly-varying layer of a viscous fluid on a solid substrate, lubrication theory can be invoked to derive an equation for the evolution of the height of the free surface of the fluid, $h = h(x, y, t)$ [77, 79]. The fluid velocity, $\mathbf{u}$, is obtained from a balance of viscous and pressure gradient terms, subject to appropriate boundary conditions. Using the depth-averaged mass flux in the continuity equation yields

$$\frac{\partial h}{\partial t} + \nabla \cdot \left(\int_0^h \mathbf{u} \, dz \right) = 0. \tag{2.7}$$

Expressing the velocity $\mathbf{u}$ in terms of the gradient of the local pressure produces the nondimensionalized lubrication equation

$$\frac{\partial h}{\partial t} - \nabla \cdot (h^3 \nabla p) = 0. \tag{2.8}$$

To complete the model we need a constitutive relation to describe the pressure $p(x, y, t)$ in terms of the film thickness. Following the work of Williams and Davis [97], we model the pressure as a sum of contributions from a disjoining pressure due to attractive van der Waals forces and a linearized curvature term corresponding to surface tension effects,

$$p = \tfrac{1}{3} h^{-3} - \nabla^2 h. \tag{2.9}$$

Using this form of the pressure, equation (2.8) yields the fourth-order nonlinear lubrication model for $h(x, y, t)$,

$$\frac{\partial h}{\partial t} = \nabla \cdot \left(h^3 \nabla \left[\tfrac{1}{3} h^{-3} - \nabla^2 h \right] \right). \tag{2.10}$$

This equation has an energy integral given in terms of gradients of the surface and van der Waals contributions,

$$\mathcal{E} = \int \tfrac{1}{2}|\nabla h|^2 - \tfrac{1}{6}h^{-2}\, dA. \tag{2.11}$$

This energy is dissipated until all pressure gradients in the solution vanish,

$$\frac{d\mathcal{E}}{dt} = -\int h^3|\nabla p|^2\, dA \leq 0. \tag{2.12}$$

Since the original work of Williams and Davis, equation (2.10) and related models have been used in numerous studies of linear and nonlinear stability of liquid films and have included other effects like evaporation, surface reactions, inertia, and repulsive van der Waals forces [28, 78, 86, 87, 93, 94]. More complete references are given in recent review articles [44, 75, 79], and see [99].

Equation (2.10) can also be written as

$$\frac{\partial h}{\partial t} = -\nabla \cdot (h^{-1}\nabla h) - \nabla \cdot (h^3 \nabla\nabla^2 h). \tag{2.13}$$

This is a member of a class of problems called unstable generalized thin film equations. Bertozzi and Pugh [14, 15, 16, 17] have done analysis of the behavior of contact lines, instabilities, and weak solutions of the nonlinear degenerate parabolic PDE,

$$\frac{\partial h}{\partial t} = -\nabla \cdot (h^m \nabla h) - \nabla \cdot (h^n \nabla\nabla^2 h), \tag{2.14}$$

and Laugesen and Pugh have done extensive studies of the properties of the equilibria for this class of PDE [65, 66, 67, 68]. Equations of this form occur as models of lubrication flows in many problems, for example, with $n = m = 3$, (2.14) is a model for fluid dripping from a ceiling, see [16].

In (2.13), the first term on the right gives the influences of the van der Waals forces, $\mathcal{V}(h) \equiv -\nabla \cdot (h^{-1}\nabla h)$, while the second term corresponds to surface tension effects, $\mathcal{S}(h) \equiv -\nabla \cdot (h^3 \nabla\nabla^2 h)$. In [99] this model was studied for a fluid layer confined in a small shallow container, with no flux and no meniscus boundary conditions,

$$\mathbf{n} \cdot \nabla p = 0, \qquad \mathbf{n} \cdot \nabla h = 0. \tag{2.15}$$

We will focus on (2.13) specialized to two fundamental geometries (see Fig. 2)—one-dimensional "line rupture" ($d = 1$), with $h = h(x,t)$ governed by

$$h_t = -(h^{-1}h_x)_x - (h^3 h_{xxx})_x, \tag{2.16}$$

and axisymmetric "point rupture" ($d = 2$), with $h = h(r,t)$ governed by

$$\frac{\partial h}{\partial t} = -\frac{1}{r}\frac{\partial}{\partial r}\left(\frac{r}{h}\frac{\partial h}{\partial r}\right) - \frac{1}{r}\frac{\partial}{\partial r}\left(rh^3 \frac{\partial}{\partial r}\left\{\frac{1}{r}\frac{\partial}{\partial r}\left[r\frac{\partial h}{\partial r}\right]\right\}\right). \tag{2.17}$$

Zhang and Lister [103] have studied rupture in this problem associated with a singularity forming in (2.10) as $h \to 0$ at a finite time t_c. They showed that similarity solutions leading to rupture exist for both planar line rupture and axisymmetric point rupture. They found the beginnings of a discrete, countably infinite set of solutions, the first of which was observed in direct simulations of the PDE (2.10). Here we review these results and the stability analysis of the similarity solutions given by Witelski and Bernoff [98, 99].

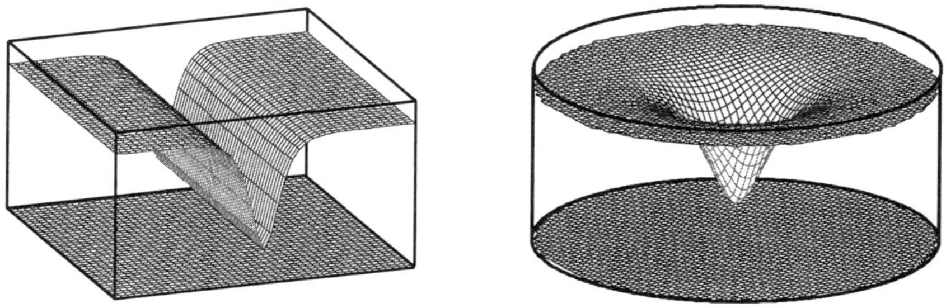

Figure 2: The geometry of (left) line rupture, $d = 1$, and (right) axisymmetric point rupture, $d = 2$, in van der Waals driven thin films (2.10).

3 Instabilities in near-equilibrium dynamics

While rigorous proof of blow-up in these fourth-order nonlinear equations can be of great technical challenge [17], it is straightforward to note the presence of strongly destabilizing terms in these PDEs. These terms can destabilize equilibrium solutions and lead to complicated dynamics, providing a potential mechanism for singularity formation [88]. Stabilizing and destabilizing contributions in (2.1) can be examined in terms of the dominance of different physical effects for different limits in the underlying models. We now illustrate these effects for problems (2.5) and (2.16).

In the surface diffusion problem, to leading order, for a very slender, slowly-varying cylinder the curvature is dominated by the hoop curvature, $\mathcal{H} \sim \kappa_1 \sim 1/h$, and (2.5) reduces to

$$\frac{\partial h}{\partial t} \sim -\frac{1}{h}\frac{\partial}{\partial x}\left(\frac{1}{h}\frac{\partial h}{\partial x}\right). \tag{3.1}$$

This is a second-order nonlinear backward heat equation, which is ill-posed and leads to potentially unbounded growth of gradients of the solution. Hence, the solution will not remain slowly-varying in space, and therefore the influence of higher-gradients terms is crucial. If we consider surface diffusion evolution for a finite-radius cylinder with variations in the radius on very short length-scales, then the curvature is dominated by the axial curvature, $\mathcal{H} \sim \kappa_2 \sim -h_{xx}$, and (2.5) reduces to

$$\frac{\partial h}{\partial t} \sim -\frac{1}{h}\frac{\partial}{\partial x}\left(h\frac{\partial^3 h}{\partial x^3}\right). \tag{3.2}$$

This is a fourth-order nonlinear diffusion equation that is well-posed for positive solutions and serves to regularize the influence of (3.1).

Similarly for thin films governed by (2.16), sources of instability and regularization can be directly tied to different physical effects. In the limit of a slowly-varying, very thin coating flow, van der Waals forces will dominate and, to leading order, (2.16) reduces to

$$\frac{\partial h}{\partial t} \sim -\frac{\partial}{\partial x}\left(\frac{1}{h}\frac{\partial h}{\partial x}\right). \tag{3.3}$$

Like (3.1), this is an ill-posed nonlinear diffusion equation with unstable amplification of short-wavelength perturbations to the solution. Fortunately, for any finite film thickness, the influence of surface tension,

$$\frac{\partial h}{\partial t} \sim -\frac{\partial}{\partial x}\left(h^3 \frac{\partial^3 h}{\partial x^3}\right), \tag{3.4}$$

will eventually dominate to regularize these instabilities.

In (2.5), (2.16), and more general problems of the form (2.1), the presence of destabilizing terms in the governing equation (2.1) can yield non-trivial dynamics, depending on the regularizing influence of the other terms in the equation:

(a) If there are no higher-order regularizing terms, then the problem is ill-posed and the solution can break down on arbitrarily short time-scales.

(b) If there is sufficiently strong regularization, then singularities will be suppressed and a solution will exist for all times; this is sometimes called sub-critical behavior [17].

(c) In other cases, called critical or super-critical behavior, the destabilizing and regularizing influences can balance. The solution may not exist forever, but it will not break-down in an "uncontrolled" manner. The formation of singularities in finite-time occurs as the limit of a well-defined dynamic process balancing the regularizing and destabilizing effects. This is the focus of this paper.

3.1 Numerical methods for nonlinear parabolic PDEs

While analysis of the full problem (2.1), or more amenable simplified models, can lead to a lot of insight into these problems, ultimately these findings must be supported by results from numerical simulations of the full problem. Analysis of singularities in second-order nonlinear PDEs has lead to a thorough understanding of many aspects of those problems [2, 49, 69, 85]. For higher-order equations, fewer analytical techniques are available [10, 11] and direct numerical simulations necessarily play a more important role. We begin by introducing a standard framework for numerical solution of well-posed PDE problems [1]. This approach will be applicable for finite times before the formation of the singularity, and later we will discuss how to extend it up to the singularity time.

For ease of presentation, we will focus on the simplest numerical method – finite differences, but most of the discussion will be equally applicable to other numerical methods for PDEs, such as finite element and spectral methods. In all of these numerical methods, we calculate the evolution of a discretized approximation of the continuous solution $h(x,t)$ on $0 \le x \le L$ for $t \ge 0$. In finite difference methods, we represent the discretized solution at each discrete time as a vector, $\mathbf{h}^m \in \mathbb{R}^{N+1}$, where

$$h(x_n, t^m) \approx \mathbf{h}_n^m, \qquad x_n = n\Delta x, \qquad t^m = m\Delta t, \tag{3.5}$$

with $\Delta x = L/N$, where $n = 0, 1, 2, \ldots, N$ and $m = 0, 1, 2, \ldots$. Practical numerical solution of stiff problems for parabolic PDEs like (2.5) and (2.16) requires the use of implicit time-stepping methods [58]. The use of an explicit method for fourth-order problems such as (2.5) and (2.16)) would yield a stability constraint that forces the use of a very small time-step $\Delta t = O(\Delta x^4)$. The simplest implicit method for (2.1) is the backward Euler scheme,

$$\frac{\mathbf{h}_n^{m+1} - \mathbf{h}_n^m}{\Delta t} = \mathcal{R}_n(\mathbf{h}^{m+1}), \quad n = 0, 1, 2, \ldots, N, \tag{3.6}$$

with the initial condition $\mathbf{h}_n^0 \approx h(x_n, 0)$. Here we use $\mathcal{R}(\mathbf{h})$ to represent an appropriate discretization of the spatial operator in (2.1), including boundary conditions. This method is only first order accurate in time, but for our purposes it is sufficient; higher order schemes can be treated similarly. With only minor changes, (3.6) could be made into a second-order accurate Crank-Nicholson scheme.

Solution of (3.6) at each time-step involves a system of $(N+1)$ nonlinear algebraic equations for $\mathbf{h}^{m+1}$, namely

$$\mathcal{F}(\mathbf{h}^{m+1}) \equiv \mathbf{h}^{m+1} - \mathbf{h}^m - \Delta t \mathcal{R}(\mathbf{h}^{m+1}) = \mathbf{0}. \tag{3.7}$$

We use Newton's method to solve $\mathcal{F}(\mathbf{h}^{m+1}) = \mathbf{0}$, that is, we calculate successive corrections, denoted by $\boldsymbol{\epsilon}_{(k)}$ for $k = 0, 1, 2, \ldots$, to an initial guess for the solution, $\mathbf{h}_{(0)}^{m+1}$, from

$$\mathbf{J}_{(k)} \boldsymbol{\epsilon}_{(k)} = -\mathcal{F}(\mathbf{h}_{(k)}^{m+1}), \tag{3.8}$$

where $\mathbf{J}$ is the Jacobian matrix for the system. This operator is given by the derivative of the discretized system (3.7), or equivalently in terms of a discretization of the functional derivative of equation (2.1),

$$\mathbf{J}_{(k)} \equiv \frac{\delta \mathcal{F}(\mathbf{h}_{(k)}^{m+1})}{\delta \mathbf{h}^{m+1}} = \mathbf{I} - \Delta t \frac{\delta \mathcal{R}}{\delta h}. \tag{3.9}$$

Under appropriate conditions on $\mathbf{h}_{(k)}^{m+1}$ [60], the corrections $\boldsymbol{\epsilon}_{(k)}$ produced by (3.8) will yield quadratic convergence to the solution of (3.7), $\mathbf{h}_{(k)}^{m+1} \to \mathbf{h}^{m+1}$,

$$\mathbf{h}_{(k+1)}^{m+1} = \mathbf{h}_{(k)}^{m+1} + \boldsymbol{\epsilon}_{(k)}, \quad k = 0, 1, 2, \ldots, \tag{3.10}$$

starting from the initial guess, $\mathbf{h}_{(0)}^{m+1} = \mathbf{h}^m$. While the calculation of the functional derivative of (2.16) and especially (2.5) may appear algebraically cumbersome, this process can be completely automated through the use of symbolic algebra programs (such as Maple or Mathematica) to calculate an analytical form of the Jacobian (3.9) in terms of derivatives of the finite difference equations (3.7).

For one-dimensional problems, finite difference discretizations yield a sparse banded structure for the Jacobian matrix. The linear algebra for the solution of each iteration of (3.8) can be done directly using an *LU* factorization, requiring $O(N)$ work and memory-space for the calculation of each time-step. We will return to the basic material constructed here several times in the course of this article as a key building block [92] allowing us to calculate linear stability of equilibria, self-similar solutions, their stability, and the dynamics of multi-dimensional problems.

3.2 The role of unstable equilibrium solutions

One of the most fundamental questions for problems of the form (2.1) is whether or not singularities will develop from given initial data [49]. In general this is a difficult question, but for initial data given by small perturbations from equilibria, we can gain some insight from linear stability analysis.

While in general the equilibrium solutions $h = \bar{h}(x)$ of (2.1) are given by the solutions of a fourth-order nonlinear elliptic problem, $\mathcal{R}(\bar{h}) = 0$, for the problems considered here we can take advantage of the form of $\mathcal{R}$ to obtain the equilibria as the solutions of second-order problems. For the surface diffusion problem, the equilibria have the surface Laplacian of their mean curvature equal to zero, $\nabla_s^2 \mathcal{H}(\bar{h}) = 0$. Subject to the periodic boundary conditions considered in Section 2.1, this forces the equilibria to have constant mean curvature, $\mathcal{H}(\bar{h}(x)) = \bar{\mathcal{H}}$, (2.4), namely they solve the second order ordinary differential equation,

$$\frac{1}{\bar{h}\sqrt{1+\bar{h}_x^2}} - \frac{\bar{h}_{xx}}{(1+\bar{h}_x^2)^{3/2}} = \bar{\mathcal{H}}. \tag{3.11}$$

There are three classes of solutions of this equation (see Fig. 3a); (i) constant radius cylinders, $\bar{h}_c(x) = \underline{h}$, (ii) spheres, $\bar{h}_s(x) = \sqrt{\underline{h}^2 - x^2}$ for $|x| \leq \underline{h}$, and (iii) the Delaunay unduloids, $\bar{h}_u(x)$. The sphere and cylinder are actually limiting cases for the family of unduloid solutions. The cylinder has zero longitudinal curvature, $\kappa_2 \equiv 0$, the sphere has constant κ_2, and there is a one-parameter family of unduloids with non-uniform $\kappa_2(x)$ connecting these limits [9] (see Fig. 3b). In the sphere-limit, the minimum of $\bar{h}_u(x)$ on the domain becomes arbitrarily small, hence smooth, simply-connected equilibrium solutions exist all the way down to the singular limit where the topology change occurs, when $\min_x \bar{h} = 0$.

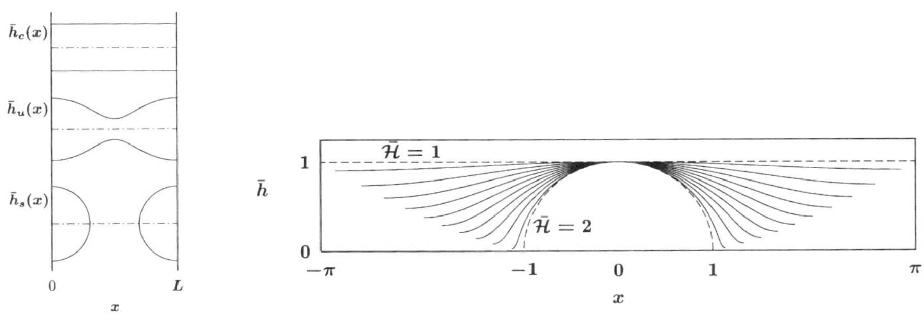

Figure 3: Constant mean curvature axisymmetric equilibria of the surface diffusion problem—(left) cross-sectional views of the cylinder, unduloid and sphere solutions, (right) these solutions as elements of a continuous family of solutions of (3.11).

For van der Waals driven thin films, the equilibria satisfy the fourth-order elliptic problem $\partial_x(\bar{h}^3 \partial_x p(\bar{h})) = 0$. Subject to the boundary conditions (2.15), any positive equilibrium solution $\bar{h}(x)$ must have a uniform constant pressure, $p(\bar{h}(x)) = \bar{p}$, (2.9),

$$\frac{1}{3\bar{h}^3} - \frac{d^2\bar{h}}{dx^2} = \bar{p}. \tag{3.12}$$

There are no solutions of (3.12) with disjoint support (with $h = 0$ over finite intervals) analogous to the $\bar{h}_s(x)$ sphere solution of the surface diffusion problem. Otherwise, there are very close parallels in the sets of equilibrium solutions; like the $\bar{h}_c(x)$ solutions, uniform flat films of any positive thickness are solutions, $\bar{h}(x) = \underline{h}$, and there is a one-parameter family of non-uniform solutions, $\bar{h}(x) = \bar{h}_u(x)$ (see Fig. 4). This family of solutions bifurcates from the flat film and ends in a singular limit (where $h = 0$ at one point), $\bar{h}_u(x) \sim C(1-(1-|x|)^2)$ [99].

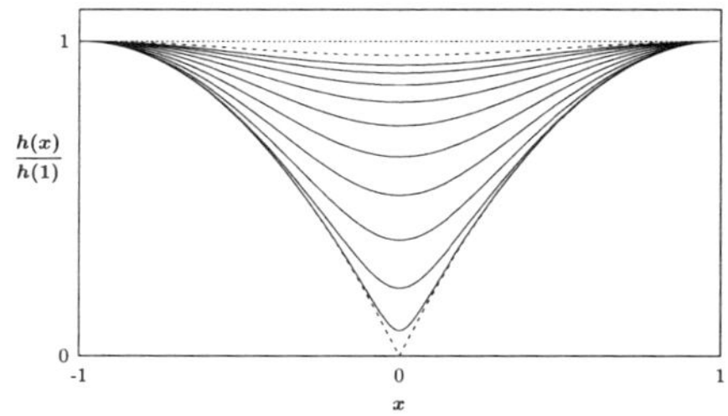

Figure 4: Constant pressure equilibrium solutions of the van der Waals driven thin film problem given as a continuous family of solutions of (3.12).

Since the solutions of problems (2.5) and (2.16) can exist in equilibrium very close to rupture, we must examine if the influence of small perturbations can drive the systems to the topology-change singularity. Asymptotic stability of an equilibrium solution $\bar{h}(x)$ with respect to infinitesimal perturbations can be investigated using the ansatz

$$h(x,t) = \bar{h}(x) + \epsilon \hat{h}(x) e^{\lambda t}, \quad \epsilon \ll 1. \tag{3.13}$$

Substituting into (2.1) and linearizing for $\epsilon \to 0$ yields the linear stability problem,

$$\lambda \hat{h} = \mathcal{L}(\bar{h})\hat{h}, \tag{3.14}$$

where the linear operator is $\mathcal{L} = \delta\mathcal{R}/\delta h$. For the constant equilibria, $\bar{h}(x) = \underline{h}$, the eigenvalues and eigenfunctions can be determined analytically [9, 13, 99]. For the nontrivial equilibria $\bar{h}_u(x)$, the spectrum must be calculated numerically; this can be accomplished using a simple extension of the numerical approach developed in Section 3.1. Since the equilibria solve $\mathcal{R}(\bar{h}) = 0$, we can apply Newton's method, with $\mathcal{F} = \mathcal{R}$, to solve the corresponding discretized problem to obtain $\bar{h}(x_n) \approx \bar{\mathbf{h}}_n$. After the solution has converged, the Jacobian matrix is a discretized representation of the linear operator, $\mathcal{L} \approx \bar{\mathbf{J}} = \delta\mathcal{R}(\bar{\mathbf{h}})/\delta\mathbf{h}$. Therefore the spectrum of the linear stability problem for $\bar{h}(x)$ can be determined numerically by calculating the eigenvalues and eigenvectors of the matrix $\bar{\mathbf{J}}$. The inverse power method [53, 81, 92] from numerical linear algebra can be efficiently used to calculate the first few modes in the spectrum [9, 98, 99].

The numerical calculations yield that the $\bar{h}_u(x)$ solutions are unstable. For the surface diffusion problem, we showed in [9] that the unduloids are unstable with respect to a single eigenmode, and the exponential growth rate for this instability increases for unduloids approaching the singular limit (the sphere). Therefore, in a restricted sense, the equilibria

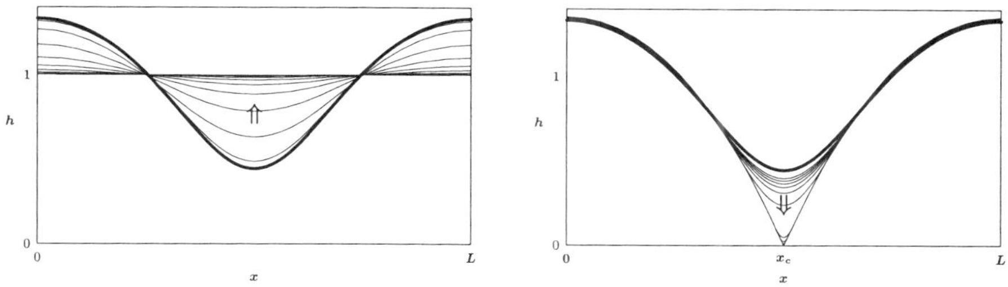

Figure 5: Evolution in the surface diffusion problem starting from opposite infinitesimal perturbations of the unstable equilibrium $\bar{h}_u(x)$; (left) evolution to the uniform cylinder, and (right) evolution to finite-time rupture.

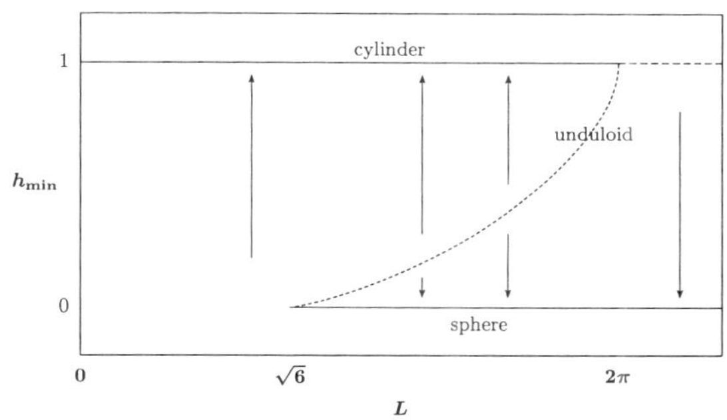

Figure 6: The equilibria and the near-equilibrium dynamics for axisymmetric surface diffusion as a function of the domain length L; solid curves show the regimes where the sphere and cylinder are stable, arrows indicate the evolution for solutions starting near the unstable unduloid equilibria.

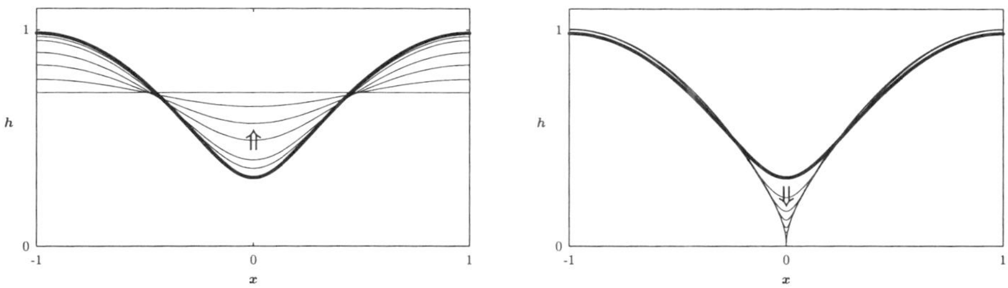

Figure 7: Evolution in the van der Waals thin film problem starting from opposite infinitesimal perturbations of the unstable equilibrium $\bar{h}_u(x)$; (left) evolution to the uniform film, and (right) evolution to finite-time rupture.

closer to rupture (with smaller $\min_x \bar{h}_u(x)$) are indeed "more unstable," since they destabilize faster.

To examine the nonlinear dynamics beginning near the unstable equilibria, we make use of direct numerical simulations of the PDEs. Fig. 5 shows numerical solutions of the surface diffusion PDE (2.5) starting from unduloids perturbed by infinitesimal perturbations of opposite signs. In both cases, the unduloid destabilizes and the solution evolves toward a stable equilibrium. Analysis shows that both the cylinder and sphere are linearly stable [9]. Fig. 5a shows convergence to the cylinder following evolution within the set of simply-connected solutions. In contrast, for the opposite perturbation, if the dynamics are to eventually lead to convergence to the stable, disconnected chain of spheres, then there must be a change in topology, where the cylinder pinches off, $h(x_c, t) \to 0$ (see Fig. 5b). This behavior is described further in [9], where for the near-equilibrium initial value problem, it is argued that the unstable equilibrium solutions serve to separate basins of attraction for the simply-connected and disconnected stable equilibria. This structure is illustrated schematically in Fig. 6. The projection of the perturbation from the unstable equilibrium solution onto the unstable eigenmode of $\bar{h}_u(x)$ can be used to predict if a rupture singularity will occur [9, 99],

$$\langle h(x,t) - \bar{h}_u(x), \hat{h}(x)\rangle \lessgtr 0 \quad \Longrightarrow \quad \text{rupture or not.} \tag{3.15}$$

The near-equilibrium dynamics in the thin film problem (2.16) have the same qualitative structure [99] (see Fig. 7).

Therefore, at least for a restricted class of initial data, we can answer the question of whether or not singularities will develop. It is notable that the answer depends sensitively on the form of the perturbation to the unstable equilibrium $\bar{h}_u(x)$. Drawing from the fact that the branch of unstable solutions connects to the uniform flat state $\bar{h}(x) = \underline{h}$ and extends to a singular limit with $\bar{h}(x_c) \to 0$ (see Figs. 3 and 4), certain conditions can lead to the following seemingly counter-intuitive behaviors:

(a) An infinitesimal perturbation can cause an $\bar{h}_u(x)$ solution that is arbitrarily close to rupture (with $\min_x \bar{h}_u(x) \to 0$) to redistribute its mass and converge to a uniform flat state, $h(x,t) \to \underline{h}$.

(b) An infinitesimal perturbation can cause an $\bar{h}_u(x)$ solution that is arbitrarily close to a stable uniform positive flat solution $\underline{h}$ (with $||\bar{h}_u(x) - \underline{h}|| \to 0$) to become unstable and rupture.

4 Self-similar dynamics

Having identified the possible destabilizing influences in our problems and having given a condition to determine if the evolution can lead to singularity formation, we now describe the details of the dynamics resulting in the finite-time singularity. In (2.5), (2.16), and a large class of other problems, these dynamics are given by self-similar solutions [3].

As described by Barenblatt [3], self-similar solutions occur in systems where, under appropriate rescalings of the dependent and independent variables, the structure of the solution remains unchanged as the system evolves towards a singular limit. In the context of problems with finite-time singularities, we will show that the structure of the solution as the singularity is approached takes a well-defined form that is independent of space and time scales. Our analysis of this behavior in fourth-order nonlinear PDEs follows the techniques for studying similarity solutions in analogous second-order problems given in the works of Barenblatt [3], Giga and Kohn [48, 50, 51], and others [26, 49].

4.1 Dimensional analysis and first-type similarity solutions

Dimensional analysis can be used to demonstrate the existence of possible similarity solutions leading to rupture in (2.5) and (2.16) by showing that the PDEs are scale-invariant. To make the description of the rupture problem more concrete, we assume that the time t_c and position x_c where the singularity will form are known, that is,

$$h(x_c, t) \to 0 \quad \text{as} \quad t \to t_c. \tag{4.1}$$

We then rescale the solution and the space and time variables in a neighborhood of the critical point (x_c, t_c),

$$h = \mathrm{H}\hat{h}, \qquad x - x_c = \mathrm{X}\hat{x}, \qquad t - t_c = \mathrm{T}\hat{t}. \tag{4.2}$$

Describing the overall structure of the rupture process then is reduced to obtaining relations between the lengthscale X and the timescale T of the solution and the height-scale H in the limit that $\mathrm{H} \to 0$. If such relations can be found that map the PDE (2.1) onto itself (in terms of the hat variables), then the PDE has a scale-invariant symmetry and supports a self-similar solution [42].

Applying the scalings (4.2) to the van der Waals problem yields

$$\left[\frac{\mathrm{H}}{\mathrm{T}}\right]\frac{\partial \hat{h}}{\partial \hat{t}} = -\left[\frac{1}{\mathrm{X}^2}\right]\frac{\partial}{\partial \hat{x}}\left(\frac{1}{\hat{h}}\frac{\partial \hat{h}}{\partial \hat{x}}\right) - \left[\frac{\mathrm{H}^4}{\mathrm{X}^4}\right]\frac{\partial}{\partial \hat{x}}\left(\hat{h}^3\frac{\partial^3 \hat{h}}{\partial \hat{x}^3}\right). \tag{4.3}$$

Since we know that influences of the van der Waals forces and surface tension effects balance as $\mathrm{H} \to 0$, the scales of the two terms on the right side must be of the same order, $\mathrm{H} = \mathrm{X}^{1/2}$. Moreover, since the formation of the singularity is a fast dynamic process, the time derivative on the left must also be of the same order, hence $\mathrm{T} = \mathrm{X}^{5/2}$. Consequently we obtain the scalings $\mathrm{X} = \mathrm{T}^{2/5}$ and $\mathrm{H} = \mathrm{T}^{1/5}$. Exactly the same scalings result from the corresponding analysis of the radially symmetric rupture problem (2.17).

Similarly, in rescaling the surface diffusion problem, to retain the balance of destabilizing and regularizing terms in the curvature, we must scale $\mathrm{H} = \mathrm{X}$,

$$\mathcal{H} = \left[\frac{1}{\mathrm{H}}\right]\frac{1}{\hat{h}\sqrt{1+\hat{h}_{\hat{x}}^2}} - \left[\frac{\mathrm{H}}{\mathrm{X}^2}\right]\frac{\hat{h}_{\hat{x}\hat{x}}}{(1+\hat{h}_{\hat{x}}^2)^{3/2}} = \left[\frac{1}{\mathrm{H}}\right]\hat{\mathcal{H}}. \tag{4.4}$$

The time-derivative term can then be balanced with the spatial operator,

$$\left[\frac{\mathrm{H}}{\mathrm{T}}\right]\frac{\partial \hat{h}}{\partial \hat{t}} = \left[\frac{\mathrm{H}}{\mathrm{H}^2\mathrm{X}^2}\right]\frac{1}{\hat{h}}\frac{\partial}{\partial \hat{x}}\left(\frac{\hat{h}}{\sqrt{1+\hat{h}_{\hat{x}}^2}}\frac{\partial \hat{\mathcal{H}}}{\partial \hat{x}}\right), \tag{4.5}$$

to yield $\mathrm{T} = \mathrm{H}^4$. Consequently, we obtain the length-scales as $\mathrm{X} = \mathrm{T}^{1/4}$ and $\mathrm{H} = \mathrm{T}^{1/4}$.

We use these scaling relations to identify the form of the self-similar solution in terms of combinations of h, x, and t that are scale-invariant. From the scalings found above, note that as the rupture singularity is approached, H $\to 0$, the time-scale also vanishes, T $\to 0$. Therefore, without loss of generality, we can choose this time-scale to be the time remaining until rupture, $\tau \equiv t_c - t \geq 0$. For contrast, in infinite-time rupture solutions, as H $\to 0$ we approach T $\to \infty$. Writing the scaling relations for finite-time self-similar solutions in general as H $=$ T$^\alpha$, X $=$ T$^\beta$, we can then determine a scale-independent spatial variable η, and similarity solution profile $\bar{H}(\eta)$ in the form

$$h(x,t) = \tau^\alpha \bar{H}(\eta), \qquad \eta = (x - x_c)/\tau^\beta, \qquad \tau = t_c - t. \tag{4.6}$$

This is a very general ansatz and encompasses the forms of similarity solutions for a wide range of problems [3]. Here the scaling exponents (α, β) describe the qualitative behavior of the solution as $\tau \to 0$:

- If $\alpha < 0$ then the solution blows up, $h \to \infty$.
- If $\alpha > 0$ then the solution ruptures, $h \to 0$.
- If $\beta > 0$ then the length-scale of the solution vanishes; this is a focusing-type solution.
- If $\beta < 0$ then the length-scale of the solution diverges; this is a spreading-type solution.

We note that in the context of population dynamics in mathematical biology, rupture/pinch-off singularities describe local extinction events [85], where h represents a population density.

In both of the problems (2.5) and (2.16), we obtain focusing-type rupture solutions; for van der Waals driven rupture $\alpha = 1/5$ and $\beta = 2/5$,

$$h(x,t) = \tau^{1/5} \bar{H}(\eta), \qquad \eta = (x - x_c)/\tau^{2/5}, \tag{4.7}$$

and for the surface diffusion problem $\alpha = \beta = 1/4$,

$$h(x,t) = \tau^{1/4} \bar{H}(\eta), \qquad \eta = (x - x_c)/\tau^{1/4}. \tag{4.8}$$

In cases, like these, where all of the scaling exponents can be determined explicitly from dimensional analysis, the solutions are called *first-type* similarity solutions [3]. If at least one of the scaling parameters can not be determined from dimensional analysis, then the similarity solutions are said to be second-type [3]; for these problems the determination of the scaling exponents and the the solution profile $\bar{H}(\eta)$ takes the form of a nonlinear eigenvalue problem.

4.2 Observing self-similar behavior in PDE simulations

Having determined the form expected for the similarity solutions of our problems, we turn to the question of determining whether these results are born out by direct numerical simulations of the PDEs. For finite times before the formation of the singularity, these simulations can be carried out using the scheme described in Section 3.1. Moreover, with a few additional considerations, the basic numerical scheme can be adapted to follow the evolution up to times arbitrarily close to the singularity time t_c, allowing observation of the self-similar dynamics in detail.

In order to accurately resolve the structure of the solution as it evolves to a finite-time singularity, the numerical representation of the solution (3.5) must be modified to capture the influence of the changing length and time scales. First, an obvious flaw with using any constant time-step Δt, however small, in (3.6) is that eventually the discrete time will exceed the critical time, $t^m = m\Delta t > t_c$, for some value of m. For most problems, analysis of the PDE can show that no classical solution exists for $t > t_c$, so the numerical solution $\mathbf{h}^m$ will certainly be spurious. Moreover, a numerical solution from a scheme like (3.6) can be hoped to converge to the solution of the underlying continuous problem only when the change in the solution is small during each discrete time-step Δt. During the formation of a singularity, there will be significant changes in the solution and its gradients in a neighborhood of the singularity location x_c, as $t \to t_c$. Consequently, adaptive time-stepping, $t^{m+1} = t^m + \Delta t^m$, is needed as $t^m \to t_c$. For problems with finite-time rupture, we must adjust Δt^m to control the rate of decrease of the minimum of the solution over each time-step. One strategy is to select Δt^m so that the minimum of the solution decreases by no more than a controlled small fraction in each time-step. For example we can select Δt^m such that the solution satisfies

$$\frac{\min_n \mathbf{h}_n^{m+1}}{\min_n \mathbf{h}_n^m} \geq 1 - \delta, \quad \delta \ll 1. \tag{4.9}$$

To correctly represent the dynamics, the changes in spatial scales for the solution must also be considered—adaptive spatial regridding near x_c is also necessary. A simple approach to this problem is to use insight from the expected form of the self-similar solution (4.6) to adapt x_n to keep the profile $\bar{H}(\eta)$ well-resolved. We assume that the position where the singularity will occur, x_c, is known, and define $h_c(t) \equiv h(x_c, t)$. To accurately resolve the similarity solution $\bar{H}(\eta)$, in a neighborhood of x_c, we should discretize $h(x_n, t^m) \approx \mathbf{h}_n^m$ not in terms of uniform Δx, but in terms of uniform $\Delta \eta$, so that (4.6) implies the choice

$$x_n = x_c + n h_c^{\beta/\alpha}(t) \Delta \eta. \tag{4.10}$$

These are the simplest approaches; related ideas have been used for other fourth-order problems [8, 10, 12], and there is extensive literature on numerical methods for studying blow-up in second-order problems [2, 7, 22, 23, 25, 27, 61].

Once the issues connected with calculating the numerical solution of the PDE problem are resolved, we can turn to the question of how to present evidence of the self-similar behavior of the solution. The problem is that the similarity solution (4.6) is very sensitive to the value of t_c as $t \to t_c$, and t_c is a priori unknown. As will be shown later, plotting $\mathbf{h}/\tau^\alpha$ to demonstrate the presence of self-similar behavior is an inherently unstable process, and small errors in the value of t_c will lead to large errors in the graph as the singularity is approached. The resolution to this difficulty is to use (4.6) to write expressions derived from the solution that are independent of τ. For example, consider the evolution of the solution at the critical position x_c,

$$h_c(t) \equiv h(x_c, t) = \tau^\alpha \bar{H}(0), \qquad \frac{dh_c}{dt} = -\beta \tau^{\beta - 1} \bar{H}(0), \tag{4.11}$$

where $\bar{H}(0)$ is an as-yet unknown constant to be determined from the similarity solution. These two relations can be used to eliminate τ and yield an expected relation between $h_c(t)$ and its derivative as rupture is approached, $\tau \to 0$,

$$\frac{dh_c}{dt} \propto -h_c^{-(1-\beta)/\alpha}. \tag{4.12}$$

Fig. 8a shows that the numerical simulation of pinch-off in surface diffusion from Fig. 5b follows this scaling relation as $h \to 0$ [9]. A stronger test for the presence of self-similar evolution is to demonstrate that the spatial structure of the solution in a neighborhood of the singularity position converges to a scale-independent profile. Again, working from (4.6), given $h(x,t)$ and $h_c(t)$, we can eliminate τ to write the solution in a scale invariant form as

$$\frac{x - x_c}{h_c^{\beta/\alpha}(t)} = \frac{\eta}{\bar{H}(0)}, \qquad \frac{h(x,t)}{h_c(t)} = \frac{\bar{H}(\eta)}{\bar{H}(0)}. \tag{4.13}$$

Fig. 8b shows the convergence to a re-normalized self-similar profile of six rescaled solution curves from the simulation of pinch-off in surface diffusion shown in Fig. 5b.

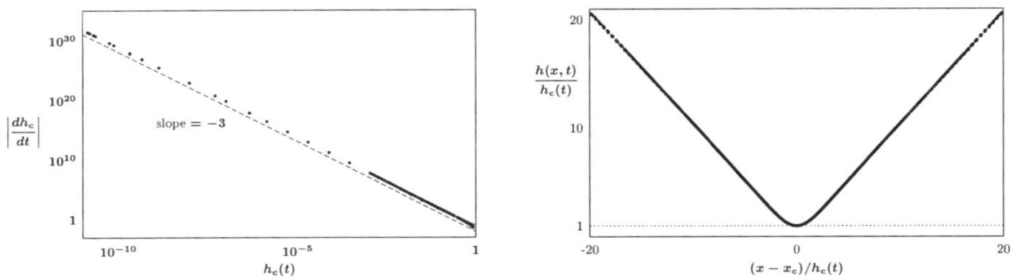

Figure 8: Observing self-similar evolution toward rupture in the surface diffusion problem, with $\alpha = \beta = 1/4$: using data from the simulation in Fig. 5b, the expected dynamics are obtained for $h_c(t)$ (left), and rescaling the solution yields a self-similar profile in the neighborhood of x_c (right).

4.3 Self-similar solutions

Further insight into the nature of the dynamics leading to singularity formation can be gained from thorough analysis of the properties of the similarity solution $\bar{H}(\eta)$. Substituting the similarity solutions found above, (4.7) and (4.8), into their corresponding PDEs yields nonlinear ordinary differential equations for the respective self-similar profiles $\bar{H}(\eta)$; for line rupture in the van der Waals problem,

$$\frac{1}{5}(2\eta \bar{H}_\eta - \bar{H}) = -(\bar{H}^{-1}\bar{H}_\eta + \bar{H}^3 \bar{H}_{\eta\eta\eta})_\eta, \tag{4.14}$$

and for surface diffusion,

$$\frac{1}{4}(\eta \bar{H}_\eta - \bar{H}) = \frac{1}{\bar{H}}\frac{d}{d\eta}\left(\frac{\bar{H}}{\sqrt{1+\bar{H}_\eta^2}}\frac{d}{d\eta}\left\{\frac{1}{\bar{H}\sqrt{1+\bar{H}_\eta^2}} - \frac{\bar{H}_{\eta\eta}}{(1+\bar{H}_\eta^2)^{3/2}}\right\}\right). \tag{4.15}$$

We observe the nature of (4.6) as a separation-of-variables ansatz, reducing the the PDE (2.1) to an ODE for $\bar{H}(\eta)$ for a special class of solutions. To provide a full description of the stability of these special solutions, we extend these similarity differential equations to capture the dynamics of all solutions of the original PDEs.

As described by Giga and Kohn [51], the self-similar ansatz (4.6) suggests a change of variables allowing the representation of all solutions of (2.1) in the form

$$h(x,t) = \tau^{\alpha} H(\eta, s) \qquad \eta = (x - x_c)/\tau^{\beta}, \tag{4.16}$$

where the new stretched "similarity time" variable s is given by

$$s = -\ln \tau, \qquad \tau = t_c - t. \tag{4.17}$$

Note that as the finite-time singularity is approached, $\tau \to 0$, we have the limit $s \to \infty$. Taking the time derivative of (4.16) yields the form

$$\frac{\partial h}{\partial t} = \tau^{\alpha-1}\left(\frac{\partial H}{\partial s} - \mathcal{T}(H)\right), \tag{4.18}$$

where the additional term on the right is given by

$$\mathcal{T}(H) \equiv \alpha H - \beta \eta H_{\eta}. \tag{4.19}$$

This linear operator reflects the time-dependent scaling of the dependent and independent variables in (4.16). From (4.16), every solution $h(x,t)$ of PDE (2.1) corresponds to a solution $H(\eta, s)$ of the similarity PDE

$$\frac{\partial H}{\partial s} = \mathcal{N}(H), \qquad \mathcal{N}(H) \equiv \mathcal{R}(H) + \mathcal{T}(H), \tag{4.20}$$

where in order to support a similarity solution of the form (4.6), the spatial operator $\mathcal{R}$ must satisfy $\mathcal{R}(h) = \tau^{\alpha-1}\mathcal{R}(H)$. Here, we have effectively done a separation of variables between the fast scales connected to the finite-time singularity, (η, s) and the slow time-scale τ.

In the framework of (4.20), self-similar solutions, $H = \bar{H}(\eta)$, are s-independent steady states on $-\infty < \eta < \infty$, satisfying the ODE

$$\mathcal{N}(\bar{H}(\eta)) = 0. \tag{4.21}$$

We need to specify boundary conditions for (4.21) to complete the problem to determine $\bar{H}(\eta)$. These conditions are determined by the requirement that the influence of the singularity be *localized in space* to a vanishingly small neighborhood of x_c as $\tau \to 0$. That is, the singularity should not influence the far-field of the solution, $x \neq x_c$, as $t \to t_c$. For a focusing similarity solution (with $\beta > 0$), while the solution, given by the self-similar form (4.6), is becoming singular at x_c as $t \to t_c$, at any finite distance away from x_c, the solution $h(x,t)$ will continue to evolve smoothly in space and time. To enforce this condition, that $\partial_t h = O(1)$ for $x \neq x_c$, for any blow-up solution ($\alpha < 0$) and rupture solutions with $\alpha < 1$ using (4.18) yields an asymptotic boundary condition for the similarity solution,

$$\mathcal{T}(\bar{H}) \to 0 \quad \text{as} \quad |\eta| \to \infty. \tag{4.22}$$

Consequently, using (4.19), the leading order asymptotic behavior of $\bar{H}(\eta)$ is given by

$$\bar{H}(|\eta| \to \infty) \sim A|\eta|^{\alpha/\beta}, \tag{4.23}$$

where A is an unknown positive constant. To observe this structure of the similarity solutions, note the linear far-field asymptotics of the surface diffusion similarity solution, with $\alpha/\beta = 1$, shown in Fig. 8b, and similarly for the square-root asymptotics of the van der Waals similarity solutions, with $\alpha/\beta = 1/2$, shown in Fig. 9.

In addition to its interpretation as a localization condition allowing for the asymptotic matching [62] of the inner similarity solution at x_c to the slowly varying outer solution, (4.23) can also be shown to determine locally unique solutions of the fourth-order singular boundary value problem for (4.21). WKB analysis [5] of the linearized far-field behavior of (4.21) shows that specifying the asymptotic behavior (4.23) suppresses unwanted exponential modes and uniquely determines a linearized solution [9, 103].

4.4 Numerical calculation of self-similar solutions

Having identified (4.14) and (4.15) as special cases of the ODE boundary value problem for focusing-type finite-time self-similar solutions,

$$\mathcal{N}(\bar{H}(\eta)) = 0, \qquad \bar{H}(|\eta| \to \infty) \sim A|\eta|^{\alpha/\beta}, \tag{4.24}$$

we turn to the problem of obtaining the solution $\bar{H}(\eta)$ of (4.24). For our two case studies and most other problems, closed-form analytic solutions of (4.24) can not be found, so the solutions must be obtained in numerical form. We now outline two numerical methods that can be used to solve this problem.

4.4.1 The shooting method

The shooting method is a classical technique for finding solutions of a boundary value problem from a set of trial solutions for a related family of initial value problems [90]; any solution of the initial value problem that also satisfies the boundary conditions solves the full boundary value problem. An important feature of the solutions of (4.14), (4.15), and many other problems that makes the shooting method easily applicable is that they yield symmetric solutions for $\bar{H}(\eta)$. All symmetric solutions of (4.24) satisfy the initial conditions at $\eta = 0$,

$$\bar{H}(0) = C_1, \qquad \bar{H}'(0) = 0, \qquad \bar{H}''(0) = C_2, \qquad \bar{H}'''(0) = 0, \tag{4.25}$$

where $C_1 > 0$ and C_2 are unknown constants. Hence, we can solve (4.24) as an initial value problem from $\eta = 0$, subject to (4.25), and then check if the trial solution satisfies the asymptotic boundary condition, (4.23), for $\eta \to \infty$. For appropriate values of (C_1, C_2) a solution can be found that satisfies (4.23). To find all possible solutions of the boundary value problem, an exhaustive search through the two-parameter space (C_1, C_2) must be made.

For the surface diffusion problem, the $\bar{H}_1(\eta)$ solution was first calculated in [101] by Wong, Miksis, Voorhees, and Davis using a shooting method for a formulation of (4.15) in arclength coordinates. For the van der Waals rupture problem, Zhang and Lister [103], carried out an exhaustive two-parameter shooting method search for both line rupture (2.16) and point rupture (2.17) problems and found the first six of an infinite sequence of different similarity solutions, $\bar{H}_m(\eta)$ for $m = 1, 2, 3, \ldots$ (see Fig. 9). See [103] for tabulated values of the shooting parameters $\bar{H}(0)$ and $\bar{H}''(0)$ and the corresponding values of the asymptotic parameter A in (4.23).

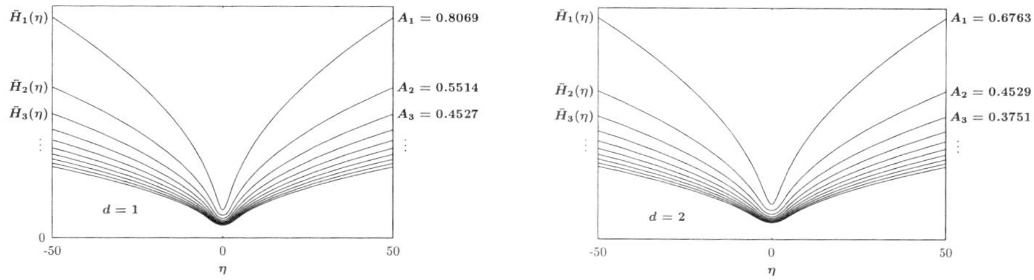

Figure 9: Discrete families of self-similar solutions for van der Waals driven rupture; (left) line rupture solutions, (right) point rupture solutions.

Shooting methods have also been used effectively to demonstrate the existence of multiple finite-time similarity solutions in a number of other problems in fluid dynamics [21, 95, 96]. The existence of multiple similarity solutions has also been noted in a wide range of other problems including the complex Ginzburg-Landau equation [80], the nonlinear Schrödinger equation [24], and chemotactic models in mathematical biology [20], just to name a few.

4.4.2 Newton's method

While the shooting method is attractive based on the ease with which initial value problems for ODEs can be implemented, it does have some shortcomings that can make it difficult to use. For many problems, (4.24) can lead to a stiff initial value problem, where the behavior of the solution is very sensitive to small changes in the initial conditions and very careful numerical implementation is necessary to accurately calculate the asymptotic behavior of the solution as $\eta \to \infty$. In [96], this made calculating each additional similarity solution $\bar{H}_m(\eta)$ in the sequence successively more difficult and computationally intensive. Thus, we consider Newton's method, which does not have the same restrictions. Newton's method offers a distinctly different and independent numerical approach for solving the boundary value problem (4.24) that overcomes some of the limitations of the shooting method and fits in with the overall program for the numerical calculation of (2.1) started in Section 3.1.

A direct numerical solution of the singular boundary value problem (4.24) using Newton's method or similar iterative relaxation methods [81] can be implemented using spectral methods on infinite domains [19]. However, for simplicity, we will consider the calculation of symmetric solutions of (4.24) using finite differences on a large but finite domain, $0 \le \eta \le L$, with $L \gg 1$, and the solution represented by

$$\bar{H}(\eta_n) \approx \bar{\mathbf{H}}_n, \qquad \eta_n = n\Delta\eta, \qquad n = 0, 1, 2, \dots, N \tag{4.26}$$

with $\Delta\eta = L/N$. Symmetry is imposed by application of the boundary conditions $\bar{H}'(0) = 0$, $\bar{H}'''(0) = 0$, and the far-field asymptotic behavior (4.22) can be approximated using the Robin boundary condition at the edge of the domain,

$$\mathcal{T}(\bar{H}) \equiv \alpha\bar{H} - \beta\eta\bar{H}_\eta = 0 \quad \text{at} \quad \eta = L. \tag{4.27}$$

This yields a system of $(N+1)$ nonlinear algebraic equations for $\bar{\mathbf{H}}$, namely,

$$\mathcal{F}(\bar{\mathbf{H}}) \equiv \mathcal{N}(\bar{\mathbf{H}}) = \mathbf{0}. \tag{4.28}$$

Using Newton's method to solve this system, we arrive at a problem of the same form as (3.8), (3.10), where the Jacobian operator is now given by

$$\mathbf{J}_{(k)} \equiv \frac{\delta \mathcal{F}(\bar{\mathbf{H}}_{(k)})}{\delta \bar{\mathbf{H}}} = \frac{\delta \mathcal{R}}{\delta \bar{H}} + \mathcal{T}. \tag{4.29}$$

Note that this problem, $\mathcal{N}(\bar{\mathbf{H}}) = \mathbf{0}$, differs from the equilibrium problem, $\mathcal{R}(\bar{\mathbf{h}}) = \mathbf{0}$, considered in Section 3.2, only in the boundary condition (4.27) and in the addition of the linear operator $\mathcal{T}$. Consequently, only minor changes to the numerical scheme developed in Section 3.1 are needed to calculate the self-similar solutions.

An important issue for the calculation of similarity solutions is the fact that Newton's method is only locally convergent and hence "good" initial iterates $\bar{\mathbf{H}}_{(0)}$ are needed to ensure convergence, $\bar{\mathbf{H}}_{(k)} \to \bar{\mathbf{H}}$. Since multiple solutions exist to $\mathcal{N}(\bar{H}) = 0$ (as suggested by the results obtained by shooting methods, as shown in Fig. 9), we must qualify the previous statement by specifying convergence to which of the many possible solutions, $\bar{H}_m(\eta)$ for $m = 1, 2, 3, \ldots$. When multiple solutions exist to nonlinear problems, Newton's method partitions the space of trial solutions into basins of attraction for each solution, possibly with fractal basin boundaries [81, 45]. This means that for poorly chosen initial guesses, the iterates, if they converge at all, will converge almost randomly to different $\bar{H}_m(\eta)$ solutions.

As in the van der Waals rupture problem, multiple similarity solutions exist for the surface diffusion problem, and they can be ordered according to their far-field asymptotic constant A (4.23), with the sequence of A_m satisfying,

$$A_1 > A_2 > A_3 > \cdots \to 0. \tag{4.30}$$

Bernoff et al. [9] suggested an approach to this problem that yielded systematic, well-ordered convergence to all of the similarity solutions of (4.24) for the surface diffusion problem. If we consider the asymptotic limit for the sequence of solutions, $m \to \infty$, then we can rescale the solutions in terms of the small parameter $\varepsilon \equiv A_m \to 0$,

$$\bar{H}(\eta) = \varepsilon^{1-\gamma\alpha/\beta} U(z) \qquad \eta = \varepsilon^{-\gamma} z, \tag{4.31}$$

where γ is a scaling constant, to be determined later. The form (4.31) eliminates the unknown constant A from the asymptotic behavior of the solution to yield,

$$U(|z| \to \infty) \sim |z|^{\alpha/\beta}. \tag{4.32}$$

The scaling parameter ε now appears in the rescaled version of the differential equation (4.21), which now takes the form of a nonlinear eigenvalue problem, $\tilde{\mathcal{N}}(U, \varepsilon) = 0$. For $\varepsilon \to 0$ there is a distinguished limit that balances the linear terms from $\mathcal{T}$ with second-order terms from the destabilizing influences described in Section 3 (see (3.1) and (3.3)). For the van der Waals problem, this limit selects the value $\gamma = 2/5$,

$$\bar{H}(\eta) = \varepsilon^{4/5} U(z), \qquad \eta = \varepsilon^{-2/5} z. \tag{4.33}$$

For the surface diffusion problem, we find that $\gamma = 1/2$,

$$\bar{H}(\eta) = \varepsilon^{1/2} U(z), \qquad \eta = \varepsilon^{-1/2} z. \tag{4.34}$$

A leading-order outer solution for (4.24) can then be found by writing $U(z) \sim U_0(z)$ plus higher order terms as $\varepsilon \to 0$, where U_0 satisfying a second-order nonlinear equation, $\mathcal{N}_2(U_0) = 0$. For the van der Waals problem for line rupture ($d = 1$) or point rupture ($d = 2$), this yields

$$\mathcal{N}_2(U_0) \equiv \frac{1}{5}\left(U_0 - 2z\frac{dU_0}{dz}\right) - \frac{1}{z^{d-1}}\frac{d}{dz}\left(z^{d-1}\frac{1}{U_0}\frac{dU_0}{dz}\right) = 0. \tag{4.35}$$

In this case, there is a simple separation of the terms in (4.21), with $\mathcal{N}_2(U) \equiv \mathcal{T}(U) + \mathcal{V}(U)$, and the full ODE taking the form $\mathcal{N}_2(U) + \varepsilon^4\mathcal{S}(U) = 0$. For the surface diffusion problem, the terms in (4.15) do not decouple as simply, but the corresponding second-order problem is

$$\mathcal{N}_2(U_0) \equiv \frac{1}{4}\left(U_0 - 2z\frac{dU_0}{dz}\right) - \frac{1}{U_0}\frac{d}{dz}\left(\frac{1}{U_0}\frac{dU_0}{dz}\right) = 0. \tag{4.36}$$

An important feature of both second-order problems, (4.35) and (4.36), is that they have a continuous scale-invariant symmetry with respect to (4.31) [42]; that is, for $\bar{H} \to U_0$ in the corresponding form of (4.31), we obtain a continuous one-parameter family of solutions of $\mathcal{N}_2(\bar{H}) = 0$ for any $\varepsilon > 0$ from the unique solution $U_0(z)$ satisfying (4.32). As we have seen, in the full problem (4.24), the solutions $\bar{H}_m(\eta)$ occur as a discrete, countable sequence, not as a continuous family. Hence we can conclude that the influence of the higher-order regularizing terms is crucial in selecting the discrete set of similarity solutions. However, for the purposes of computing a convergent sequence via Newton's method, the scale-invariant solutions (4.33) and (4.34) with $U = U_0(z)$ and $\varepsilon = A$ provide a very good set of initial guesses for $\bar{\mathbf{H}}_{(0)}$ in a one-parameter search. This approach was used in [9, 98, 99] to obtain large numbers of similarity solutions $\bar{H}_m(\eta)$, with m up to twenty or so, with piecewise convergence from initial guesses with A in a finite neighborhood of each A_m. See Fig. 9 for the first ten similarity solutions for finite-time rupture in the van der Waals problem [98, 99].

4.5 Stability analysis

While multiple finite-time self-similar solutions exist for our problems, results from all numerical simulations of the PDEs always show a single solution, $\bar{H}_1(\eta)$, being selected by the evolution under all conditions where rupture occurs. This situation is consistent with $\bar{H}_1(\eta)$ being distinguished from the other $\bar{H}_m(\eta)$ in terms of its stability to perturbations. We will show that $\bar{H}_1(\eta)$ is the only stable similarity solution.

First, we must address the question of how to characterize the stability of the dynamics leading to finite-time singularity formation. In fact, there are four classes of perturbations that must be considered to evaluate stability,

Perturbations of:

(4.37)

(a) the time of the singularity, $t_c \to t_c + \epsilon$;

(b) the position of the singularity, $x_c \to x_c + \epsilon$;

(c) the mode of self-similar evolution, $\bar{H}_m(\eta) \to \bar{H}_k(\eta)$;

(d) the geometry or dimension of the singularity, $d_m \to d_k$.

We discuss all of these instabilities in the framework of linear stability analysis of the self-similar solutions. While the study the stability of the singularity formation process is a complicated problem with respect to the original PDE (2.1), the use of similarity variables (4.16) considerably simplifies the difficulties. With respect to the similarity PDE (4.20), the self-similar solutions $\bar{H}_m(\eta)$ are steady state solutions, hence classical linear stability theory can be applied to the problem in similarity coordinates [26, 51].

Consider a infinitesimal perturbation to a self-similar solution of the form

$$H(\eta, s) = \bar{H}_m(\eta) + \epsilon \hat{H}(\eta) e^{\lambda s}, \tag{4.38}$$

with $\epsilon \to 0$. Using the definition of the similarity time variable (4.17), $s = -\ln \tau$, and (4.16), (4.38) corresponds to $h(x,t)$ of the form,

$$h(x,t) = \tau^\alpha \left(\bar{H}_m(\eta) + \epsilon \hat{H}(\eta) \tau^{-\lambda} \right). \tag{4.39}$$

Consequently, as the singular time is approached ($\tau \to 0$) an instability growing relative to the dynamics of the assumed self-similar solution occurs if $\mathrm{Re}(\lambda) > 0$. Hence, using (4.38) the condition defining linear stability of finite-time self-similar solutions is the same as the classical result for stability of equilibria.

In writing the similarity variables (4.16), (4.17), it is assumed that the position x_c and time t_c of the singularity are known exactly. In general, this is not the case, and errors in the values assumed for (t_c, x_c) will lead to instabilities associated with (4.37(a),(b)). This result can be demonstrated using the invariance of the PDE (2.1) with respect to spatial translations and time-shifts of localized solutions. Consider (4.37(a)), that is, suppose that there is an error in the value of the critical time, $t_c \to t_c + \epsilon$. This corresponds to a transformation from one solution of (2.1), to a time-shifted version, $h(x,t) \to h(x, t-\epsilon)$, which is also a solution of the PDE. Translating this shift into similarity variables, $\tau \to \tau(1+\epsilon e^s)$, and applying it to the self-similar solution $\bar{H}(\eta)$ yields

$$\bar{H}(\eta) \to H(\eta, s) = (1+\epsilon e^s)^\alpha \bar{H}(\eta[1+\epsilon e^s]^{-\beta}). \tag{4.40}$$

Linearizing the action of this symmetry transformation for $\epsilon \to 0$, we can identify the unstable eigenmode connected with time-shifts of the finite-time self-similar solution $\bar{H}(\eta)$,

$$H(\eta, s) \sim \bar{H}(\eta) + \epsilon \mathcal{T}(\bar{H}) e^s, \qquad \hat{H}_T = \mathcal{T}(\bar{H}), \qquad \lambda_T = 1. \tag{4.41}$$

A similar description applies to errors in the assumed position of the singularity, $x_c \to x_c + \epsilon$ (4.37(b)). This case corresponds to spatial shifts of solutions of (2.1), $h(x,t) \to h(x-\epsilon, t)$, and in terms of similarity variables,

$$\bar{H}(\eta) \to H(\eta, s) = \bar{H}(\eta + \epsilon e^{\beta s}). \tag{4.42}$$

Linearizing the action of this symmetry for infinitesimal spatial translations of the focusing similarity solution $\bar{H}(\eta)$ yields an unstable eigenmode,

$$H(\eta, s) \sim \bar{H}(\eta) + \epsilon \bar{H}'(\eta) e^{\beta s}, \qquad \hat{H}_X = \bar{H}'(\eta), \qquad \lambda_X = \beta. \tag{4.43}$$

Similarly, eigenmodes can also be derived for the other symmetry modes that can be applied to solutions of the PDE (2.1), such as rotations [99].

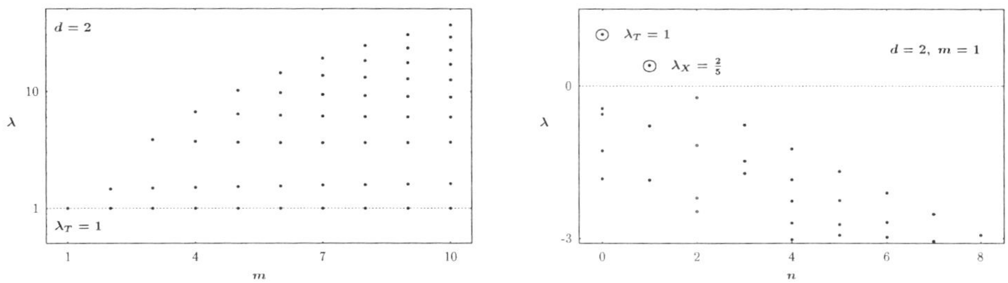

Figure 10: Eigenvalues for the linear stability of the self-similar solutions for axisymmetric van der Waals driven thin film rupture: (left) for axisymmetric perturbations of $\bar{H}_m(\eta)$ with $m = 1, 2, \ldots, 10$ (4.37(c)), and (right) for non-axisymmetric perturbations that could lead to a change in geometry of the point rupture solution $\bar{H}_1(\eta)$ (4.37(d)).

Having accounted for possible instabilities due to the definitions of the similarity variables, we assume that (t_c, x_c) are known exactly and turn to the question of stability of $\bar{H}_m(\eta)$ with respect to perturbations that change the shape of the self-similar profile, (4.37(c)). Substituting (4.38) into (4.20), and linearizing for $\epsilon \to 0$ yields

$$\lambda \hat{H} = \mathcal{L}(\bar{H}_m)\hat{H} \tag{4.44}$$

on $-\infty < \eta < \infty$, where the linear operator $\mathcal{L}$ is given by

$$\mathcal{L} \equiv \frac{\delta \mathcal{N}}{\delta H} = \frac{\delta \mathcal{R}}{\delta H} + \mathcal{T}. \tag{4.45}$$

To specify the appropriate boundary conditions for $|\eta| \to \infty$, we recall the localization condition (4.22). For solutions of the form (4.38), we use (4.18) to obtain the condition $\lambda \hat{H} = \mathcal{T}(\hat{H})$ as $|\eta| \to \infty$. Therefore, the complete form of the linear operator with boundary conditions is

$$\hat{\mathcal{L}}\hat{H} \equiv \begin{cases} \mathcal{L}(\bar{H}_m)\hat{H} & |\eta| < \infty, \\ \mathcal{T}(\hat{H}) & |\eta| \to \infty, \end{cases} \tag{4.46}$$

and the eigenvalue problem for the stability analysis takes the form $\hat{\mathcal{L}}\hat{H} = \lambda \hat{H}$ [99]. As in the stability analysis of equilibrium solutions of (2.1), in general, the spectrum of $\bar{H}_m(\eta)$ must be calculated numerically. Noting that the discretized form of the linear operator (4.45) is identical with the Jacobian matrix at the converged solution $\bar{H}_m(\eta)$, (4.29), we observe that this stability calculation via the inverse power method is a simple extension of the Newton's method calculation of the similarity solutions given in Section 4.4.2 [92].

Fig. 10a shows the positive eigenvalues in spectra for axisymmetric perturbations of the $\bar{H}_m(\eta)$ point-rupture similarity solutions (see Fig. 9b) of the $d = 2$ axisymmetric van der Waals thin film problem (2.17) [99]. Note that the time-shift symmetry eigenmode $\lambda_T = 1$ is present in the spectra of all of the similarity solutions. Only the first similarity solution $\bar{H}_1(\eta)$ has no other positive eigenvalues, corresponding to unstable modes; hence $\bar{H}_1(\eta)$ is the only stable axisymmetric self-similar solution. Being the only stable solution, $\bar{H}_1(\eta)$ will be the global attractor among solutions going to a self-similar finite-time singularity. Therefore we

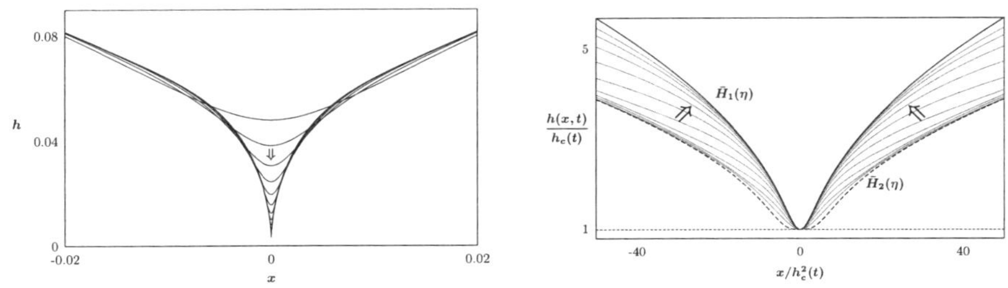

Figure 11: Perturbations leading to a change of the self-similar profile (4.37(c)): (left) for initial data $h(x,0)$ very close to the $\bar{H}_2(\eta)$ profile, the solution of the van der Waals problem (2.16) approaches finite-time rupture, but destabilizes and converges to the $\bar{H}_1(\eta)$ solution as $\tau \to 0$ (right).

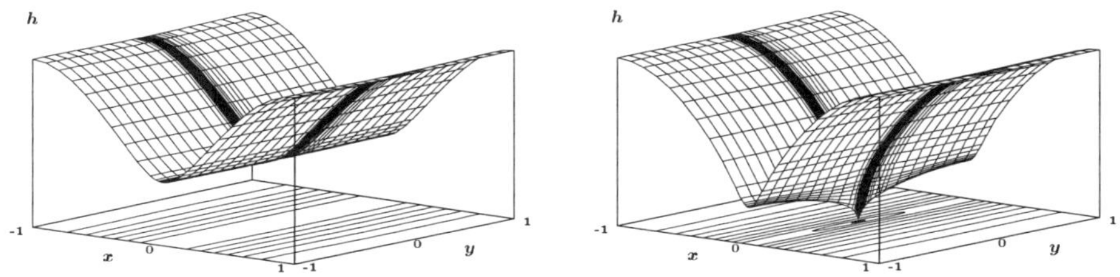

Figure 12: Perturbations leading to a change in the rupture geometry (4.37(d)): (left) small perturbations of initial data leading to line rupture in the van der Waals problem (2.10) yield a point rupture singularity (right).

can restate (4.37(c)) to reflect this universal asymptotic behavior for singularity formation, starting from almost any axisymmetric initial data, $\bar{H}_m(\eta) \to \bar{H}_1(\eta)$ (see Fig. 11).

Finally, we address the question of perturbations that can change the geometric structure of the singularity, (4.37(d)). Fig. 10b shows that, apart from symmetry modes, the $\bar{H}_1(\eta)$ axisymmetric van der Waals rupture solution is stable with respect to non-axisymmetric perturbations, $H(\eta, s) \sim \bar{H}_1(\eta) + \epsilon \hat{H}(\eta) e^{in\theta} e^{\lambda s}$ [99]. Consequently, problems for (2.10) starting from nearly-axisymmetric data will converge to axisymmetric point-rupture singularities. In contrast, the $\bar{H}_1(\eta)$ line rupture solution ($d = 1$) of (2.16) is unstable with respect to transverse perturbations. Fig. 12 shows a simulation of the dynamics of (2.10), where a small perturbation of a line rupture solution destabilizes to ultimately yield a point rupture singularity, $d = 1 \to d = 2$.

5 ADI methods for multi-dimensional problems

Calculating more complicated dynamics like the geometric instability shown in Fig. 12, pattern formation [38], and other spatial structuring, requires accurate, efficient numerical schemes to tackle these more computationally intensive problems. To solve a two-dimensional version of (2.1) for $h(x, y, t)$, naively, we can proceed exactly as in Section 3.1, with

$$h(x_i, y_j, t^m) \approx \mathbf{h}_{i,j}^m, \qquad x_i = i\Delta x, \qquad y_j = j\Delta y, \qquad t^m = m\Delta t, \tag{5.1}$$

where $\Delta x = L/N$, $\Delta y = L/N$, and then the backward Euler method for (2.1) is

$$\frac{\mathbf{h}_{i,j}^{m+1} - \mathbf{h}_{i,j}^m}{\Delta t} = \mathcal{R}_{i,j}(\mathbf{h}^{m+1}). \tag{5.2}$$

In the one-dimensional case, say for the van der Waals rupture problem (2.16), the solution at each time-step can be obtained using Newton's method, where the Jacobian matrix $\mathbf{J}$ was pentadiagonal, and the associated linear system could be solved directly with a sparse LU factorization. In two dimensions, (5.2) yields a system of $(N+1)^2$ nonlinear equations to be solved using Newton's method. The corresponding Jacobian is block-pentadiagonal (with only thirteen nonzero diagonals), but attempting to solve this problem directly with an LU factorization would yield a factored matrix with $O(N^3)$ entries. To achieve a desired level of spatial resolution with large N, the computer memory required for this approach can rapidly become prohibitive. Therefore, alternative methods have been used; (a) iterative methods (conjugate gradients, GMRES, and others) to solve the linear systems, or (b) alternating direction implicit (ADI) methods [36, 39, 40] to approximate the solution of the two-dimensional problem (5.2) in terms of direct solutions of a series of related one-dimensional sub-problems (3.6). We will describe how ADI schemes derived from approximate factorizations can be constructed for two-dimensional fourth-order problems like (2.10) as extensions of the numerical methods for one-dimension problems given in Section 3.1.

To focus attention on the new aspects of implementing ADI methods for higher-order nonlinear equations with mixed derivatives, we consider a model problem for the thin film equation [100],

$$\frac{\partial h}{\partial t} + \nabla \cdot (f(h)\nabla\nabla^2 h) = 0. \tag{5.3}$$

The system of nonlinear algebraic equations corresponding to (5.2) is then given by

$$\mathcal{F}(\mathbf{h}^{m+1}) \equiv \mathbf{h}^{m+1} - \mathbf{h}^m + \Delta t \nabla \cdot (f(\mathbf{h}^{m+1})\nabla\nabla^2 \mathbf{h}^{m+1}) = \mathbf{0}. \tag{5.4}$$

Rather than attacking this nonlinear problem directly, we begin by constructing ADI methods for the linear problem with $f(h) \equiv 1$, that is $\partial_t h + \nabla^4 h = 0$. Here the spatial operator is the two-dimensional biharmonic operator, which intrinsically includes mixed derivatives, $\nabla^4 h \equiv h_{xxxx} + 2h_{xxyy} + h_{yyyy}$. The biharmonic operator is a fundamental part of many models in applied mathematics for fluid dynamics [6], elastic waves [35, 36, 37], combustion, and other applications, see [100] and references therein.

The derivation of the ADI scheme begins by separating implicit and explicit terms in (5.4),

$$(\mathbf{I} + \Delta t \nabla^4)\mathbf{h}^{m+1} = \mathbf{h}^m. \tag{5.5}$$

To avoid inverting the two-dimensional operator $I + \Delta t \nabla^4$, we use an approximate factorization of this operator [71]. Specifically, we replace the implicit term $(\mathbf{I} + \Delta t \nabla^4)\mathbf{h}^{m+1}$ with a product of one-dimensional operators acting on $\mathbf{h}^{m+1}$ plus other explicit terms (operators acting on $\mathbf{h}^m$) without loss of time-accuracy in the scheme. We can write the operator on the left side of (5.5) as

$$\mathbf{I} + \Delta t \nabla^4 = \mathbf{L}_x \mathbf{L}_y + 2\Delta t \partial_{xxyy} - \Delta t^2 \mathbf{D}_x \mathbf{D}_y, \tag{5.6}$$

where the one-dimensional operators are defined by

$$\mathbf{L}_x \equiv \mathbf{I} + \Delta t \mathbf{D}_x, \qquad \mathbf{L}_y \equiv \mathbf{I} + \Delta t \mathbf{D}_y, \qquad \mathbf{D}_x \equiv \partial_{xxxx}, \qquad \mathbf{D}_y \equiv \partial_{yyyy}. \tag{5.7}$$

Using the fact that $\mathbf{h}^{m+1} = \mathbf{h}^m + O(\Delta t)$, we obtain

$$\begin{aligned}(\mathbf{I} + \Delta t \nabla^4)\mathbf{h}^{m+1} &= \mathbf{L}_x \mathbf{L}_y \mathbf{h}^{m+1} + 2\Delta t \partial_{xxyy} \mathbf{h}^{m+1} - \Delta t^2 \mathbf{D}_x \mathbf{D}_y \mathbf{h}^{m+1} \\ &= \mathbf{L}_x \mathbf{L}_y \mathbf{h}^{m+1} + [2\Delta t \partial_{xxyy} - \Delta t^2 \mathbf{D}_x \mathbf{D}_y](\mathbf{h}^m + O(\Delta t)).\end{aligned} \tag{5.8}$$

This yields the first-order accurate approximately-factored ADI scheme,

$$\mathbf{L}_x \mathbf{L}_y \mathbf{h}^{m+1} = (\mathbf{I} - 2\Delta t \partial_{xxyy} + \Delta t^2 \mathbf{D}_x \mathbf{D}_y)\mathbf{h}^m. \tag{5.9}$$

If mixed derivatives are present in the problem, then non-iterative ADI schemes (like this one) can be at most first-order accurate in time, as shown by [4].

Equation (5.9) can be written in a more compact form by subtracting $\mathbf{L}_x \mathbf{L}_y \mathbf{h}^m$ from both sides to obtain an equation for the change between time-steps, $\boldsymbol{\epsilon} = \mathbf{h}^{m+1} - \mathbf{h}^m$,

$$\mathbf{L}_x \mathbf{L}_y \boldsymbol{\epsilon} = -\Delta t \nabla^4 \mathbf{h}^m. \tag{5.10}$$

Writing this in factored form yields the scheme

$$\left.\begin{aligned}\mathbf{L}_x \boldsymbol{\mu} &= -\Delta t \nabla^4 \mathbf{h}^m \\ \mathbf{L}_y \boldsymbol{\epsilon} &= \boldsymbol{\mu}\end{aligned}\right\} \quad \mathbf{h}^{m+1} = \mathbf{h}^m + \boldsymbol{\epsilon}. \tag{5.11}$$

Here $\boldsymbol{\epsilon} = \boldsymbol{\epsilon}_{i,j}$ and $\boldsymbol{\mu} = \boldsymbol{\mu}_{i,j}$ are intermediate variables needed to determine the solution at the next time-step, $\mathbf{h}^{m+1}$. The virtue of (5.11) is that $\boldsymbol{\mu}$ satisfies a set of $(N+1)$ independent one-dimensional problems on the lines $y = y_j$ for $j = 0, 1, 2, \ldots, N$, and similarly for $\boldsymbol{\epsilon}$ once $\boldsymbol{\mu}$ is determined. Hence the two-dimensional problem (5.5) is reduced to a series of one-dimensional problems and efficient direct methods can be used for the remaining linear algebra.

Returning our focus to the solution of the nonlinear problem (5.3), we first consider a "pseudo-linear" ADI scheme [100] extending (5.11) for cases where $f(h)$ depends weakly on h. If we neglect the dependence of $f(\mathbf{h}^{m+1})$ on the unknown $\mathbf{h}^{m+1}$ in (5.4), and instead approximately evaluate that term using some explicit estimate, $\mathbf{h}^{m+1} \approx \mathbf{h}^{m+1}_{(k)}$, then the rest of the linear ADI scheme follows with $\mathbf{L}_x \equiv \mathbf{I} + \Delta t \partial_x (f(\mathbf{h}^{m+1}_{(k)})\partial_{xxx})$ and similarly for the y-operator, $\mathbf{L}_y$. This yields an iterative scheme for $k = 0, 1, 2, \ldots$:

$$\left.\begin{aligned}\mathbf{L}_x \boldsymbol{\mu} &= -[\mathbf{h}^{m+1}_{(k)} - \mathbf{h}^m + \Delta t \nabla \cdot (f(\mathbf{h}^{m+1}_{(k)})\nabla\nabla^2 \mathbf{h}^{m+1}_{(k)})] \\ \mathbf{L}_y \boldsymbol{\epsilon} &= \boldsymbol{\mu}\end{aligned}\right\} \quad \mathbf{h}^{m+1}_{(k+1)} = \mathbf{h}^{m+1}_{(k)} + \boldsymbol{\epsilon}. \tag{5.12}$$

Here we begin with an initial guess for the solution, $\mathbf{h}^{m+1}_{(0)} = \mathbf{h}^m$, and under appropriate assumptions the scheme will converge as $k \to \infty$, $\mathbf{h}^{m+1}_{(k)} \to \mathbf{h}^{m+1}$. Noting that scheme (5.12) can be written in the form $\mathbf{L}_x\mathbf{L}_y\boldsymbol{\epsilon} = -\mathcal{F}(\mathbf{h}^{m+1})$, we observe that this ADI method is actually a linear iterative method for the solution of the full two-dimensional backward Euler problem (5.4).

For strongly nonlinear operators and problems with more complicated spatial dependence, where the convergence of the pseudo-linear scheme might be very limited, we consider an approximate Newton-ADI scheme. As in Section 3.1, we use Newton's method to iteratively solve (5.4),

$$\mathbf{J}_{(k)}\boldsymbol{\epsilon}_{(k)} = -\mathcal{F}(\mathbf{h}_{(k)}^{m+1}), \qquad \mathbf{h}_{(k+1)}^{m+1} = \mathbf{h}_{(k)}^{m+1} + \boldsymbol{\epsilon}_{(k)}, \tag{5.13}$$

where the Jacobian, $\mathbf{J}(\mathbf{h}) = \delta\mathcal{F}/\delta\mathbf{h}$, is given by

$$\mathbf{J}(\mathbf{h})\boldsymbol{\epsilon} = \boldsymbol{\epsilon} + \Delta t\nabla \cdot (\boldsymbol{\epsilon} f'(\mathbf{h})\nabla\nabla^2\mathbf{h} + f(\mathbf{h})\nabla\nabla^2\boldsymbol{\epsilon}). \tag{5.14}$$

To avoid the computationally intensive linear algebra associated with the exact Jacobian, consider the use of an approximate factorization $\mathbf{J} \sim \mathbf{L}_x\mathbf{L}_y$ with

$$\mathbf{L}_x\boldsymbol{\mu} \equiv \boldsymbol{\mu} + \Delta t\partial_x[\boldsymbol{\mu} f'(\mathbf{h})\partial_x\nabla^2\mathbf{h} + f(\mathbf{h})\partial_{xxx}\boldsymbol{\mu}], \tag{5.15}$$

and similarly for $\mathbf{L}_y$. This yields the approximate-Newton iterative ADI scheme,

$$\left.\begin{aligned} \mathbf{L}_x\boldsymbol{\mu} &= -\mathcal{F}(\mathbf{h}_{(k)}^{m+1}) \\ \mathbf{L}_y\boldsymbol{\epsilon} &= \boldsymbol{\mu} \end{aligned}\right\} \quad \mathbf{h}_{(k+1)}^{m+1} = \mathbf{h}_{(k)}^{m+1} + \boldsymbol{\epsilon}. \tag{5.16}$$

If the solution of problem (5.2) is uniform in one direction, i.e. $h = h(x,t)$ or $h = h(y,t)$, then (5.16) reduces to Newton's method (3.8), (3.10) for the remaining one-dimensional problem.

To obtain a second-order accurate method, we can replace the backward Euler scheme (5.2) for (2.1) with a second-order midpoint/Crank-Nicholson single-step scheme [100],

$$\mathcal{F}(\mathbf{h}^{m+1}) \equiv \mathbf{h}^{m+1} - \mathbf{h}^m + \Delta t\mathcal{R}(\tfrac{1}{2}[\mathbf{h}^{m+1} + \mathbf{h}^m]) = \mathbf{0}. \tag{5.17}$$

Then a similar approximate-Newton ADI iterative method (5.16) can be constructed with appropriate modifications to (5.15). Such a scheme is not in conflict with the result of Beam and Warming [4] limiting single-step ADI methods for problem with mixed derivatives to first-order accuracy in time. Their study applied to non-iterative schemes, where $\mathbf{h}^{m+1}$ is calculated in one-step. The consequence is that for (5.17), (5.16), the first iterate, $\mathbf{h}_{(1)}^{m+1}$, is limited to be first-order accurate at best, but for larger k, the iterates $\mathbf{h}_{(k)}^{m+1}$ converges to a second-order accurate solution.

6 Further directions

We conclude by briefly touching on two problems that extend the ideas described above; finite-time singularities in systems of equations, and regularization of singularities.

6.1 Similarity solutions for systems of equations

One natural generalization of the scalar problem (2.1) is the study of singularity formation in systems of equations. For many problems in fluid mechanics and other physical systems, the best suited mathematical models take the form of systems of coupled nonlinear evolution equations. This is the case for long wave models of fluid jets and the flow of sheets of fluid in free space [44]. Finite-time break-up in these free-surface flows has been the focus of much recent work [21, 46, 57, 82, 89, 95, 96].

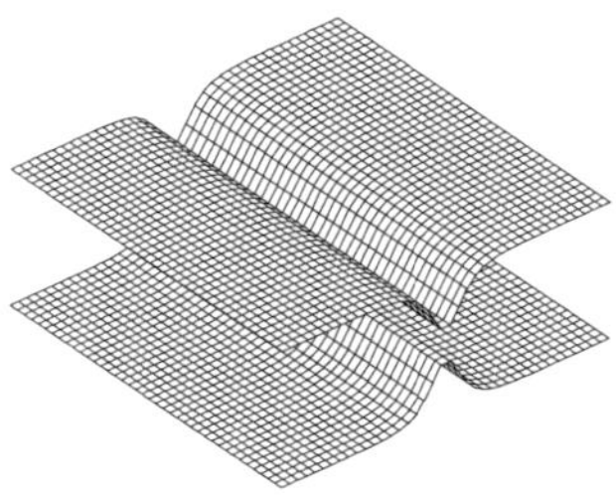

Figure 13: The geometry of line rupture in a two-dimensional sheet of viscous fluid in free space.

We briefly review the work of Vaynblat et al. [95, 96] on van der Waals driven rupture of free sheets (see Fig. 13) as an example of a system of equations with a finite-time self-similar solution. The long wave equations for this problem govern the height, or half thickness, of the fluid film $h = h(x,t)$, and the depth-averaged horizontal velocity in the film, $u = u(x,t)$. The statement of conservation of mass for the flow takes the form

$$\frac{\partial h}{\partial t} + \frac{\partial (hu)}{\partial x} = 0, \tag{6.1a}$$

and conservation of momentum yields

$$\frac{\partial u}{\partial t} + u\frac{\partial u}{\partial x} = \frac{4}{h}\frac{\partial}{\partial x}\left(h\frac{\partial u}{\partial x}\right) - \frac{\partial (h^{-3})}{\partial x}, \tag{6.1b}$$

where the van der Waals intermolecular forces, given by the last term in (6.1b), provide the destabilizing mechanism. In [95] it was shown that a uniform flat film at rest, $h = \underline{h}, u = 0$, is unstable. From the momentum balance, a local thinning of the film, $h \to 0$, increases the local outflow velocity, $u^2 \propto h^{-3} \to \infty$, and from (6.1a) hence promotes rupture, $h \to 0$.

The same program of analysis used in Section 4 can be applied to this system to obtain the self-similar rupture solutions. Dimensional analysis of (6.1a) yields the relation between the length, velocity and timescales, $\mathrm{X} = \mathrm{UT}$. By balancing inertial and viscous terms in (6.1b) we determine that $\mathrm{U} = \mathrm{T}^{-1/2}$ and hence $\mathrm{X} = \mathrm{T}^{1/2}$. Then by additionally balancing the van der Waals forces with these terms, we find that $\mathrm{H} = \mathrm{T}^{1/3}$. As $\mathrm{H} \to 0$, these scalings correspond to a finite-time self-similar rupture solution of the form

$$h(x,t) = \tau^{1/3}\bar{H}(\eta), \qquad u(x,t) = \tau^{-1/2}\bar{U}(\eta), \qquad \eta = (x - x_c)/\tau^{1/2}. \tag{6.2}$$

Substituting (6.2) into (6.1) yields the system of ordinary differential equations for $\bar{H}(\eta)$ and $\bar{U}(\eta)$,

$$-\frac{1}{3}\bar{H} + \frac{1}{2}\eta\bar{H}_\eta + \bar{U}\bar{H}_\eta + \bar{H}\bar{U}_\eta = 0, \tag{6.3a}$$

$$\frac{1}{2}\bar{U} + \frac{1}{2}\eta\bar{U}_\eta + \bar{U}\bar{U}_\eta = \frac{4}{\bar{H}}\left(\bar{H}\bar{U}_\eta\right)_\eta - (\bar{H}^{-3})_\eta. \tag{6.3b}$$

In Figs. 14a, 15a we plot results from a numerical simulation of (6.1), showing rupture of the thin film, $h \to 0$, with the corresponding finite-time blow-up of the local velocity, $u \to \infty$. Using the approach described in Section 4.2 with the similarity solution given by (6.2), Figs. 14b, 15b show that these results do indeed converge to self-similar profiles $\bar{H}(\eta), \bar{U}(\eta)$. As described in [95, 96], these profiles are in fact the first solutions of the system (6.3), $\bar{H}_1(\eta), \bar{U}_1(\eta)$, and there are an infinite set of other rupture solutions, as observed in Section 4.4 and in other systems [21].

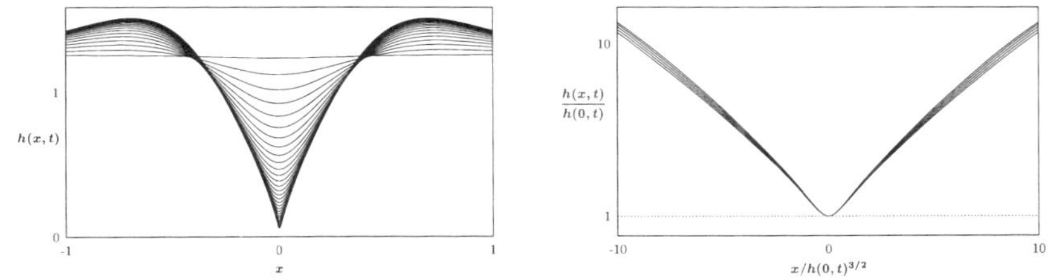

Figure 14: The dynamics of the thickness of the fluid film, $h(x, t)$ for (6.1) (left), and convergence of this data to a finite-time self-similar rupture profile using (6.2) (right).

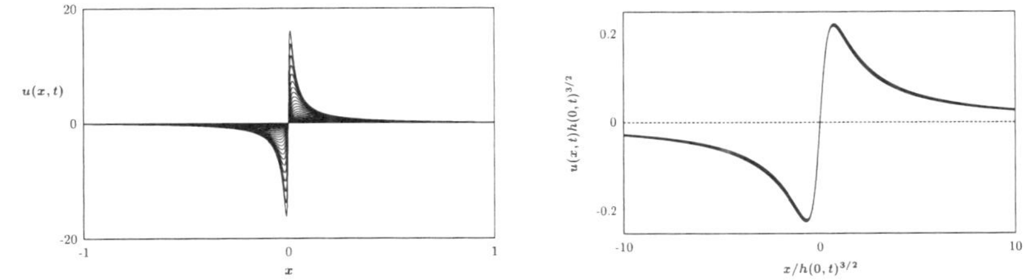

Figure 15: The dynamics of the fluid velocity $u(x, t)$ corresponding to Fig. 14. As finite-time rupture of the film approaches, the velocity approaches finite-time blow-up (left) in the consistent self-similar form (6.2) (right).

6.2 Regularization of problems with finite-time singularities

As described earlier, the formation of finite-time singularities in mathematical models for physical systems indicates break-down of the mathematical model as the physical system undergoes a fundamental change in its character. In the equations for surface diffusion (2.5), unstable thin fluid films (2.10), and many other problems, once the singularity has formed, the mathematical solution can not be continued for later times. However, the original physical problems being modeled do continue to exist, and generally go on to develop even more interesting and complex behavior following the first singularity. For example, consider the intricate cascade of behavior in the break-up of a fluid jet [89]. It is very desirable to set up mathematical models that can capture the essential features of the transitions corresponding to the finite-time singularities and also describe the evolution of the physical system for later times.

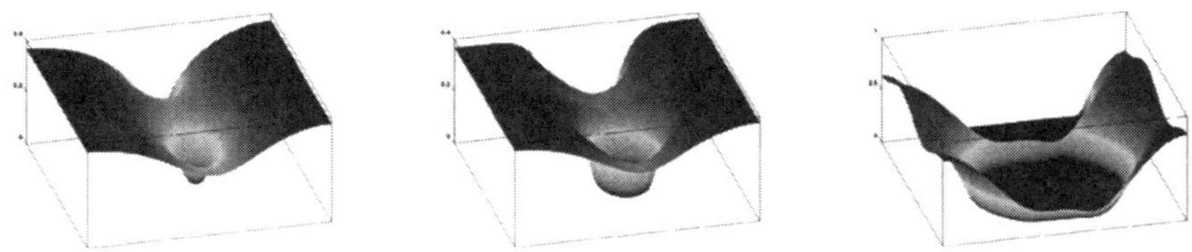

(a) finite-time formation of a rupture point

(b) nucleation of a dry-spot or "hole"

(c) further dewetting producing a ridge network

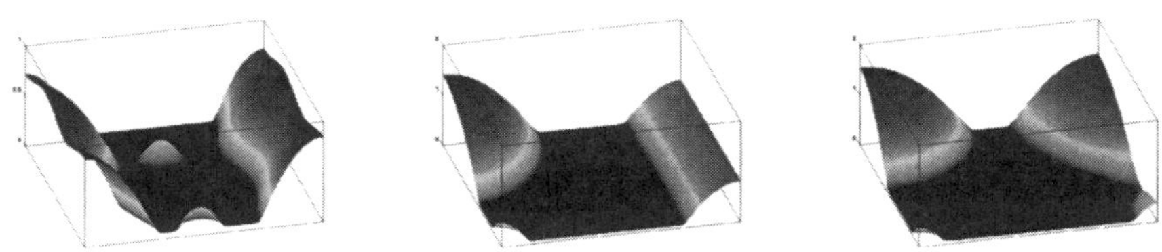

(d) break-up of some ridges

(e) co-existence of fluid droplets and ridges

(f) droplet coarsening

Figure 16: Stages in the dynamics of dewetting films (6.4).

The techniques used to construct such regularized models can vary greatly depending on details of the physical systems and the forms of the mathematical models. Hence, we will specialize our remarks to the surface diffusion and thin film rupture problems. For the surface diffusion problem, Coleman et al. [32, 33] use (2.5) until pinch-off occurs, and then restart the problem after the change in topology has occurred, using appropriate representations of (2.2) for each newly-formed separate surface. There are some open mathematical questions connected to such "surface surgery" techniques [9], and Cahn and Taylor [30] discuss other methods for continuing solutions past the singularity.

For the problem of rupture of thin films by van der Waals forces, regularization of the model to provide solutions valid for all times can be accomplished by using a better physical model for the pressure in the fluid. The model (2.9) used for the pressure included only "disjoining" van der Waals forces, which serve to promote rupture. For finite thickness films, these are indeed the dominant intermolecular forces, but in the limit of rupture, $h \to 0$, higher order "conjoining" forces should also be included to yield the modified pressure,

$$p = \frac{1}{3}h^{-3}\left(1 - \frac{\epsilon}{h}\right) - \nabla^2 h, \qquad \frac{\partial h}{\partial t} - \nabla \cdot (h^3 \nabla p) = 0, \tag{6.4}$$

where $h = \epsilon \ll 1$ defines the scale at which the conjoining forces become significant. For $\epsilon = 0$, (6.4) reduces to (2.10) and finite-time rupture singularities can occur. For $\epsilon > 0$, the solution has a positive lower bound determined by ϵ, so complete rupture ($h \to 0$) can not occur, consequently solutions of (6.4) exist for all times. While finite-time singularities do not occur, this model does approximately preserve the self-similar dynamics associated with the rupture process until $h_{\min} = O(\epsilon)$ is reached. For (6.4) this behavior is called "dewetting." In [13] it was shown that the balance of disjoining and conjoining forces leads to a more complicated set of equilibrium states, including solutions that are topologically disconnected (not counting an ultra-thin film with $h \sim \epsilon$), like the $\bar{h}_s(x)$ equilibrium solution of the surface diffusion problem (see Fig. 3a). Fig. 16 shows a numerical simulation of a two-dimensional dewetting film, calculated using the ADI methods given in Section 5 [100]. The first stage of the dewetting process corresponds to finite-time rupture, the later stages describe a complex series of coarsening processes currently under active study [52].

Acknowledgments I wish to give thanks to all of my research collaborators, especially Andy Bernoff and Andrea Bertozzi. I also acknowledge the invaluable aid of Mark Bowen in giving many helpful suggestions on preparing this manuscript. The author also gives thanks to the organizers of the ASI for the opportunity to present this work. The author was supported by a fellowship from the Alfred P. Sloan foundation and NSF FRG grant DMS–0074049.

References

[1] W. F. Ames, *Numerical Methods for Partial Differential Equations*, 3rd ed., Academic Press, Boston, MA, 1992.

[2] C. Bandle and H. Brunner, Blowup in diffusion equations: a survey, *J. Comput. Appl. Math.* **97** (1-2) (1998), 3–22.

[3] G. I. Barenblatt, *Scaling, Self-Similarity, and Intermediate Asymptotics*, Cambridge University Press, Cambridge, 1996.

[4] R. M. Beam and R. F. Warming, Alternating direction implicit methods for parabolic equations with a mixed derivative, *SIAM J. Sci. Statist. Comput.* **1** (1) (1980), 131–159.

[5] C. M. Bender and S. A. Orszag, *Advanced Mathematical Methods for Scientists and Engineers*, McGraw-Hill, New York, 1978.

[6] D. J. Benney, Long waves on liquid films, *J. Math. and Phys.* **45** (1966), 150–155.

[7] M. Berger and R. V. Kohn, A rescaling algorithm for the numerical calculation of blowing-up solutions, *Comm. Pure Appl. Math.* **41** (6) (1988), 841–863.

[8] A. J. Bernoff and A. L. Bertozzi, Singularities in a modified Kuramoto-Sivashinsky equation describing interface motion for phase transition, *Phys. D* **85** (3) (1995), 375–404.

[9] A. J. Bernoff, A. L. Bertozzi, and T. P. Witelski, Axisymmetric surface diffusion: Dynamics and stability of self-similar pinchoff, *J. Statist. Phys.* **93** (3–4) (1998), 725–776.

[10] A. L. Bertozzi, Symmetric singularity formation in lubrication-type equations for interface motion, *SIAM J. Appl. Math.* **56** (3) (1996), 681–714.

[11] A. L. Bertozzi, The mathematics of moving contact lines in thin liquid films, *Notices Amer. Math. Soc.* **45** (6) (1998), 689–697.

[12] A. L. Bertozzi, M. P. Brenner, T. F. Dupont, and L. P. Kadanoff, Singularities and similarities in interface flows, in: *Trends and Perspectives in Applied Mathematics*, Springer, New York, 1994, 155–208.

[13] A. L. Bertozzi, G. Grün, and T. P. Witelski, Dewetting films: bifurcations and concentrations, *Nonlinearity* **14** (2001), 1569–1592.

[14] A. L. Bertozzi and M. C. Pugh, The lubrication approximation for thin viscous films: the moving contact line with a "porous media" cut-off of van der Waals interactions, *Nonlinearity* **7** (6) (1994), 1535–1564.

[15] A. L. Bertozzi and M. C. Pugh, The lubrication approximation for thin viscous films: regularity and long-time behavior of weak solutions, *Comm. Pure Appl. Math.* **49** (2) (1996), 85–123.

[16] A. L. Bertozzi and M. C. Pugh, Long-wave instabilities and saturation in thin film equations, *Comm. Pure Appl. Math.* **51** (6) (1998), 625–661.

[17] A. L. Bertozzi and M. C. Pugh, Finite-time blow-up of solutions of some long-wave unstable thin film equations, *Indiana Univ. Math. J.* **49** (4) (2000), 1323–1366.

[18] J. Bischof, D. Scherer, S. Herminghaus, and P. Leiderer, Dewetting modes of thin metallic films: Nucleation of holes and spinodal dewetting, *Phys. Rev. Lett.* **77** (8) (1996), 1536–1539.

[19] J. P. Boyd, Spectral methods using rational basis functions on an infinite interval, *J. Comput. Phys.* **69** (1) (1987), 112–142.

[20] M. P. Brenner, P. Constantin, L. P. Kadanoff, A. Schenkel, and S. C. Venkataramani, Diffusion, attraction and collapse, *Nonlinearity* **12** (4) (1999), 1071–1098.

[21] M. P. Brenner, J. R. Lister, and H. A. Stone, Pinching threads, singularities and the number 0.0304..., *Phys. Fluids* **8** (11) (1996), 2827–2836.

[22] C. Budd, B. Dold, and A. Stuart, Blowup in a partial differential equation with conserved first integral, *SIAM J. Appl. Math.* **53** (3) (1993), 718–742.

[23] C. J. Budd, J. Chen, W. Huang, and R. D. Russell, Moving mesh methods with applications to blow-up problems for PDEs, in: *Numerical Analysis 1995 (Dundee, 1995)*, Longman, Harlow, 1996, 1–18.

[24] C. J. Budd, S. Chen, and R. D. Russell, New self-similar solutions of the nonlinear Schrödinger equation with moving mesh computations, *J. Comput. Phys.* **152** (2) (1999), 756–789.

[25] C. J. Budd, G. J. Collins, and V. A. Galaktionov, An asymptotic and numerical description of self-similar blow-up in quasilinear parabolic equations, *J. Comput. Appl. Math.* **97** (1-2) (1998), 51–80.

[26] C. J. Budd and V. Galaktionov, Stability and spectra of blow-up in problems with quasi-linear gradient diffusivity, *Proc. Roy. Soc. London Ser. A* **454** (1977) (1998), 2371–2407.

[27] C. J. Budd, W. Huang, and R. D. Russell, Moving mesh methods for problems with blow-up, *SIAM J. Sci. Comput.* **17** (2) (1996), 305–327.

[28] J. P. Burelbach, S. G. Bankoff, and S. H. Davis, Nonlinear stability of evaporating/condensing liquid films, *J. Fluid Mech.* **195** (1988), 463–494.

[29] R. E. Caflisch and G. C. Papanicolaou, eds., *Singularities in Fluids, Plasmas and Optics*, NATO ASI Ser. C **404**, Kluwer, Dordrecht, 1993.

[30] J. W. Cahn and J. E. Taylor, Surface motion by surface diffusion, *Acta Metallica Materiala* **42** (4) (1994), 1045–1063.

[31] D. L. Chopp and J. A. Sethian, Flow under curvature: singularity formation, minimal surfaces, and geodesics, *Experiment. Math.* **2** (4) (1993), 235–255.

[32] B. D. Coleman, R. S. Falk, and M. Moakher, Stability of cylindrical bodies in the theory of surface diffusion, *Phys. D* **89** (1–2) (1995), 123–135.

[33] B. D. Coleman, R. S. Falk, and M. Moakher, Space-time finite element methods for surface diffusion with application to the theory of the stability of cylinders, *SIAM J. Sci. Comput.* **17** (6) (1996), 1434–1448.

[34] P. Constantin, T. F. Dupont, R. E. Goldstein, L. P. Kadanoff, M. J. Shelley, and S.-M. Zhou, Droplet breakup in a model of the Hele-Shaw cell, *Phys. Rev. E (3)* **47** (6) (1993), 4169–4181.

[35] S. D. Conte, Numerical solution of vibration problems in two space variables, *Pacific J. Math.* **7** (1957), 1535–1544.

[36] S. D. Conte and R. T. Dames, An alternating direction method for solving the biharmonic equation., *Math. Tables Aids Comput.* **12** (1958), 198–205.

[37] S. D. Conte and R. T. Dames, On an alternating direction method for solving the plate problem with mixed boundary conditions, *J. Assoc. Comput. Mach.* **7** (1960), 264–273.

[38] M. C. Cross and P. C. Hohenberg, Pattern formation outside of equilibrium, *Rev. Modern Phys.* **65** (3) (1993), 854–1112.

[39] J. Douglas, Jr. and J. E. Gunn, A general formulation of alternating direction methods. I. Parabolic and hyperbolic problems, *Numer. Math.* **6** (1964), 428–453.

[40] J. Douglas, Jr. and H. H. Rachford, Jr., On the numerical solution of heat conduction problems in two and three space variables, *Trans. Amer. Math. Soc.* **82** (1956), 421–439.

[41] P. G. Drazin and W. H. Reid, *Hydrodynamic Stability*, Cambridge Monogr. Mech. Appl. Math., Cambridge University Press, Cambridge, 1981.

[42] L. Dresner, *Applications of Lie's Theory of Ordinary and Partial Differential Equations*, Institute of Physics Publishing, Bristol, 1999.

[43] T. F. Dupont, R. E. Goldstein, L. P. Kadanoff, and S.-M. Zhou, Finite-time singularity formation in Hele-Shaw systems, *Phys. Rev. E (3)* **47** (6) (1993), 4182–4196.

[44] J. Eggers, Nonlinear dynamics and breakup of free-surface flows, *Rev. Modern Phys.* **69** (3) (1997), 865–929.

[45] B. I. Epureanu and H. S. Greenside, Fractal basins of attraction associated with a damped Newton's method, *SIAM Rev.* **40** (1) (1998), 102–109 (electronic).

[46] T. Erneux and S. H. Davis, Nonlinear rupture of free films, *Phys. Fluids A* **5** (1993), 1117–1122.

[47] L. C. Evans and J. Spruck, Motion of level sets by mean curvature. I, *J. Differential Geom.* **33** (3) (1991), 635–681.

[48] S. Filippas and R. V. Kohn, Refined asymptotics for the blowup of $u_t - \Delta u = u^p$, *Comm. Pure Appl. Math.* **45** (7) (1992), 821–869.

[49] V. A. Galaktionov and J. L. Vazquez, The problem of blow-up in nonlinear parabolic equations, in: *Proc. Summer Course on PDEs*, Pitman, 2000.

[50] Y. Giga and R. V. Kohn, Asymptotically self-similar blow-up of semilinear heat equations, *Comm. Pure Appl. Math.* **38** (3) (1985), 297–319.

[51] Y. Giga and R. V. Kohn, Characterizing blowup using similarity variables, *Indiana Univ. Math. J.* **36** (1) (1987), 1–40.

[52] K. Glasner and T. P. Witelski, Coarsening dynamics of dewetting films, preprint, Duke University, 2001.

[53] G. H. Golub and C. F. Van Loan, *Matrix Computations*, 3rd ed., Johns Hopkins University Press, Baltimore, MD, 1996.

[54] K. Große-Brauckmann and K. Polthier, Compact constant mean curvature surfaces with low genus, *Experiment. Math.* **6** (1) (1997), 13–32.

[55] S. Herminghaus, K. Jacobs, K. Mecke, J. Bischof, A. Fery, M. Ibn-Elhaj, and S. Schlagowski, Spinodal dewetting in liquid crystal and liquid metal films, *Science* **282** (5390) (1998), 916–919.

[56] G. Huisken, Asymptotic behavior for singularities of the mean curvature flow, *J. Differential Geom.* **31** (1) (1990), 285–299.

[57] M. P. Ida and M. J. Miksis, Thin film rupture, *Appl. Math. Lett.* **9** (3) (1996), 35–40.

[58] A. Iserles, *A First Course in the Numerical Analysis of Differential Equations*, Cambridge University Press, Cambridge, 1996.

[59] L. P. Kadanoff, Singularities and blowups, *Physics Today* **50** (9) (1997), 11.

[60] H. B. Keller, *Lectures on Numerical Methods in Bifurcation Problems*, published for the Tata Institute of Fundamental Research, Bombay, by Springer, Berlin-New York, 1987.

[61] J. B. Keller and J. S. Lowengrub, Asymptotic and numerical results for blowing-up solutions to semilinear heat equations, in: *Singularities in Fluids, Plasmas and Optics* (R. E. Caflisch and G. C. Papanicolaou, eds.), NATO ASI Ser. C **404**, Kluwer, Dordrecht, 1993, 111–129.

[62] J. Kevorkian and J. D. Cole, *Multiple Scale and Singular Perturbation Methods*, Appl. Math. Sci. **114**, Springer-Verlag, New York, 1996.

[63] J. R. King, Emerging areas of mathematical modelling, *R. Soc. Lond. Philos. Trans. Ser. A Math. Phys. Eng. Sci.* **358** (1765) (2000), 3–19.

[64] L. Kulinsky, J. D. Powers, and A. M. Glaeser, Morphological evolution of pre-perturbed pore channels in sapphire, *Acta Materialia* **44** (10) (1996), 4115–4130.

[65] R. S. Laugesen and M. C. Pugh, Energy levels of steady states for thin film type equations, preprint, 2000, to appear in *J. Differential Equations.*

[66] R. S. Laugesen and M. C. Pugh, Heteroclinic orbits, mobility parameters and stability for thin film type equations, preprint, 2000; `http://xxx.lanl.gov.abs/math.AP/0003209`.

[67] R. S. Laugesen and M. C. Pugh, Linear stability of steady states for thin film and Cahn-Hilliard type equations, *Arch. Rational Mech. Anal.* **154** (1) (2000), 3–51.

[68] R. S. Laugesen and M. C. Pugh, Properties of steady states for thin film equations, *European J. Appl. Math.* **11** (3) (2000), 293–351.

[69] H. A. Levine, The role of critical exponents in blowup theorems, *SIAM Rev.* **32** (2) (1990), 262–288.

[70] A. J. Majda and A. L. Bertozzi, *Vorticity and Incompressible Flow*, Cambridge University Press, Cambridge, 2002.

[71] G. I. Marchuk, Splitting and alternating direction methods, in: *Handbook of Numerical Analysis* (P. G. Ciarlet and J.-L. Lions, eds.), vol. I, North-Holland, Amsterdam, 1990, 197–462.

[72] J. A. Moriarty and L. W. Schwartz, Dynamic considerations in the closing and opening of holes in thin liquid films, *J. Colloid Interface Sci.* **161** (1993), 335–342.

[73] W. W. Mullins, Theory of thermal grooving, *J. Appl. Phys.* **28** (3) (1957), 333–339.

[74] W. W. Mullins, Mass transport at interfaces in single component systems, *Metall. Trans. A* **26** (1995), 1917–1929.

[75] T. G. Myers, Thin films with high surface tension, *SIAM Rev.* **40** (3) (1998), 441–462.

[76] F. A. Nichols and W. W. Mullins, Morphological changes of a surface of revolution due to capillarity-induced surface diffusion, *J. Appl. Phys.* **36** (6) (1965), 1826–1835.

[77] J. R. Ockendon and H. Ockendon, *Viscous Flow*, Cambridge University, Cambridge, 1995.

[78] A. Oron and S. G. Bankoff, Dewetting of a heated surface by an evaporating liquid film under conjoining/disjoing pressures, *J. Colloid Interface Sci.* **218** (1999), 152–166.

[79] A. Oron, S. H. Davis, and S. G. Bankoff, Long-scale evolution of thin liquid films, *Rev. Modern Phys.* **69** (3) (1997), 931–980.

[80] P. Plecháč and V. Šverák, On self-similar singular solutions of the complex Ginzburg-Landau equation, *Comm. Pure Appl. Math.* **54** (10) (2001), 1215–1242.

[81] W. H. Press, S. A. Teukolsky, W. T. Vetterling, and B. P. Flannery, *Numerical Recipes in C*, 2nd ed., Cambridge University Press, Cambridge, 1992.

[82] M. C. Pugh and M. J. Shelley, Singularity formation in thin jets with surface tension, *Comm. Pure Appl. Math.* **51** (7) (1998), 733–795.

[83] C. Redon, F. Brochard-Wyart, and F. Rondelez, Dynamics of dewetting, *Phys. Rev. Lett.* **66** (6) (1991), 715–718.

[84] G. Reiter, Dewetting of thin polymer-films, *Phys. Rev. Lett.* **68** (1) (1992), 75–78.

[85] A. A. Samarskii, V. A. Galaktionov, S. P. Kurdyumov, and A. P. Mikhailov, *Blow-up in Quasilinear Parabolic Equations*, de Gruyter, Berlin, 1995.

[86] A. Sharma and R. Khanna, Pattern formation in unstable thin liquid films, *Phys. Rev. Lett.* **81** (16) (1998), 3463–3466.

[87] A. Sharma and R. Khanna, Pattern formation in unstable thin liquid films under the influence of antagonistic short- and long-range forces, *J. Chem. Phys.* **110** (10) (1999), 4929–4936.

[88] M. J. Shelley, R. E. Goldstein, and A. I. Pesci, Topological transitions in Hele-Shaw flow, in: *Singularities in Fluids, Plasmas and Optics* (R. E. Caflisch and G. C. Papanicolaou, eds.), NATO ASI Ser. C **404**, Kluwer, Dordrecht, 1993, 167–188.

[89] X. D. Shi, M. P. Brenner, and S. R. Nagel, A cascade of structure in a drop falling from a faucet, *Science* **265** (5169) (1994), 219–222.

[90] J. Stoer and R. Bulirsch, *Introduction to Numerical Analysis*, 2nd ed., Texts Appl. Math. **12**, Springer-Verlag, New York, 1993.

[91] B. Straughan, *Explosive Instabilities in Mechanics*, Springer-Verlag, Berlin, 1998.

[92] L. S. Tuckerman and D. Barkley, Bifurcation analysis for timesteppers, in: *Numerical Methods for Bifurcation Problems and Large-Scale Dynamical Systems (Minneapolis, MN, 1997)*, Springer, New York, 2000, 453–466.

[93] S. J. VanHook, M. F. Schatz, W. D. McCormick, J. B. Swift, and H. L. Swinney, Long-wavelength instability in surface-tension-driven Bénard convection, *Phys. Rev. Lett.* **75** (24) (1995), 4397–4400.

[94] S. J. VanHook, M. F. Schatz, J. B. Swift, W. D. McCormick, and H. L. Swinney, Long-wavelength surface-tension-driven Bénard convection: experiment and theory, *J. Fluid Mech.* **345** (1997), 45–78.

[95] D. Vaynblat, J. R. Lister, and T. P. Witelski, Rupture of thin viscous films by van der Waals forces: Evolution and self-similarity, *Phys. Fluids* **13** (5) (2001), 1130–1140.

[96] D. Vaynblat, J. R. Lister, and T. P. Witelski, Symmetry and self-similarity in rupture and pinchoff: A geometric bifurcation, *European J. Appl. Math.* **12** (3) (2001), 209–232.

[97] M. B. Williams and S. H. Davis, Nonlinear theory of film rupture, *J. Colloid Interface Sci.* **90** (1982), 220.

[98] T. P. Witelski and A. J. Bernoff, Stability of self-similar solutions for van der Waals driven thin film rupture, *Phys. Fluids* **11** (9) (1999), 2443–2445.

[99] T. P. Witelski and A. J. Bernoff, Dynamics of three-dimensional thin film rupture, *Phys. D* **147** (1-2) (2000), 155–176.

[100] T. P. Witelski and M. Bowen, ADI schemes for higher-order nonlinear diffusion equations, preprint, Duke University, 2001.

[101] H. Wong, M. J. Miksis, P. W. Voorhees, and S. H. Davis, Universal pinch off of rods by capillarity-driven surface diffusion, *Scripta Materialia* **39** (1) (1998), 55–60.

[102] R. Xie, A. Karim, J. Douglas, C. Han, and R. Weiss, Spinodal dewetting of thin polymer films, *Phys. Rev. Lett.* **81** (6) (1998), 1251–1254.

[103] W. W. Zhang and J. R. Lister, Similarity solutions for van der Waals rupture of a thin film on a solid substrate, *Phys. Fluids* **11** (9) (1999), 2454–2462.

Index